This book is an up-to-date introduction to univariate spectral analysis at the graduate level, which reflects a new scientific awareness of spectral complexity, as well as the widespread use of spectral analysis on digital computers with considerable computational power. The authors view the subject from a practical perspective while giving a firm underpinning of theory. The text provides theoretical and computational guidance on the available techniques, emphasizing those that work in practice.

Spectral analysis finds extensive application in the analysis of data arising in many of the physical sciences, ranging from electrical engineering and physics to geophysics and oceanography. Throughout the text the authors give all formulae in terms that can be immediately and easily coded for a computer, and a thorough review of the necessary Fourier theory is presented. Algorithmic innovations and statistical properties are also presented. A valuable feature of the text is that many examples are given showing the application of spectral analysis to real data sets.

Special emphasis is placed on the multitaper technique, because of its practical success in handling spectra with intricate structure, and its power to handle data with or without spectral lines. A thorough discussion of conventional nonparametric and parametric (autoregressive) spectral analysis is included, along with an examination of potential problems with their application. The text contains a large number of exercises of varying levels of difficulty, together with an extensive bibliography. The emphasis throughout the book is on providing insight and guidance on the many techniques available.

This book will be of interest to advanced undergraduates, graduate students and researchers in mathematics, statistics and the whole of the physical sciences with an interest in spectral analysis.

SPECTRAL ANALYSIS FOR PHYSICAL APPLICATIONS

SPECTRAL ANALYSIS
FOR
PHYSICAL APPLICATIONS

MULTITAPER AND CONVENTIONAL
UNIVARIATE TECHNIQUES

DONALD B. PERCIVAL

Senior Mathematician, Applied Physics Laboratory
University of Washington, Seattle

and

ANDREW T. WALDEN

Senior Lecturer, Mathematics Department
Imperial College of Science, Technology and Medicine, London

CAMBRIDGE
UNIVERSITY PRESS

Scholium: May 1995   90.40

Published by the Press Syndicate of the University of Cambridge
The Pitt Building, Trumpington Street, Cambridge CB2 1RP
40 West 20th Street, New York, NY 10011-4211, USA
10 Stamford Road, Oakleigh, Melbourne 3166, Australia

First published 1993

Printed in Great Britain at the University Press, Cambridge

A catalogue record for this book is available from the British Library

Library of Congress cataloguing in publication data
Percival, Donald B.
Spectral analysis for physical applications : multitaper and conventional
univariate techniques/Donald B. Percival and Andrew T. Walden.
p.   cm.
Includes bibliographical references and indexes.
ISBN 0-521-35532-X – ISBN 0-521-43541-2 (pbk)
1. Spectral theory (Mathematics)   I. Walden, Andrew T.   II. Title.
III. Title: Multitaper and conventional univariate techniques.
QA320.P434   1993
515′.7222–dc20   92-45862   CIP

ISBN 0 521 35532 x hardback
ISBN 0 521 43541 2 paperback

To Piroska, Hilary, Adam, Jane, Dancsi, Robin and Clare,
and in memory of Helen

Contents

Preface

All theory, my friend, is gray,
But green is the lustrous tree of life.
Mephistopheles, as quoted in Goethe's *Faust*

In many areas of the physical sciences, spectral analysis finds frequent and extensive use. Examples abound in oceanography, electrical engineering, geophysics, astronomy and hydrology. Spectral analysis is a well-established standard – its use in so many areas in fact facilitates the exchange of ideas across a broad array of scientific endeavors. As old and as well-established as it is, however, spectral analysis is still an area of active on-going research (indeed, one of the problems for both practitioners and theorists alike is that important methodological advances are spread out over many different disciplines and literally dozens of scientific journals). Since the 1960s, developments in spectral analysis have had to take into account three new factors:

[1] *Digital data explosion.* The amount of data routinely collected in the form of time series is staggering by 1960s standards. Examples include exploration seismic data, continuous recordings of the earth's magnetic field, large seismology networks, real-time processing of sonar signals and remote sensing from satellites. The impact on spectral analysis is that *fast* and *digital* processing has become very important.

[2] *Enormous increase in computational power.* In retrospect, it is clear that much of the spectral analysis methodology in vogue in the 1960s was influenced heavily by what *could* be calculated with commonly available computers. With the computational power now widely available on today's computers, we are finally in a position to concentrate on what *should* be calculated.

[3] *New awareness of spectral complexity.* In the 1960s spectral analysis was designed to be applied primarily to processes with relatively simple spectra. It has become clear since then – particularly from geophysical applications – that real data can have spectra with intricate detail and an enormous dynamic range. Hence even small improvements in methodology can be important for interpretations.

Our intent in writing this book is to provide a graduate-level introduction to spectral analysis that covers both well-accepted methodology and some important recent advances addressing these new factors. We place special emphasis on the multitaper technique due to Thomson (1982) because of its potential for routinely handling spectra with intricate structure. At the risk of appearing to agree with Mephistopheles, our treatment is slanted away from theory and toward practical applications. However, it is our experience that successful application of spectral analysis requires a solid understanding of its theoretical underpinnings. We have thus endeavored to present a satisfying balance between the underlying theory and its applications. Examples are given of the application of spectral analysis techniques to real data sets, as well as to synthetic data when pedagogically appropriate. We have tried to weave together the best of the algorithmic innovations from the electrical engineering literature with the more useful statistical inference results from the statistics literature. There are several distinctive features to this book.

- A thorough review of the necessary Fourier theory is given. A major source of confusion to those learning or using spectral methods is meshing discrete time computing with theory developed for continuous time or frequency. Hence, in Chapter 3 we look at results for continuous time and frequency, discrete time and continuous frequency, and discrete time and discrete frequency. This review also includes spectral concentration measures, providing the background to the nonparametric multitaper spectral analysis method of Chapter 7.

- Computational considerations are extremely important to the practitioner. We have given formulae in terms that can be immediately and easily coded (for example, the sampling interval is explicitly included). Choosing bandwidths for smoothing depends on finding computable bandwidth measures for both smoothing windows and the time series of interest. We give such measures and demonstrate their utility. Tapering is often given little coverage in books on time series and spectral analysis. We have found it to be extremely useful in practice, and Chapter 6 develops the structure and behavior of tapered and smoothed spectral estimators, with several examples.

- The multitaper spectral analysis method – currently proving to be so powerful in many areas of science where spectra with high dynamic range are commonplace – is studied in Chapter 7, with some necessary computational details supplied in Chapter 8. Many scientists are initially intimidated by this method due to its *apparent* complexity. In reality, it is conceptually simple, as we show with a simple heuristic interpretation. In addition, we provide formal justification via both matrix algebra and projection methods.

- In general most significant developments in harmonic analysis of data containing sinusoids (spectral lines) in noise have taken place in the electrical engineering literature. Chapter 10 reflects this, but builds in newer statistical tests for sinusoids. It considers in detail problems of spectral line splitting and frequency shifting, and gives a method for reshaping the spectral continuum around the location of a spectral line.

- A large number of exercises, of varying levels of difficulty, are provided at the end of the chapters. Many of these exercises support results in the main text (instructors wishing to obtain a solution guide for the exercises should contact the authors for details).

This book grew out of a graduate course given by each of us to students of statistics, electrical engineering and the physical sciences at the University of Washington. The book is similarly best suited to graduate students and researchers. The emphasis is, naturally enough, very strongly in the frequency domain, but time domain methods are introduced as a natural complement where necessary and/or useful. While the more traditional forms of spectral analysis have been necessarily included in detail, the emphasis of the book is on methodology developed in the 1970s on up to the current time. Because of their statistical tractability, we also stress nonparametric spectral analysis techniques. While parametric autoregressive moving average (ARMA) models are often convenient for mathematical purposes, they are often very inconvenient or unsatisfactory when used by themselves for analysis of spectra with high dynamic range and a complicated shape, as often occur in studies in the physical sciences. As we discuss in Chapter 9, simple parametric models (such as pure autoregressive) can, however, often be used to good effect for prewhitening, prior to nonparametric analysis.

Most of the algorithms discussed in this book have been coded in Common Lisp and can be run under the various implementations of this language available on a number of different computers (e.g., Macintosh Common Lisp^{TM}). This code and most of the data sets used in this book can be obtained through e-mail. Details are given on page xxvii.

A complementary volume to this one is currently under preparation and is scheduled to include topics such as spectral analysis of multivariate time series, higher order spectra, spectral estimation for irregularly sampled time series, robust spectral estimation, estimation of power-law spectra, spectral ratios and the simulation of stationary processes.

The book was written using Donald Knuth's superb typesetting system TeX as implemented by Blue Sky Research in their product TeXtures for Apple Macintosh^{TM} computers. With one exception, all of the figures in this book were created using the plotting system GPL written by W. Hess, who provided us with excellent technical support. The majority of the computations necessary for the various examples and figures were carried out using $P_I TSS_A$, a Lisp-based object-oriented program for interactive time series and signal analysis that was developed in part by one of us (Percival) with support from the Office of Naval Research, the Naval Observatory and the Naval Research Laboratory. Without these software tools, this book would certainly never have existed.

We thank S. Murphy, J. Harlett, and R. Spindel of the Applied Physics Laboratory, University of Washington, for providing discretionary funding at critical times to help us start and finish this book. We are much indebted to those who have commented on the manuscript or supplied data to us, namely, B. Bell, W. Dunlap, J. Filliben, W. Fox, P. Guttorp, A. Jessup, R. D. Martin, D. McCarthy, E. McCoy, H. Nhu, C. Siedenburg and B. Walter. Particular thanks are due to C. Greenhall, whose careful and thorough critiques substantially improved this manuscript in a number of areas. We thank C. Andersen and P. Magassy for carefully proofreading earlier versions of the manuscript. We are also very grateful to the many graduate students who suffered through incomplete early versions of various chapters, giving us valuable critiques of the manuscript and exercises and spotting numerous errors. Any remaining errors are, of course, our responsibility, and we would be pleased to hear from any reader who finds a mistake (our 'paper' and electronic mailing addresses are listed below). Finally we thank our families for their patience over the seven year period this book was being written.

Don Percival
Applied Physics Laboratory
HN-10
University of Washington
Seattle, WA 98195
dbp@apl.washington.edu
October, 1992

Andrew Walden
Department of Mathematics
Imperial College of Science,
  Technology and Medicine
London SW7 2BZ, U.K.
a.walden@ic.ac.uk

Conventions and Notation

- *Important conventions*

(5)	refers to the single displayed equation on page 5
(10a), (10b)	refers to different displayed equations on page 10
Figure 9	refers to the figure on page 9
Table 248	refers to the table on page 248
Exercise [1.4]	refers to the fourth exercise at the end of Chapter 1
$S(\cdot)$	refers to a function
$S(f)$	refers to the value of the function $S(\cdot)$ at f
$\{h_t\}$	refers to a sequence of values indexed by t
h_t	refers to a single value of a sequence

In the following lists, the numbers at the end of the brief descriptions are page numbers where more information about – or an example of the use of – an abbreviation or symbol can be found.

- *Abbreviations used frequently*

acf	autocorrelation function	37
acls	approximate conditional least squares	502
acs	autocorrelation sequence	37, 39
acvf	autocovariance function	36
acvs	autocovariance sequence	36, 39
AR(p)	pth order autoregressive process	44, 392, 168

ARMA(p, q)	autoregressive moving average process	
	of orders p and q	46, 168
cpdf	cumulative probability distribution function	33
dB	decibels, i.e., $10 \log_{10}(\cdot)$	
DFT	discrete Fourier transform	110
dpss	discrete prolate spheroidal sequence	104, 378
dpswf	discrete prolate spheroidal wave function	104
FFT	fast Fourier transform	110, 114
FIR	finite impulse response	170
FPE	final prediction error	436
Hz	Hertz: 1 Hz = 1 cycle per second	
IIR	infinite impulse response	170
ls	least squares	426
LTI	linear time-invariant	155
MA(q)	qth order moving average process	43, 167, 443
ml	maximum likelihood	429
mle	maximum likelihood estimate or estimator	429
mse	mean square error	191
pdf	probability density function	228, 54
pswf	prolate spheroidal wave function	77
rv	random variable	31
sdf	spectral density function	132
SVD	singular value decomposition	533
WOSA	Welch's (or weighted) overlapped segment	
	averaging	289

- *Non-Greek notation used frequently*

$\arg(z)$	argument of complex-valued number z	65
A_l	real-valued amplitude associated	
	with $\cos(2\pi f_l t \, \Delta t)$	463
$b\{\cdot\}$	bias	356
$b^{(l)}\{\cdot\}$	local bias	356
$b^{(b)}\{\cdot\}$	broad-band bias	356
$b_k(f)$	weight associated with kth eigenspectrum	
	at frequency f	366
$b_W(\cdot)$	bias due to smoothing window only	244
B_l	real-valued amplitude associated	
	with $\sin(2\pi f_l t \, \Delta t)$	463
B_S	spectral bandwidth	274, 277

$\{Z(t)\}$	complex-valued continuous parameter stochastic process	32
$\{Z(f)\}$	orthogonal process	129

- *Greek notation used frequently*

α	intercept parameter of a linear regression, level of a test, or a scalar	51, 233
$\alpha^2(T)$	fraction of signal's energy lying in $[-T/2, T/2]$	75
β	slope parameter of a linear regression	51
β_W	Grenander's measure of bandwidth of a smoothing window	241
$\beta^2(W)$	fraction of signal's energy lying in $[-W, W]$	75
γ	Euler's constant $(0.5772\ldots)$	228, 281
$\Gamma(\cdot)$	gamma function	493
Γ_p	Toeplitz covariance matrix	38, 394
$\delta_{j,k}$	Kronecker delta function	79
$\delta(\cdot)$	Dirac delta function	144
Δt	sampling interval	87, 144
ϵ_k	1 if k even, i if k odd	104, 362
$\{\epsilon_t\}$	white noise or error sequence	43, 460
$\overrightarrow{\epsilon_t}(k)$	forward prediction error: $X_t - \overrightarrow{X_t}(k)$	398
$\overleftarrow{\epsilon_t}(k)$	backward prediction error: $X_t - \overleftarrow{X_t}(k)$	400
$\{\eta_t\}$	colored noise or error sequence	467
$\theta(\cdot)$	phase function corresponding to transfer function $G(\cdot)$	158
$\theta_{1,q}, \ldots, \theta_{q,q}$	coefficients of MA(q) model	43, 443
λ	constant to define different logarithmic scales	257
$\lambda_k(c)$	eigenvalue associated with pswf, order k	77
$\lambda_k(N, W)$	eigenvalue associated with dpswf, order k	103
μ	expected value of a stationary process	36
$\tilde{\mu}$	estimator of μ used with tapered data	217, 501
ν	(equivalent) degrees of freedom of spectral estimator	255
$\{\rho_\tau\}$	autocorrelation sequence (acs)	37
$\rho(\cdot)$	autocorrelation function (acf)	37
σ^2	variance	36
σ_ϵ^2	white noise variance	43, 461
σ_η^2	colored noise variance	467
σ_p^2	innovations variance for an AR(p) process	392

$\bar{\sigma}_p^2$	Burg estimate of σ_p^2	416
$\tilde{\sigma}_p^2$	Yule–Walker estimate of σ_p^2	395, 404
τ	lag value	36
ϕ_l	phase of a sinusoid	46
$\phi_{1,p}, \ldots, \phi_{p,p}$	coefficients of an AR(p) model	44, 392
$\bar{\phi}_{1,p}, \ldots, \bar{\phi}_{p,p}$	Burg estimates of AR(p) coefficients	416, 452
$\tilde{\phi}_{1,p}, \ldots, \tilde{\phi}_{p,p}$	Yule–Walker estimates of AR(p) coefficients	404, 452
$\varphi_{1,p}, \ldots, \varphi_{p,p}$	coefficients of a pseudo-AR(p) model	515
$\mathbf{\Phi}_p$	$[\phi_{1,p}, \phi_{2,p}, \ldots, \phi_{p,p}]^T$	394
$\Phi^{-1}(p)$	$p \times 100\%$ percentage point of standard Gaussian distribution	256
χ_ν^2	chi-square distribution with ν degrees of freedom	221
$\psi_k(\cdot; c)$	prolate spheroidal wave function, order k	77
ω	angular frequency	10

- *Other mathematical symbols used frequently*

$\approx$	approximately equal to	
z^*	complex conjugate of z	39
A^H	complex-conjugate (Hermitian) transpose of matrix A	351, 533
$\in, \notin$	contained in, not contained in	
$*$	convolution operator	117–9
$\star$	cross-correlation operator	84
$10 \log_{10}(\cdot)$	decibel scale for power	
$|L_N|$	determinant of matrix L_N	430
$\equiv$	equal by definition	
$\overset{\text{d}}{=}$	equal in distribution	222
$\overset{\text{ms}}{=}$	equal in mean square sense	58
$\hat{\cdot}$	estimator or estimate; e.g., $\hat{S}_j$ is an estimator of S_j	19
$\longleftrightarrow$	Fourier transform pair relationship	116–9
$R^{\#}$	generalized inverse of matrix R	535
$\lfloor N \rfloor$	greatest integer $\leq N$	10
$\langle \cdot, \cdot \rangle$	inner product	419
$(a)_+$	max $(a, 0)$	492
$\|\cdot\|^2$	squared norm	382, 419
$\mathbf{a}^T, L_N^T$	transpose of vector $\mathbf{a}$, transpose of matrix L_N	37

Data and Software

Many of the time series used in this book have been deposited in the datasets library of the e-mail based retrieval system StatLib. To use StatLib it is necessary to be able to send electronic mail to an internet host. (Those with BITNET or UUCP mail access should ask their local system administrator how to send internet mail.) To get more information on StatLib, send the single-line electronic mail message

<code>send index</code>

to

<code>statlib@lib.stat.cmu.edu</code>

To obtain the time series used in this book, send the message

<code>send sapa from datasets</code>

to

<code>statlib@lib.stat.cmu.edu</code>

Common Lisp code was developed for many of the algorithms discussed in this book and has also been deposited with StatLib. It can be run using, e.g., Macintosh Common Lisp^{TM}. To obtain an index and details of how to obtain the algorithms, send the message

<code>send index from sapaclisp</code>

to

<code>statlib@lib.stat.cmu.edu</code>

1

Introduction to Spectral Analysis

1.0 Introduction

This chapter provides a quick introduction to the subject of spectral analysis. Except for some later references to the exercises of Section 1.6, this material is independent of the rest of the book and can be skipped without loss of continuity. Our intent is to use some simple examples to motivate the key ideas. Since our purpose is to view the forest before we get lost in the trees, the particular analysis techniques we use here have been chosen for their simplicity rather than their appropriateness.

1.1 Some Aspects of Time Series Analysis

Spectral analysis is part of time series analysis, so the natural place to start our discussion is with the notion of a time series. The quip (attributed to R. A. Fisher) that a time series is 'one damned thing after another' is not far from the truth: loosely speaking, a time series is a set of observations made sequentially in time. Examples abound in the real world, and Figures 2 and 3 show plots of small portions of four actual time series:

[1] the speed of the wind in a certain direction at a certain location, measured every 0.025 second;

[2] the monthly average measurements related to the flow of water in the Willamette River at Salem, Oregon;

[3] the daily record of a quantity (to be precise, the change in average daily frequency) that tells how well an atomic clock keeps time on a day to day basis (a constant value of 0 would indicate that the clock agreed perfectly with a time scale maintained by the U. S. Naval Observatory); and

1

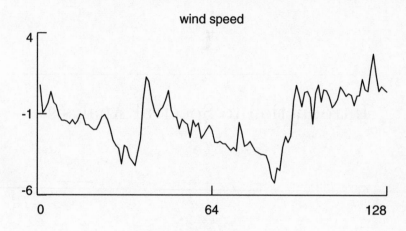

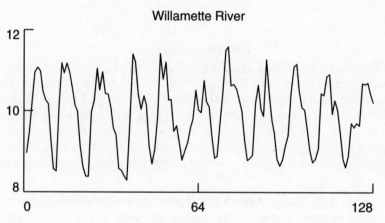

Figure 2. Plots of portions of the first two time series. For both series the vertical axis is the value of the time series (in unspecified units), while the horizontal axis is time (measured at 0.025 second intervals for the wind speed series and in months for the Willamette River series).

[4] the change in the level of ambient noise in the ocean from one second to the next.

For each of these plots, the values of the time series at 128 successive times are connected by lines to help the eye follow the variations in the series. The visual appearances of these four series are quite different.

The chief aim of time series analysis is to develop quantitative means to allow us to characterize time series, e.g., to say quantitatively how one series differs from another or how two series are related. There are two broad classes of characterizations, namely, time domain techniques and frequency domain techniques. Spectral analysis is the prime

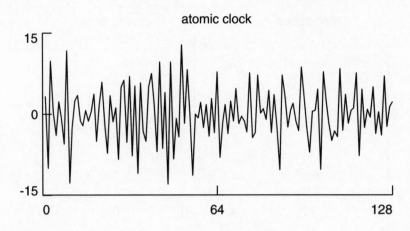

atomic clock

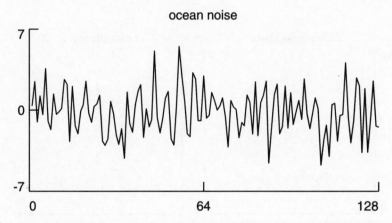

ocean noise

Figure 3. Plots of portions of the last two time series. The horizontal axes are again time (measured in days for the atomic clock data and in seconds for the ocean noise series).

example of a frequency domain technique. Before we introduce it, we will first consider a popular time domain technique. We contend that this latter technique is not completely satisfactory and that spectral analysis is a useful and complementary alternative to it.

Let us concentrate for the moment on the wind speed and atomic clock data (top plots of Figures 2 and 3, respectively). How do these two series differ? One way is that in the wind speed series adjacent points of the time series tend to be close in value, while in the atomic clock series positive values tend to be followed by negative values and vice versa. To see this effect graphically, we can plot x_{t+1} versus x_t as t varies from 1 to $N-1$, where we let x_1, x_2, ..., x_N represent any

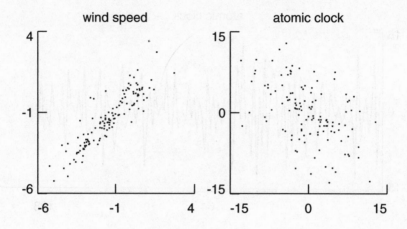

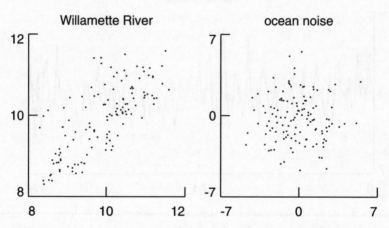

Figure 4. Lag 1 scatter plots for the time series in Figures 2 and 3. In each of these plots, the value of the time series at time $t + 1$ is plotted on the vertical axis versus the value at time t on the horizontal axis (for t ranging from 1 to 127).

one of our series and let N represent the sample size, i.e., the number of data points in a time series, 128 in our case. Such a plot is called a lag 1 scatter plot, and Figure 4 shows this plot for each of our four series. We note the following:

[1] For the wind speed series, the points tend to fall about a line of positive slope. Thus a wind speed with a certain value tends to be followed by one near that same value.

[2] The plot for the Willamette River data resembles that of the wind speed series except that the points are more spread out.

[3] For the atomic clock data, the points fall loosely about a line with a negative slope.

[4] For the ocean noise data, it is not obvious that there is a tendency of the points to cluster about a line in one direction or another.

We could create a lag k scatter plot by plotting x_{t+k} versus x_t, but, while such plots are informative, they are unwieldy to work with. To summarize the information in scatter plots similar to those in Figure 4, note that these plots indicate a roughly linear relationship between x_{t+1} and x_t; i.e., with $k = 1$, we can write

$$x_{t+k} = \alpha_k + \beta_k x_t + \epsilon_{t,k}$$

for some intercept α_k and slope β_k (possibly equal to 0), where $\epsilon_{t,k}$ represents an 'error' term that models deviations from strict linearity. If we make the assumption that a linear relationship holds approximately between x_{t+k} and x_t for all k, we can use as a summary statistic a well-known measure of the strength of the linear relationship between two ordered collections of variables $\{y_t\}$ and $\{z_t\}$, namely, the Pearson product moment correlation coefficient:

$$\hat{\rho} = \frac{\Sigma(y_t - \bar{y})(z_t - \bar{z})}{[\Sigma(y_t - \bar{y})^2 \Sigma(z_t - \bar{z})^2]^{1/2}},$$

where $\bar{y}$ and $\bar{z}$ are the sample means of the y_t and z_t terms, respectively. If we let $y_t = x_{t+k}$ and $z_t = x_t$ and if we adjust the summations in the denominator to make use of all available data, we are led to the lag k sample autocorrelation for a time series:

$$\hat{\rho}_k = \frac{\sum_{t=1}^{N-k}(x_{t+k} - \bar{x})(x_t - \bar{x})}{\sum_{t=1}^{N}(x_t - \bar{x})^2}. \tag{5}$$

Note that $\hat{\rho}_0 = 1$. As a sequence indexed by the lag k, the quantity $\{\hat{\rho}_k\}$ is called the sample autocorrelation sequence (sample acs) for the time series x_t.

The sample acs up to lag 32 is plotted for our four time series in Figures 6 and 7. A careful study of these plots can reveal a lot about our four time series. For example, in the Willamette River data, we see that x_t and x_{t+6} are negatively correlated, while x_t and x_{t+12} are positively correlated. This pattern is consistent with the visual evidence in Figure 2 that the river flow varies with a period of roughly 12 months.

Let us now assume that the time series $x_1, x_2, \ldots, x_N$ can be regarded as observed values (i.e., realizations) of corresponding random variables (rv's) $X_1, X_2, \ldots, X_N$. We use the term 'modeling of a time series' for the procedure by which we specify the properties of these

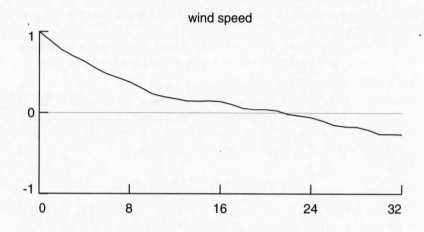

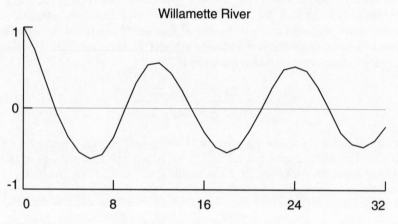

Figure 6. Sample autocorrelation sequences for the time series of Figure 2. The value of the acs at lag k is plotted versus k for k ranging from 0 to 32. By definition the acs for lag 0 is 1.

N rv's. For a class of models reasonable for time series such as those in Figures 2 and 3, $\hat{\rho}_k$ is an estimate of a corresponding population quantity called the lag k theoretical autocorrelation, defined as

$$\rho_k = E\{(X_t - \mu)(X_{t+k} - \mu)\}/\sigma^2\,,$$

where $E\{Z\}$ is our notation for the expectation operator applied to the random variable Z; $\mu = E\{X_t\}$ is the population mean of the time series; and $\sigma^2 = E\{(X_t - \mu)^2\}$ is the corresponding population variance. (Note, in particular, that ρ_k, μ and σ^2 do not depend on t. As we shall see later, models for which this is true play a central role in spectral analysis and are called stationary.) Moreover, if we make an additional

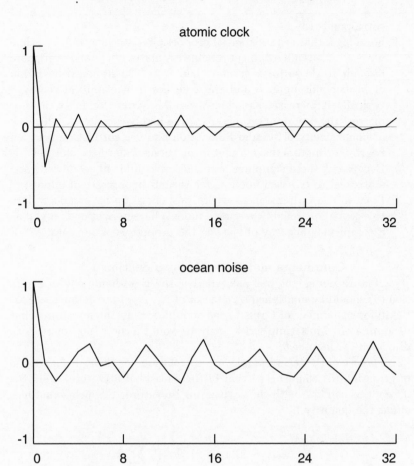

Figure 7. Sample autocorrelation sequences for the time series of Figure 3.

assumption, namely, that the X_t terms follow a multivariate Gaussian (normal) distribution, knowledge of the ρ_k terms, σ^2 and μ completely specifies our model. Thus, to fit such a model to a time series, we need only estimate ρ_k, σ^2 and μ from the available data.

As a set of parameters, the ρ_k terms, σ^2 and μ constitute a time domain characterization of a model. Since a model is completely specified by these parameters in the Gaussian case and since these parameters can all be estimated from a time series, why would we want to consider other characterizations? There are several reasons:

[1] The parameters of a model should ideally make it easy for us to visualize typical time series that can be generated by the model. Unfortunately, it takes a fair amount of experience to be able to look at a theoretical acs and visualize what kind of time series it

corresponds to.

[2] For a lag k that is a substantial proportion of the length N of a time series, it is often hard to get reliable estimates of ρ_k (and even more difficult to do so for k greater than N). The standard deviation of $\hat{\rho}_k$ depends upon k and the true acs in a complicated way – typically it increases as k increases. Moreover, for most cases of interest the estimators $\hat{\rho}_k$ and $\hat{\rho}_{k+1}$ are highly correlated. This lack of homogeneity of standard deviation and the correlation between nearby estimators make a plot of $\hat{\rho}_k$ versus k hard to interpret.

[3] Because of these sampling problems, it is difficult to devise good statistical tests based upon $\hat{\rho}_k$ for various hypotheses of interest.

[4] Even in the rare instances where we believe we have enough data to estimate ρ_k reliably, a second model characterization is useful as a complementary way of viewing the properties of our data.

Comments and Extensions to Section 1.1

[1]   The reader might well ask whether the right-hand side of Equation (5) should be multiplied by a factor of $N/(N-k)$ to compensate for the different number of terms in the summations in the numerator and denominator. Most time series analysts would answer 'no' for reasons discussed in Chapter 6.

[2]   The use of the term 'sample autocorrelation' for the right-hand side of Equation (5) conforms to that of the statistical literature. Unfortunately this conflicts with the engineering literature, in which sometimes either the quantity

$$\frac{1}{N} \sum_{t=1}^{N-k} (x_{t+k} - \bar{x})(x_t - \bar{x}) \text{ or } \frac{1}{N} \sum_{t=1}^{N-k} x_{t+k}x_t$$

is called the lag k sample autocorrelation. This latter notation can lead to unnecessary confusion between correlations and covariances and cause nonzero means to be ignored.

[3]   We do not want to leave the impression that lag k scatter plots for time series always indicate an approximately linear relationship between x_{t+k} and x_t. As a simple counterexample, Figure 9 shows the first 24 years of a time series of monthly average temperatures at St. Paul, Minnesota, and the lag 6 and 9 scatter plots for the entire time series (this extends from 1820 to 1983), both of which are highly nonlinear.

[4]   While it is reasonable that $\mu = E\{X_t\}$ is independent of time for the wind speed, atomic clock and ocean noise series, it would seem to be an unreasonable assumption for the Willamette River data. A more natural assumption is that $E\{X_t\}$ is a function of which month t occurs in. As we shall see, the key concept of stationarity assumes that certain quantities – including $E\{X_t\}$ – are independent of time. It would appear

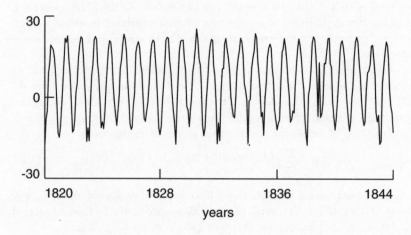

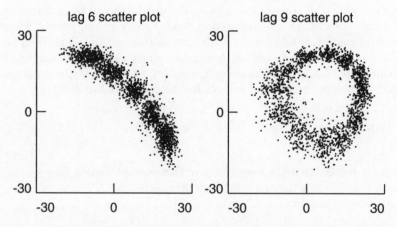

Figure 9. Plots of the first 24 years of the St. Paul temperature time series and the lag 6 and 9 scatter plots for the entire series. The temperature series is measured in degrees centigrade. For the lag k scatter plot ($k = 6$, 9), the value x_{t+k} is plotted on the vertical axis versus x_t on the horizontal axis.

at first that we can not assume a stationary model for the Willamette River data as we have implied above. In fact, as we shall discuss later, there is a mathematical trick that can be used to force such data into a stationary model (it involves assuming that the time origin of a periodic phenomenon can be regarded as being picked at random).

1.2 Spectral Analysis for a Simple Time Series Model
Some of the problems of estimation and interpretation that are associated with the acs are lessened (but not completely alleviated) when

we deal with a frequency domain characterization called the spectrum. The spectrum is simply a second way of characterizing models for time series. The objective of spectral analysis is to study and estimate the spectrum.

How exactly we define the spectrum depends upon what class of models we assume for a time series. A detailed definition for a useful class of models is presented in Chapter 4, but the key idea behind the spectrum is based upon a model for a time series consisting of a linear combination of sines and cosines with different frequencies; i.e.,

$$X_t = \mu + \sum_f \left[A(f) \cos\left(2\pi f t\right) + B(f) \sin\left(2\pi f t\right) \right] . \tag{10a}$$

The summation in Equation (10a) is a rather special one. To say what it means for the class of stationary processes is the subject of the spectral representation theorem (see Section 4.1). Fortunately, if we deal with a particularly simple (but unrealistic) model, we can say exactly what the summation means, define the spectrum in terms of elements involved in the summation, and thereby get an idea of what spectral analysis is all about. Let us assume that our time series can be modeled by a sum of a constant term μ and sinusoids with different fixed frequencies $\{f_j\}$ and random amplitudes $\{A_j\}$ and $\{B_j\}$ (the notation $\lfloor N/2 \rfloor$ refers to the greatest integer less than or equal to $N/2$):

$$X_t = \mu + \sum_{j=1}^{\lfloor N/2 \rfloor} \left[A_j \cos\left(2\pi f_j t\right) + B_j \sin\left(2\pi f_j t\right) \right], \qquad t = 1, 2, \ldots, N.$$

$$\tag{10b}$$

Here we require that the frequencies of the sinusoids have a very special form, namely, that they be related to the sample size N in the following way:

$$f_j \equiv j/N, \qquad 1 \le j \le \lfloor N/2 \rfloor.$$

The frequency f_j is often called the jth standard (or Fourier) frequency; it is a cyclical frequency measured in cycles per unit time as opposed to an angular frequency $\omega_j \equiv 2\pi f_j$ measured in radians per unit time (here and throughout this book the symbol '$\equiv$' means 'equal by definition'). For example, f_j is measured in cycles per 0.025 second for the wind speed series, while its units are cycles per month for the Willamette River series. We also assume that the amplitudes $\{A_j\}$ and $\{B_j\}$ are random variables with the following stipulations: for all j

$$E\{A_j\} = E\{B_j\} = 0 \text{ and } E\{A_j^2\} = E\{B_j^2\} = \sigma_j^2.$$

Thus the variance of the amplitudes associated with the jth standard frequency is just σ_j^2. We further assume that the A_j and B_j rv's are all mutually uncorrelated; i.e.,

$$E\{A_j A_k\} = E\{B_j B_k\} = 0 \text{ for } j \ne k$$

and
$$E\{A_j B_k\} = 0 \text{ for all } j, k.$$

It can now be argued (see Exercise [1.1]) that $E\{X_t\} = \mu$,

$$\sigma^2 = E\{(X_t - \mu)^2\} = \sum_{j=1}^{\lfloor N/2 \rfloor} \sigma_j^2, \tag{11a}$$

and

$$\rho_k = \frac{\sum_{j=1}^{\lfloor N/2 \rfloor} \sigma_j^2 \cos(2\pi f_j k)}{\sum_{j=1}^{\lfloor N/2 \rfloor} \sigma_j^2}. \tag{11b}$$

(We emphasize that we are considering models defined by (10b) for pedagogical purposes only. These have a number of undesirable features, not the least of which is an explicit dependence of the component frequencies f_j on the sample size N.)

For this model we *define* the spectrum by

$$S_j \equiv \sigma_j^2, \qquad 1 \le j \le \lfloor N/2 \rfloor.$$

A plot of S_j versus f_j merely shows us the variances of the random variables that determine the amplitudes of the sinusoidal terms at the standard frequencies. From Equation (11a), we have the following fundamental relationship:

$$\sum_{j=1}^{\lfloor N/2 \rfloor} S_j = \sigma^2.$$

Thus, for a time series generated by the model in Equation (10b), the population variance, σ^2, can be regarded as being composed of a sum of a number of components, each of which is associated with a different nonzero standard frequency. The contribution to the variance due to the sinusoidal terms with frequency f_j is given by S_j. A study of S_j versus f_j indicates where the variability in a time series is likely to come from.

Equation (11b) and the definition of the spectrum tell us that we can determine the acs and σ^2 if we know the spectrum. Conversely, it can be shown (see Exercise [1.5]) that we can determine the spectrum if we know the acs and σ^2. The spectrum is a frequency domain characterization for a model of a time series and is fully equivalent to the time domain characterization given by the acs and σ^2.

For a model given by Equation (10b), it is easy to simulate a typical time series: we use a random number generator on a computer to pick values for A_j and B_j and plug these into (10b) to generate a simulated time series. To illustrate this procedure, we will generate four such series using four different spectra. This exercise will show how a spectrum

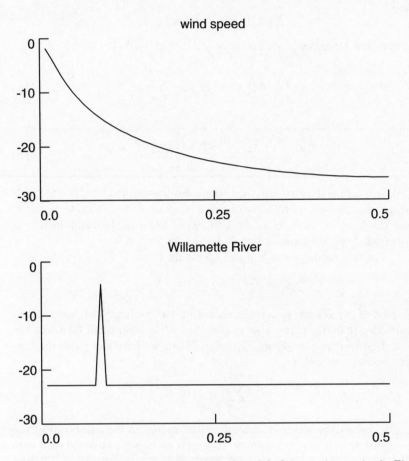

Figure 12. Plots of theoretical spectra of models for two time series in Figure 2. The 64 values that determine each spectra are connected by solid lines. The horizontal axis represents frequency measured in cycles per unit time. The vertical axis represents $10 \log_{10}(S_j)$, i.e., S_j expressed in decibels (dB).

can be used to tell us something about the structure of an associated time series. The four spectra that we will use are actually rough models for the four time series in Figures 2 and 3 (for the moment we ignore the question of where these models came from). Figures 12 and 13 show the four theoretical spectra; Figures 14 and 15 show the corresponding acs's (calculated via Equation (11b)); and Figures 16 and 17 show a simulated time series that corresponds to each of the four spectra. If a proposed spectrum is a reasonable model for a time series, the corresponding theoretical acs should resemble the sample acs for the series, and simulated time series from that spectrum should have roughly the

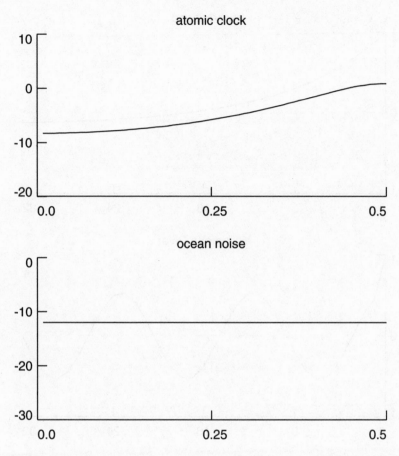

Figure 13. Plots of theoretical spectra of models for two time series in Figure 3.

same visual properties as the actual time series. Here are some specifics about our four time series and these figures.

[1] For the wind speed data, we assume that S_j is large for $j = 1$ and then tapers off rapidly as j gets large. Thus, the low frequency terms in Equation (10b) – these correspond to sinusoids with long periods – should predominate. The theoretical acs in Figure 14 is positive until lag 18. This picture agrees fairly well with the corresponding sample acs in Figure 6, which is positive until lag 22. The appearance of the simulated time series is one of rather broad swoops together with some choppiness (evidently due to the higher frequencies in Equation (10b)). The wind speed series and the corresponding simulated series appear to have the same kind of

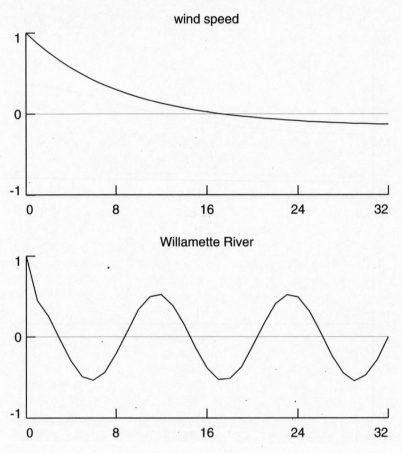

Figure 14. Plots of theoretical autocorrelation sequences of models for two time series in Figure 2 (cf. Figure 6).

bumpiness.

[2] For the Willamette River data, we assume a spectrum that is constant except for a spike at $j = 11$. Since $f_{11} = 11/128$, this frequency corresponds to a period of $1/f_{11} = 128/11 \approx 11.6$ months. This is the frequency with a period closest to 1 year in our model (the next closest is f_{10} with a corresponding period of 12.8 months). We would thus expect to see terms with about this period predominant in Equation (10b). The generated time series should have a tendency to fluctuate with this period. This is roughly true for both the simulated series (Figure 16) and the Willamette River series. The sample acs for the river flow data (Figure 6) and the theoretical acs (Figure 14) look fairly similar. (Here one of the limitations

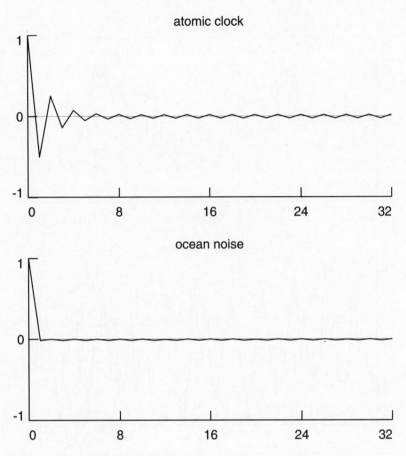

Figure 15. Plots of theoretical autocorrelation sequences of models for two time series in Figure 3 (cf. Figure 7).

of our simple model is apparent: from physical considerations, it would make more sense to have a term corresponding to a frequency of one cycle per year in our model, but to include such a term would destroy some of the nice mathematical properties of our model that we will need shortly.)

[3] For the atomic clock data, we assume S_j is large for $j = \lfloor N/2 \rfloor = 64$ and then tapers off rapidly as j decreases. Thus the high frequency terms (i.e., sinusoids with short periods) should predominate. The theoretical acs oscillates between positive and negative values with an amplitude that decreases rapidly for the first few lags. The sample acs in Figure 7 for these data shows more variability than this theoretical acs (particularly for the higher lags), but the dis-

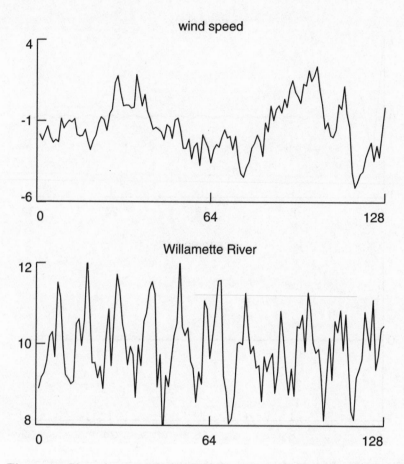

Figure 16. Plots of two simulated time series with statistical properties similar to series in Figure 2.

crepancy can be due to sampling variation. The appearance of the simulated time series in Figure 17 is one of choppiness as the series swings back and forth from positive to negative values. The atomic clock data and the simulated series have the same 'feel' to them.

[4] Finally, for the ocean noise data, we assume a spectrum S_j that is constant for all j. A time series generated from such a spectrum is often called 'white noise' in analogy to white light, which is composed of equal contributions of a whole range of colors. The theoretical acs is close to 0 for all $|k| > 0$ (see Exercise [1.6]). There should be no discernible patterns in a time series generated from a white noise spectrum, and indeed there appears to be little in the ocean noise data and none in its simulation. (We will return to

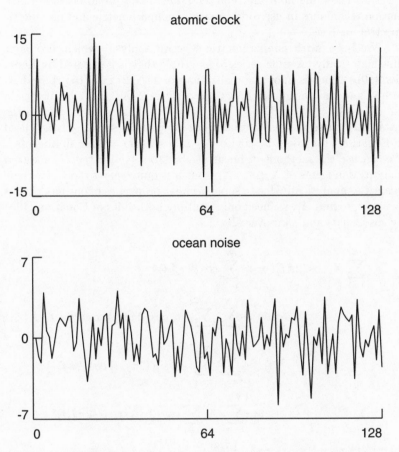

Figure 17. Plots of two simulated time series with statistical properties similar to series in Figure 3.

this example in Chapter 10, where we will find that, using a test for white noise, we can reject the hypothesis that this time series is white noise! This is somewhat evident from its sample acs in Figure 7, which shows a tendency to oscillate with a period of five time units. Nonetheless, for the purposes of this chapter, the ocean noise series is close enough to white noise for us to use it as an example of such.)

1.3 Nonparametric Estimation of the Spectrum from Data

The estimation of spectra from a given time series is a complicated subject and is the main concern of this book. For the simple model described by Equation (10b), we will give two methods for estimating spectra. These methods are representative of two broad classes of estimation techniques in use today, namely, nonparametric and parametric spectral analysis.

We begin with nonparametric spectral analysis, which also came first historically. A time series of length N that is generated by Equation (10b) depends upon the realizations of $2\lfloor N/2\rfloor$ rv's (the A_j and B_j terms) and the parameter μ, a total of $M \equiv 2\lfloor N/2\rfloor + 1$ quantities in all. Now, $M = N$ for N odd, and $M = N + 1$ for N even, but in the latter case there are also actually just N quantities: $B_{N/2}$ is not used in Equation (10b) because $\sin(2\pi f_{N/2}t) = \sin(\pi t) = 0$ for all integers t. We can use the methods of linear algebra to solve for the N unknown quantities in terms of $X_1, \ldots, X_N$, but it is quite easy to solve for these quantities analytically due to some peculiar properties of our model. For example, to find A_j, we need only multiply both sides of Equation (10b) by $\cos(2\pi f_j t)$ and sum over all t:

$$\sum_{t=1}^{N} X_t \cos(2\pi f_j t) = \mu \sum_{t=1}^{N} \cos(2\pi f_j t)$$
$$+ \sum_{t=1}^{N} \sum_{k=1}^{\lfloor N/2\rfloor} A_k \cos(2\pi f_k t) \cos(2\pi f_j t)$$
$$+ \sum_{t=1}^{N} \sum_{k=1}^{\lfloor N/2\rfloor} B_k \sin(2\pi f_k t) \cos(2\pi f_j t)$$
$$= \sum_{k=1}^{\lfloor N/2\rfloor} A_k \sum_{t=1}^{N} \cos(2\pi f_k t) \cos(2\pi f_j t)$$
$$+ \sum_{k=1}^{\lfloor N/2\rfloor} B_k \sum_{t=1}^{N} \sin(2\pi f_k t) \cos(2\pi f_j t), \quad (18)$$

since it can be shown that $\sum_{t=1}^{N} \cos(2\pi f_j t) = 0$ (see Exercise [1.3c]). For $1 \leq j < N/2$, the following 'orthogonality relationships' hold:

$$\sum_{t=1}^{N} \cos(2\pi f_k t) \cos(2\pi f_j t) = \begin{cases} 0, & \text{if } k \neq j; \\ N/2, & \text{if } k = j, \end{cases}$$

and

$$\sum_{t=1}^{N} \sin(2\pi f_k t) \cos(2\pi f_j t) = 0 \qquad \text{for all } j \text{ and } k$$

(see Exercise [1.4c]). If N is even and $j = N/2$, the only difference is that $N/2$ is replaced by N in the second line above. Thus it follows from Equation (18) that

$$A_j = \frac{2}{N} \sum_{t=1}^{N} X_t \cos(2\pi f_j t)$$

if $1 \leq j < N/2$ and, if N is even,

$$A_{N/2} = \frac{1}{N} \sum_{t=1}^{N} X_t \cos(2\pi f_{N/2} t).$$

Likewise, it can be shown that for $1 \leq j < N/2$

$$B_j = \frac{2}{N} \sum_{t=1}^{N} X_t \sin(2\pi f_j t).$$

Finally, by simply summing both sides of Equation (10b) with respect to t, we find that

$$\bar{X} \equiv \frac{1}{N} \sum_{t=1}^{N} X_t = \mu$$

exactly. (Perfect estimation of model parameters rarely occurs in statistics: the simple model we are using here for pedagogical purposes has some special – and implausible – properties.)

Since $E\{A_j^2\} = E\{B_j^2\} = \sigma_j^2$, for a time series $x_1, x_2, \ldots, x_N$ that is a realization of a model given by (10b), it is natural to estimate S_j by

$$\hat{S}_j \equiv \frac{A_j^2 + B_j^2}{2}$$

$$= \frac{2}{N^2} \left[\left(\sum_{t=1}^{N} x_t \cos(2\pi f_j t) \right)^2 + \left(\sum_{t=1}^{N} x_t \sin(2\pi f_j t) \right)^2 \right] \quad (19a)$$

for $1 \leq j < N/2$ and, if N is even,

$$\hat{S}_{N/2} \equiv \frac{1}{N^2} \left(\sum_{t=1}^{N} x_t \cos(2\pi f_{N/2} t) \right)^2. \quad (19b)$$

As examples of this estimation procedure, the thin curves in Figures 20 and 21 are graphs of $\hat{S}_j$ versus f_j for the four time series in Figures 2 and 3. We can now see some justification for the theoretical spectra for these time series given previously (shown in these figures by the thick curves): the theoretical spectra are smoothed versions of the

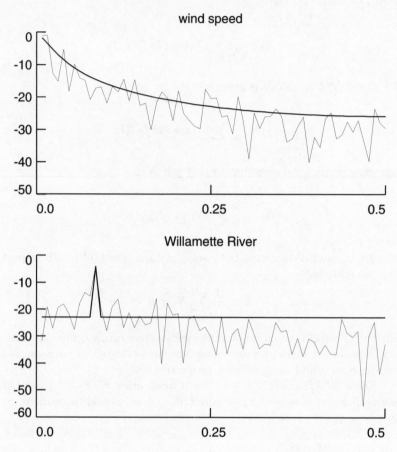

Figure 20. Comparison of theoretical and estimated spectra for two time series in Figure 2. The thick curves are the theoretical spectra (copied from Figure 12), while the thin curves are the estimated spectra. The units of the axes are the same as those of Figure 12.

estimated spectra. Here are some points to note about the nonparametric spectral estimates:

[1] Since $\hat{S}_j$ involves just A_j^2 and B_j^2 and since the A_j and B_j rv's are mutually uncorrelated, it can be argued that the $\hat{S}_j$ rv's should be approximately uncorrelated (the assumption of Gaussianity for A_j and B_j makes this statement true without approximation). This property should be contrasted to that of the sample acs, which is highly correlated for values with lags close to one another.

[2] Because of this approximate uncorrelatedness, it is possible to derive good statistical tests of hypotheses by basing the tests on some

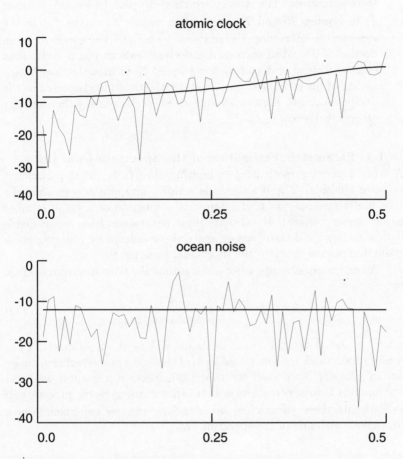

Figure 21. Comparison of theoretical and estimated spectra for two time series in Figure 3.

form of the $\hat{S}_j$.

[3] Since $\hat{S}_j$ is only a 'two degrees of freedom' estimate of σ_j^2, it follows that there should be considerable variability in $\hat{S}_j$ as a function of j even if the true underlying spectrum changes slowly with j. The bumpiness of the estimated spectra in Figures 20 and 21 can be attributed to this sampling variability. If we can assume that S_j varies slowly with j, we can smooth the $\hat{S}_j$ locally to come up with a more reasonable estimate of S_j.

[4] It can be shown (see Chapter 6) that, if we consider $\log(\hat{S}_j)$ instead of $\hat{S}_j$, the variance of the former is approximately the same for all f_j. Thus a plot of $\log(\hat{S}_j)$ versus f_j is easier to interpret than plots of the sample acs, for which there is no known 'variance stabilizing'

transformation. This is why we chose to plot $10 \log_{10}(\hat{S}_j)$ versus f_j in Figures 20 and 21 rather than simply $\hat{S}_j$ versus f_j. In the engineering literature, the units of $10 \log_{10}(\hat{S}_j)$ are said to be in decibels (dB). We can regard a decibel scale as just a convenient equivalent way of depicting a log scale. Note that the 'local' variability of the estimated spectra about the theoretical spectra on the decibel scales of Figures 20 and 21 is approximately the same for all spectral levels.

1.4 Parametric Estimation of the Spectrum from Data

A second popular method of estimating spectra is called parametric spectral analysis. The basic idea is simple: first, we assume that the true underlying spectrum of interest is a function of a small number of parameters; second, we estimate these parameters from the available data somehow; and third, we estimate the spectrum by plugging these estimated parameters into the functional form for S.

As an example, suppose we assume that the true spectrum is given by .

$$S_j(\alpha, \beta) = \frac{\beta}{1 + \alpha^2 - 2\alpha \cos{(2\pi f_j)}}, \tag{22}$$

where α and β are the parameters we will need to estimate. (The reader might think the functional form of this spectrum is rather strange, but in Chapter 9 we shall see that (22) arises in a natural way and corresponds to a spectrum of a first-order autoregressive process.) If we estimate these parameters by $\hat{\alpha}$ and $\hat{\beta}$, say, we can then form a parametric estimate of the spectrum by

$$\hat{S}_j(\hat{\alpha}, \hat{\beta}) = \frac{\hat{\beta}}{1 + \hat{\alpha}^2 - 2\hat{\alpha} \cos{(2\pi f_j)}}.$$

Now it can be shown that, if a time series has an underlying spectrum given by (22), then $\rho_1 \approx \alpha$ to a good approximation so that we can estimate α by $\hat{\alpha} = \hat{\rho}_1$. By analogy to the relationship

$$\sum_{j=1}^{\lfloor N/2 \rfloor} S_j = \sigma^2,$$

we require that

$$\sum_{j=1}^{\lfloor N/2 \rfloor} \hat{S}_j(\hat{\alpha}, \hat{\beta}) = \frac{1}{N} \sum_{t=1}^{N} (x_t - \bar{x})^2 \equiv \hat{\sigma}^2$$

and hence estimate β by

$$\hat{\beta} = \hat{\sigma}^2 \left(\sum_{j=1}^{\lfloor N/2 \rfloor} \frac{1}{1 + \hat{\alpha}^2 - 2\hat{\alpha} \cos(2\pi f_j)} \right)^{-1}.$$

In fact, we have already shown two examples of parametric spectral analysis using Equation (22): what we called the theoretical spectra for the wind speed data and the atomic clock data (the thick curves in Figures 20 and 21) are actually just parametric spectral estimates using the procedure we just described. The parametric estimates of the spectra are much smoother than their corresponding nonparametric estimates. This might or might not be desirable depending upon the physical process that generates a particular time series.

There are two difficulties with the parametric approach that we need to mention. One is that it is hard to say much about the statistical properties of the estimated spectrum. Without making some additional assumptions, we cannot, for example, calculate a 95% confidence interval for S_j. Also we must be careful about which class of functional forms we assume initially. If there are too many parameters to be estimated (relative to the length of the time series), all the parameters cannot be reliably estimated; if there are too few parameters, the true underlying spectrum might not be well approximated by any spectrum specified by that number of parameters (in fact, this happens here: an sdf of form (22) is not a good approximation for the Willamette River data).

1.5 Uses of Spectral Analysis

In summary, spectral analysis is an analysis of variance technique. It is based upon our ability to represent a time series in some fashion as a sum of sines and cosines of different frequencies and amplitudes. The variance of a time series is broken down into a number of components, the totality of the components being called the spectrum. Each component is associated with a particular frequency and represents the contribution that frequency makes to the total variability of the series.

In the remaining chapters of this book we extend the basic ideas discussed here to the class of stationary stochastic processes (Chapter 2). We discuss the theory of spectral analysis for deterministic functions (Chapter 3), which is important in its own right and also serves as a motivating example for the corresponding theory for stationary processes. We motivate and state the spectral representation theorem for stationary processes (Chapter 4). This theorem essentially defines what Equation (10a) means for stationary processes, which in turn allows us to define their spectra. We next discuss the central role that the theory of linear filters plays in spectral analysis (Chapter 5); in particular, this

theory allows us to easily determine the spectra for an important class
of processes. The bulk of the remaining chapters is devoted to various
aspects of nonparametric and parametric spectral estimation theory for
stationary processes.

. We close this introductory chapter with some examples of the many
practical uses for spectral analysis. The following quote from Kung and
Arun (1987) indicates a few of the areas in which spectral analysis has
been used:

> Some typical applications where the problem of spectrum estima-
> tion is encountered are: interference spectrometry; the design of
> Wiener filters for signal recovery and image restoration; the design
> of channel equalizers in communications systems; the determina-
> tion of formant frequencies (location of spectral peaks) in speech
> analysis; the retrieval of hidden periodicities from noisy data (lo-
> cating spectral lines); the estimation of source-energy distribution
> as a function of angular direction in passive underwater sonar; the
> estimation of the brightness distribution (of the sky) using aperture
> synthesis telescopy in radio astronomy; and many others.

Here we list some of the uses for spectral analysis.

[1] *Testing theories*

It is sometimes possible to derive the spectrum for a certain physical phenomenon based upon a theoretical model. One way to test
the assumed theory is to collect data concerning the phenomenon,
estimate the spectrum from the observed time series, and compare
it with the theoretical spectrum. For example, scientists have been
able to derive the spectrum for thermal noise produced by the random motion of free electrons in a conductor. The average current
is zero, but the fluctuations produce a noise voltage across the conductor. It is easy to see how one could test the physical theory by
an appropriate experiment. A second example concerns the wind
speed data we have examined above. Some physical theories predict that there should be a well-defined peak in the spectrum in the
low frequency range for wind speed data collected near coastlines.
One can use spectral analysis to test this theory on actual data.

[2] *Investigating data*

We have seen that knowledge of a spectrum allows us to describe
in broad outline what a time series should look like that is drawn
from a process with such a spectrum. Conversely, spectral analysis
may reveal certain features of a time series that are not obvious
from other analyses. It has been used in this way in many of the
physical sciences. Two examples are geophysics (for the study of
intraplate volcanism in the Pacific Ocean basin and attenuation of
compressional wav s) and oceanography (for examining tidal be-

havior and the effect of pressure and wind stress on the ocean). As a further example, Walker (1985) in the 'Amateur Scientist' column of *Scientific American* describes the efforts of researchers to answer the question 'Does the rate of rainfall in a storm have a pattern or is it random?' – it evidently has a pattern which spectral analysis can discern.

[3] *Discriminating data*

Spectral analysis is also a popular means of clearly demonstrating the qualitative differences between different time series. For example, Jones *et al.* (1972) investigated the spectra of brain wave measurements of babies before and after they were subjected to a flashing light. The two sets of estimated spectra showed a clear difference between the 'before' and 'after' time series. As a second example, Jones *et al.* (1987) investigated electrical waves generated during the contraction of human muscles and found that waves typical of normal, myopathic and neurogenic subjects could be classified in terms of their spectral characteristics.

[4] *Performing diagnostic tests*

Spectral analysis is often used in conjunction with model fitting. For example, Box and Jenkins (1976) describe a procedure for fitting autoregressive, integrated, moving average (ARIMA) models to time series. Once a particular ARIMA model has been fit to some data, an analyst can tell how well the model represents the data by examining what are called the estimated innovations. For ARIMA models, the estimated innovations play the same role that residuals play in linear least squares analysis. Now the estimated innovations are just another time series, and if the ARIMA model for the original data is a good one, the spectrum of the estimated innovations should approximate a white noise spectrum. Thus spectral analysis can be used to perform a goodness of fit test for ARIMA models.

[5] *Assessing the predictability of a time series*

One of the most popular uses of time series analysis is to predict future values of a time series. For example, one measure of the performance of atomic clocks is based upon how predictable they are as time keepers. Spectral analysis plays a key role in assessing the predictability of a time series.

1.6 Exercises

[1.1] Verify Equations (11a) and (11b).

[1.2] Use the Euler relationship

$$e^{i2\pi f} = \cos{(2\pi f)} + i\sin{(2\pi f)},$$

where $i \equiv \sqrt{-1}$, to show that Equations (19a) and (19b) can be rewritten in the following form:

$$\hat{S}_j = \frac{k}{N^2} \left| \sum_{t=1}^{N} x_t e^{-i2\pi f_j t} \right|^2$$

for $1 \le j \le \lfloor N/2 \rfloor$, where $k = 2$ if $j < N/2$, and $k = 1$ if N is even and $j = N/2$.

[1.3]   a) If z is any complex number not equal to 1, show that

$$\sum_{t=1}^{N} z^t = \frac{z - z^{N+1}}{1 - z}.$$

b) Show (using Euler's relationship in Exercise [1.2]) that

$$\cos(2\pi f) = \frac{e^{i2\pi f} + e^{-i2\pi f}}{2} \quad \text{and} \quad \sin(2\pi f) = \frac{e^{i2\pi f} - e^{-i2\pi f}}{2i}.$$

c) Show that

$$\sum_{t=1}^{N} e^{i2\pi f t} = \begin{cases} N e^{i(N+1)\pi f} \mathcal{D}_N(f), & \text{if } f \ne 0, \pm 1, \pm 2, \ldots; \\ N, & \text{if } f = 0, \pm 1, \pm 2, \ldots, \end{cases}$$

where

$$\mathcal{D}_N(f) \equiv \frac{\sin(N\pi f)}{N \sin(\pi f)}$$

is one form of Dirichlet's kernel (this plays a prominent role in Chapters 3 and 6). Use the above to show that, for integer j such that $1 \le j < N$,

$$\sum_{t=1}^{N} \cos(2\pi f_j t) = 0 \quad \text{and} \quad \sum_{t=1}^{N} \sin(2\pi f_j t) = 0,$$

where $f_j \equiv j/N$.

d) Use part c) to show that

$$\sum_{t=0}^{N-1} e^{i2\pi f t} = \begin{cases} N e^{i(N-1)\pi f} \mathcal{D}_N(f), & \text{if } f \ne 0, \pm 1, \pm 2, \ldots; \\ N, & \text{if } f = 0, \pm 1, \pm 2, \ldots. \end{cases}$$

e) Use part c) to show that

$$\sum_{t=-N}^{N} e^{i2\pi f t} = \begin{cases} (2N+1)\mathcal{D}_{2N+1}(f), & \text{if } f \ne 0, \pm 1, \pm 2, \ldots; \\ 2N+1, & \text{if } f = 0, \pm 1, \pm 2, \ldots. \end{cases}$$

[1.4] The following trigonometric relationships are used in Chapters 6 and 10.

a) Use Exercise [1.3c] and the fact that

$$e^{i2\pi(f\pm f')t}$$

$$= e^{i2\pi ft}e^{\pm i2\pi f't}$$
$$= [\cos{(2\pi ft)} + i\sin{(2\pi ft)}][\cos{(2\pi f't)} \pm i\sin{(2\pi f't)}]$$
$$= \cos{(2\pi ft)}\cos{(2\pi f't)} \mp \sin{(2\pi ft)}\sin{(2\pi f't)}$$
$$\quad + i[\sin{(2\pi ft)}\cos{(2\pi f't)} \pm \cos{(2\pi ft)}\sin{(2\pi f't)}]$$

to evaluate

$$\sum_{t=1}^{N}\cos{(2\pi ft)}\cos{(2\pi f't)}, \quad \sum_{t=1}^{N}\cos{(2\pi ft)}\sin{(2\pi f't)},$$

and

$$\sum_{t=1}^{N}\sin{(2\pi ft)}\sin{(2\pi f't)}.$$

In particular show that, if $f \pm f' \neq 0, \pm 1, \pm 2, \ldots,$

$$\sum_{t=1}^{N}\cos{(2\pi ft)}\cos{(2\pi f't)} = C_N(f - f') + C_N(f + f'),$$

$$\sum_{t=1}^{N}\cos{(2\pi ft)}\sin{(2\pi f't)} = S_N(f - f') + S_N(f + f'),$$

and

$$\sum_{t=1}^{N}\sin{(2\pi ft)}\sin{(2\pi f't)} = C_N(f - f') - C_N(f + f'),$$

where

$$C_N(f) \equiv \frac{N}{2}\mathcal{D}_N(f)\cos{[(N+1)\pi f]}$$

and

$$S_N(f) \equiv \frac{N}{2}\mathcal{D}_N(f)\sin{[(N+1)\pi f]},$$

with $\mathcal{D}_N(\cdot)$ defined as in Exercise [1.3c].

b) Show that for $f \neq 0, \pm 1/2, \pm 1, \ldots,$

$$\sum_{t=1}^{N}\cos^2{(2\pi ft)} = \frac{N}{2} + \frac{\sin{(N2\pi f)}}{2\sin{(2\pi f)}}\cos{[(N+1)2\pi f]},$$

$$\sum_{t=1}^{N}\cos{(2\pi ft)}\sin{(2\pi ft)} = \frac{\sin{(N2\pi f)}}{2\sin{(2\pi f)}}\sin{[(N+1)2\pi f]},$$

and

$$\sum_{t=1}^{N} \sin^2(2\pi ft) = \frac{N}{2} - \frac{\sin(N2\pi f)}{2\sin(2\pi f)} \cos\left[(N+1)2\pi f\right].$$

c) Show that

$$\sum_{t=1}^{N} \cos^2(2\pi f_j t) = \sum_{t=1}^{N} \sin^2(2\pi f_j t) = \frac{N}{2},$$

$$\sum_{t=1}^{N} \cos(2\pi f_j t)\sin(2\pi f_j t) = \sum_{t=1}^{N} \cos(2\pi f_j t)\sin(2\pi f_k t) = 0,$$

and

$$\sum_{t=1}^{N} \cos(2\pi f_j t)\cos(2\pi f_k t) = \sum_{t=1}^{N} \sin(2\pi f_j t)\sin(2\pi f_k t) = 0,$$

where $f_j \equiv j/N$ and $f_k \equiv k/N$ with j and k both integers such that $j \neq k$ and $1 \leq j, k < N/2$. Show that, for N even,

$$\sum_{t=1}^{N} \cos^2(2\pi f_{N/2} t) = N$$

and

$$\sum_{t=1}^{N} \cos(2\pi f_{N/2} t)\sin(2\pi f_{N/2} t) = \sum_{t=1}^{N} \sin^2(2\pi f_{N/2} t) = 0.$$

[1.5] Show how σ_j^2 in Equation (11b) can be expressed in terms of $\{\rho_k\}$ and σ^2.

[1.6]   a) Show that the acs for the 'white noise' defined in Section 1.2 as a model for the ocean noise data is close to – but not exactly – zero for all lags k such that $0 < |k| < N$ (see also Figure 15; here we do *not* assume that N is necessarily 128). This is another peculiarity of the model defined by Equation (10b): the usual definition for white noise (see Chapters 2 and 4) implies both a constant spectrum (to be precise, a constant spectral density function) and an acs that is *exactly* zero for all $|k| > 0$, whereas a constant spectrum in (10b) has a corresponding acs that is *not* exactly zero at all nonzero lags.

   b) Now consider the model

$$X_t = \mu + \sum_{j=0}^{\lfloor N/2 \rfloor} \left[A_j \cos(2\pi f_j t) + B_j \sin(2\pi f_j t)\right],$$

$t = 1, 2, \ldots, N$. This differs from Equation (10b) only in that the summation starts at $j = 0$ instead of $j = 1$. We assume that the statistical properties of A_j and B_j are as described below Equation (10b). Since $\cos(2\pi f_0 t) = 1$ and $\sin(2\pi f_0 t) = 0$ for all t, this modification just introduces randomness in the constant term in our model. For this new model, show that $E\{X_t\} = \mu$ and that the variance σ^2 and the acs $\{\rho_k\}$ are given, respectively, by Equations (11a) and (11b) if we replace '$j = 1$' with '$j = 0$' in the summations. Show that, if

$$\sigma_j^2 = \begin{cases} 2\nu^2, & 1 \le j < N/2; \\ \nu^2, & \text{otherwise,} \end{cases}$$

with $\nu^2 > 0$, then we have

$$\rho_k = \begin{cases} 1, & k = 0; \\ 0, & 0 < |k| < N. \end{cases}$$

Thus, if we introduce randomness in the constant term, we can produce a model that agrees with the usual definition of white noise for $|k| < N$ in terms of its acs, but we no longer have a constant spectrum.

2

Stationary Stochastic Processes

2.0 Introduction

Spectral analysis almost invariably deals with a class of models called stationary stochastic processes. The material in this chapter is a brief review of the theory behind such processes. The reader is referred to Chapter 3 of Priestley (1981), Chapter 10 of Papoulis (1991) or Chapter 1 of Yaglom (1987) for complementary discussions.

2.1 Stochastic Processes

Consider the following experiment (see Figure 31): we hook up a resistor to an oscilloscope in such a way that we can examine the voltage variations across the resistor as a function of time. Every time we press a 'reset' button on the oscilloscope, it displays the voltage variations for the 1 second interval following the 'reset.' Since the voltage variations are presumably caused by such factors as small temperature variations in the resistor, each time we press the 'reset' button, we will observe a different display on the oscilloscope. Owing to the complexity of the factors that influence the display, there is no way that we can use the laws of physics to predict what will appear on the oscilloscope. However, if we repeat this experiment over and over, we soon see that, although we view a different display each time we press the 'reset' button, the displays resemble each other: there is a characteristic 'bumpiness' shared by all the displays.

We can model this experiment by considering a large bowl in which we have placed pictures of all the oscilloscope displays that we could possibly observe. Pushing the 'reset' button corresponds to reaching into the bowl and choosing 'at random' one of the pictures. Loosely speaking, we call the bowl of all possible pictures together with the

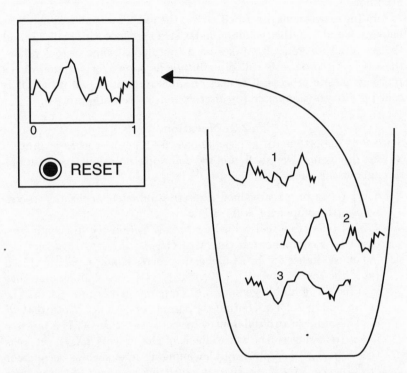

Figure 31. The oscilloscope experiment and a simplified bowl model. Here the ensemble contains only three possible realizations, the second of which was chosen.

mechanism by which we select the pictures a *stochastic process*. The one particular picture we actually draw out at a given time is called a *realization* of the stochastic process. The collection of all possible realizations is called the *ensemble*.

A more precise definition utilizes the concept of a random variable (rv), defined as a function, or mapping, from the sample space of possible outcomes of a random experiment to the real line (for a real-valued rv), the complex plane (for a complex-valued rv) or m-dimensional Euclidean space (for a vector-valued rv of dimension m). If we let the sample space be the ensemble of all possible realizations for an experiment such as the one described above, then for any fixed time t we can define an rv $X(t)$ that describes the outcome of the experiment at time t (this would be the height of the picture on the oscilloscope at time t in our example). We can now give the following formal definition of a stochastic process: a stochastic (or random) process, denoted by $\{\, X(t) : t \in T \,\}$, is a family of rv's indexed by t, where t belongs to some given index set T. For convenience, we refer to t as representing time, although in general it need not.

In the experiment described above, the index set is all real numbers between 0 and 1. Other common index sets are the entire real axis and the set of all integers. If t takes on a continuous range of real values, the stochastic process is called a *continuous parameter* (or *continuous time*) stochastic process; if t takes on a discrete set of real values, it is called a *discrete parameter* (or *discrete time*) stochastic process.

2.2 Notation

Since it is important to be clear about the type of stochastic process we are discussing, we shall use the following notational conventions throughout this book:

[1] $\{X_t\}$ refers to a real-valued *discrete parameter* stochastic process whose tth component is X_t, while

[2] $\{X(t)\}$ refers to a real-valued *continuous parameter* stochastic process whose component at time t is $X(t)$.

[3] When the index set for a stochastic process is not explicitly stated (as is the case above with $\{X_t\}$ and $\{X(t)\}$), we shall assume that it is the set of all integers for a discrete parameter process and the entire real axis for a continuous parameter process. Note that 't' is being used in two different ways here: the 't' in X_t is a unitless index (referring to the tth element of the process $\{X_t\}$), whereas the 't' in $X(t)$ has physically meaningful units such as seconds or days (hence $X(t)$ is the element occurring at time t of the process $\{X(t)\}$).

[4] On occasion we will need to discuss more than one stochastic process at a time. To distinguish among them, we will either introduce another symbol besides X (such as in $\{Y_t\}$) or add a second index following the time index. For example, $\{X_{t,j}\}$ and $\{X_{t,k}\}$ refer to the jth and kth discrete parameter processes, while $\{X(t,j)\}$ and $\{X(t,k)\}$ refer to two continuous parameter processes.

[5] We reserve the symbol Z for a complex-valued rv whose real and imaginary components are real-valued rv's. With an index added, $\{Z_t\}$ is a complex-valued discrete parameter stochastic process with a tth component formed from, say, the real-valued rv's $X_{t,1}$ and $X_{t,2}$:

$$Z_t = X_{t,1} + iX_{t,2},$$

where $i \equiv \sqrt{-1}$ (the symbol '$\equiv$' means 'equal by definition'). Likewise, $\{Z(t)\}$ is a complex-valued continuous parameter stochastic process with a tth component formed from two real-valued rv's, say, $X(t,1)$ and $X(t,2)$:

$$Z(t) = X(t,1) + iX(t,2).$$

2.3 Basic Theory for Stochastic Processes

Let us first consider the real-valued discrete parameter stochastic process $\{X_t\}$. Since, for t fixed, X_t is an rv, it has an associated cumulative probability distribution function (cpdf) given by

$$F_t(a) = \mathbf{P}[X_t \le a],$$

where the notation $\mathbf{P}[A]$ is the probability that the event A will occur. If we assume that they exist, the first moment and the second central moment for this cpdf are given, respectively, by

$$E\{X_t\} = \int_{-\infty}^{\infty} x\,dF_t(x) \equiv \mu_t \tag{33}$$

and

$$\mathrm{var}\{X_t\} = \int_{-\infty}^{\infty} (x - \mu_t)^2\,dF_t(x) \equiv \sigma_t^2.$$

In general, μ_t and σ_t^2 depend upon the index t. (The two integrals above are examples of Riemann–Stieltjes integrals; these are briefly reviewed in the Comments and Extensions below.)

Since $\{X_t\}$ is a stochastic process, we are interested in the relationships between the various rv's that are part of it. These are expressed by various higher order cpdf's. For example, for any t_1 and t_2 in the index set T,

$$F_{t_1,t_2}(a_1, a_2) = \mathbf{P}[X_{t_1} \le a_1, X_{t_2} \le a_2]$$

gives the bivariate cpdf for X_{t_1} and X_{t_2}. More generally, for any integer $n \ge 1$ and any $t_1, t_2, \ldots, t_n$ in the index set, we can define the n-dimensional cpdf by

$$F_{t_1,t_2,\ldots,t_n}(a_1, a_2, \ldots, a_n) = \mathbf{P}[X_{t_1} \le a_1, X_{t_2} \le a_2, \ldots, X_{t_n} \le a_n].$$

These higher order cpdf's completely specify the joint statistical properties of the rv's in the stochastic process.

The above discussion also holds for a real-valued continuous parameter stochastic process $\{X(t)\}$ and for complex-valued discrete and continuous parameter stochastic processes if we make some obvious extensions. For example, for a complex-valued continuous parameter stochastic process $\{Z(t)\}$ with

$$Z(t) = X(t, 1) + iX(t, 2),$$

we can define a univariate cpdf by

$$F_t(a, b) = \mathbf{P}[X(t, 1) \le a, X(t, 2) \le b]$$

and a bivariate cpdf by

$$F_{t_1,t_2}(a_1, b_1, a_2, b_2)$$
$$= \mathbf{P}[X(t_1, 1) \le a_1, X(t_1, 2) \le b_1, X(t_2, 1) \le a_2, X(t_2, 2) \le b_2].$$

Comments and Extensions to Section 2.3

[1] We state here one definition for the Riemann–Stieltjes integral and a few facts concerning it (we will need a stochastic version of this integral when we introduce the spectral representation theorem for stationary processes in Section 4.1); for more details, see Section 18.9 of Taylor and Mann (1972) or Section 4.2 of Greene and Knuth (1990), from which the following material is adapted. Let $g(\cdot)$ and $H(\cdot)$ be two real-valued functions defined over the interval $[L, U]$ with $L < U$, and let P_n be a partition of this interval of size $n + 1$; i.e., P_n is a set of $n + 1$ points x_j such that

$$L = x_0 < x_1 < \cdots < x_{n-1} < x_n = U.$$

Define the 'mesh fineness' of the partition P_n as

$$|P_n| \equiv \max\left(x_1 - x_0, x_2 - x_1, \ldots, x_{n-1} - x_{n-2}, x_n - x_{n-1}\right).$$

Let x_j' be any point in the interval $[x_{j-1}, x_j]$, and consider the summation

$$\mathcal{S}(P_n) \equiv \sum_{j=1}^{n} g(x_j') \left[H(x_j) - H(x_{j-1})\right].$$

The Riemann–Stieltjes integral is defined as the limit

$$\int_L^U g(x)\, dH(x) \equiv \lim_{|P_n| \to 0} \mathcal{S}(P_n) \tag{34}$$

provided that $\mathcal{S}(P_n)$ converges to a unique limit as the mesh fineness decreases to zero. Extension of the above to allow $L = -\infty$ and $U = \infty$ is done by the same limiting argument as is used in the case of a Riemann integral.

We now note the following facts (all of which assume that $g(\cdot)$ and $H(\cdot)$ are such that the Riemann–Stieltjes integral over $[L, U]$ exists).

a) If $H(x) = x$, the Riemann–Stieltjes integral reduces to the ordinary Riemann integral $\int_L^U g(x)\, dx$.

b) If $H(\cdot)$ is differentiable everywhere over the interval $[L, U]$ with derivative $h(\cdot)$, the Riemann–Stieltjes integral reduces to the ordinary Riemann integral $\int_L^U g(x)h(x)\, dx$.

c) Suppose that a is such that $L < a < U$ and that

$$H(x) = \begin{cases} c, & L \leq x < a; \\ b + c, & a \leq x \leq U; \end{cases}$$

i.e., $H(\cdot)$ is a step function with a single step of size b at $x = a$. Then

$$\int_L^U g(x)\, dH(x) = bg(a).$$

In general, if $H(\cdot)$ is a step function with steps of sizes b_1, b_2, ..., b_N at points a_1, a_2, ..., a_N (all of which are distinct and satisfy $L < a_k < U$), then

$$\int_L^U g(x)\,dH(x) = \sum_{k=1}^N b_k g(a_k).$$

The last fact shows that many ordinary summations can be expressed as Riemann–Stieltjes integrals. This gives us a certain compactness in notation. For example, in many elementary textbooks on statistics, Equation (33) would have to be handled as three separate cases: a summation if $F_t(\cdot)$ were a step function (i.e., the rv X_t has a probability mass function); an ordinary Riemann integral if $F_t(\cdot)$ were differentiable (i.e., X_t has a probability density function); or as a summation plus an ordinary Riemann integral if $F_t(\cdot)$ were a 'mixed' cpdf. The Riemann–Stieltjes integral thus gives us some – but by no means all – of the advantages of the Lebesgue integral commonly used in advanced texts on probability and statistics.

2.4 Real-Valued Stationary Processes

The class of all stochastic processes is 'too large' to work with in practice. In spectral analysis, we consider only a special subclass of these called stationary processes. Basically, stationarity requires that certain statistical properties of a stochastic process be invariant with respect to time.

There are two common types of stationarity. The first type is *complete* stationarity (sometimes referred to as *strong* stationarity or *strict* stationarity): the process $\{X_t\}$ is said to be completely stationary if, for all $n \geq 1$, for any t_1, t_2, ..., t_n contained in the index set, and for any τ such that $t_1 + \tau$, $t_2 + \tau$, ..., $t_n + \tau$ are also contained in the index set, the joint cpdf of X_{t_1}, X_{t_2}, ..., X_{t_n} is the same as that of $X_{t_1+\tau}$, $X_{t_2+\tau}$, ..., $X_{t_n+\tau}$; i.e.,

$$F_{t_1,t_2,\ldots,t_n}(a_1, a_2, \ldots, a_n) = F_{t_1+\tau, t_2+\tau, \ldots, t_n+\tau}(a_1, a_2, \ldots, a_n).$$

In other words, the probabilistic structure of a completely stationary process is invariant under a shift in time.

Unfortunately, completely stationary processes are too difficult to work with as models for most time series of interest since they have to be specified by using n-dimensional cpdf's. A simplifying assumption leads to the second common type of stationarity: the process $\{X_t\}$ is said to be *second-order* stationary (sometimes called *weakly* stationary or *covariance* stationary) if, for all $n \geq 1$, for any t_1, t_2, ..., t_n contained in the index set, and for any τ such that $t_1 + \tau$, $t_2 + \tau$, ..., $t_n + \tau$ are also contained in the index set, all the joint moments of orders 1 and 2 of $X_{t_1}, X_{t_2}, \ldots, X_{t_n}$ exist, are finite and are equal to the corresponding joint moments of $X_{t_1+\tau}, X_{t_2+\tau}, \ldots, X_{t_n+\tau}$. As immediate consequences of this definition, we have

$$E\{X_t\} \equiv \mu \text{ and } E\{X_t^2\} \equiv \mu_2',$$

where μ and μ_2' are constants independent of t. This implies that

$$\text{var}\{X_t\} = \mu_2' - \mu^2 \equiv \sigma^2$$

is also a constant independent of t. If we allow the shift $\tau = -t_1$, we see that

$$E\{X_{t_1} X_{t_2}\} = E\{X_0 X_{t_2-t_1}\}$$

is a function of the difference $t_2 - t_1$ only. Actually, it is a function of the absolute difference $|t_2 - t_1|$ only, since, if we now let $\tau = -t_2$, we have

$$E\{X_{t_1} X_{t_2}\} = E\{X_{t_2} X_{t_1}\} = E\{X_0 X_{t_1-t_2}\}.$$

Recall that, if U and V are two real-valued rv's, then their covariance is defined as

$$\text{cov}\{U, V\} \equiv E\{(U - E\{U\})(V - E\{V\})\}.$$

The above implies that the covariance between X_{t_1} and X_{t_2} is also a function of the absolute difference $|t_2 - t_1|$, since

$$\text{cov}\{X_{t_1}, X_{t_2}\} = E\{[X_{t_1} - \mu][X_{t_2} - \mu]\} = E\{X_{t_1} X_{t_2}\} - \mu^2.$$

For a discrete parameter second-order stationary process $\{X_t\}$, we define the *autocovariance sequence* (acvs) by

$$s_\tau \equiv \text{cov}\{X_t, X_{t+\tau}\} = \text{cov}\{X_0, X_\tau\}.$$

Likewise, for a continuous parameter second-order stationary process $\{X(t)\}$, we define the *autocovariance function* (acvf) by

$$s(\tau) \equiv \text{cov}\{X(t), X(t+\tau)\} = \text{cov}\{X(0), X(\tau)\}.$$

Both s_τ and $s(\tau)$ measure the covariance between members of a process which are separated by τ units. The variable τ is called the *lag*.

Here are some further properties of the acvs and acvf.

[1] For a discrete parameter process with an index set given by the integers, τ can assume any integer value; for a continuous parameter

stationary process with an index set given by the entire real axis, τ can assume any real value.

[2] Note that, in the discrete parameter case, $s_0 = \sigma^2$ and $s_{-\tau} = s_\tau$; likewise, in the continuous parameter case, $s(0) = \sigma^2$ and $s(-\tau) = s(\tau)$.

[3] We can define $\rho_\tau = s_\tau/s_0$ as the *autocorrelation sequence* (acs) for $\{X_t\}$; likewise, we can define $\rho(\tau) = s(\tau)/s(0)$ as the *autocorrelation function* (acf) for $\{X(t)\}$. Since we have

$$\rho_\tau = \frac{\text{cov}\{X_t, X_{t+\tau}\}}{[\text{var}\{X_t\}\,\text{var}\{X_{t+\tau}\}]^{1/2}} = \frac{\text{cov}\{X_t, X_{t+\tau}\}}{\text{var}\{X_t\}},$$

ρ_τ is the correlation coefficient between pairs of rv's from the process $\{X_t\}$ that are τ units apart. There is an analogous interpretation for $\rho(\tau)$. (As already mentioned in Chapter 1, the definitions we have given for the acvs, acs, acvf and acf are standard in the statistical literature, but other definitions for these terms are sometimes used in the engineering literature.)

[4] Since ρ_τ and $\rho(\tau)$ are correlation coefficients and hence constrained to lie between -1 and 1, it follows that

$$|s_\tau| \le s_0 \quad \text{and} \quad |s(\tau)| \le s(0) \quad \text{for all } \tau.$$

[5] The sequence $\{s_\tau\}$ is *positive semidefinite*; i.e., for all $n \ge 1$, for any $t_1, t_2, \ldots, t_n$ contained in the index set, and for any set of nonzero real numbers $a_1, a_2, \ldots, a_n$,

$$\sum_{j=1}^{n}\sum_{k=1}^{n} s_{t_j - t_k} a_j a_k \ge 0$$

(if this double summation is strictly greater than 0, then $\{s_\tau\}$ is said to be *positive definite*). To see this, let us define the following two column vectors of size n:

$$\mathbf{a} \equiv (a_1, a_2, \ldots, a_n)^T \quad \text{and} \quad \mathbf{V} \equiv [X_{t_1}, X_{t_2}, \ldots, X_{t_n}]^T,$$

where superscript T denotes the operation of vector transposition. Let Σ be the variance-covariance matrix of the vector $\mathbf{V}$. Its (j, k)th element is given by

$$E\{[X_{t_j} - \mu][X_{t_k} - \mu]\} = s_{t_j - t_k}.$$

Define the rv

$$W = \sum_{j=1}^{n} a_j X_{t_j} = \mathbf{a}^T \mathbf{V}.$$

Then we have

$$0 \leq \text{var}\{W\} = \text{var}\{\mathbf{a}^T\mathbf{V}\} = \mathbf{a}^T\Sigma\mathbf{a} = \sum_{j=1}^{n}\sum_{k=1}^{n} s_{t_j - t_k} a_j a_k$$

(see Exercise [2.8]). The important point here is that an acvs cannot just be any arbitrary sequence, a fact that will become important when we discuss estimators for $\{s_\tau\}$. The same comments hold – with obvious modifications – for the function $s(\cdot)$. (For a fascinating discussion on how severely the requirement of positive semidefiniteness constrains a sequence of numbers, see Makhoul, 1990.)

[6] If we now consider the variance-covariance matrix Γ_N of the vector $[X_1, X_2, \ldots, X_N]^T$, we find that

$$\Gamma_N = \begin{bmatrix} s_0 & s_1 & \cdots & s_{N-2} & s_{N-1} \\ s_1 & s_0 & \cdots & s_{N-3} & s_{N-2} \\ \vdots & \vdots & \ddots & \vdots & \vdots \\ s_{N-2} & s_{N-3} & \cdots & s_0 & s_1 \\ s_{N-1} & s_{N-2} & \cdots & s_1 & s_0 \end{bmatrix}. \tag{38}$$

A two-dimensional matrix whose elements depend only upon the *difference* between the row and column indices rather than the individual values of the indices is known as a *Toeplitz matrix*. We thus see that the variance-covariance matrix of a subsequence of a discrete parameter real-valued stationary process is a symmetric Toeplitz matrix.

It follows from the definitions of complete and second-order stationarity that, if $\{X_t\}$ is a completely stationary process with finite variance, then it is also second-order stationary. In general, second-order stationarity does not imply complete stationarity. An important exception is a *Gaussian process* (also called a *normal process*), defined as follows: the stochastic process $\{X_t\}$ is said to be Gaussian if, for all $n \geq 1$ and for any $t_1, t_2, \ldots, t_n$ contained in the index set, the joint cpdf of $X_{t_1}, X_{t_2}, \ldots, X_{t_n}$ is multivariate Gaussian. A second-order stationary Gaussian process is also completely stationary due to the fact that the multivariate Gaussian distribution is completely characterized by its moments of first and second order.

Hereinafter the unadorned term 'stationarity' will be taken to mean 'second-order stationarity.'

In summary, a discrete parameter stochastic process $\{X_t\}$ is said to be second-order stationary if $E\{X_t\} = \mu$ and $\text{cov}\{X_t, X_{t+\tau}\} = s_\tau$ for $\tau = 0, \pm 1, \pm 2, \ldots$, where μ and s_τ are finite numbers independent of the index t.

2.5 Complex-Valued Stationary Processes

We say that a discrete parameter complex-valued process $\{Z_t\}$, defined via

$$Z_t = X_{t,1} + iX_{t,2},$$

is (second-order) stationary if, for all $n \geq 1$, for any $t_1, t_2, \ldots, t_n$ contained in the index set, and for any τ such that $t_1+\tau, t_2+\tau, \ldots, t_n+\tau$ are also contained in the index set, all the joint first- and second-order moments of $X_{t_1,1}, X_{t_1,2}, X_{t_2,1}, X_{t_2,2}, \ldots, X_{t_n,1}$ and $X_{t_n,2}$ exist, are finite and are equal to the corresponding joint moments of $X_{t_1+\tau,1}$, $X_{t_1+\tau,2}, X_{t_2+\tau,1}, X_{t_2+\tau,2}, \ldots, X_{t_n+\tau,1}$ and $X_{t_n+\tau,2}$. This definition immediately implies that the component processes $\{X_{t,1}\}$ and $\{X_{t,2}\}$ are stationary. We thus have

$$E\{Z_t\} = E\{X_{t,1}\} + iE\{X_{t,2}\} \equiv \mu_1 + i\mu_2 \equiv \mu,$$

where μ_1 and μ_2 are real-valued constants and μ is a complex-valued constant, all of which are independent of t. We now *define*

$$\mathrm{cov}\{Z_{t_1}, Z_{t_2}\} \equiv E\{[Z_{t_1} - \mu]^*[Z_{t_2} - \mu]\},$$

where the asterisk indicates the operation of complex conjugation. By arguments analogous to those used in the previous section, it is possible to show that

$$\mathrm{cov}\{Z_t, Z_{t+\tau}\} = E\{[Z_t - \mu]^*[Z_{t+\tau} - \mu]\} = s_\tau$$

is independent of t and that the acvs has the property

$$s_{-\tau} = s_\tau^*.$$

Note that the above also holds for a real-valued stationary process – the complex conjugate of a real number is just the real number itself. For $\tau = 0$, we have

$$s_0 = E\{|Z_t - \mu|^2\} \equiv \mathrm{var}\{Z_t\}.$$

The acs is defined in the same way as before, so we must have

$$\rho_{-\tau} = \rho_\tau^*.$$

The positive semidefinite property of $\{s_\tau\}$ is now defined slightly differently: $\{s_\tau\}$ is said to be such if, for all $n \geq 1$, for any $t_1, t_2, \ldots, t_n$ contained in the index set, and for any set of *complex* numbers $c_1, c_2, \ldots, c_n$,

$$\mathrm{var}\left\{\sum_{j=1}^n c_j Z_{t_j}\right\} = \sum_{j=1}^n \sum_{k=1}^n s_{t_j - t_k} c_j c_k^* \geq 0.$$

While the variance-covariance matrix of a subsequence of a discrete parameter complex-valued stationary process is a Toeplitz matrix, it is not necessarily a symmetric Toeplitz matrix (as in the real-valued case) since $s_{-\tau} \neq s_\tau$ in general; however, because of the condition $s_{-\tau} = s_\tau^*$, the covariance matrix falls in the class of *Hermitian Toeplitz matrices*.

A continuous parameter complex-valued stationary process $\{Z(t)\}$ is defined in a similar way. In particular, we note that its acvf satisfies

$$s(-\tau) = s^*(\tau)$$

and is positive semidefinite.

We note the following definitions and results for later use.

[1] A complex-valued rv Z_t is said to have a *complex Gaussian* (or *normal*) distribution if its (real-valued) real and imaginary components $X_{t,1}$ and $X_{t,2}$ are bivariate Gaussian; likewise, a collection of n complex-valued rv's Z_1, ..., Z_n is said to follow a complex Gaussian distribution if all of its real and imaginary components are multivariate Gaussian (of dimension $2n$).

[2] The complex-valued process $\{Z_t\}$ is said to be complex Gaussian if, for all $n \geq 1$ and for any $t_1, t_2, \ldots, t_n$ contained in the index set, the joint cpdf of the real and imaginary components $X_{t_1,1}$, $X_{t_1,2}$, $X_{t_2,1}$, $X_{t_2,2}$, ..., $X_{t_n,1}$ and $X_{t_n,2}$ is multivariate Gaussian. (For a comprehensive discussion of complex-valued stochastic processes, see Miller, 1974.)

[3] If Z_1, Z_2, Z_3 and Z_4 are any four complex-valued Gaussian rv's with zero means, then the *Isserlis theorem* (Isserlis, 1918) states that

$$\begin{aligned} \mathrm{cov}\,&\{Z_1 Z_2, Z_3 Z_4\} \\ &= \mathrm{cov}\,\{Z_1, Z_3\}\,\mathrm{cov}\,\{Z_2, Z_4\} + \mathrm{cov}\,\{Z_1, Z_4\}\,\mathrm{cov}\,\{Z_2, Z_3\} \end{aligned} \quad (40)$$

(in fact, the above also holds if we replace Z_1, Z_2, Z_3 and Z_4 with any four real-valued Gaussian rv's with zero means).

Although the main focus of this book is on the application of real-valued stationary processes to time series data, complex-valued processes arise indirectly in several contexts, including the spectral representation theorem (Section 4.1) and the statistical properties of spectral estimators (Section 6.6). Complex-valued stationary processes are used directly to model certain bivariate radar, sonar and medical time series; Loupas and McDicken (1990) discuss an interesting example concerning time series collected by a Doppler ultrasound system for monitoring blood flow.

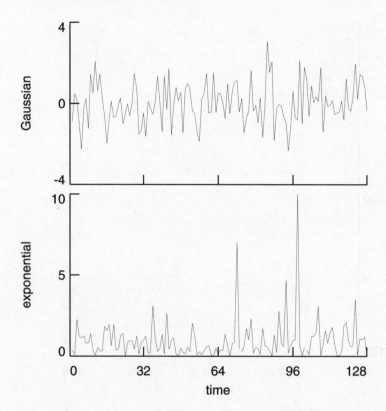

Figure 41. Realizations of size 128 of two white noise processes. The distribution for the process in the top plot is Gaussian (with zero mean and unit variance), while the one for the bottom plot is chi-square with two degrees of freedom (exponential) – this distribution is one-sided (i.e., an rv with this distribution can assume only nonnegative values) and has a heavier tail than the Gaussian distribution. This gives the time series the appearance of having outliers (discordant values).

2.6 Examples of Discrete Parameter Stationary Processes

- *White noise process*

This is also called a *purely random* process. Let $\{X_t\}$ be a sequence of uncorrelated rv's such that $E\{X_t\} = \mu$ and var $\{X_t\} = \sigma^2$ for all t. Since uncorrelatedness means that cov $\{X_t, X_{t+\tau}\} = 0$ for all t and $\tau \neq 0$, it follows that $\{X_t\}$ is a stationary process with acvs

$$s_\tau = \begin{cases} \sigma^2, & \text{if } \tau = 0; \\ 0, & \text{if } \tau \neq 0. \end{cases}$$

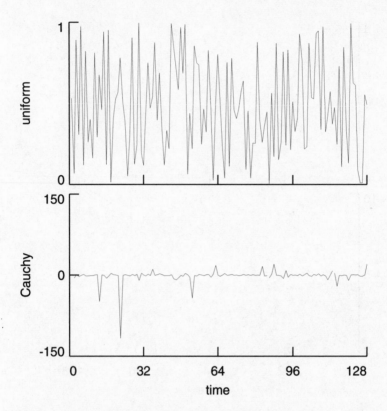

Figure 42. Realizations of two more white noise processes. The process in the top plot has a uniform distribution over the interval $[0, 1]$, while the one in the bottom plot has a Cauchy distribution that has been truncated at $\pm 10^{10}$ to produce a distribution with finite mean and variance – a requirement for second-order stationarity. The tails for the truncated Cauchy distribution are heavier than for both distributions in Figure 41, while the tails for the uniform distribution are lighter.

This implies that

$$\rho_\tau = \begin{cases} 1, & \text{if } \tau = 0; \\ 0, & \text{if } \tau \neq 0. \end{cases}$$

Despite the simplicity of this process, it plays a central role in the theory of stationary processes since many other processes can be created from white noise by manipulating it properly (see the examples which follow). Four examples of realizations of white noise processes (each with a different distribution) are shown in Figures 41 and 42. Although the acvs's for these processes are identical, their realizations are quite different – the characterization of a process by its first- and second-order moments can gloss over potentially important features.

• *Moving average process*

The process $\{X_t\}$ is called a qth order moving average process – denoted by MA(q) – if it can be expressed in the form

$$X_t = \mu - \theta_{0,q}\epsilon_t - \theta_{1,q}\epsilon_{t-1} - \cdots - \theta_{q,q}\epsilon_{t-q}, \tag{43a}$$

where μ and $\theta_{j,q}$ are constants ($\theta_{0,q} \equiv -1$ and $\theta_{q,q} \neq 0$) and $\{\epsilon_t\}$ is a white noise process with zero mean and variance σ_ϵ^2 (the notation $\theta_{j,k}$ refers to the jth coefficient associated with a kth order process). Note first that $E\{X_t\} = \mu$, a constant independent of t – we assume it to be 0 in what follows. Second, since $E\{\epsilon_t\epsilon_{t+\tau}\} = 0$ for all $\tau \neq 0$, it follows from Equation (43a) that, for $\tau \geq 0$,

$$\text{cov}\{X_t, X_{t+\tau}\} = \sum_{j=0}^{q}\sum_{k=0}^{q}\theta_{j,q}\theta_{k,q}E\{\epsilon_{t-j}\epsilon_{t+\tau-k}\}$$

$$= \sigma_\epsilon^2 \sum_{j=0}^{q-\tau}\theta_{j,q}\theta_{j+\tau,q} \equiv s_\tau$$

depends only on the lag τ. Here we interpret the last summation to be equal to 0 when $q - \tau < 0$. Since it is easy to show that

$$\text{cov}\{X_t, X_{t-\tau}\} = \text{cov}\{X_t, X_{t+\tau}\},$$

it follows that $\{X_t\}$ is a stationary process with acvs given by

$$s_\tau = \begin{cases} \sigma_\epsilon^2 \sum_{j=0}^{q-|\tau|}\theta_{j,q}\theta_{j+|\tau|,q}, & |\tau| \leq q; \\ 0, & |\tau| > q. \end{cases} \tag{43b}$$

Note that we did not need to place any restrictions on the $\theta_{j,q}$ terms to ensure stationarity. Note also that the variance of $\{X_t\}$ is given by

$$s_0 = \sigma_\epsilon^2 \sum_{j=0}^{q}\theta_{j,q}^2,$$

which can be used to compute the acs $\{\rho_\tau\}$.

Figure 44 shows realizations for two different Gaussian MA(1) processes of the form

$$X_t = \epsilon_t - \theta\epsilon_{t-1}.$$

For the first process, $\theta = 1$; for the second, $\theta = -1$. Since $\rho_1 = -\theta/(1+\theta^2)$ here, we see that adjacent values of the first process are negatively correlated ($\rho_1 = -1/2$), while adjacent values of the second process are positively correlated ($\rho_1 = 1/2$). The realizations in Figure 44 agree with this description.

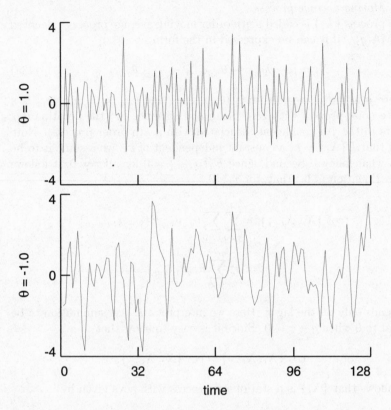

Figure 44. Realizations of size 128 of two first-order Gaussian moving average processes. The top and bottom plots are for processes with $\theta = 1.0$ and -1.0, respectively.

An interesting example of a physical process that can be modeled as a moving average process is the thickness of textile slivers (an intermediate stage in converting flax fibers into yarn) as a function of displacement along a sliver (Spencer-Smith and Todd, 1941). Note that the 'time' variable here is a displacement (distance) rather than physical time.

- *Autoregressive process*

The process $\{X_t\}$ with zero mean is called a pth order autoregressive process – denoted by $\mathrm{AR}(p)$ – if it satisfies an equation such as

$$X_t = \phi_{1,p}X_{t-1} + \phi_{2,p}X_{t-2} + \cdots + \phi_{p,p}X_{t-p} + \epsilon_t,$$

where $\phi_{1,p}$, $\phi_{2,p}$, $\ldots$, $\phi_{p,p}$ are constants (with $\phi_{p,p} \neq 0$) and $\{\epsilon_t\}$ is a white noise process with zero mean and variance σ_p^2. In words, X_t is assumed to be a linear combination of X_{t-1}, $\ldots$, X_{t-p} plus a random

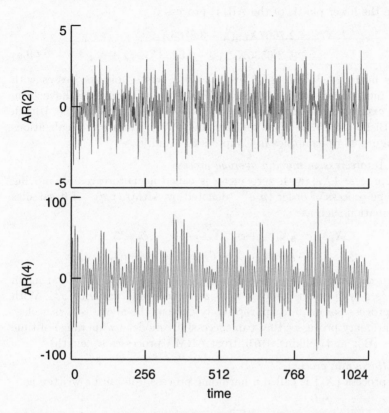

Figure 45. Realizations of size 1024 of two Gaussian autoregressive processes. The top plot is a realization of the AR(2) process of Equation (45), and the bottom plot is a realization of the AR(4) process of Equation (46a).

component (ϵ_t). In contrast to the parameters of an MA(q) process, the $\phi_{k,p}$ terms must satisfy certain conditions for $\{X_t\}$ to be a stationary process; i.e., not all AR(p) processes are stationary. If a stationary AR(p) process is nondeterministic (i.e., X_t cannot be perfectly predicted from $X_{t-1}, X_{t-2}, \ldots, X_{t-p}$), it can be written as an infinite-order moving average process:

$$X_t = -\sum_{j=0}^{\infty} \theta_j \epsilon_{t-j},$$

where θ_j can be determined from the $\phi_{k,p}$ terms. AR(p) processes play an important role in modern spectral analysis and are discussed in detail in Chapter 9.

Figure 45 shows realizations from two AR processes; the upper plot is of the AR(2) process

$$X_{t,2} = 0.75X_{t-1,2} - 0.5X_{t-2,2} + \epsilon_{t,2}, \tag{45}$$

while the lower plot is of the AR(4) process

$$X_{t,4} = 2.7607X_{t-1,4} - 3.8106X_{t-2,4}$$
$$+ 2.6535X_{t-3,4} - 0.9238X_{t-4,4} + \epsilon_{t,4} \qquad (46a)$$

(here both $\{\epsilon_{t,2}\}$ and $\{\epsilon_{t,4}\}$ are Gaussian white noise processes with zero means and unit variances). These two AR processes have been used extensively in the literature as test cases (see, for example, Ulrych and Bishop, 1975, and Box and Jenkins, 1976). We use the realizations in Figure 45 as examples in Chapters 6, 7 and 9.

• *Autoregressive moving average process*
The process $\{X_t\}$ with zero mean is called an autoregressive moving average process of order (p, q) – denoted by ARMA(p,q) – if it satisfies an equation such as

$$X_t = \phi_{1,p}X_{t-1} + \phi_{2,p}X_{t-2} + \cdots + \phi_{p,p}X_{t-p}$$
$$+ \epsilon_t - \theta_{1,q}\epsilon_{t-1} - \theta_{2,q}\epsilon_{t-2} - \cdots - \theta_{q,q}\epsilon_{t-q},$$

where $\phi_{j,p}$ and $\theta_{j,q}$ are constants ($\phi_{p,p} \neq 0$ and $\theta_{q,q} \neq 0$) and again $\{\epsilon_t\}$ is a white noise process with zero mean and variance σ_ϵ^2. With the process parameters appropriately chosen, the above is a rich class of stationary processes that can successfully model a wide range of time series (Box and Jenkins, 1976, treat ARMA processes at length).

• *Harmonic process*
The process $\{X_t\}$ is called a harmonic process if it can be written as

$$X_t = \mu + \sum_{l=1}^{L} D_l \cos\left(2\pi f_l t + \phi_l\right), \qquad (46b)$$

where μ, L, D_l and f_l are real-valued constants and the ϕ_l terms are independent rv's, each having a rectangular (or uniform) distribution on the interval $[-\pi, \pi]$.

A harmonic process is a stationary process, as can be seen from the following arguments. First, note that

$$E\{X_t\} = \mu + \sum_{l=1}^{L} D_l E\left\{\cos\left(2\pi f_l t + \phi_l\right)\right\}$$

$$= \mu + \sum_{l=1}^{L} D_l \int_{-\pi}^{\pi} \cos\left(2\pi f_l t + \phi_l\right)\frac{1}{2\pi}\, d\phi_l = \mu, \qquad (46c)$$

a constant independent of t, which we take to be 0. Next, note that

$$\text{cov}\{X_t, X_{t+\tau}\}$$
$$= \sum_{k=1}^{L}\sum_{l=1}^{L} D_k D_l E\left\{\cos\left(2\pi f_k t + \phi_k\right)\cos\left[2\pi f_l(t+\tau) + \phi_l\right]\right\}. \qquad (46d)$$

In the above double summation, if $k \neq l$, it follows from the independence of the ϕ_l rv's that

$$E\left\{\cos\left(2\pi f_k t + \phi_k\right)\cos\left[2\pi f_l(t+\tau)+\phi_l\right]\right\}$$
$$= E\left\{\cos\left(2\pi f_k t + \phi_k\right)\right\}E\left\{\cos\left[2\pi f_l(t+\tau)+\phi_l\right]\right\} = 0$$

by the same integration used in (46c). If $k = l$, the expectation in (46d) is equal to

$$\frac{1}{4\pi}\int_{-\pi}^{\pi}\cos\left[(4\pi f_l t + 2\pi f_l\tau)+2\phi_l\right]+\cos\left(2\pi f_l\tau\right)d\phi_l = \cos\left(2\pi f_l\tau\right)/2.$$

Equation (46d) thus reduces to

$$\operatorname{cov}\left\{X_t, X_{t+\tau}\right\} = \sum_{l=1}^{L} D_l^2\cos\left(2\pi f_l\tau\right)/2 \equiv s_\tau, \tag{47a}$$

which shows that a harmonic process is stationary with acvs $\{s_\tau\}$. Because

$$s_0 = \sum_{l=1}^{L} D_l^2/2, \tag{47b}$$

it follows that the acs is given by

$$\rho_\tau = \sum_{l=1}^{L} D_l^2\cos\left(2\pi f_l\tau\right)\bigg/\sum_{l=1}^{L} D_l^2.$$

If we compare Equations (46b) and (47a), we see that both the harmonic process and its acvs consist of sums of cosine waves with exactly the same frequencies, but that for the acvs all of the cosine terms are 'in phase;' i.e., the ϕ_l terms are all 0. Note that the sequence $\{s_\tau\}$ does not damp down to 0 as τ gets large.

If we do not regard the phases ϕ_l in Equation (46b) as rv's, we have a purely deterministic model, i.e., there are no random elements. Any given realization from a harmonic process can be regarded as a deterministic function since each ϕ_l is constant with respect to time. Knowledge of any segment of such a realization with a length at least as long as the number of unknown parameters is enough to fully specify the entire realization. In one sense, randomizing the phases is a mathematical trick to allow us to treat models like (46b) within the context of the theory of stationary processes. In another sense, however, *all* stationary processes can be written as a generalization of a harmonic process with an infinite number of terms – as we shall see, this is the essence of the spectral representation theorem to be discussed in Chapter 4.

It should be noted that the independence of the ϕ_l rv's is critical to ensure stationarity – see Exercise [2.6]. Although independence of phases is often assumed for convenience when the model described by (46b) is fit to actual data, this assumption can lead to subtle problems. Walden and Prescott (1983) modeled tidal elevations using the harmonic process approach. The disagreement (often small) between the calculated and measured tidal probability density functions is largely attributable to the lack of phase independence of the constituents of the tide (see Section 10.1 for details).

By using the trigonometric identity $\cos(x + y) = \cos(x)\cos(y) - \sin(x)\sin(y)$, we can rewrite (46b) as

$$X_t = \mu + \sum_{l=1}^{L} \left[A_l \cos\left(2\pi f_l t\right) + B_l \sin\left(2\pi f_l t\right) \right], \qquad (48a)$$

where

$$A_l \equiv D_l \cos(\phi_l) \text{ and } B_l \equiv -D_l \sin(\phi_l).$$

Since, for all l,

$$E\{A_l\} = E\{B_l\} = 0, \quad \text{var}\{A_l\} = \text{var}\{B_l\} = D_l^2/2 \qquad (48b)$$

and

$$\text{cov}\{A_l, B_l\} = -D_l^2 \int_{-\pi}^{\pi} \cos(\phi_l)\sin(\phi_l)\frac{1}{2\pi}\,d\phi_l = 0, \qquad (48c)$$

and since, for all $k \neq l$,

$$\text{cov}\{A_k, A_l\} = \text{cov}\{B_k, B_l\} = \text{cov}\{A_k, B_l\} = 0 \qquad (48d)$$

due to the independence of the ϕ_l rv's, we can regard a harmonic process as involving either independent random phases – Equation (46b) – or uncorrelated random amplitudes – Equation (48a). Conversely, if $A_1, A_2, \ldots, A_L, B_1, B_2, \ldots, B_L$ are any set of rv's that satisfy Equations (48b), (48c) and (48d), the process formed by Equation (48a) is a stationary process with mean μ and an acvs given by (47a); however, without additional stipulations on A_l and B_l, it cannot in general also be written in the form of Equation (46b) – this equation implies that, for each realization, the sinusoidal component associated with frequency f_l has a fixed amplitude D_l, which will not be the case if, say, A_l and B_l are drawn from a Gaussian distribution. The formulation with uncorrelated random amplitudes is thus more general than that with independent random phases. In practical applications, however, the harmonic processes of Equation (46b) are often sufficient – particularly since we would need to have multiple realizations of a process in order to distinguish between the two formulations. (The model we considered in Chapter 1, namely Equation (10b), is a special case of a process generated using uncorrelated random amplitudes.)

2.7 Comments on Continuous Parameter Processes

All of the stationary processes mentioned in Section 2.6 – with the important exception of harmonic processes – can be constructed by taking (possibly infinite) linear combinations of discrete parameter white noise. One might suppose that we can likewise construct numerous examples of continuous parameter stochastic processes by taking linear combinations of continuous parameter white noise. There is a fundamental technical difficulty with this approach – a continuous parameter white noise process with properties similar to those of a discrete parameter process does not exist; i.e., a continuous parameter process $\{\epsilon(t)\}$ with an acvf given by

$$s(\tau) = \begin{cases} \sigma^2 > 0, & \text{if } \tau = 0; \\ 0, & \text{if } \tau \neq 0, \end{cases}$$

does not exist! It is possible to circumvent these technical problems and to deal with a 'fictitious' continuous parameter white noise process to construct a wide range of continuous parameter stationary processes. This fiction is useful in much the same way as the Dirac delta function is a useful fiction (as long as certain precautions are taken). We do not deal with it directly in this book since our primary concern is analyzing time series using techniques appropriate for a digital computer. We thus do not need to construct continuous parameter stationary processes directly – though we do deal with them indirectly through the process of sampling (Section 4.5). For our purposes the spectral representation theorem for continuous parameter stationary processes will serve as the necessary foundation for future study (see item [3] in the Comments and Extensions to Section 4.1).

2.8 Use of Stationary Processes as Models for Data

The chief use of stationary processes in this book is to serve as models for various time series. Although the concept of stationarity is defined for models and not for data, it is proper to ask whether certain time series can be usefully modeled by stationary processes. We should be careful about drawing conclusions about a time series based upon analysis techniques (such as spectral analysis) that assume stationarity when in fact this assumption is suspect for the data at hand.

There are many ways in which the assumption of stationarity can fail. Fortunately, some common violations are relatively easy to patch up. For example, a process can have infinite variance and yet have statistical properties that are invariant with respect to translations in time. Here we examine two other types of nonstationarity by considering two physical time series for which use of a stationary process as a model is suspect.

The first example is the time series shown in the top plot of Figure 50. These data concern a spinning rotor which is used to measure

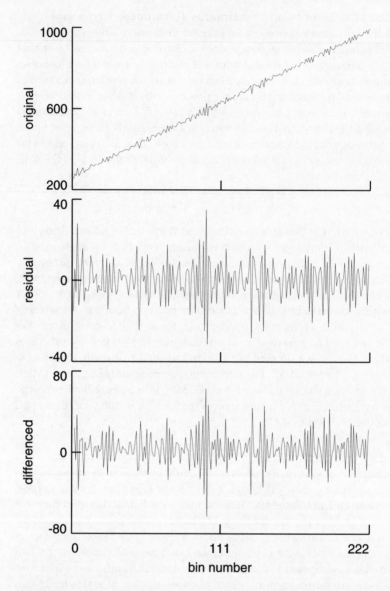

Figure 50. Spinning rotor time series. From top to bottom, the plots are:
the original data; the residuals from a linear least squares fit; and the first
difference of the original data. The data are measured in units of microseconds.
(This data set was collected by S. Wood at the National Bureau of Standards
and made available to the authors by J. Filliben.)

air density (i.e., pressure). During the period over which the data were collected, the rotor was always slowing down – the more molecules the rotor hit, the faster it slowed down. The data plotted are the amount of time (measured in microseconds) that it took the rotor to make 400 revolutions. The x-axis is the bin number: the first bin corresponds to the first group of 400 observed revolutions, and the tth bin, to the tth such group. (These data are a good example of a 'time series' for which the 'time index' is not actually time – however the values of the series are measured in time!)

As the spinning rotor slowed down, the time it took to make 400 revolutions necessarily increased. If we use a stochastic process $\{X_t\}$ as a model for these data, it is unrealistic to assume that it is stationary. This requires that $E\{X_t\}$ be a constant independent of time, which from physical considerations (and Figure 50) is unrealistic. However, the time dependence of $E\{X_t\}$ appears to be linear, so a reasonable model might be

$$X_t = \alpha + \beta t + Y_t, \tag{51}$$

where α and β are unknown parameters and $\{Y_t\}$ is a stationary process with zero mean.

How we proceed now depends upon what questions we want answered about the data. For heuristic purposes, let us assume that we want to examine the covariance structure of $\{Y_t\}$. Two important ways of recovering $\{Y_t\}$ from $\{X_t\}$ are shown in the middle and bottom plots of Figure 50. The middle plot shows the residuals from an ordinary linear least squares fit to the parameters α and β; i.e.,

$$\hat{Y}_t \equiv X_t - \hat{\alpha} - \hat{\beta}t,$$

where $\hat{\alpha}$ and $\hat{\beta}$ are the linear least squares estimates of α and β. The residual $\hat{Y}_t$ only approximates the unknown Y_t, but nonetheless we can analyze $\{\hat{Y}_t\}$ to get some idea of the covariance structure of $\{Y_t\}$.

A second way of analyzing $\{Y_t\}$ is to examine the first difference of the data:

$$X_t^{(1)} \equiv X_t - X_{t-1} = \beta + Y_t^{(1)}, \quad \text{where } Y_t^{(1)} \equiv Y_t - Y_{t-1}.$$

This is shown in the bottom plot of Figure 50 (the sample mean of the $X_t^{(1)}$ values is an estimator of β and is equal to 3.38 here). It can be shown (by arguments similar to those needed for Exercise [2.5]) that the first difference of a stationary process is also a stationary process – hence, if $\{Y_t\}$ is stationary, so is $\{Y_t^{(1)}\}$. Moreover, by using the theory of linear filters (see Chapter 5), it is possible to relate the covariance (or spectral) properties of $\{Y_t^{(1)}\}$ to those of $\{Y_t\}$. Because of this, differencing is

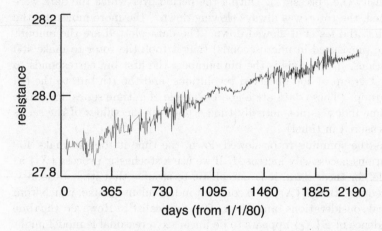

Figure 52. Standard resistor time series. The plot shows the value of a resistor measured over approximately a six-year period. (This data set was collected by S. Dziuba at the National Bureau of Standards and made available to the authors through the courtesy of J. Filliben.)

often preferred over linear least squares to effectively remove a linear trend in a time series.

An extension to simple differencing for removing a trend can be applied when there is seasonality, or periodicity, in the data with period p (Box and Jenkins, 1976). If p is a multiple of the sampling interval, we could choose to examine $X_t - X_{t-p}$. Nature might choose to be less cooperative as in the case of tidal analysis where the dominant tidal period is 12.42 hours and the sampling interval is often 1 hour. Conradsen and Spliid (1981) showed that in this case the periodicity can be largely removed by using $X_t - (1 - \delta)X_{t-n} - \delta X_{t-n-1}$, where $n = \lfloor p \rfloor$ and $\delta = p - n$ (these are 12 and 0.42, respectively, in the tidal example). However, as explained in Section 2.6, harmonic processes incorporating such periodicities can be examined within the framework of stationary processes if desired (see also Chapter 10).

The second example of a time series for which a stationary process is a questionable model is shown in Figure 52. The data are the resistance of a standard resistor over approximately a six-year period. Here we see a linear or, perhaps, quadratic drift over time. If we assume only a linear drift, an appropriate model for these data might be the same stochastic process considered for the spinning rotor data (Equation (51)). However, here it might be questionable to assume that $\{Y_t\}$, the deviations from the linear drift, is a stationary process. If we compare the data from about days 1095 to 1460 with the rest of the series, it appears that the series has a variance that changes with time

– a violation of the stationarity assumption. There are no simple transformations that correct this problem. One common approach is to use different stationary processes to model various chunks of the data. Here two models might work – one for the deviations about a linear drift for days 1095 to 1460, and a second for the data before and after.

2.9 Exercises

[2.1] Let $\{X_{t,1}\}, \{X_{t,2}\}, \ldots, \{X_{t,m}\}$ be m stationary processes with zero means and acvs's $\{s_{\tau,1}\}, \{s_{\tau,2}\}, \ldots, \{s_{\tau,m}\}$, respectively; i.e.,

$$s_{\tau,j} = \text{cov}\{X_{t,j}, X_{t+\tau,j}\}.$$

If the processes $\{X_{t,j}\}$ and $\{X_{t,k}\}$ are uncorrelated for all $j \neq k$ (i.e., $E\{X_{t,j}X_{t+\tau,k}\} = 0$ for all t, τ and $j \neq k$), what is the acvs of the stationary process

$$X_t \equiv \sum_{j=1}^{m} X_{t,j}?$$

(The procedure of adding together processes to form a new process is sometimes called *aggregation*. A simple extension to this exercise shows that finite linear combinations of stationary processes are also stationary.)

[2.2] Let the real-valued sequence $\{x_{t,1}\}$ be defined by

$$x_{t,1} = \begin{cases} +1, & \text{if } t = 0, -1, -2, \ldots; \\ -1, & \text{if } t = 1, 2, 3, \ldots, \end{cases}$$

and let the real-valued sequence $\{x_{t,2}\}$ be defined by $x_{t,2} = -x_{t,1}$. Suppose we have a discrete parameter stochastic process whose ensemble consists of just $\{x_{t,1}\}$ and $\{x_{t,2}\}$. We can generate realizations of this process by picking either $\{x_{t,1}\}$ or $\{x_{t,2}\}$ at random with probability 1/2 each. Let $\{X_t\}$ represent the process itself.
   a) What is $E\{X_t\}$?
   b) What is $\text{var}\{X_t\}$?
   c) What is $\text{cov}\{X_t, X_{t+\tau}\}$?
   d) Is $\{X_t\}$ a (second-order) stationary process?

[2.3] In the defining Equation (43a) for a qth order moving average process, we specified that $\theta_{0,q} \equiv -1$. In terms of the covariance structure for moving average processes, why do we not gain more generality by letting $\theta_{0,q}$ be any arbitrary number?

[2.4] Show that the first-order moving average process defined by

$$X_t = \epsilon_t - \theta\epsilon_{t-1}$$

can be written in terms of previous values of the process as

$$X_t = \epsilon_t - \sum_{j=1}^{p} \theta^j X_{t-j} - \theta^{p+1} \epsilon_{t-p-1}$$

for any positive integer p. In the above, what conditions on θ must hold in order that X_t can be expressed as an infinite-order autoregressive process, i.e.,

$$X_t = \epsilon_t - \sum_{j=1}^{\infty} \theta^j X_{t-j}?$$

(A moving average process that can be written as an infinite-order autoregressive process is called *invertible*.)

[2.5] Suppose that
$$Y_t = \alpha + \beta t + \gamma t^2 + X_t,$$

where α, β and γ are nonzero constants and $\{X_t\}$ is a stationary process with acvs $\{s_{\tau,X}\}$. Show that the first difference of the first difference of Y_t, i.e.,

$$Y_t^{(2)} \equiv Y_t^{(1)} - Y_{t-1}^{(1)}, \quad \text{where } Y_t^{(1)} \equiv Y_t - Y_{t-1},$$

is a stationary process, and find its acvs $\{s_{\tau,Y^{(2)}}\}$ in terms of $\{s_{\tau,X}\}$.

[2.6] In definition (46b) for a harmonic process, we stipulated that the ϕ_l rv's be independent and uniformly distributed over the interval $[-\pi, \pi]$. Suppose that the probability density function (pdf) of ϕ_l is not that of a uniformly distributed rv, but rather is of the form

$$f(\phi_l) \equiv \frac{1}{2\pi} \left(1 + \cos \phi_l \right), \qquad |\phi_l| \leq \pi.$$

A realization of ϕ_l from this pdf is more likely to be close to 0 than to $\pm\pi$. With this new stipulation, are both $E\{X_t\}$ and $\text{cov}\{X_t, X_{t+\tau}\}$ still independent of t for the process defined by Equation (46b)?

[2.7] a) Suppose that $\{X_t\}$ and $\{Y_t\}$ are both stationary processes with zero means and with acvs's $\{s_{\tau,X}\}$ and $\{s_{\tau,Y}\}$, respectively. If $\{X_t\}$ and $\{Y_t\}$ are uncorrelated with each other, show that the acvs $\{s_{\tau,Z}\}$ of the complex-valued process defined by

$$Z_t = X_t + iY_t$$

must be real-valued. Can $\{s_{\tau,X}\}$ and $\{s_{\tau,Y}\}$ be determined if $\{s_{\tau,Z}\}$ is known?

b) Suppose that $Y_t = X_{t+k}$ for some integer k; i.e., the process $\{Y_t\}$ is just a shifted version of $\{X_t\}$. What is $\{s_{\tau,Z}\}$ now?

[2.8] Show that, if $W_1, W_2, \ldots, W_n$ are any set of n real-valued rv's with finite variances, and if $a_1, a_2, \ldots, a_n$ are any set of n real numbers, then

$$\mathrm{var}\left\{\sum_{j=1}^{n} a_j W_j\right\} = \sum_{j=1}^{n} a_j^2 \, \mathrm{var}\,\{W_j\} + 2\sum_{j<k} a_j a_k \, \mathrm{cov}\,\{W_j, W_k\}.$$

(55)

If we further let

$$\mathbf{a} \equiv (a_1, a_2, \ldots, a_n)^T \text{ and } \mathbf{W} \equiv [W_1, W_2, \ldots, W_n]^T,$$

and if we let Σ be the variance-covariance matrix of the vector $\mathbf{W}$ (i.e., the (j, k)th element of Σ is $\mathrm{cov}\,\{W_j, W_k\}$), show that the above implies that

$$\mathrm{var}\,\{\mathbf{a}^T \mathbf{V}\} = \mathbf{a}^T \Sigma \mathbf{a}.$$

[2.9] Suppose that $\{X_t\}$ is a stationary process with zero mean and acvs $\{s_{\tau,X}\}$ and that C is a random variable with zero mean and variance σ_C^2. Suppose that C and X_t are uncorrelated for all t. Show that $Y_t \equiv X_t + C$ is a stationary process and determine its acvs. (This is an example of a *nonergodic* stationary process – see [2] of the Comments and Extensions to Section 6.1.)

[2.10] Suppose that $\{\epsilon_t\}$ is a white noise process with zero mean and variance σ_ϵ^2. Define the process $\{X_t\}$ via

$$X_t = \begin{cases} \epsilon_0, & t = 0; \\ X_{t-1} + \epsilon_t, & t = 1, 2, \ldots. \end{cases}$$

Is $\{X_t\}$ a stationary process?

[2.11] Suppose that $\{X_t\}$ is a discrete parameter white noise process with zero mean, and consider the process defined by $Y_t = X_t \cos(2\pi f_0 t + \phi)$, where $0 < f_0 < 1/2$ is a fixed frequency.

a) Show that, if we regard ϕ as a constant, the process $\{Y_t\}$ is nonstationary because its variance is a function of t.

b) Show that, if ϕ is an rv that is uniformly distributed over the interval $[-\pi, \pi]$ and is independent of $\{X_t\}$, then $\{Y_t\}$ is a white noise process.

3

Deterministic Spectral Analysis

3.0 Introduction

In Chapter 1 we modeled four different time series of length N using a stochastic model of the form

$$X_t = \mu + \sum_{j=1}^{\lfloor N/2 \rfloor} [A_j \cos(2\pi f_j t) + B_j \sin(2\pi f_j t)], \qquad f_j \equiv j/N, \quad (56)$$

where the A_j and B_j rv's are uncorrelated with zero mean and variance σ_j^2 (this is Equation (10b)). We defined a variance spectrum

$$S_j = \sigma_j^2, \qquad 1 \leq j \leq \lfloor N/2 \rfloor$$

and found that

$$\sum_{j=1}^{\lfloor N/2 \rfloor} S_j = \sigma^2,$$

the variance of the process. The variance spectrum thus decomposes the process variance into $\lfloor N/2 \rfloor$ components, each associated with the expected squared amplitude of sinusoids of a particular frequency. It gives us a concise way of summarizing some of the important statistical properties of $\{X_t\}$.

In Chapter 2 we introduced the concept of stationary processes. The task that awaits us is to define an appropriate spectrum for these processes. We do so in Chapter 4. We devote this chapter to the definition and study of various spectra for deterministic (nonrandom) functions (or sequences) of time. Our rationale for doing so is three-fold. First, since every *realization* of a stochastic process is a deterministic function of time, we shall use the material here to motivate the definition of a spectrum for stationary processes. Second, deterministic functions and sequences appear quite naturally in many aspects of the study of the spectra of stationary processes. The concepts we discuss here will be used repeatedly in subsequent chapters when we discuss, for example, the relationship between the spectrum and the acvs for a stationary process, the relationship between a linear filter and its transfer function, and the use of data tapers in the estimation of spectra. Third, just as the spectrum for stationary processes is a way of summarizing some important features of those processes, the various spectra that can be defined for deterministic functions can be used to express certain properties of these functions easily.

In model (56) we dealt with a discrete time (or discrete parameter) process $\{X_t\}$ and defined a variance spectrum for it over a set of discrete frequencies, namely, $f_j = j/N$ with $1 \leq j \leq \lfloor N/2 \rfloor$. Now for a real or complex-valued deterministic *function* $g(\cdot)$, it might be possible to define a reasonable spectrum over a discrete set of frequencies or a continuum of frequencies; the same remark also holds for a deterministic *sequence* $\{g_t\}$. There are four possible combinations of interest: continuous time with continuous frequency; continuous time with discrete frequency; discrete time with continuous frequency; and discrete time with discrete frequency. The first case arises in much of the historical theoretical framework on spectral analysis; the second provides a description for periodic signals often encountered in electrical engineering; the third appears in the relationship between the spectrum and acvs of a stationary process; and the fourth is important as the route into digital computer techniques. The remaining sections in this chapter are an investigation of some of the important time/frequency relationships in each of these categories. For heuristic reasons, we begin with the case of a continuous time deterministic function with a spectrum defined over a discrete set of frequencies.

3.1 Fourier Theory – Continuous Time/Discrete Frequency

Since $\cos(2\pi nt/T)$ defines a periodic function of t with period $T > 0$ for any integer n, i.e.,

$$\cos[2\pi n(t+T)/T] = \cos(2\pi nt/T) \text{ for all times } t,$$

and since the same holds for $\sin(2\pi nt/T)$, it follows that

$$\tilde{g}_p(t) = \frac{a_0}{2} + \sum_{n=1}^{\infty} a_n \cos(2\pi nt/T) + b_n \sin(2\pi nt/T) \qquad (57)$$

also defines a periodic function of t with period T. Here $\{a_n\}$ and $\{b_n\}$ are arbitrary sequences of constants (either real or complex-valued) subject only to the condition that the summation above must converge for all t. We can rewrite (57) in a more compact form by expressing the cosines and sines as complex exponentials (see Exercise [1.3b]):

$$\tilde{g}_p(t) = \sum_{n=-\infty}^{\infty} G_n e^{i2\pi f_n t} \tag{58a}$$

where

$$G_n \equiv \begin{cases} (a_n - ib_n)/2, & \text{for } n \geq 1; \\ a_0/2, & \text{for } n = 0; \\ (a_{|n|} + ib_{|n|})/2, & \text{for } n \leq -1, \end{cases} \tag{58b}$$

and $f_n \equiv n/T$. Equation (58a) involves both positive and negative frequencies. Negative frequencies are a somewhat strange concept at first sight, but they are a useful mathematical fiction that allows us to simplify the mathematical theory below considerably. We note that, if a_n and b_n are real-valued, $G_{-n}^* = G_n$ so that $|G_{-n}| = |G_n|$. Exercise [3.1] is to express a_n and b_n in terms of the G_n (and to examine a third common way of expressing (57)).

Let $g_p(\cdot)$ be a deterministic (nonrandom) real or complex-valued function of t which is periodic with period T and is square integrable; i.e.,

$$\int_{-T/2}^{T/2} |g_p(t)|^2 \, dt < \infty.$$

Can $g_p(\cdot)$ be represented by a series such as in Equation (58a)? The answer is 'yes, in a certain sense.' The exact result is as follows: define

$$g_{p,m}(t) \equiv \sum_{n=-m}^{m} G_n e^{i2\pi f_n t}, \tag{58c}$$

where

$$G_n \equiv \frac{1}{T} \int_{-T/2}^{T/2} g_p(t) e^{-i2\pi f_n t} \, dt. \tag{58d}$$

Then, as $m \to \infty$, it can be shown that $g_{p,m}(\cdot)$ converges to $g_p(\cdot)$ in the mean square sense; i.e.,

$$\lim_{m \to \infty} \int_{-T/2}^{T/2} |g_p(t) - g_{p,m}(t)|^2 \, dt = 0.$$

Hence we can write

$$g_p(t) \stackrel{\text{ms}}{=} \sum_{n=-\infty}^{\infty} G_n e^{i2\pi f_n t} \tag{58e}$$

if we define the symbol $\overset{\text{ms}}{=}$ to mean 'equal in the mean square sense.'

Let us now offer some justification for Equation (58d). Assume that (58e) is true, multiply both sides by $\exp(-i2\pi f_m t)$, integrate them from $-T/2$ to $T/2$, and interchange the order of summation and integration on the right-hand side to get

$$\int_{-T/2}^{T/2} g_p(t)e^{-i2\pi f_m t}\,dt = \sum_{n=-\infty}^{\infty} G_n \int_{-T/2}^{T/2} e^{i2\pi(f_n-f_m)t}\,dt. \qquad (59a)$$

Invoke the following orthogonality relationship:

$$\int_{-T/2}^{T/2} e^{i2\pi(f_n-f_m)t}\,dt = \begin{cases} 0, & m \neq n; \\ T, & m = n. \end{cases}$$

Equation (59a) now reduces to

$$\int_{-T/2}^{T/2} g_p(t)e^{-i2\pi f_m t}\,dt = TG_m,$$

from which (58d) follows immediately.

Equation (58e) is called the *Fourier series representation* of a periodic function; G_n is referred to as the nth *Fourier coefficient*.

By analogy with the case of a real-valued function where we might regard $g_p(\cdot)$ as representing an electrical signal such as a periodic fluctuation in the voltage in a circuit, we can define

$$\text{energy of } g_p(\cdot) \text{ over } [-T/2, T/2] \equiv \int_{-T/2}^{T/2} |g_p(t)|^2\,dt.$$

We can multiply both sides of (58e) by the complex conjugate of each side, integrate from $-T/2$ to $T/2$, interchange the order of summation and integration and apply the orthogonality relationship to get

$$\int_{-T/2}^{T/2} |g_p(t)|^2\,dt = \sum_{n=-\infty}^{\infty} \sum_{m=-\infty}^{\infty} G_n G_m^* \int_{-T/2}^{T/2} e^{i2\pi(f_n-f_m)t}\,dt$$

$$= T \sum_{n=-\infty}^{\infty} |G_n|^2. \qquad (59b)$$

This equation is referred to as *Parseval's theorem* (or Rayleigh's theorem) for Fourier series. Because $g_p(\cdot)$ is periodic, the energy in it over the interval from $-\infty$ to ∞ would be infinite. A related concept is that of power, which is defined as energy per unit time interval:

$$\text{power over } (-T/2, T/2) \equiv \frac{\text{energy over } (-T/2, T/2)}{T}$$

$$= \frac{1}{T} \int_{-T/2}^{T/2} |g_p(t)|^2\,dt = \sum_{n=-\infty}^{\infty} |G_n|^2. \qquad (59c)$$

Whereas the energy in $g_p(\cdot)$ from $-\infty$ to ∞ is infinite, the power over this infinite interval is finite since

$$\lim_{k\to\infty} \frac{1}{kT} \int_{-kT/2}^{kT/2} |g_p(t)|^2 \, dt = \sum_{n=-\infty}^{\infty} |G_n|^2 .$$

Equation (59c) decomposes the power into an infinite sum of terms, $|G_n|^2$, the nth term of which is the contribution to the power from the term in the Fourier series for $g_p(\cdot)$ with frequency n/T. We can now define the *discrete power spectrum* for $g_p(\cdot)$ as the sequence whose nth element is

$$S_n \equiv |G_n|^2 , \qquad n = 0, \pm 1, \pm 2, \dots .$$

A plot of S_n versus f_n shows us how the power is distributed over the various frequency components of $g_p(\cdot)$. Note, however, that we can recover $|G_n|$ – but not G_n itself – from S_n. This makes it impossible to reconstruct $g_p(\cdot)$ from knowledge of its discrete power spectrum alone.

As a simple example of the above theory, consider the 2π periodic, even and real-valued function given by

$$g_p(t) \equiv \frac{1-\phi^2}{1+\phi^2 - 2\phi \cos(t)}. \tag{60a}$$

It can be shown that, if $|\phi| < 1$, then $g_p(\cdot)$ is square integrable with Fourier coefficients $G_n = \phi^{|n|}$. Hence

$$g_{p,m}(t) = \sum_{n=-m}^{m} G_n e^{i2\pi f_n t} = 1 + 2\sum_{n=1}^{m} \phi^n \cos(nt) \tag{60b}$$

should converge in mean square to $g_p(\cdot)$ as $m \to \infty$. Figure 61 shows plots of $g_p(\cdot)$ compared with $g_{p,m}(\cdot)$ for $m = 4, 8, 16$ and 32 and $\phi = 0.9$. For this example, the discrete power spectrum is given by

$$S_n = \phi^{2|n|}.$$

Figure 62 shows a plot of $10 \log_{10}(S_n)$ – again with $\phi = 0.9$ – versus $f_n = n/2\pi$ for $n = 0, \pm 1, \dots, \pm 32$.

It is interesting that $g_{p,m}(\cdot)$ in (58c) has the following least squares interpretation as an approximation to $g_p(\cdot)$. Let

$$h_{p,m}(t) \equiv \sum_{n=-m}^{m} H_n e^{i2\pi f_n t}$$

be any function whose discrete power spectrum is identically zero for $|n| > m$, where we now regard m as a fixed positive integer. Suppose

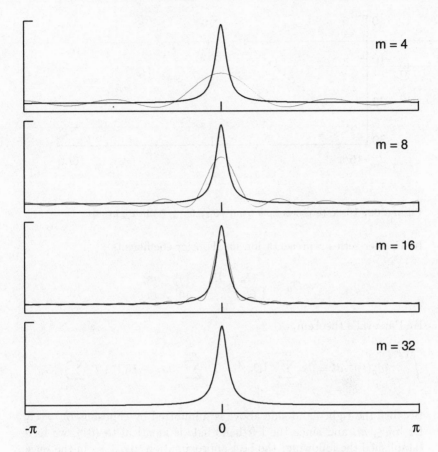

Figure 61. Approximation of a function by a truncated Fourier series. The thick curve in each of the four plots above is $g_p(\cdot)$ in Equation (60a) (with $\phi = 0.9$). The thin curves are $g_{p,m}(\cdot)$ in Equation (60b) for $m = 4$, 8, 16 and 32 – the approximation is sufficiently good in the latter case that there is little discernible difference between it and $g_p(\cdot)$.

we want to find a function of this form that minimizes the integrated square error given by

$$\int_{-T/2}^{T/2} |g_p(t) - h_{p,m}(t)|^2 \, dt. \tag{61}$$

Then the function $h_{p,m}(\cdot)$ that minimizes the above is in fact $g_{p,m}(\cdot)$. To see this, consider the function defined by

$$d_p(t) \equiv g_p(t) - h_{p,m}(t).$$

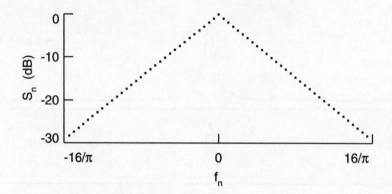

Figure 62. Discrete power spectrum $\{S_n\}$ for $g_p(\cdot)$ in Figure 61.

Its Fourier series representation has Fourier coefficients

$$D_n = \begin{cases} G_n - H_n, & |n| \le m; \\ G_n, & |n| > m. \end{cases}$$

By Parseval's theorem,

$$\int_{-T/2}^{T/2} |d_p(t)|^2 \, dt = T \sum_{n=-\infty}^{\infty} |D_n|^2 = T \sum_{n=-m}^{m} |G_n - H_n|^2 + T \sum_{|n|>m} |G_n|^2.$$

Because the right-hand side above is minimized by choosing $H_n = G_n$ for $|n| \le m$, and since the left-hand side is identical to (61), we have established the following: the best approximation to $g_p(\cdot)$ – in the sense of minimizing (61) – in the class of functions that have discrete power spectra with nonzero elements only for $|n| \le m$ is in fact $g_{p,m}(\cdot)$, the function formed by truncating the Fourier series representation for $g_p(\cdot)$ at indices $\pm m$.

Comments and Extensions to Section 3.1

[1] There are important differences between equality in the mean square sense and equality in the pointwise sense. Consider the following two periodic functions with period $2T$:

$$g_p(t) = \begin{cases} e^{-t}, & 0 < t \le T; \\ 0, & -T < t \le 0, \end{cases}$$

and

$$h_p(t) = \begin{cases} e^{-t}, & 0 \le t \le T; \\ 0, & -T < t < 0. \end{cases}$$

These functions are not equal in the pointwise sense since $g_p(0) = 0$ while $h_p(0) = 1$; however, they are equal in the mean square sense because

$$\int_{-\infty}^{\infty} |g_p(t) - h_p(t)|^2 \, dt = 0.$$

It should also be emphasized that convergence in mean square is computed by integration (or summation) and hence is very different from pointwise convergence. For example, consider the periodic function with period $T > 2$ defined by

$$g_p(t) = \begin{cases} 1, & |t| \leq 1; \\ 0, & 1 < |t| \leq T/2. \end{cases}$$

Its nth Fourier coefficient is

$$G_n = \frac{1}{T} \int_{-1}^{1} e^{-i2\pi f_n t} \, dt = \frac{\sin(2\pi f_n)}{T\pi f_n},$$

where we interpret the latter quantity to be $2/T$ when $n = 0$. Since $g_p(\cdot)$ is square integrable, evidently the sequence of functions defined by

$$g_{p,m}(t) = \sum_{n=-m}^{m} \frac{\sin(2\pi f_n)}{T\pi f_n} e^{i2\pi f_n t}$$

converges to $g_p(\cdot)$ in the mean square sense as $m \to \infty$; however, from Theorem 15.1 of Champeney (1987), it can be shown that, while we do have

$$\lim_{m \to \infty} g_{p,m}(t) = g_p(t) \text{ for all } t \neq \pm 1,$$

in fact we do *not* have pointwise convergence where $g_p(\cdot)$ is discontinuous since

$$\lim_{m \to \infty} g_{p,m}(1) = [g_p(1-) + g_p(1+)]/2 = 1/2 \neq g_p(1) = 1,$$

with a similar result for $t = -1$; here we define

$$g_p(t-) \equiv \lim_{\substack{\epsilon \to 0 \\ \epsilon > 0}} g_p(t - \epsilon) \text{ and } g_p(t+) \equiv \lim_{\substack{\epsilon \to 0 \\ \epsilon > 0}} g_p(t + \epsilon).$$

[2] In this chapter we use the symbol '$\overset{\text{ms}}{=}$' whenever equality in the mean square sense is meant, but in latter chapters we follow standard mathematical practice and use '$=$' to stand for both types of equality – we hope the distinction will be clear from the context.

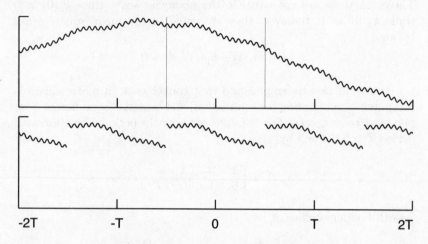

-2T   -T   0   T   2T

Figure 64. Periodic extension of a function. The upper plot shows a nonperiodic real-valued function of time. The bottom plot shows a new function that results from taking the values in the top plot from $[-T/2, T/2]$ – i.e., those values between the thin vertical lines – and replicating them periodically.

3.2 Fourier Theory – Continuous Time and Frequency

Suppose now that $g(\cdot)$ is a deterministic real or complex-valued function of t that is not periodic. Since it does not possess a periodic structure, we cannot express it in the form of a Fourier series (the right-hand side of (58e)) that will be valid for all t. However, we can do the following (cf. Figure 64). For any positive real number T, define a new function $g_T(\cdot)$ with period T first by letting

$$g_T(t) \equiv g(t) \text{ for } -T/2 \leq t \leq T/2$$

and then by letting

$$g_T(t + pT) \equiv g_T(t) \text{ for } -T/2 \leq t \leq T/2 \text{ and } p = \pm 1, \pm 2, \ldots.$$

If $g_T(\cdot)$ is square integrable over $[-T/2, T/2]$, the discussion in the previous section says that we can write

$$g_T(t) \stackrel{\text{ms}}{=} \sum_{n=-\infty}^{\infty} G_{n,T} e^{i2\pi f_n t},$$

where

$$G_{n,T} = \frac{1}{T} \int_{-T/2}^{T/2} g_T(t) e^{-i2\pi f_n t} \, dt = \frac{1}{T} \int_{-T/2}^{T/2} g(t) e^{-i2\pi f_n t} \, dt,$$

since $g(t) = g_T(t)$ over the interval $[-T/2, T/2]$. Hence, for $-T/2 \le t \le T/2$, we have

$$g(t) \equiv g_T(t) \stackrel{ms}{=} \sum_{n=-\infty}^{\infty} \left(\int_{-T/2}^{T/2} g(t) e^{-i2\pi f_n t} \, dt \right) e^{i2\pi f_n t} \, \Delta f,$$

where

$$\Delta f \equiv f_n - f_{n-1} = 1/T.$$

Now as $T \to \infty$, we have $\Delta f \to 0$, and the summation above becomes an integral, so that (formally at least) for all t

$$g(t) \stackrel{ms}{=} \int_{-\infty}^{\infty} G(f) e^{i2\pi f t} \, df, \tag{65a}$$

where

$$G(f) \equiv \int_{-\infty}^{\infty} g(t) e^{-i2\pi f t} \, dt. \tag{65b}$$

The heuristic argument above can be made rigorous under the assumption that

$$\int_{-\infty}^{\infty} |g(t)|^2 \, dt < \infty.$$

The set of all functions that satisfy this property of square integrability is denoted as $L^2(-\infty, \infty)$, where the superscript reflects the square, and the bounds on the integral are given in the parentheses. Equations (65a) and (65b) are thus well-defined for all functions in $L^2(-\infty, \infty)$. The right-hand side of Equation (65a) is called the *Fourier integral representation* of $g(\cdot)$. The function $G(\cdot)$ in (65b) is said to be the *Fourier transform* of $g(\cdot)$. The function of t that is defined by the right-hand side of (65a) is called the *inverse Fourier transform* of $G(\cdot)$ (or, sometimes, the *Fourier synthesis* of $G(\cdot)$). The functions $g(\cdot)$ and $G(\cdot)$ are said to be a *Fourier transform pair* – a fact commonly denoted by

$$g(\cdot) \longleftrightarrow G(\cdot).$$

The inverse Fourier transform on the right-hand side of (65a) represents $g(\cdot)$ (in the mean square equality sense) as the 'sum' of complex exponentials $\exp(i2\pi f t)$ having amplitude $|G(f)|$ and phase $\arg(G(f))$, where

$$\arg(G(f)) \equiv \theta(f) \text{ if we write } G(f) = |G(f)| \, e^{i\theta(f)}.$$

The function $|G(\cdot)|$ is often referred to as the *amplitude spectrum* for $g(\cdot)$.

If we regard $g(\cdot)$ as a signal, its energy over the interval $(-\infty, \infty)$ is just

$$\int_{-\infty}^{\infty} |g(t)|^2 \, dt.$$

This quantity is related to the amplitude spectrum $|G(\cdot)|$ through the following version of Parseval's theorem:

$$\int_{-\infty}^{\infty} |g(t)|^2 \, dt = \int_{-\infty}^{\infty} |G(f)|^2 \, df. \qquad (66)$$

This result is easily proved:

$$\begin{aligned}
\int_{-\infty}^{\infty} |g(t)|^2 \, dt &= \int_{-\infty}^{\infty} g^*(t) \left(\int_{-\infty}^{\infty} G(f) e^{i2\pi ft} \, df \right) dt \\
&= \int_{-\infty}^{\infty} G(f) \left(\int_{-\infty}^{\infty} g^*(t) e^{i2\pi ft} \, dt \right) df \\
&= \int_{-\infty}^{\infty} G(f) G^*(f) \, df = \int_{-\infty}^{\infty} |G(f)|^2 \, df
\end{aligned}$$

(for an extension, see Exercise [3.3]). Note that the above shows that the Fourier transform of a square integrable function is itself a square integrable function (Champeney, 1987, p. 62). The function $|G(\cdot)|^2$ is called the *energy spectral density function* for $g(\cdot)$ by analogy to, say, a probability density function. This is appropriate terminology since Parseval's relationship implies that $|G(f)|^2 \, df$ represents the contribution to the energy from those components in $g(\cdot)$ whose frequencies lie between f and $f + df$. Note that for this case the power (energy per unit time interval) is given by

$$\lim_{T \to \infty} \frac{\text{energy over } [-T/2, T/2]}{T} = 0,$$

since the energy from $-\infty$ to ∞ is finite.

Because we can recover (at least in the mean square sense) $g(\cdot)$ if we know $G(\cdot)$ and vice versa, the time and frequency domain representations of a deterministic function contain equivalent information. A quote from Bracewell (1978, Chapter 8) is appropriate here:

> We may think of functions and their transforms as occupying two domains, sometimes referred to as the upper and the lower, as if functions circulated at ground level and their transforms in the underworld. There is a certain convenience in picturing a function as accompanied by a counterpart in another domain, a kind of shadow which is associated uniquely with the function through the Fourier transformation, and which changes as the function changes.

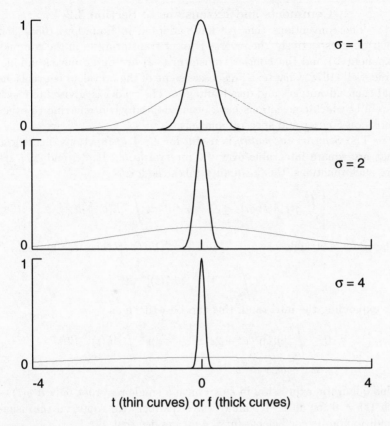

Figure 67. Gaussian probability density functions (thin curves) and their Fourier transforms (thick curves). The standard deviations σ for the pdf's are (from top to bottom) 1, 2 and 4.

As an example of the above theory, consider the Gaussian probability density function with zero mean:

$$g_\sigma(t) = \frac{1}{(2\pi\sigma^2)^{1/2}} e^{-t^2/(2\sigma^2)}.$$

Since the Fourier transform of the function defined by $\exp(-\pi t^2)$ is the function itself (Bracewell, 1978, p. 386), it follows that the Fourier transform of the above is given by

$$G_\sigma(f) = e^{-2\pi^2 f^2 \sigma^2}.$$

Plots of portions of $g_\sigma(\cdot)$ and $G_\sigma(\cdot)$ for $\sigma = 1$, 2 and 4 are shown in Figure 67. Since the energy spectral density function is just $|G_\sigma(f)|^2$, we can readily infer its properties from the plot of $G_\sigma(\cdot)$.

Comments and Extensions to Section 3.2

[1]   The conventions that we have adopted in Equations (65a) and (65b) for, respectively, the inverse Fourier transform ($+i$ in the complex exponential) and the Fourier transform ($-i$) are by no means unique. Bracewell (1978, Chapter 2) discusses some of the prevalent conventions and their advantages and disadvantages. The reader is advised to check carefully which conventions have been adopted when referring to other books on Fourier and spectral analysis.

[2]   The *Schwarz inequality* is useful for $L^2(a, b)$ functions, i.e., those that are square integrable over the interval $[a, b]$. If $g(\cdot)$ and $h(\cdot)$ are two such functions, this inequality takes the form

$$\left| \int_a^b g(t) h(t) \, dt \right|^2 \leq \int_a^b |g(t)|^2 \, dt \int_a^b |h(t)|^2 \, dt. \tag{68}$$

The proof is straightforward. If $p \equiv \int_a^b g(t) h(t) \, dt$, then for any real ϵ

$$0 \leq \int_a^b |g(t) + \epsilon p h^*(t)|^2 \, dt.$$

By expanding the integrand, this can be written as

$$0 \leq \int_a^b |g(t)|^2 \, dt + 2\epsilon |p|^2 + \epsilon^2 |p|^2 \int_a^b |h(t)|^2 \, dt$$

$$\equiv c + b\epsilon + a\epsilon^2.$$

This quadratic expression in ϵ can have a single real root only if $g(t) + \epsilon p h^*(t) = 0$ for all t. It cannot have distinct real roots, so the usual condition on its coefficients ($b^2 \leq 4ac$) implies that

$$|p|^4 \leq |p|^2 \int_a^b |g(t)|^2 \, dt \int_a^b |h(t)|^2 \, dt$$

or, equivalently (even if $p = 0$),

$$|p|^2 = \left| \int_a^b g(t) h(t) \, dt \right|^2 \leq \int_a^b |g(t)|^2 \, dt \int_a^b |h(t)|^2 \, dt.$$

The frequency domain version of (68) is obviously

$$\left| \int_c^d G(f) H(f) \, df \right|^2 \leq \int_c^d |G(f)|^2 \, df \int_c^d |H(f)|^2 \, df,$$

provided that $G(\cdot)$ and $H(\cdot)$ are functions in $L^2(c, d)$. If we set $a = c = -\infty$ and $b = d = \infty$, both the above and (68) hold for $g(\cdot)$, $G(\cdot)$, $h(\cdot)$ and $H(\cdot)$ such that

$$g(\cdot) \longleftrightarrow G(\cdot) \quad \text{and} \quad h(\cdot) \longleftrightarrow H(\cdot)$$

since in this case all these functions are in $L^2(-\infty, \infty)$.

3.3 Band-Limited and Time-Limited Functions

We have just defined the amplitude spectrum and energy spectral density function for a function in $L^2(-\infty, \infty)$. In practice, most physical signals (which can be regarded as functions in $L^2(-\infty, \infty)$) have amplitude spectra with finite support (i.e., the amplitude spectra are nonzero only over a finite range of frequencies). Slepian (1976, 1983) gives some nice illustrations. A pair of solid copper wires will not propagate electromagnetic waves at optical frequencies, so received signals would not be expected to contain energy at frequencies greater than, say, 10^{20} Hz (cycles per second). Recorded male speech gives an amplitude spectrum that is zero for frequencies exceeding 8000 Hz, while orchestral music has no frequencies higher than 20 000 Hz. These band limitations are all at the high-frequency end, and the cutoff frequencies are called high-cut frequencies. It is possible to record signals deficient in both low and high frequencies if the recording instrument has some sort of built-in filter that does so. An example is equipment used in seismic prospecting for oil and gas, which contains a low-cut filter to eliminate certain deleterious noise forms and a high-cut *antialias* filter (the necessity for which is due to effects described in Section 4.5).

If there is no energy above a frequency $|f| = W$, say, then the finite energy signal $g(\cdot)$ is said to be *band-limited* to the band $[-W, W]$. In this case the signal has the following Fourier integral representation:

$$g(t) \overset{\text{ms}}{=} \int_{-W}^{W} G(f) e^{i2\pi ft} \, df.$$

Slepian (1983) notes that band-limited signals are necessarily 'smooth' in the following sense. If we replace t in the right-hand side of the above by the complex number z, the resulting function is defined over the complex plane. Its kth derivative is given by

$$\int_{-W}^{W} G(f) (i2\pi f)^k e^{i2\pi fz} \, df.$$

The Schwarz inequality and Parseval's relationship show that the above exists and is finite for all k (see Exercise [3.10]). A complex-valued function that can be differentiated an arbitrary number of times in the finite complex plane is called an *entire function* in complex analysis. It has a Taylor series expansion about every point with an infinite radius of convergence – these properties make it 'smooth' in the eyes of mathematicians.

Just as some physical signals are band-limited and hence smooth, others can be time-limited and possibly 'rough': we say that $g(\cdot)$ is time-limited if, for some $T > 0$, $g(t)$ is zero for all $|t| > T/2$. Examples of time-limited signals abound in the real world. The seismic trace due to a particular earthquake is necessarily time-limited, as is the amount of gas released by a rocket.

Can a signal be both band-limited and time-limited? The answer is 'yes,' but only in a trivial sense. Slepian (1983) notes that the only $L^2(-\infty, \infty)$ signal with both these properties is zero for all t. The argument for this is simple. If $g(\cdot)$ is time-limited and band-limited, we can express it as a Taylor series about any particular point. If we pick the point to be in the region where $g(\cdot)$ is zero (i.e., $|t| > T/2$), all its derivatives are necessarily 0, and hence the Taylor series representation for $g(\cdot)$ shows that it must necessarily be identically zero everywhere.

3.4 Continuous/Continuous Reciprocity Relationships

We now explore in more detail the relative behavior of the representation of a signal (here taken to be an $L^2(-\infty, \infty)$ function) in the time and frequency domains, i.e., the signal itself $g(\cdot)$ and its Fourier transform $G(\cdot)$. We shall do this by examining three different measures of reciprocity between $g(\cdot)$ and $G(\cdot)$, the most important of which is the last one, the so-called *fundamental uncertainty relationship*.

● *Similarity theorem*

It can be shown (see Exercise [3.13]) that, if $g(\cdot)$ and $G(\cdot)$ are a Fourier transform pair and a is a nonzero real-valued number, then the functions defined by

$$|a|^{1/2}g(at) \quad \text{and} \quad \frac{1}{|a|^{1/2}}G(f/a) \tag{70a}$$

form a Fourier transform pair also. Hence if one member of the transform pair contracts horizontally and expands vertically, the other member expands horizontally and contracts vertically. This is illustrated in Figure 71 for the Fourier transform pair

$$g(t) = \frac{1}{(2\pi)^{1/2}}e^{-t^2/2} \quad \text{and} \quad G(f) = e^{-2\pi^2 f^2}$$

for $a = 1, 2$ and 4 (Figure 67 is also an example of this).

● *Equivalent width*

Suppose that $g(0) \neq 0$. We can define the width of such a signal as the width of the rectangle whose height is equal to $g(0)$ and whose area is the same as that under the curve of $g(\cdot)$; i.e.,

$$\text{width}_e\{g(\cdot)\} \equiv \int_{-\infty}^{\infty} g(t)\,dt \Big/ g(0). \tag{70b}$$

This measure of signal width makes sense for a $g(\cdot)$ that is real-valued, positive everywhere, peaked about 0 and continuous at 0 (see Figure 72), but it is less satisfactory for other types of functions. Now

$$\int_{-\infty}^{\infty} g(t)\,dt = G(0) \quad \text{and} \quad g(0) = \int_{-\infty}^{\infty} G(f)\,df$$

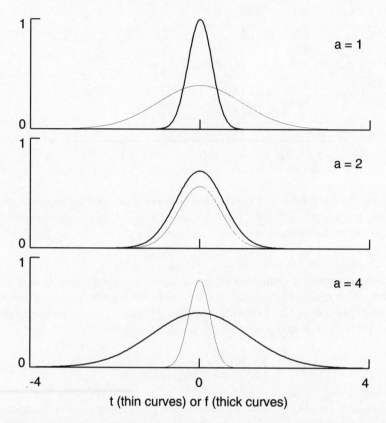

t (thin curves) or f (thick curves)

Figure 71. Similarity theorem. As $g(\cdot)$ (thin curve in top plot) is contracted horizontally and expanded vertically by letting $a = 2$ and 4 (middle and bottom plots, respectively), its Fourier transform $G(\cdot)$ (thick curve in top plot) is expanded horizontally and contracted vertically.

(if $g(\cdot)$ is continuous at 0), so we have

$$\text{width}_\text{e}\,\{g(\cdot)\} = \int_{-\infty}^{\infty} g(t)\,dt \Big/ g(0)$$

$$= G(0) \Big/ \int_{-\infty}^{\infty} G(f)\,df = 1/\text{width}_\text{e}\,\{G(\cdot)\}.$$

Note that the equivalent width of a signal is the reciprocal of the equivalent width of its transform; i.e.,

equivalent width (signal) × equivalent width (transform) = 1.

Thus, as the equivalent width of a signal increases, the equivalent width of its transform necessarily decreases. (We will meet other measures of width in Chapter 6 when we discuss ways of measuring bandwidth.)

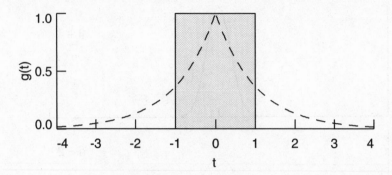

Figure 72. Equivalent width. The dashed curve shows the function defined by $g(t) = \exp(-|t|)$. Since $\int_{-\infty}^{\infty} g(t)\, dt = 2$ and $g(0) = 1$, the equivalent width is 2, which is the width of the shaded rectangle.

- *Fundamental uncertainty relationship*

A useful alternative definition of signal width for the special case of a nonnegative real-valued signal can be based on matching it to a rectangular function in the following way. Let $r(\cdot\,; \mu_r, W_r)$ be a rectangular function with unit area centered at μ_r with width $2W_r$; i.e.,

$$r(t; \mu_r, W_r) \equiv \begin{cases} 1/(2W_r), & \mu_r - W_r \leq t \leq \mu_r + W_r; \\ 0, & \text{otherwise.} \end{cases} \tag{72}$$

Note that $r(\cdot\,; \mu_r, W_r)$ can be regarded as the probability density function (pdf) of a random variable (rv) uniformly distributed over the interval $[\mu_r - W_r, \mu_r + W_r]$ – in this interpretation μ_r is the expected value of the rv. A measure of spread of this pdf is its variance (second central moment):

$$\sigma_r^2 \equiv \int_{-\infty}^{\infty} (t - \mu_r)^2\, r(t; \mu_r, W_r)\, dt = \frac{W_r^2}{3}.$$

Thus we can express the width of $r(\cdot\,; \mu_r, W_r)$ in terms of its variance, namely, $2W_r = 2\sigma_r\sqrt{3}$. This suggests that we take the following steps to define the width for a nonnegative real-valued signal $g(\cdot)$ such that

$$\int_{-\infty}^{\infty} g(t)\, dt \equiv C \ \text{ with } 0 < C < \infty.$$

First, we renormalize $g(\cdot)$ as $\tilde{g}(t) \equiv g(t)/C$ so that $\tilde{g}(\cdot)$ can be regarded as a pdf. Second, we calculate the second central moment

$$\sigma_{\tilde{g}}^2 \equiv \int_{-\infty}^{\infty} (t - \mu_{\tilde{g}})^2\, \tilde{g}(t)\, dt, \ \text{ where } \mu_{\tilde{g}} \equiv \int_{-\infty}^{\infty} t\tilde{g}(t)\, dt$$

(we assume both these quantities are finite). Finally we *define*

$$\text{width}_v \left\{ g(\cdot) \right\} \equiv 2\sigma_{\tilde{g}}\sqrt{3}. \tag{73a}$$

For the important special case in which $C = 1$ and $\mu_{\tilde{g}} = 0$, we have

$$\text{width}_v \left\{ g(\cdot) \right\} = \left(12 \int_{-\infty}^{\infty} t^2 g(t) \, dt \right)^{1/2} \tag{73b}$$

(we make use of this formula in Chapter 6 as a measure of the bandwidth of a smoothing window).

As we have defined it, the above measure of width is not meaningful for a complex-valued signal $g(\cdot)$; however, it can be applied to its modulus $|g(\cdot)|$ or squared modulus $|g(\cdot)|^2$ since both are necessarily nonnegative and real-valued. The latter case leads to a more precise statement about the relationship between the width of a signal – defined now as $\text{width}_v \left\{ |g(\cdot)|^2 \right\}$ using (73a) – and the width of its Fourier transform $\text{width}_v \left\{ |G(\cdot)|^2 \right\}$ via a version of Heisenberg's uncertainty principle (Slepian, 1983; Bracewell, 1978). Let the signal $g(\cdot)$ have unit energy; i.e.,

$$\int_{-\infty}^{\infty} |g(t)|^2 \, dt = \int_{-\infty}^{\infty} |G(f)|^2 \, df = 1.$$

Then both $|g(\cdot)|^2$ and $|G(\cdot)|^2$ act like probability density functions, and their associated variances (a measure of spread or width) are given by

$$\sigma_g^2 = \int_{-\infty}^{\infty} (t - \mu_g)^2 \, |g(t)|^2 \, dt, \quad \text{where} \quad \mu_g \equiv \int_{-\infty}^{\infty} t \, |g(t)|^2 \, dt,$$

and

$$\sigma_G^2 = \int_{-\infty}^{\infty} (f - \mu_G)^2 \, |G(f)|^2 \, df, \quad \text{where} \quad \mu_G \equiv \int_{-\infty}^{\infty} f \, |G(f)|^2 \, df.$$

For convenience assume that $\mu_g = \mu_G = 0$. Then we have

$$\sigma_g^2 \sigma_G^2 = \int_{-\infty}^{\infty} t^2 \, |g(t)|^2 \, dt \int_{-\infty}^{\infty} f^2 \, |G(f)|^2 \, df.$$

Since

$$g(t) \overset{\text{ms}}{=} \int_{-\infty}^{\infty} G(f) e^{i2\pi f t} \, df,$$

it can be argued (under a regularity condition) that

$$g'(t) \equiv \frac{dg(t)}{dt} \overset{\text{ms}}{=} \int_{-\infty}^{\infty} i2\pi f G(f) e^{i2\pi f t} \, df,$$

so that $g'(\cdot)$ and the function defined by $i2\pi f G(f)$ are a Fourier transform pair. By Parseval's theorem for a Fourier transform pair (66),

$$\int_{-\infty}^{\infty} |g'(t)|^2 \, dt = 4\pi^2 \int_{-\infty}^{\infty} f^2 \, |G(f)|^2 \, df.$$

Hence

$$\sigma_g^2 \sigma_G^2 = \frac{1}{4\pi^2} \int_{-\infty}^{\infty} t^2 \, |g(t)|^2 \, dt \int_{-\infty}^{\infty} |g'(t)|^2 \, dt.$$

By the Schwarz inequality (Equation (68)) we have

$$\int_{-\infty}^{\infty} t^2 \, |g(t)|^2 \, dt \int_{-\infty}^{\infty} |g'(t)|^2 \, dt \geq \left| \int_{-\infty}^{\infty} t g^*(t) g'(t) \, dt \right|^2$$

$$\geq \left[\Re \left(\int_{-\infty}^{\infty} t g^*(t) g'(t) \, dt \right) \right]^2 = 1/4 \left[\int_{-\infty}^{\infty} t g^*(t) g'(t) + t g(t) \, [g'(t)]^* \, dt \right]^2$$

(here $\Re(z)$ is the real part of the complex number z, and we use the facts that $|z|^2 \geq (\Re(z))^2$ and $\Re(z) = (z + z^*)/2$). We thus have

$$16\pi^2 \sigma_g^2 \sigma_G^2 \geq \left[\int_{-\infty}^{\infty} t g^*(t) g'(t) + t g(t) \, [g'(t)]^* \, dt \right]^2 ;$$

however, since

$$\frac{d \, |g(t)|^2}{dt} = g^*(t) g'(t) + g(t) \, [g'(t)]^* ,$$

we can conclude that

$$16\pi^2 \sigma_g^2 \sigma_G^2 \geq \left[\int_{-\infty}^{\infty} t \frac{d \, |g(t)|^2}{dt} \, dt \right]^2 = \left[\int_{-\infty}^{\infty} |g(t)|^2 \, dt \right]^2 = 1,$$

where we have used integration by parts and the fact that $g(\cdot)$ has unit energy. This is Heisenberg's uncertainty principle, namely,

$$\sigma_g^2 \sigma_G^2 \geq \frac{1}{16\pi^2}. \tag{74}$$

Note for example that, if σ_g^2 is very small, then σ_G^2 must be large to ensure that (74) is satisfied.

Is the lower bound in (74) attainable? Consider the Gaussian shaped function

$$g(t) = b e^{-\pi a^2 t^2}, \quad \text{where } b^2 = a\sqrt{2}$$

(this condition on b forces $g(\cdot)$ to have unit energy). Then

$$\sigma_g^2 = \int_{-\infty}^{\infty} t^2 g^2(t)\, dt = a\sqrt{2} \int_{-\infty}^{\infty} t^2 e^{-2\pi a^2 t^2}\, dt = 1/(4\pi a^2).$$

The Fourier transform of $g(\cdot)$ is given by

$$G(f) = ba^{-1} e^{-\pi f^2/a^2}, \quad \text{so} \quad |G(f)|^2 = b^2 a^{-2} e^{-2\pi f^2/a^2}.$$

Thus

$$\sigma_G^2 = b^2 a^{-2} \int_{-\infty}^{\infty} f^2 e^{-2\pi f^2/a^2}\, df = a^2/(4\pi) \quad \text{and} \quad \sigma_g^2 \sigma_G^2 = 1/(16\pi^2),$$

which is the minimum value obtainable under the uncertainty relationship. (Some further mathematical insights into equivalent width and the fundamental uncertainty relationship are given by Champeney, 1987, pp. 75–6.)

3.5 Concentration Problem – Continuous/Continuous Case

Three ways of measuring time and frequency concentrations of signals belonging to $L^2(-\infty, \infty)$ have been discussed in the previous section. An alternative way of measuring concentration is discussed by Slepian (1983). Concentration in time is measured by the ratio

$$\alpha^2(T) \equiv \int_{-T/2}^{T/2} |g(t)|^2\, dt \Big/ \int_{-\infty}^{\infty} |g(t)|^2\, dt, \qquad (75a)$$

i.e., the fraction of the signal's energy lying in a time interval of length T centered about 0. Analogously,

$$\beta^2(W) \equiv \int_{-W}^{W} |G(f)|^2\, df \Big/ \int_{-\infty}^{\infty} |G(f)|^2\, df \qquad (75b)$$

is a measure of concentration in the frequency domain.

These measures of concentration have considerable intuitive appeal. They have also had a profound effect on current research in spectral analysis because an analytic solution to the following problem leads to a way of characterizing time-limited and band-limited signals. The problem is: among all signals, say $g_W(\cdot)$, that are band-limited to $[-W, W]$, find all those such that $\alpha^2(T)$ is as large as possible, i.e., that have the greatest concentration of their energy in the time interval $[-T/2, T/2]$. Since a nontrivial band-limited signal cannot be time-limited also, we must have $1 > \alpha^2(T) > 0$ for those signals that are solutions to this maximization problem. We now outline the steps that can be taken to solve this problem by following Slepian (1983).

In definition (75a) we can express $g_W(\cdot)$ in terms of its Fourier transform $G_W(\cdot)$ to show that

$$\alpha^2(T, W) = \frac{\int_{-W}^{W} \int_{-W}^{W} G_W(f) K(f, f'; T) G_W^*(f') \, df' \, df}{\int_{-W}^{W} |G_W(f)|^2 \, df}, \qquad (76a)$$

where

$$K(f, f'; T) \equiv \frac{\sin (\pi T[f - f'])}{\pi[f - f']};$$

here we have added a second argument to $\alpha^2(\cdot)$ to emphasize its dependence on W as well as T. It can be shown from the calculus of variations (see, for example, Courant and Hilbert, 1953) that all functions $G_W(\cdot)$ that maximize $\alpha^2(T, W)$ must satisfy the following integral equation:

$$\int_{-W}^{W} K(f, f'; T) G_W(f') \, df' = \alpha^2(T, W) G_W(f), \quad |f| \leq W. \qquad (76b)$$

An integral equation of the above form is known in the literature as a *Fredholm equation of the second kind*; the quantity $K(\cdot, \cdot; T)$ is called the *kernel* of the integral equation. Engineers will recognize this kernel as the function defined by $T \operatorname{sinc}[T(f - f')]$, where $\operatorname{sinc}(\cdot)$ is the sinc function:

$$\operatorname{sinc}(t) \equiv \frac{\sin (\pi t)}{\pi t}.$$

The unknown quantities that a solution to this equation gives us are the functions $G_W(\cdot)$ and the corresponding scalar $\alpha^2(T, W)$. The theory for integral equations such as (76b) is dependent on the nature of the associated kernel. Here the kernel is such that there is a countably infinite number of solutions to (76b) (this statement regards functions that differ only by a multiplicative constant – such as $G_W(\cdot)$, $2G_W(\cdot)$, $\pi G_W(\cdot)$, etc. – as being a single solution). Not all of these solutions are also solutions to our maximization problem. If we let

$$G_{W,k}(\cdot) \text{ and } \alpha_k^2(T, W), \quad k = 0, 1, \ldots,$$

represent these solutions, where

$$1 > \alpha_0^2(T, W) \geq \alpha_1^2(T, W) \geq \cdots,$$

one solution to our maximization problem is given by $G_{W,0}(\cdot)$, and the degree of concentration of energy is given by $\alpha_0^2(T, W)$.

This maximization problem that we have just described reappears in another form when we discuss filters and data tapers in Chapters 5 and 7. The set of all solutions to the integral equation (76b) is also of considerable interest. We summarize some of their properties here. We first make the following substitutions:

$$x \equiv f/W; \quad y \equiv f'/W; \quad \text{and} \quad c \equiv \pi WT.$$

Equation (76b) now becomes

$$\int_{-1}^{1} \frac{\sin(c[x-y])}{\pi[x-y]} G_W(Wy) \, dy = \alpha^2(T, W) G_W(Wx), \quad |x| \leq 1.$$

The theory of integral equations says that the solutions to the above equation depend only upon the properties of the associated kernel. The kernel depends upon T and W only through $c = \pi WT$, so we define

$$\psi(y; c) \equiv G_W(Wy) \quad \text{and} \quad \lambda(c) \equiv \alpha^2(T, W)$$

to reflect this dependence. Our integral equation now takes the form

$$\int_{-1}^{1} \frac{\sin(c[x-y])}{\pi[x-y]} \psi(y; c) \, dy = \lambda(c)\psi(x; c), \quad |x| \leq 1. \tag{77}$$

We denote the countably infinite set of functions that solve the above by

$$\psi_0(\cdot; c), \psi_1(\cdot; c), \psi_2(\cdot; c), \ldots,$$

and the corresponding proportions of energy by

$$\lambda_0(c), \lambda_1(c), \lambda_2(c), \ldots.$$

The solution $\psi_k(\cdot; c)$ is called the kth *eigenfunction*, and $\lambda_k(c)$ is the associated *eigenvalue*. The eigenfunction corresponding to each eigenvalue is unique except for a multiplicative constant. These eigenfunctions are called *prolate spheroidal wave functions* (pswf's). The pswf $\psi_0(\cdot; 4)$ is shown in Figure 78 – it is the portion of the thick curve between the two thin vertical lines located at $x = \pm 1$. Here are some of the important properties of these eigenfunctions and eigenvalues.

[1] The eigenfunctions are real-valued and orthogonal on $[-1, 1]$; i.e.,

$$\int_{-1}^{1} \psi_j(x; c)\psi_k(x; c) \, dx = 0, \quad \text{for } j \neq k.$$

[2] The eigenvalues have the following properties:

$$\lambda_k(c) > 0 \text{ for all } k; \quad 1 > \lambda_0(c) > \lambda_1(c) > \cdots; \quad \text{and} \quad \lim_{k \to \infty} \lambda_k(c) = 0.$$

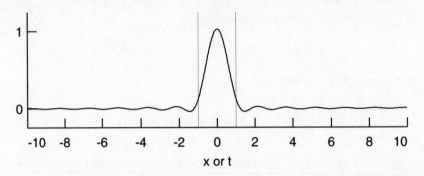

Figure 78. Prolate spheroidal wave function $\psi_0(\cdot; 4)$ (pswf) and associated rescaled Fourier transform. The pswf is the portion of the thick curve between the two thin vertical lines, while its rescaled Fourier transform is the entire thick curve.

[3] Equation (77) only defines $\psi(\cdot; c)$ for $|x| \leq 1$. The left-hand side of that equation, however, is well-defined for all x, so we can *define*

$$\psi(x; c) \equiv \lambda(c)^{-1} \int_{-1}^{1} \frac{\sin(c[x-y])}{\pi[x-y]} \psi(y; c) \, dy, \quad |x| > 1. \tag{78}$$

With this definition, it can be shown that the eigenfunctions are orthogonal on $(-\infty, \infty)$ as well as on $[-1, 1]$. Moreover, the Fourier transform of $\psi_k(\cdot; c)$ restricted to $|x| \leq 1$ has the same form as $\psi_k(\cdot; c)$ except for a scale change; i.e.,

$$\int_{-1}^{1} \psi_k(x; c) e^{-i2\pi xt} \, dx \propto \psi_k(2\pi t/c; c), \quad -\infty < t < \infty.$$

The thick curve in Figure 78 shows $\psi_0(\pi t/2; 4)$ versus t for $|t| \leq 10$ scaled to the same values as $\psi_0(\cdot; 4)$ (recall that this is shown by the segment of the solid curve between the thin vertical lines). Similarly,

$$\int_{-\infty}^{\infty} \psi_k(x; c) e^{-i2\pi xt} \, dx \propto \begin{cases} \psi_k(2\pi t/c; c), & |t| \leq c/(2\pi); \\ 0, & \text{otherwise.} \end{cases}$$

[4] If we normalize the eigenfunctions to have an energy of unity over $(-\infty, \infty)$, it follows that

$$\int_{-\infty}^{\infty} \psi_j(x; c) \psi_k(x; c) \, dx = \delta_{j,k}$$

($\delta_{j,k}$ is Kronecker's delta function; i.e., $\delta_{j,k} = 0$ if $j \neq k$ and $= 1$ if $j = k$) and that

$$\int_{-1}^{1} \psi_j(x;c)\psi_k(x;c)\,dx = \lambda_k(c)\delta_{j,k}.$$

The pswf's are thus orthogonal over $[-1,1]$ and orthonormal over $(-\infty,\infty)$. The value $\lambda_k(c)$ gives the proportion of energy in the interval $[-1,1]$.

[5] At this point we can return to the solution to the concentration problem posed in this section. Recall that $g_W(\cdot)$ is to be a signal whose Fourier transform $G_W(\cdot)$ is zero outside $[-W,W]$. If we make

$$G_{W,k}(f) \propto \begin{cases} \psi_k(f/W;c), & |f| \leq W; \\ 0, & \text{otherwise,} \end{cases}$$

where $c = \pi WT$, then $G_{W,k}(\cdot)$ is zero outside $[-W,W]$, and its inverse Fourier transform $g_{W,k}(\cdot)$ is such that $g_{W,k}(t) \propto \psi_k(2t/T;c)$. Thus there is only one solution to the concentration problem, namely, the band-limited signal $g_{W,0}(\cdot) = \psi_0(2\cdot/T;c)$ (if we count solutions that differ only by a multiplicative constant as just one solution). For this solution $\lambda_0(c)$ represents the degree of concentration of $g_{W,0}(\cdot)$ in $[-T/2,T/2]$. Among all functions that are band-limited to $[-W,W]$ and are orthogonal to $g_{W,0}(\cdot)$, the signal $g_{W,1}(\cdot) = \psi_1(2\cdot/T;c)$ is most concentrated in $[-T/2,T/2]$ with a degree of concentration of $\lambda_1(c) < \lambda_0(c)$. This pattern continues – among all functions that are band-limited to $[-W,W]$ and are orthogonal to $g_{W,0}(\cdot)$, $g_{W,1}(\cdot)$, ..., $g_{W,k-1}(\cdot)$, the function $g_{W,k}(\cdot)$ is most concentrated in $[-T/2,T/2]$ with a degree of concentration of $\lambda_k(c)$.

[6] For large c the eigenvalue series $\lambda_k(c)$ drops sharply from approximately unity to nearly zero at a value of k known as the *Shannon number*, namely,

$$k = 2c/\pi = 2WT.$$

This quantity is fundamental in electrical engineering and has a vital role in spectral analysis (see Chapter 7). Loosely speaking, if $2WT$, the so-called duration–bandwidth product is large (and hence also c), the space of signals of approximate duration T and approximate bandwidth $2W$ has approximate dimension (or complex degrees of freedom) $2WT$. Since there are no nontrivial signals that are both time-limited and band-limited, this statement is necessarily vague (Slepian, 1976, has made it mathematically precise).

[7] The eigenfunctions $\psi_k(\cdot;c)$ are even or odd as k is even or odd; $\psi_k(\cdot;c)$ has exactly k zeros in the interval $[-1,1]$.

[8] With $c = 2\pi W$ the eigenfunctions $\psi_k(\cdot; c)$ form a complete basis for the class of all functions that are band-limited on $[-W, W]$. This means that, if $g_W(\cdot)$ is *any* $L^2(-\infty, \infty)$ function in that band-limited class, it can be represented as

$$g_W(t) \overset{\text{ms}}{=} \sum_{k=0}^{\infty} \gamma_k \psi_k(t; c), \qquad -\infty < t < \infty, \qquad (80a)$$

where

$$\gamma_k \equiv \int_{-\infty}^{\infty} g_W(t) \psi_k(t; c) \, dt.$$

It follows from the symmetry between time and frequency that these eigenfunctions also form a complete basis for the class of $L^2(-1, 1)$ functions.

As an application of the above theory, let us consider the problem of extrapolating a band-limited function (Slepian and Pollak, 1961). Suppose that we only know $g_W(\cdot)$ over the interval $[-1, 1]$ and that we wish to extrapolate it outside this interval. Since $g_W(\cdot)$ can be represented as in Equation (80a), we can multiply both sides of that equation by $\psi_j(\cdot; c)$ and integrate over $[-1, 1]$ to get (after exchanging the order of integration and summation on the right-hand side)

$$\int_{-1}^{1} \psi_j(t; c) g_W(t) \, dt = \sum_{k=0}^{\infty} \gamma_k \int_{-1}^{1} \psi_j(t; c) \psi_k(t; c) \, dt$$

$$= \gamma_j \lambda_j(c)$$

(from property [4] above), or, equivalently,

$$\gamma_k = \lambda_k^{-1}(c) \int_{-1}^{1} g_W(t) \psi_k(t; c) \, dt.$$

Thus, given $g_W(\cdot)$ over the interval $[-1, 1]$, we can extrapolate to values outside this interval by using

$$g_W(t) \overset{\text{ms}}{=} \sum_{k=0}^{\infty} \lambda_k^{-1}(c) \left[\int_{-1}^{1} g_W(t') \psi_k(t'; c) \, dt' \right] \psi_k(t; c); \qquad (80b)$$

i.e., it is possible to extrapolate the band-limited function $g_W(\cdot)$ *perfectly* (at least in the mean square sense) just from knowledge of its form over the interval $[-1, 1]$. By repeating the above argument with a suitable rescaling of the time axis, it follows that a band-limited function can be reconstructed perfectly (in mean square) from knowledge of its values over *any* interval of nonzero length.

Now suppose that we truncate the infinite summation in (80b) at, say, $k = m$, and use

$$g_m(t) \equiv \sum_{k=0}^{m} \gamma_k \psi_k(t; c)$$

to approximate $g_W(\cdot)$. The energy in the error of fit of $g_m(\cdot)$ to $g_W(\cdot)$ in $[-1, 1]$ is

$$\int_{-1}^{1} |g_W(t) - g_m(t)|^2 \, dt = \sum_{k=m+1}^{\infty} \gamma_k^2 \lambda_k(c).$$

From property [5] above, $\lambda_k(c)$ is close to 0 if m exceeds the Shannon number $2c/\pi$. Hence the error in $[-1, 1]$ will be small if m is chosen thus. However, the energy in the error of fit in $(-\infty, \infty)$ is

$$\int_{-\infty}^{\infty} |g_W(t) - g_m(t)|^2 \, dt = \sum_{k=m+1}^{\infty} \gamma_k^2,$$

which does not depend on $\lambda_k(c)$.

3.6 Convolution Theorem – Continuous Time and Frequency

We state and prove here one version of the widely used convolution theorem. This theorem is often paraphrased as 'convolution in the time domain is equivalent to multiplication in the frequency domain.' Let $g(\cdot)$ and $h(\cdot)$ be two real or complex-valued functions. The convolution of $g(\cdot)$ and $h(\cdot)$ is the function of t defined by

$$\int_{-\infty}^{\infty} g(u)h(t - u) \, du,$$

provided the integral exists. Convolution involves reflecting one of the functions about the time axis, shifting it by t units, multiplying it by the corresponding coordinate of the other function, and integrating this product from $-\infty$ to ∞. The function defined above is conveniently denoted by $g * h(\cdot)$, and its value at t by $g * h(t)$; i.e.,

$$g * h(t) \equiv \int_{-\infty}^{\infty} g(u)h(t - u) \, du. \tag{81}$$

A change of variable in the integral shows that $h * g(\cdot) = g * h(\cdot)$.

The Fourier transform of $g * h(\cdot)$ is given by

$$\int_{-\infty}^{\infty} g * h(t)e^{-i2\pi ft} \, dt.$$

If we substitute (81) into the above, we have

$$\int_{-\infty}^{\infty}\int_{-\infty}^{\infty} g(u)h(t-u)e^{-i2\pi ft}\,du\,dt$$

$$= \int_{-\infty}^{\infty} g(u)e^{-i2\pi fu}\int_{-\infty}^{\infty} h(t-u)e^{-i2\pi f(t-u)}\,dt\,du.$$

Let $y = t - u$ to get

$$\int_{-\infty}^{\infty} g(u)e^{-i2\pi fu}\,du\int_{-\infty}^{\infty} h(y)e^{-i2\pi fy}\,dy = G(f)H(f).$$

Hence the Fourier transform of the function $g*h(\cdot)$ is given by the function defined by $G(f)H(f)$; i.e., the Fourier transform of the convolution of $g(\cdot)$ and $h(\cdot)$ is the product of the Fourier transforms of $g(\cdot)$ and $h(\cdot)$. This fact is often written as

$$g * h(\cdot) \longleftrightarrow G(\cdot)H(\cdot).$$

There are a variety of convolution theorems that stipulate under what conditions the above makes sense. Exercise [3.6] concerns one of these theorems; see Champeney (1987) for others.

Convolution is often regarded as a smoothing operation (this is used extensively in Chapter 6). In this interpretation, $g(\cdot)$ is regarded as a signal we desire to smooth, while $h(\cdot)$ is called a *smoothing kernel*. The convolution $g * h(\cdot)$ is then a smoothed version of $g(\cdot)$ whose degree of smoothness is determined by the properties of $h(\cdot)$. As a concrete example, let us assume that

$$g(t) = \sum_{l=1}^{L} A_l \cos\left(2\pi f_l t + \phi_l\right), \tag{82}$$

so that our signal is just a sum of sinusoids with different amplitudes, frequencies and phases. For the smoothing kernel let us pick the Gaussian probability density function

$$h(t) = \frac{1}{(2\pi\sigma^2)^{1/2}}e^{-t^2/(2\sigma^2)}$$

with zero mean and standard deviation $\sigma > 0$ – here σ plays the role of an adjustable smoothing parameter (the boundedness of $g(\cdot)$ in (82) and the integrability of $h(\cdot)$ is sufficient for the convolution integral (81) to exist). Now the Fourier transform of the function defined by $\exp\left(-\pi t^2\right)$ is the function itself, which implies that

$$\int_{-\infty}^{\infty} e^{-\pi t^2}\cos\left(2\pi ft\right)dt = e^{-\pi f^2} \quad\text{and}\quad \int_{-\infty}^{\infty} e^{-\pi t^2}\sin\left(2\pi ft\right)dt = 0$$

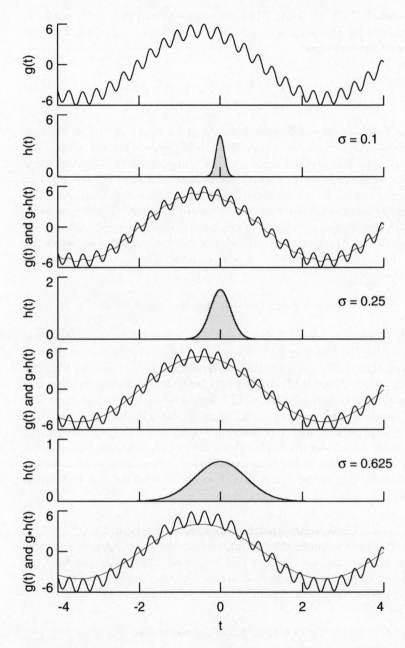

Figure 83. Convolution as a smoothing operation. Convolution of $g(\cdot)$ (thick curves) with different smoothing kernels $h(\cdot)$ (curves above shaded areas) produces functions (thin curves in plots below shaded areas) that are smoothed versions of $g(\cdot)$.

(Bracewell, 1978, p. 386). These integral expressions – along with the trigonometric identity $\cos{(a+b)} = \cos{(a)}\cos{(b)} - \sin{(a)}\sin{(b)}$ – can be used to show that

$$g * h(t) = \sum_{l=1}^{L} e^{-(\sigma 2\pi f_l)^2/2} A_l \cos{(2\pi f_l t + \phi_l)}. \qquad (84a)$$

Note that the only difference between $g(\cdot)$ and $g * h(\cdot)$ is in the amplitudes assigned to the sinusoids – their frequencies and phases are unchanged. The original amplitude A_l is multiplied by the attenuation factor $\exp{[-(\sigma 2\pi f_l)^2/2]}$, which lies between 0 and 1. For fixed σ, this factor is smaller for an amplitude associated with a high frequency term than for one associated with a low frequency term. This attenuation of the high frequency contributions to $g(\cdot)$ makes $g * h(\cdot)$ smoother in appearance. The degree of smoothness can be controlled by adjusting σ – the larger σ is, the more A_l is attenuated. Figure 83 illustrates this for the case

$$g(t) = 5\cos{(2\pi f_1 t + 0.5)} + \cos{(2\pi f_2 t + 1.1)} \qquad (84b)$$

with $f_1 = 1/6$ and $f_2 = 3$. For $\sigma = 0.1$, 0.25 and 0.625, the attenuation factors for the low frequency term f_1 are, respectively to two decimal places, 0.99, 0.97 and 0.81; the corresponding factors for the high frequency term f_2 are 0.17, 0.0 and 0.0. If we define the proper degree of smoothing here to mean that $g * h(\cdot)$ should be as close as possible to the low frequency term in $g(\cdot)$, then Figure 83 shows that the choice of 0.25 for σ is preferable to 0.1 (undersmoothing) or 0.625 (oversmoothing). (Chapter 5 on linear filters provides some more insight into this example and discusses some of the properties that $h(\cdot)$ must have to be a reasonable smoothing kernel – see also [3] in the Comments and Extensions below.)

Comments and Extensions to Section 3.6

[1]   There are no fewer than 20 different versions of the convolution theorem in wide use (see Chapter 6 of Bracewell, 1978). We note here two variants called the (complex) cross-correlation theorem and the (complex) autocorrelation theorem. The cross-correlation of $g(\cdot)$ and $h(\cdot)$ is defined by

$$g^* \star h(t) \equiv \int_{-\infty}^{\infty} g^*(u) h(u+t) \, du,$$

where '$\star$' is defined by

$$a \star b(t) = \int_{-\infty}^{\infty} a(u) b(u+t) \, du.$$

This is similar to – but not the same as – the convolution integral since

$$a * b(t) = \int_{-\infty}^{\infty} a(u)b(t - u)\, du.$$

The cross-correlation of $g(\cdot)$ and $h(\cdot)$ is in fact equal to the convolution of $g^*(\cdot)$ with the *time reversed* version of $h(\cdot)$. Now

$$g^* \star h(t) = \int_{-\infty}^{\infty} g^*(u)h(u+t)\, du = \int_{-\infty}^{\infty} h(u)g^*(u-t)\, du = [h^* \star g(-t)]^*,$$

which in general is *not* equal to $h^* \star g(t)$ – unlike ordinary convolution, cross-correlation is not commutative. Note that

$$g^* \star h(\cdot) \longleftrightarrow G^*(\cdot)H(\cdot) \tag{85a}$$

(this is Exercise [3.7a]). If we put $h(\cdot) = g(\cdot)$, the cross-correlation becomes an autocorrelation:

$$g^* \star g(t) = \int_{-\infty}^{\infty} g^*(u)g(u + t)\, du,$$

from which we have

$$g^* \star g(\cdot) \longleftrightarrow G^*(\cdot)G(\cdot) = |G(\cdot)|^2. \tag{85b}$$

[2] The autocorrelation above can be used to define another measure of the width of a signal $g(\cdot)$ (in addition to $\text{width}_e\{g(\cdot)\}$ and $\text{width}_v\{g(\cdot)\}$ defined in Equations (70b) and (73a)). This measure is simply the equivalent width of the autocorrelation of $g(\cdot)$:

$$\text{width}_a\{g(\cdot)\} \equiv \text{width}_e\{g^* \star g(\cdot)\}$$

$$= \frac{\int_{-\infty}^{\infty} g^* \star g(t)\, dt}{g^* \star g(0)} = \frac{\left|\int_{-\infty}^{\infty} g(t)\, dt\right|^2}{\int_{-\infty}^{\infty} |g(t)|^2\, dt}. \tag{85c}$$

The rectangular function of Equation (72) has an autocorrelation given by

$$r \star r(t; \mu_r, W_r) = \begin{cases} 1 - [|t|/(2W_r)], & -2W_r < t < 2W_r; \\ 0, & \text{otherwise}, \end{cases}$$

and hence has an autocorrelation width given by

$$\text{width}_a\{r(\cdot; \mu_r, W_r)\} = \int_{-2W_r}^{2W_r} 1 - \frac{|t|}{2W_r}\, dt = 2W_r$$

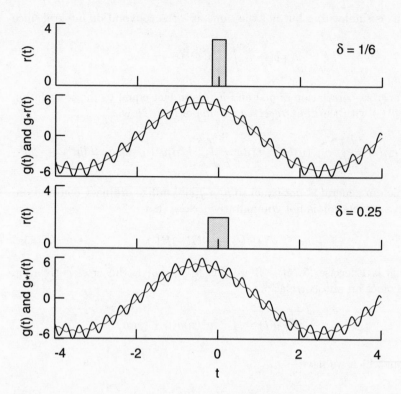

Figure 86. Convolution with a rectangular smoothing kernel (cf. Figure 83).

as intuitively required. Note that the autocorrelation width of a signal is the reciprocal of the equivalent width of its energy spectral density function since

$$\frac{\left|\int_{-\infty}^{\infty} g(t)\, dt\right|^2}{\int_{-\infty}^{\infty} |g(t)|^2\, dt} = \frac{|G(0)|^2}{\int_{-\infty}^{\infty} |G(f)|^2\, df}.$$

Use can be made of the autocorrelation width of a signal for defining the bandwidth of a smoothing window – see Chapter 6.

[3] In preparation for a discussion in Chapter 6, let us reconsider our example of convolution as a smoothing operation by looking at another widely used smoothing kernel, namely, the rectangular kernel defined by

$$r(t) \equiv \begin{cases} 1/(2\delta), & |t| \leq \delta; \\ 0, & |t| > \delta. \end{cases}$$

The parameter δ determines the degree of smoothing. Note that

$$g * r(t) = \frac{1}{2\delta} \int_{t-\delta}^{t+\delta} g(u)\, du, \tag{86}$$

a quantity that in calculus books is called the *average value* of the function $g(\cdot)$ over the interval from $t - \delta$ to $t + \delta$. With $g(\cdot)$ defined by Equation (82), we have

$$g * r(t) = \sum_{l=1}^{L} \operatorname{sinc}(2f_l\delta) A_l \cos(2\pi f_l t + \phi_l), \quad \text{where } \operatorname{sinc}(u) \equiv \frac{\sin(\pi u)}{\pi u}.$$

It is interesting to compare this smoothed version of $g(\cdot)$ with $g * h(\cdot)$ in Equation (84a), which uses a Gaussian smoothing kernel. Note that, in the latter case, the attenuation factor $\exp\left[-\left(\sigma 2\pi f_l\right)^2 / 2\right]$ for the lth term in $g(\cdot)$ decreases monotonically from 1 to 0 as the smoothing parameter σ increases from 0 to ∞. This is not true for $g * r(\cdot)$ since $\operatorname{sinc}(2f_l\delta)$ oscillates about 0 with decreasing amplitude as δ increases. The effect of varying δ in $g * r(\cdot)$ is thus less transparent than varying σ in $g * h(\cdot)$.

As an example, let us reconsider $g(\cdot)$ given by Equation (84b) (shown as the thick curves in Figure 86). Since $\operatorname{sinc}(u) = 0$ for all nonzero integers u, the high frequency term f_2 is completely eliminated for $\delta = u/(2f_2) = u/6$. This is illustrated in the top two plots of Figure 86 with $\delta = 1/6$. On the other hand, $\operatorname{sinc}(u)$ has a negative-valued local minimum for u near $(4k - 1)/2$ for positive integers k, which means that the attenuation factor is negative for $\delta = (4k - 1)/(4f_2) = (4k - 1)/12$. This is illustrated in the bottom two plots with $\delta = 1/4$, for which $g * r(\cdot)$ is seen to have small ripples in the *opposite* direction from those in $g(\cdot)$. Note that the width of the smoothing kernel is now larger than that used in the top two plots – in contrast to the Gaussian kernel, a wider rectangular kernel does not necessarily imply a smoother looking $g * r(\cdot)$. For a $g(\cdot)$ with many more high frequency components than in our simple example, it can be difficult to find a δ that yields a $g * r(\cdot)$ with the proper degree of smoothness. For these reasons smoothers that have the monotonic attenuation property of the Gaussian smoothing kernel are recommended for use with various nonparametric spectral estimates in Chapter 6.

3.7 Fourier Theory – Discrete Time/Continuous Frequency

Suppose that the continuous time $L^2(-\infty, \infty)$ signal $g(\cdot)$ is sampled at equally spaced time intervals of duration Δt to define a discrete time sequence

$$g_t \equiv g(t\,\Delta t), \qquad t = 0, \pm 1, \pm 2, \ldots$$

(note that, as opposed to its use in the previous six sections, here 't' is a unitless index and not physical time – see item [3] of Section 2.2). The Fourier transform of this sequence is *defined* to be

$$G_p(f) \equiv \Delta t \sum_{t=-\infty}^{\infty} g_t e^{-i2\pi f t \,\Delta t} \tag{87}$$

(the rationale for the subscript p will be explained below). We motivate this definition by considering the *continuous time* signal defined by the product of Δt, $g(\cdot)$ and an infinite set of equally spaced Dirac delta functions; i.e.,

$$g_\delta(t) \equiv \Delta t\, g(t) \sum_{k=-\infty}^{\infty} \delta(t - k\,\Delta t).$$

With this definition, for ϵ small enough,

$$g_t = \frac{1}{\Delta t} \int_{t\,\Delta t-\epsilon}^{t\,\Delta t+\epsilon} g_\delta(u)\, du, \qquad t = 0, \pm 1, \pm 2, \dots.$$

Let us regard $g_\delta(\cdot)$ as an $L^2(-\infty, \infty)$ signal – technically it is not, but we can manipulate it as if it were. Equation (65b) states that the Fourier transform of $g_\delta(\cdot)$ is

$$\begin{aligned}
G_\delta(f) &= \int_{-\infty}^{\infty} g_\delta(t) e^{-i2\pi ft}\, dt \\
&= \int_{-\infty}^{\infty} \left(\Delta t\, g(t) \sum_{k=-\infty}^{\infty} \delta(t - k\,\Delta t) \right) e^{-i2\pi ft}\, dt \\
&= \Delta t \sum_{k=-\infty}^{\infty} \int_{-\infty}^{\infty} g(t)\delta(t - k\,\Delta t) e^{-i2\pi ft}\, dt \\
&= \Delta t \sum_{k=-\infty}^{\infty} g_k e^{-i2\pi fk\,\Delta t},
\end{aligned}$$

which is simply the right-hand side of (87) if we relabel the dummy index t as k.

Equation (87) is effectively a rectangular integration approximation of Equation (65b) with the term Δt ensuring conservation of integrated area between the two equations as $\Delta t \to 0$. It is easy to see that

$$G_p(f) = G_p(f + k/\Delta t), \qquad k = 0, \pm 1, \pm 2, \dots,$$

so that $G_p(\cdot)$ is periodic with period $1/\Delta t$ (hence the subscript p). Suppose for the moment that $G_p(\cdot)$ is in $L^2(-1/[2\,\Delta t], 1/[2\,\Delta t])$. From our discussion of the Fourier series representation of square integrable periodic functions in Section 3.1, we know from Equation (58e) that (at least in the mean square sense)

$$G_p(f) = \sum_{t=-\infty}^{\infty} \tilde{g}_t e^{i2\pi ft\,\Delta t}, \tag{88}$$

where (from Equation (58d))

$$\tilde{g}_t \equiv \Delta t \int_{-1/(2\,\Delta t)}^{1/(2\,\Delta t)} G_p(f) e^{-i2\pi ft\,\Delta t}\,df. \tag{89a}$$

This correspondence is a little tricky, since we must interchange the roles of time and frequency between here and Section 3.1. In particular, we must note that

T corresponds to $1/\Delta t$ and $f_n = n/T$ corresponds to $t\,\Delta t$.

The necessity for this interchange emphasizes the duality between time and frequency: the cases continuous time/discrete frequency and discrete time/continuous frequency are in many ways complementary. In particular, we note that Equations (88) and (87) are identical if we define

$$\tilde{g}_t = \Delta t\, g_{-t}.$$

The inverse Fourier transform of $G_p(\cdot)$ is evident from Equation (89a):

$$g_t \equiv \int_{-1/(2\,\Delta t)}^{1/(2\,\Delta t)} G_p(f) e^{i2\pi ft\,\Delta t}\,df. \tag{89b}$$

We can now restate and expand upon some of the results we found in Section 3.1 in terms of our present case of discrete time/continuous frequency. Parseval's theorem – Equation (59b) – becomes

$$\Delta t \sum_{t=-\infty}^{\infty} |g_t|^2 = \int_{-1/(2\,\Delta t)}^{1/(2\,\Delta t)} |G_p(f)|^2\,df. \tag{89c}$$

Very often Δt is taken to be unity – as we shall do in the remainder of this section for simplicity – so that the Fourier transform and its inverse become

$$G_p(f) = \sum_{t=-\infty}^{\infty} g_t e^{-i2\pi ft} \quad \text{and} \quad g_t = \int_{-1/2}^{1/2} G_p(f) e^{i2\pi ft}\,df.$$

If we are only given g_t for $t = -m,\ \dots,\ m$, then the truncated Fourier series

$$G_{p,m}(f) \equiv \sum_{t=-m}^{m} g_t e^{-i2\pi ft}$$

minimizes the integrated square error between $G_p(\cdot)$ and functions of the form

$$H_{p,m}(f) \equiv \sum_{t=-m}^{m} h_t e^{-i2\pi ft};$$

i.e., the quantity

$$\int_{-1/2}^{1/2} |G_p(f) - H_{p,m}(f)|^2 \, df$$

is minimized by setting $h_t = g_t$ (see the discussion surrounding Equation (61)). What other properties does $G_{p,m}(\cdot)$ possess? We can rewrite it as

$$G_{p,m}(f) = \sum_{t=-m}^{m} \left(\int_{-1/2}^{1/2} G_p(f') e^{i2\pi f' t} \, df' \right) e^{-i2\pi f t}$$

$$= \int_{-1/2}^{1/2} G_p(f') \left(\sum_{t=-m}^{m} e^{i2\pi(f'-f)t} \right) df'$$

$$= (2m+1) \int_{-1/2}^{1/2} G_p(f') \mathcal{D}_{2m+1}(f - f') \, df', \qquad (90)$$

where $\mathcal{D}_{2m+1}(\cdot)$ is Dirichlet's kernel (see Exercise [1.3e]). Thus $G_{p,m}(\cdot)$ is the convolution of $G_p(\cdot)$ and the function given by $2m+1$ times Dirichlet's kernel; note, however, that here convolution is defined differently than in Section 3.6, where we introduced it for the continuous/continuous case – since $G_p(\cdot)$ and Dirichlet's kernel are both periodic functions with unit period, the limits on the integral in the convolution are now over one complete cycle instead of from $-\infty$ to ∞ (see [1d] of Section 3.11 for details). Figure 91 shows plots of $\mathcal{D}_{2m+1}(f)$ versus f for $m = 4$, 16 and 64. Since $\mathcal{D}_{2m+1}(0) = 1$ for all m, the amplitude of the central lobe of $(2m+1)\mathcal{D}_{2m+1}(\cdot)$ grows as m increases whereas, as can be seen in the plots, its width decreases; the same holds for all the sidelobes. Because

$$\int_{-1/2}^{1/2} (2m+1) \mathcal{D}_{2m+1}(f) \, df = \int_{-1/2}^{1/2} \sum_{t=-m}^{m} e^{i2\pi f t} \, df = 1,$$

the area under all the lobes of $(2m+1)\mathcal{D}_{2m+1}(\cdot)$ is constant.

Let us look in more detail at the effect of convolving a signal with Dirichlet's kernel. Both the central lobe and the sidelobes of this kernel will cause $G_{p,m}(\cdot)$ to be a distorted version of $G_p(\cdot)$. The effect of the central lobe is to smooth out shape features (peaks) in $G_p(\cdot)$ that are small compared to the width of the lobe – this is often referred to as a *loss of resolution* due to the use of a finite amount of data. This width can be measured roughly as half the distance between the two nulls of $\mathcal{D}_{2m+1}(\cdot)$ closest to 0. These occur at $f = \pm 1/(2m+1)$, so the width by this measure is $1/(2m+1)$. As an example, the thick curves in Figure 92 show

$$G_p(f) = \sum_{j=1}^{2} e^{-10000(f-f_j)^2} + e^{-10000(f+f_j)^2}$$

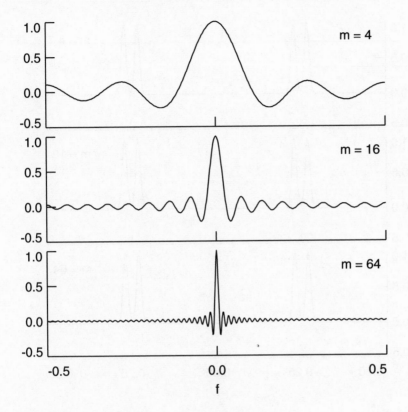

Figure 91. $\mathcal{D}_{2m+1}(\cdot)$ for $m = 4$, 16 and 64.

with $f_1 = 1/4 - 1/50$ and $f_2 = 1/4 + 1/50$. This function has twin peaks whose separation in frequency is $1/25$. The thin curves show the corresponding $G_{p,m}(\cdot)$ for $m = 4$, 16 and 64. The corresponding widths of the central lobe of Dirichlet's kernel are, respectively, $1/9$, $1/33$ and $1/129$. The twin peaks are completely smeared together in $G_{p,4}(\cdot)$, while they are faithfully resolved in $G_{p,64}(\cdot)$. For the intermediate case $m = 16$, the width of the central lobe is just slightly less than the separation in frequency between the twin peaks. In fact, a magnification of $G_{p,64}(\cdot)$ around $f = \pm 1/4$ shows a slight dip so that the twin peaks are just barely resolved.

The sidelobes of Dirichlet's kernel have positive peaks at approximately $\pm 5/(4m + 2)$, $\pm 9/(4m + 2)$, ..., and negative peaks at approximately $\pm 3/(4m + 2)$, $\pm 7/(4m + 2)$, ... Thus, we can expect a significant discrepancy between $G_{p,m}(f)$ and $G_p(f)$ if $G_p(\cdot)$ happens to have large values close to, say, $f + 3/(4m + 2)$. The integral in Equation (90) would then have significant contributions from parts of $G_p(\cdot)$ far away from $G_p(f)$. In the literature this phenomenon is often called

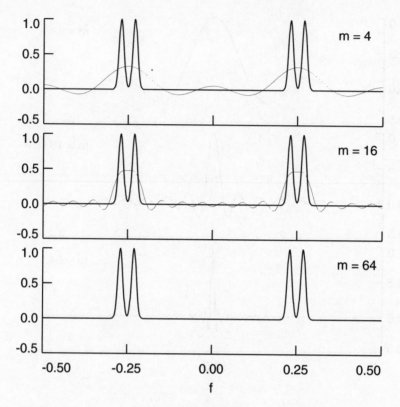

Figure 92. Illustration of loss of resolution and leakage. The function $G_p(\cdot)$ – thick curve in the above plots – has twin peaks and is approximated by the truncated Fourier series $G_{p,m}(\cdot)$ with $m = 4$, 16 and 64 (thin curves from top to bottom). Note the resolution of the approximation increases as m increases, while its leakage decreases as m increases.

leakage. Again, we look to Figure 92 for an example: the function $G_p(\cdot)$ there is essentially zero around $f = 0$, yet both $G_{p,4}(0)$ and $G_{p,16}(0)$ are non-zero due to leakage from the twin peaks in $G_p(\cdot)$. Leakage typically decreases as m increases – here $G_{p,64}(\cdot)$ shows little evidence of leakage.

We can investigate another aspect of the sidelobes of Dirichlet's kernel by considering the periodic function with unit period described over $[-1/2, 1/2]$ by

$$G_p(f) = \begin{cases} 1, & |f| \leq 1/4; \\ 0, & 1/4 < |f| \leq 1/2. \end{cases} \tag{92}$$

The result of convolving $G_p(\cdot)$ with $(2m+1)\mathcal{D}_{2m+1}(\cdot)$ is that $G_{p,m}(\cdot)$ has ripples where the discontinuities in $G_p(\cdot)$ meet the lobes of the Dirichlet

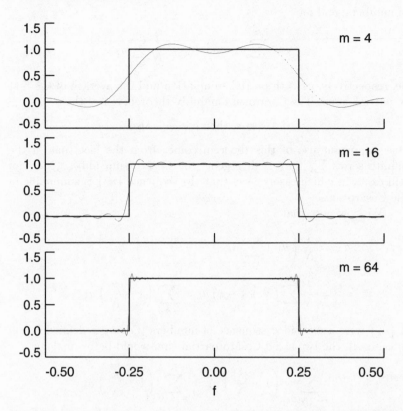

Figure 93. Illustration of the Gibbs phenomenon. The discontinuous rectangular function $G_p(\cdot)$ – thick curve in the above plots – is approximated by the truncated Fourier series $G_{p,m}(\cdot)$ with $m = 4$, 16 and 64 (thin curves from top to bottom). Note the ripples around the discontinuities at $f = \pm 1/4$.

kernel. The result of increasing m is to make the ripples occur more frequently since the lobes become narrower but their amplitudes do not decrease. This behavior is known as the *Gibbs phenomenon* and is illustrated in Figure 93. As m increases, the maximum 'overshoot' occurs closer and closer to the discontinuities at $\pm 1/4$ (toward zero) while the maximum 'undershoot' does likewise away from zero – each tends to about 9% of the amplitude of the discontinuity.

There is an interesting way of reducing both leakage and the Gibbs phenomenon – at a certain cost – by approximating $G_p(\cdot)$ with a function that, like the truncated Fourier series $G_{p,m}(\cdot)$, is based upon the available subsequence $g_{-m}, \ldots, g_m$. In the theory of partial sums of sequences of numbers, there is a concept of summability called *Cesàro summability* (Titchmarsh, 1939, p. 411). Let $u_1, u_2, \ldots$ be a sequence of numbers, and let

$$s_N \equiv \sum_{k=1}^{N} u_k \text{ and } a_N \equiv \frac{1}{N} \sum_{j=1}^{N} s_j$$

be, respectively, its Nth partial summation and the average of the first N partial sums. The Cesàro summability theorem states that,

$$\text{if } s_N \to s, \text{ then } a_N \to s \text{ also.}$$

One important use of this theorem comes from the fact that, if the infinite series $\sum_{k=1}^{\infty} u_k$ is divergent, the Cesàro sum $\lim_{N \to \infty} a_N$ can still exist, in which case we say that the sequence $\{u_k\}$ is summable in the Cesàro sense.

We now note that

$$a_N = \frac{1}{N} \left(u_1 + \sum_{k=1}^{2} u_k + \cdots + \sum_{k=1}^{N} u_k \right)$$

$$= \frac{1}{N} \sum_{k=1}^{N} (N + 1 - k) u_k = \sum_{k=0}^{N-1} \left(1 - \frac{k}{N} \right) u_{k+1}.$$

If $\{u_k\}$ were a two-sided sequence of numbers (i.e., u_k is defined for all integers k), the two-sided Cesàro partial sum would be by analogy

$$\sum_{k=-(N-1)}^{N-1} \left(1 - \frac{|k|}{N} \right) u_k = \sum_{k=-N}^{N} \left(1 - \frac{|k|}{N} \right) u_k$$

(Wiener, 1949). These results suggest that, if

$$G_{p,m}(f) = \sum_{t=-m}^{m} g_t e^{-i2\pi ft} \text{ converges as } m \to \infty,$$

then

$$G_{p,m}^{(C)}(f) \equiv \sum_{t=-m}^{m} \left(1 - \frac{|t|}{m} \right) g_t e^{-i2\pi ft}$$

should converge to the same quantity, and hence $G_{p,m}^{(C)}(\cdot)$ can also be a useful approximation to $G_p(\cdot)$. Now

$$G_{p,m}^{(C)}(f) = \sum_{t=-m}^{m} \left(1 - \frac{|t|}{m} \right) \left(\int_{-1/2}^{1/2} G_p(f') e^{i2\pi f't} \, df' \right) e^{-i2\pi ft}$$

$$= \int_{-1/2}^{1/2} G_p(f') \sum_{t=-m}^{m} \left(1 - \frac{|t|}{m} \right) e^{i2\pi(f'-f)t} \, df'$$

$$= m^2 \int_{-1/2}^{1/2} G_p(f') \mathcal{D}_m^2(f' - f) \, df',$$

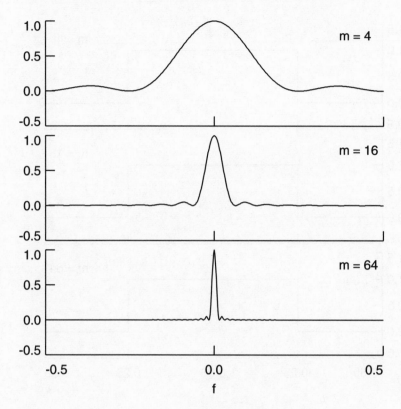

Figure 95. $\mathcal{D}_m^2(\cdot)$ for $m = 4$, 16 and 64 (cf. Figure 91).

i.e., the convolution of $G_p(\cdot)$ with a kernel that is m^2 times the square of
the Dirichlet kernel. (This result can be verified by filling in the details
in the following line of thought: (a) the sequence defined by $1 - |t|/m$ is
essentially the convolution of a rectangular sequence of length m with
itself; (b) the Fourier transform of a rectangular sequence is essentially
the Dirichlet kernel; and (c) an appropriate version of the convolution
theorem thus says that the Fourier transform of $\{1 - |t|/m\}$ should be
the square of the Dirichlet kernel.) A rescaled version of $\mathcal{D}_m^2(\cdot)$ is called
Fejér's kernel and is discussed in detail in Section 6.3. The quantity
$\mathcal{D}_m^2(\cdot)$ is shown in Figure 95. If we compare this with the Dirichlet
kernel in Figure 91, we note that $\mathcal{D}_m^2(\cdot)$ is always positive and that its
sidelobes are smaller relative to its central lobe, but at the expense of
a wider central lobe. If we again consider $G_p(\cdot)$ of Equation (92), we
obtain the results shown in Figure 96: negative overshoots cannot occur
(because of the positive nature of the kernel) and the positive overshoot
can be reduced (improved 'fidelity'). This is bought at the cost of a less
clearly 'resolved' discontinuity due to the wider central lobe.

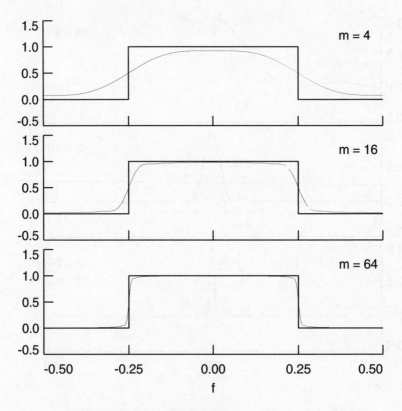

Figure 96. Alleviation of the Gibbs phenomenon (cf. Figure 93).

Note that both $G_{p,m}(\cdot)$ and $G_{p,m}^{(C)}(\cdot)$ can be defined by a summation of the form

$$\sum_{t=-m}^{m} c_t g_t e^{-i2\pi f t}$$

with an appropriate choice of c_t – 'rectangular' (or uniform) for $G_{p,m}(\cdot)$ and 'triangular' for $G_{p,m}^{(C)}(\cdot)$. This introduces the possibility of other definitions for the c_t terms, which are often called *convergence factors* in the literature on Fourier series and *windows* in the engineering literature (Rabiner and Gold, 1975, Section 3.8). These have been studied to manage the tradeoff between fidelity and resolution. We shall meet some of them in Chapter 6, where summations of the above form will reappear in our discussion of lag windows for managing the tradeoff between bias and variance in lag window spectral estimators.

3.8 Aliasing Problem – Discrete Time/Continuous Frequency

In Section 3.2 we found that a continuous time $L^2(-\infty, \infty)$ signal $g(\cdot)$ has an amplitude spectrum defined as $|G(f)|$, where

$$G(f) \equiv \int_{-\infty}^{\infty} g(t)e^{-i2\pi f t} \, dt, \qquad -\infty < f < \infty$$

(this is Equation (65b)). The motivation for calling this an amplitude spectrum is that $g(\cdot)$ has the representation

$$g(t) \overset{\text{ms}}{=} \int_{-\infty}^{\infty} G(f)e^{i2\pi f t} \, df,$$

so that $|G(f)|$ represents the amplitude associated with the complex exponential with frequency f (in what follows, we assume that $g(\cdot)$ is a continuous function so that we can replace '$\overset{\text{ms}}{=}$' with '$=$' in the above).

In the previous section we found the following representation for the sequence $\{g_t\}$:

$$g_t = \int_{-1/(2\,\Delta t)}^{1/(2\,\Delta t)} G_p(f)e^{i2\pi f t\,\Delta t} \, df$$

(this is Equation (89b)). Again we can claim that $|G_p(f)|$ defines an amplitude spectrum. Since $\{g_t\}$ is a sequence obtained by sampling the function $g(\cdot)$, the question arises as to what connection there is between $G(\cdot)$ and $G_p(\cdot)$. Now

$$g_t = g(t\,\Delta t) = \int_{-\infty}^{\infty} G(f')e^{i2\pi f' t\,\Delta t} \, df'$$

$$= \sum_{k=-\infty}^{\infty} \int_{(2k-1)/(2\,\Delta t)}^{(2k+1)/(2\,\Delta t)} G(f')e^{i2\pi f' t\,\Delta t} \, df'$$

$$= \sum_{k=-\infty}^{\infty} \int_{-1/(2\,\Delta t)}^{1/(2\,\Delta t)} G(f + k/\Delta t)e^{i2\pi (f+k/\Delta t)t\,\Delta t} \, df$$

after we make the change of variable $f \equiv f' - k/\Delta t$. Since

$$e^{i2\pi (f+k/\Delta t)t\,\Delta t} = e^{i2\pi f t\,\Delta t} \quad \text{for all } t \text{ and } k,$$

we have both

$$g_t = g(t\,\Delta t) = \int_{-1/(2\,\Delta t)}^{1/(2\,\Delta t)} \sum_{k=-\infty}^{\infty} G(f + k/\Delta t)e^{i2\pi f t\,\Delta t} \, df$$

and

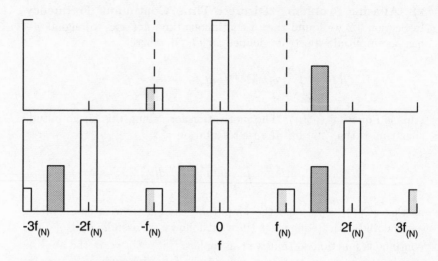

Figure 98. Example of the aliasing effect.

$$g_t = \int_{-1/(2\,\Delta t)}^{1/(2\,\Delta t)} G_p(f)e^{i2\pi ft\,\Delta t}\,df,$$

so that evidently

$$G_p(f) = \sum_{k=-\infty}^{\infty} G(f + k/\Delta t), \qquad |f| \le 1/(2\,\Delta t). \tag{98a}$$

This equation tells us that $G_p(f)$ – the Fourier transform at frequency f for the sampled sequence $\{g_t\}$ – is the sum of contributions from the Fourier transform of $g(\cdot)$ at frequencies f, $f \pm 1/\Delta t$, $f \pm 2/\Delta t$, $f \pm 3/\Delta t$, ... In general there is no way of recovering exactly the amplitude spectrum for $g(\cdot)$ given that of $\{g_t\}$. The value of $G_p(f)$ depends upon $G(\cdot)$ not only at f but also at a countably infinite set of frequencies $f + k/\Delta t$, $k = \pm 1, \pm 2, \ldots$. This phenomenon is called *aliasing*, and the frequency f is said to be aliased with each of the frequencies $f \pm 1/\Delta t$, $f \pm 2/\Delta t$, $f \pm 3/\Delta t$, ... These latter frequencies are called *aliases* of the frequency f. The highest frequency that is not an alias of a lower frequency is $1/(2\,\Delta t)$. This frequency is often called the *Nyquist frequency* or the *folding frequency*. We shall denote it by

$$f_{(N)} \equiv \frac{1}{2\,\Delta t}.$$

As a simple example, consider the real-valued function

$$G(f) = r(f + f_{(N)})/4 + r(f) + r(f - 1.5f_{(N)})/2, \tag{98b}$$

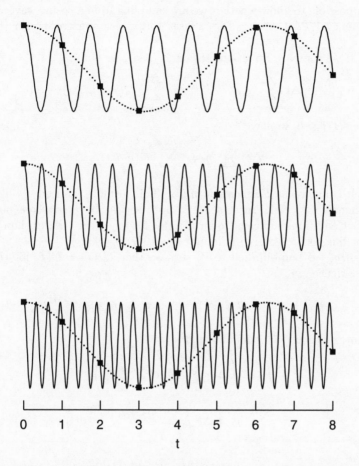

Figure 99. Illustration of the aliasing effect. The dotted curves above show $\cos(t)$ versus t. The solid curves show $\cos([1 + 2k\pi]t)$ versus t for (from top to bottom) $k = 1$, 2 and 3. The solid black squares show the common value of all four sinusoids when sampled at $t = 0, 1, \ldots, 8$.

where $r(f) \equiv 1$ for $|f| \leq f_{(N)}/8$ and 0 otherwise. This function is plotted in the top part of Figure 98, while the bottom part is its aliased version $G_p(\cdot)$ (the vertical dashed lines indicate the Nyquist frequency and its negative). Note that $G_p(\cdot)$ is periodic with period $2f_{(N)}$; that $G(\cdot)$ around $f = 1.5f_{(N)}$ is aliased around $f = -f_{(N)}/2$ in $G_p(\cdot)$; that $G(\cdot)$ for f just above $f = -f_{(N)}$ is aliased to just below $f = f_{(N)}$; and that, while $G(\cdot)$ and $G_p(\cdot)$ agree perfectly at certain frequencies (0, for example), they disagree substantially at others $(f_{(N)})$.

A graphical illustration of what causes aliasing is shown in Figure 99. By sampling at discrete times (marked by the black squares), it is impossible to know whether we are sampling from a cosine wave with frequency $1/(2\pi)$ or one of its aliases $1/(2\pi) + $ a nonzero integer (here $\Delta t = 1$).

If we compare Equations (87) and (98a), we obtain

$$\Delta t \sum_{t=-\infty}^{\infty} g_t e^{-i2\pi ft\,\Delta t} = \sum_{k=-\infty}^{\infty} G(f + k/\Delta t).$$

If we set $f = 0$, we have

$$\Delta t \sum_{t=-\infty}^{\infty} g_t = \sum_{k=-\infty}^{\infty} G(k/\Delta t),$$

which is known as *Poisson's formula*. This formula (and others like it for nonzero f) make most sense if both series above converge pointwise. Conditions sufficient to ensure this for any f are stricter than $g(\cdot)$ belonging to $L^2(-\infty, \infty)$ (Champeney, 1987, p. 163).

If $g(\cdot)$ is band-limited to $W < f_{(N)}$, then $G_p(f) = G(f)$ for $|f| \leq f_{(N)}$. Hence

$$g(t) = \int_{-f_{(N)}}^{f_{(N)}} G(f)e^{i2\pi ft}\,df = \int_{-f_{(N)}}^{f_{(N)}} G_p(f)e^{i2\pi ft}\,df.$$

From Equation (87) it follows that

$$
\begin{aligned}
g(t) &= \int_{-f_{(N)}}^{f_{(N)}} \left(\Delta t \sum_{t'=-\infty}^{\infty} g_{t'} e^{-i2\pi ft'\,\Delta t} \right) e^{i2\pi ft}\,df \\
&= \sum_{t'=-\infty}^{\infty} g_{t'} \left(\Delta t \int_{-f_{(N)}}^{f_{(N)}} e^{i2\pi f(t - t'\,\Delta t)}\,df \right) \\
&= \sum_{t'=-\infty}^{\infty} g_{t'} \operatorname{sinc} \left[2f_{(N)}(t - t'\,\Delta t) \right]
\end{aligned}
$$

(recall that $\operatorname{sinc}(t) \equiv \sin(\pi t)/(\pi t)$). This gives us an interpolation formula for recovering the continuous time signal from its sampled values (this only works for the case of a band-limited signal with W less than the Nyquist frequency).

As an example, a simple band-limited time function is $g(t) = \operatorname{sinc}^2(t)$. Its Fourier transform is band-limited to ± 1 cycles per unit time. If we sample at $\Delta t = 1/4$, then $W = 1 < f_{(N)} = 1/(2\,\Delta t) = 2$. We can reconstruct, say, $\operatorname{sinc}^2(0.1)$ from its values at $t = 0, \pm 1/4, \pm 1/2,$ ..., since

$$\operatorname{sinc}^2(0.1) = \sum_{t'=-\infty}^{\infty} \operatorname{sinc}^2(t'/4) \operatorname{sinc}(0.4 - t').$$

3.9 Concentration Problem – Discrete/Continuous Case

Our development here is quite analogous to that of Section 3.5 for the continuous time/continuous frequency case. We present it because a number of details are different and because the discrete time version is most relevant to recent new approaches to spectral estimation (see Chapter 7 on multitaper spectral estimation). The results for this case were originally derived by Slepian (1978), to whom the reader is referred for more details.

Let $\{g_t\}$ be a real or complex-valued sequence with finite energy and a sampling interval $\Delta t = 1$ (this yields a Nyquist frequency of $f_{(N)} = 1/2$). The Fourier transform of $\{g_t\}$ is

$$G_p(f) = \sum_{t=-\infty}^{\infty} g_t e^{-i2\pi f t}$$

(Equation (87)), and Parseval's theorem states that

$$\sum_{t=-\infty}^{\infty} |g_t|^2 = \int_{-1/2}^{1/2} |G_p(f)|^2 \, df$$

(Equation (89c)). The energy in the index range 0 to $N-1$ is just $\sum_{t=0}^{N-1} |g_t|^2$, and the fraction of the energy lying in this index range is

$$\alpha^2(N) \equiv \sum_{t=0}^{N-1} |g_t|^2 \Big/ \sum_{t=-\infty}^{\infty} |g_t|^2. \tag{101a}$$

As in the continuous time/continuous frequency case of Section 3.5, the energy in the frequency range $|f| \leq W < 1/2$ is $\int_{-W}^{W} |G_p(f)|^2 \, df$, and the fraction of the energy in this range is

$$\beta^2(W) \equiv \int_{-W}^{W} |G_p(f)|^2 \, df \Big/ \int_{-1/2}^{1/2} |G_p(f)|^2 \, df. \tag{101b}$$

Just as we described a signal $g(\cdot)$ as being time-limited if $g(t) = 0$ for $|t| > T/2$, so we can describe a sequence $\{g_t\}$ as being *index-limited* to the interval $[0, N-1]$ if $g_t = 0$ for $t < 0$ or $t \geq N$. There are two concentration problems of interest here. The first is to determine how large $\alpha^2(N)$ can be for $\{g_t\}$ band-limited to $|f| \leq W < 1/2$; the second, to determine how large $\beta^2(W)$ can be for $\{g_t\}$ index-limited to $[0, N-1]$.

For the first problem, we note that the concentration measure can be rewritten as

$$\alpha^2(N) =$$
$$\frac{\int_{-W}^{W} \int_{-W}^{W} G_p(f) e^{i\pi(N-1)f} N \mathcal{D}_N(f' - f) G_p^*(f') e^{-i\pi(N-1)f'} \, df \, df'}{\int_{-W}^{W} |G_p(f)|^2 \, df},$$

$$\tag{101c}$$

where $\mathcal{D}_N(\cdot)$ is Dirichlet's kernel (the proof of this is Exercise [3.9]). Note that we can simplify the above somewhat by defining

$$H_p(f) \equiv G_p(f)e^{i\pi(N-1)f}$$

to obtain

$$\alpha^2(N) = \frac{\int_{-W}^{W}\int_{-W}^{W} H_p(f)N\mathcal{D}_N(f'-f)H_p^*(f')\,df\,df'}{\int_{-W}^{W}|H_p(f)|^2\,df}.$$

Note that $G_p(\cdot)$ and $H_p(\cdot)$ differ only in phase but not magnitude. As in Section 3.5, it can be shown that all functions $H_p(\cdot)$ that maximize $\alpha^2(N)$ must satisfy the following integral equation:

$$\int_{-W}^{W} N\mathcal{D}_N(f'-f)H_p(f)\,df = \alpha^2(N)H_p(f'), \quad |f'| \leq W \qquad (102)$$

(see Equation (76b)). This again is a Fredholm integral equation of the second kind; however, now the kernel is given by

$$N\mathcal{D}_N(f'-f) = \frac{\sin\left[N\pi(f'-f)\right]}{\sin\left[\pi(f'-f)\right]}$$

instead of

$$\frac{\sin\left[T\pi(f'-f)\right]}{\pi(f'-f)} = T\,\mathrm{sinc}\left[T(f'-f)\right],$$

the kernel we encountered in the continuous time/continuous frequency case. These two kernels are both symmetric about the origin since

$$\mathcal{D}_N(-f) = \mathcal{D}_N(f) \text{ and } \mathrm{sinc}\,(-f) = \mathrm{sinc}\,(f);$$

they are in fact close in form for small $\pi(f'-f)$ since then

$$\frac{\sin\left[N\pi(f'-f)\right]}{\sin\left[\pi(f'-f)\right]} \approx \frac{\sin\left[N\pi(f'-f)\right]}{\pi(f'-f)} = N\,\mathrm{sinc}\left[N(f'-f)\right].$$

However, there is an important difference between them – whereas the kernel $N\mathcal{D}_N(\cdot)$ can be expressed as a *finite* sum of products of functions of f' and f alone, i.e.,

$$N\mathcal{D}_N(f'-f) = \sum_{t=0}^{N-1} e^{i2\pi f[t-(N-1)/2]}e^{-i2\pi f'[t-(N-1)/2]},$$

the same does not hold for the continuous time/continuous frequency kernel. This difference is summarized by calling $N\mathcal{D}_N(\cdot)$ a *degenerate* kernel. Because its kernel is degenerate, the integral equation (102) possesses only a *finite* number of eigenvalues λ and eigenfunctions $U(\cdot)$, i.e., values and functions such that

$$\int_{-W}^{W} N \mathcal{D}_N(f' - f)U(f)\,df = \lambda U(f') \tag{103a}$$

holds true (here – as before – we count functions that differ only by a nonzero scale factor as one function). There are in fact only N nonzero eigenvalues, say,

$$\lambda_0(N, W), \lambda_1(N, W), \ldots, \lambda_{N-1}(N, W).$$

These eigenvalues are distinct, real, positive, less than one and can be ordered such that

$$1 > \lambda_0(N, W) > \lambda_1(N, W) > \cdots > \lambda_{N-1}(N, W) > 0.$$

The first $2WN$ eigenvalues are extremely close to one, and then the eigenvalues fall off rapidly to zero. The maximum value of $\alpha^2(N)$ in our first concentration problem is just $\lambda_0(N, W)$, the largest eigenvalue of the integral equation.

For each eigenvalue $\lambda_k(N, W)$, there is an associated eigenfunction $U_k(\cdot; N, W)$ defined on the interval $[-W, W]$. As with the pswf (see Equation (78)), the integral equation (103a) can be used to extend $U_k(\cdot; N, W)$ to be defined over all of $[-1/2, 1/2]$. After standardization, these eigenfunctions can be taken to be *orthonormal* over $[-1/2, 1/2]$,

$$\int_{-1/2}^{1/2} U_j(f; N, W)U_k(f; N, W)\,df = \delta_{j,k},$$

and *orthogonal* over $[-W, W]$,

$$\int_{-W}^{W} U_j(f; N, W)U_k(f; N, W)\,df = \delta_{j,k}\lambda_k(N, W).$$

The band-limited sequence that solves the first concentration problem is

$$g_t = \frac{1}{\lambda_0(N, W)} \int_{-W}^{W} U_0(f; N, W)e^{i2\pi f[t-(N-1)/2]}\,df, \quad t = 0, \pm 1, \ldots. \tag{103b}$$

The normalization we have used including the term $1/\lambda_0(N, W)$ is that of Slepian (1978), but we could have used any nonzero constant (independent of t) times g_t. The function $U_k(\cdot; N, W)$ is called the kth order *discrete prolate spheroidal wave function* (dpswf). The sequence defined in Equation (103b) is called a zeroth-order *discrete prolate spheroidal sequence* (dpss).

The kth order dpss can be generated by substituting $U_k(\cdot; N, W)$ for $U_0(\cdot; N, W)$ in Equation (103b) and defining the normalizing constant now to be $(-1)^k / (\epsilon_k \lambda_k(N, W))$ instead of $1/\lambda_0(N, W)$, where $\epsilon_k \equiv 1$ for even k and $\equiv \sqrt{-1}$ for odd k (again this is done to conform to Slepian's notation). Let us now denote the kth sequence by

$$\ldots, v_{-1,k}(N, W), v_{0,k}(N, W), v_{1,k}(N, W), \ldots.$$

In this notation the band-limited sequence that solves the first concentration problem is just

$$\ldots, v_{-1,0}(N, W), v_{0,0}(N, W), v_{1,0}(N, W), \ldots.$$

There is a second way of generating the kth order dpss. In fact, Slepian (1978) *defines* this sequence via the solution to the following system of equations:

$$\sum_{t'=0}^{N-1} \frac{\sin[2\pi W(t - t')]}{\pi(t - t')} v_{t',k}(N, W) = \lambda_k(N, W) v_{t,k}(N, W), \qquad (104)$$

$t = 0, 1, \ldots, N - 1$. This is equivalent to saying that the eigenvalues $\lambda_k(N, W)$ are the eigenvalues of the $N \times N$ matrix whose (t, t')th element is

$$\frac{\sin[2\pi W(t - t')]}{\pi(t - t')}, \quad t, t' = 0, 1, \ldots, N - 1,$$

and that the N elements of the corresponding eigenvectors for this matrix are in fact subsequences of length N of the dpss's. The subsequences are in fact those elements of the dpss's with indices in the range 0 to $N - 1$. Slepian (1978) shows how the remaining elements of the dpss's can be generated based upon these subsequences. These sequences are real-valued. (Yet a third way of obtaining the dpss's involves a tridiagonal matrix. This method is often the fastest for generating dpss's on a computer. See Section 8.3 for details.)

Let us now consider the second concentration problem, namely, to determine how large $\beta^2(W)$ in Equation (101b) can be for $\{g_t\}$ index-limited to $[0, N - 1]$. By making use of the relationship

$$G_p(f) = \sum_{t=-\infty}^{\infty} g_t e^{-i2\pi ft} = \sum_{t=0}^{N-1} g_t e^{-i2\pi ft}$$

(from Equation (87) with $\Delta t = 1$), it follows that Equation (101b) can be rewritten as

$$\beta^2(W) = \sum_{t'=0}^{N-1} \sum_{t=0}^{N-1} g_t^* \frac{\sin\left[2\pi W(t'-t)\right]}{\pi(t'-t)} g_{t'} \Bigg/ \sum_{t=0}^{N-1} |g_t|^2 . \qquad (105a)$$

The sequence $\{g_t\}$ that maximizes $\beta^2(W)$ must satisfy

$$\sum_{t'=0}^{N-1} \frac{\sin\left[2\pi W(t'-t)\right]}{\pi(t'-t)} g_{t'} = \lambda_k(N,W) g_t, \quad t = 0, 1, \ldots, N-1 \quad (105b)$$

(see Exercise [3.11]). The above is equivalent to Equation (104). We can rewrite it in the compact notation

$$A\mathbf{g} = \lambda_k(N,W)\mathbf{g},$$

where A is a matrix of order $N \times N$ whose (t', t)th element is given by $\sin\left[2\pi W(t'-t)\right]/\left[\pi(t'-t)\right]$ (the rows and columns of A are labeled from 0 to $N-1$); and $\mathbf{g}$ is an N-dimensional vector whose tth element (again labeled from 0 to $N-1$) is g_t. Thus the sequence that is index-limited from 0 to $N-1$ and that has the highest concentration of energy in the frequency interval $[-W, W]$ is a vector $\mathbf{v}_0(N, W)$ whose elements are a finite subsequence of the zeroth-order dpss, namely,

$$v_{0,0}(N, W), v_{1,0}(N, W), \ldots, v_{N-1,0}(N, W).$$

We note here a few of the important features of the solutions to Equation (105b).

[1] There are N nonzero eigenvalues that satisfy (105b). These have exactly the same values and properties as the eigenvalues that satisfy the first concentration problem.

[2] The N eigenvectors that are associated with these eigenvalues can be standardized such that they are orthonormal; i.e.,

$$\mathbf{v}_j^T(N, W)\mathbf{v}_k(N, W) = \sum_{t=0}^{N-1} v_{t,j}(N, W) v_{t,k}(N, W) = \delta_{j,k}.$$

[3] These N eigenvectors form a basis for N-dimensional Euclidean space; i.e., any real-valued N-dimensional vector can be expressed as a linear combination of these eigenvectors.

[4] As we have already seen, the index-limited sequence with the highest concentration of energy in the frequency interval $[-W, W]$ is

$$v_{0,0}(N, W), v_{1,0}(N, W), \ldots, v_{N-1,0}(N, W).$$

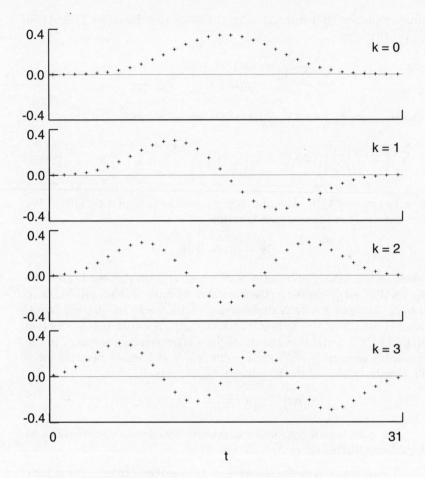

Figure 106. Plots of $\mathbf{v}_k(N, W)$ with $N = 32$ and $W = 4/N = 1/8$.

The concentration $\beta^2(W)$ is just $\lambda_0(N, W)$. The index-limited sequence that is orthogonal to this one and has the highest concentration of energy in the frequency interval $[-W, W]$ is

$$v_{0,1}(N, W), v_{1,1}(N, W), \ldots, v_{N-1,1}(N, W).$$

The concentration in this case is $\lambda_1(N, W) < \lambda_0(N, W)$. This pattern continues in an obvious way.

As an example, Figure 106 shows a plot of $\mathbf{v}_k(N, W)$ for $k = 0$, 1, 2 and 3 with $N = 32$ and $W = 4/N = 1/8$. The degree of concentration for each of these sequences is high: we have (to the first digit that differs from 9) $\lambda_0(32, 1/8) = 0.999\,999\,999\,8$, $\lambda_1(32, 1/8) = 0.999\,999\,98$, $\lambda_2(32, 1/8) = 0.999\,999\,2$ and $\lambda_3(32, 1/8) = 0.999\,98$. The square modulus of $G_p(\cdot)$ for each of these sequences is shown in Figure 107, where

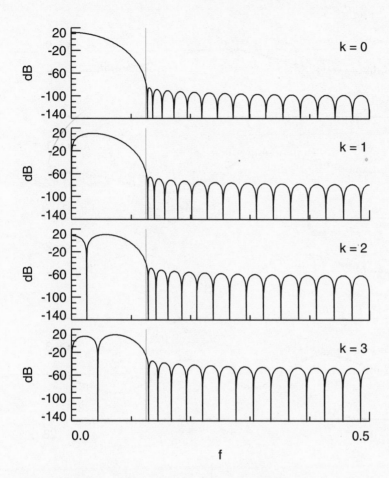

Figure 107. Plots of the $\left|G_p(\cdot)\right|^2$ for each of the sequences in Figure 106. The thin vertical lines indicate the frequency $W = 1/8$.

thin vertical lines indicate the frequency $W = 1/8$. We see that the energy is concentrated in the interval $[-W, W]$ (since the sequences in 106 are real-valued, $G_p(\cdot)$ is symmetric about 0 so we only need plot it for positive frequencies). Figures 108 and 109 show similar plots for the case $N = 99$ and $W = 4/N = 4/99$.

Figure 110 shows $\lambda_k(N, W)$ plotted versus k for the two cases illustrated in the previous four figures. The upper plot is for the case $N = 32$ and $W = 1/8$, while the lower plot shows the case $N = 99$ and $W = 4/99$. In both cases the Shannon number $2NW$ is 8, so we can expect $\lambda_k(N, W)$ to be close to 1 for $k < 8$ and close to 0 for $k > 8$. The plots show this pattern.

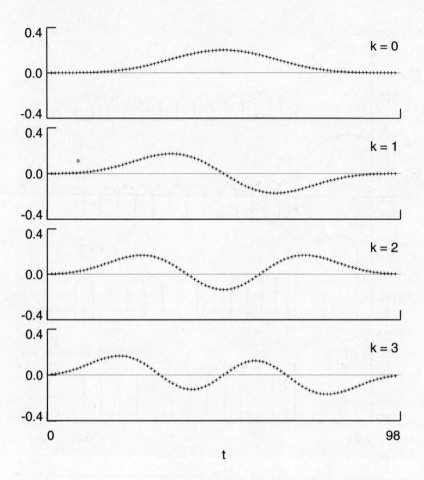

Figure 108. Plots of $\mathbf{v}_k(N, W)$ with $N = 99$ and $W = 4/N = 4/99$.

3.10 Fourier Theory – Discrete Time and Frequency

Suppose now that only a finite number of samples of a continuous time real or complex-valued signal is available, say, g_0, g_1, ..., g_{N-1}. In the case of equally spaced time sampling and continuous frequency, the Fourier transform is the periodic function $G_p(\cdot)$ defined in Equation (87) as

$$G_p(f) = \Delta t \sum_{t=-\infty}^{\infty} g_t e^{-i2\pi f t \Delta t}.$$

We could *define* a Fourier transform $G_p(\cdot; 0, N-1)$ for our subsequence of N values by setting $g_t = 0$ for $t < 0$ and $t \geq N$ in the above:

$$G_p(f; 0, N-1) \equiv \Delta t \sum_{t=0}^{N-1} g_t e^{-i2\pi f t \Delta t}. \tag{108}$$

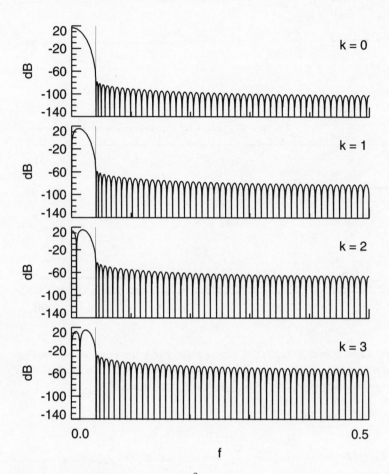

Figure 109. Plots of the $\left|G_p(\cdot)\right|^2$ for each of the sequences in Figure 108. The thin vertical lines indicate the frequency $W = 4/99$.

This definition has its uses and will appear again when we discuss the periodogram in Chapter 6; however, it involves an infinite number of frequencies. Since our subsequence has only a finite number of values, it seems plausible that we should be able to construct a representation for it that involves only a finite number of frequencies. This would be quite useful for digital computations. Accordingly, let us define a grid of N equally spaced frequencies, say,

$$f_n \equiv \frac{n}{N\,\Delta t} \ \text{ with } \ n = 0, 1, \ldots, N-1.$$

This set of N frequencies is sometimes called the *Fourier frequencies* or *standard frequencies*. For this grid of frequencies Equation (108) now

Deterministic Spectral Analysis

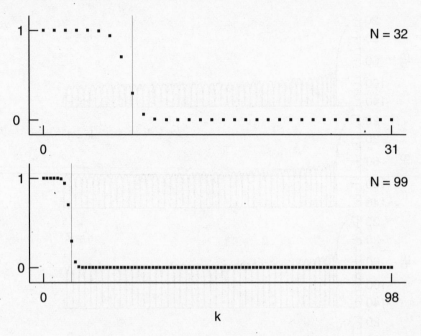

Figure 110. Plots of $\lambda_k(N, W)$ versus k for (top) the cases shown in Figures 106 and 107 and (bottom) Figures 108 and 109. The thin vertical lines indicate the position of $k = 2NW$, which is 8 in both cases.

becomes

$$G_n \equiv G_p(f_n; 0, N - 1) = \Delta t \sum_{t=0}^{N-1} g_t e^{-i2\pi f_n t \Delta t} = \Delta t \sum_{t=0}^{N-1} g_t e^{-i2\pi nt/N}.$$

(110a)

The above is known in the literature as the *discrete* Fourier transform (DFT) of the subsequence $g_0, g_1, \ldots, g_{N-1}$. The DFT plays a major role in spectral analysis and other fields of electrical engineering because it can often be computed very efficiently using an algorithm called the fast Fourier transform (FFT).

Let us now derive a representation for our subsequence in terms of the G_n in Equation (110a). If we multiply both sides of that equation by $\exp(i2\pi nt'/N)$ and sum with respect to n, we get

$$\sum_{n=0}^{N-1} G_n e^{i2\pi nt'/N} = \Delta t \sum_{n=0}^{N-1} \sum_{t=0}^{N-1} g_t e^{i2\pi n(t'-t)/N}$$

$$= \Delta t \sum_{t=0}^{N-1} g_t \sum_{n=0}^{N-1} e^{i2\pi n(t'-t)/N}.$$

(110b)

It follows from Exercise [1.3d] that

$$\sum_{n=0}^{N-1} e^{i2\pi nk/N} = \begin{cases} N, & \text{if } k = mN \text{ for integer } m; \\ 0, & \text{otherwise.} \end{cases}$$

The only time that $t' - t$ is an integer multiple of N is when $t' = t$, so (110b) reduces to

$$\sum_{n=0}^{N-1} G_n e^{i2\pi nt'/N} = g_{t'} N \Delta t.$$

We now have

$$g_t = \frac{1}{N \Delta t} \sum_{n=0}^{N-1} G_n e^{i2\pi nt/N}. \tag{111a}$$

This is known as the *inverse discrete Fourier transform* (inverse DFT) and gives us a representation for g_t. The appropriate form of Parseval's theorem here is

$$\Delta t \sum_{t=0}^{N-1} |g_t|^2 = \frac{1}{N \Delta t} \sum_{n=0}^{N-1} |G_n|^2. \tag{111b}$$

Two standard forms of the DFT and inverse DFT are in common use:

(a) $\Delta t = 1$ and (b) $\Delta t = 1/N$.

The first choice leads to

$$g_t = \frac{1}{N} \sum_{n=0}^{N-1} G_n e^{i2\pi nt/N} \text{ and } G_n = \sum_{t=0}^{N-1} g_t e^{-i2\pi nt/N}$$

and agrees with the definition in, e.g., Oppenheim and Schafer (1989). The second leads to

$$g_t = \sum_{n=0}^{N-1} G_n e^{i2\pi nt/N} \text{ and } G_n = \frac{1}{N} \sum_{t=0}^{N-1} g_t e^{-i2\pi nt/N}$$

and agrees with the definition in, e.g., Bloomfield (1976). This second form also appears to be that most often used in published FFT programs.

Note that, if we use Equation (110a) to *define* G_n for all integers n instead of just for $n = 0$ to $N-1$, the resulting infinite sequence $\{G_n\}$ is periodic with period N. Since the sequence $\{G_n \exp(i2\pi nt/N)\}$ is also periodic with period N, it follows that an alternative to Equation (111a) is

$$g_t = \frac{1}{N\,\Delta t} \sum_{n=k}^{N+k-1} G_n e^{i2\pi nt/N},$$

where k is any integer. The frequencies f_n that are associated with the G_n in the summation range from f_k up to f_{N+k-1}. If $k=0$ as in Equation (111a), the range is thus from 0 up to $(N-1)/(N\,\Delta t)$, which is just less than twice the Nyquist frequency $f_{(N)} = 1/(2\,\Delta t)$; alternatively, if $k = -\lfloor (N+1)/2 \rfloor + 1$, the f_n terms all satisfy $-f_{(N)} < f_n \le f_{(N)}$.

Note also that, if we use Equation (111a) to *define* g_t for $t < 0$ and $t \ge N$, the resulting infinite sequence $\{g_t\}$ is also periodic with period N. This supports the statement (sometimes found in the literature) that the DFT implies a periodic extension of the original subsequence; however, it is also sometimes claimed that the DFT is a result of assuming g_t is zero outside the original subsequence – this alludes to the argument we used to obtain Equation (108). Except for the trivial case of a subsequence whose terms are all zero, a periodic extension would imply a sequence such that $\sum_{t=-\infty}^{\infty} |g_t|^2 = \infty$, so the theory of Section 3.7 is not applicable. If, however, a zero extension is assumed, Equation (110a) shows that the Fourier transform of this infinite sequence is identical to the DFT of the subsequence at frequency f_n (see [1] in the Comments and Extensions below).

Suppose now, however, that our subsequence is a portion of a sequence $\{g_t\}$ whose moduli are square summable but whose terms are not necessarily zero outside the range $t = 0$ to $N-1$. What is the relationship between the Fourier transform $G_p(\cdot)$ for $\{g_t\}$ and the G_n for the subsequence? If we insert the Fourier representation for $\{g_t\}$ given by Equation (89b) into Equation (110a), we find that

$$G_n = \Delta t \sum_{t=0}^{N-1} \left(\int_{-f_{(N)}}^{f_{(N)}} G_p(f) e^{i2\pi ft\,\Delta t}\, df \right) e^{-i2\pi nt/N}$$

$$= \Delta t \int_{-f_{(N)}}^{f_{(N)}} G_p(f) \sum_{t=0}^{N-1} e^{i2\pi t(f\,\Delta t - n/N)}\, df.$$

Now G_n is associated with frequency f_n, so we can rewrite the above as follows:

$$G_n = \Delta t \int_{-f_{(N)}}^{f_{(N)}} G_p(f) \sum_{t=0}^{N-1} e^{i2\pi t\,(f-f_n)\,\Delta t}\, df.$$

If we appeal to Exercises [1.3c] and [1.3d], the above becomes

$$G_n = \Delta t \int_{-f_{(N)}}^{f_{(N)}} G_p(f) e^{-i(N-1)\pi(f_n-f)\,\Delta t} N\mathcal{D}_N \left([f_n - f]\,\Delta t \right)\, df,$$

where $\mathcal{D}_N(\cdot)$ is Dirichlet's kernel (this kernel has been plotted in Figure 91 for $N = 9, 33$ and 129). If we define

$$P(f) \equiv \Delta t \, e^{-i(\tilde{N}-1)\pi f \, \Delta t} N \mathcal{D}_N(f \, \Delta t),$$

we obtain

$$G_n = \int_{-f_{(N)}}^{f_{(N)}} G_p(f) P(f_n - f) \, df \equiv G_p * P(f_n).$$

We see that $P(\cdot)$ acts as a blurring function since it is convolved with $G_p(\cdot)$. If we are calculating G_n to obtain an approximation to $G_p(f_n)$, this blurring is obviously undesirable. We also note that the above equation is clearly related to Equation (90), which shows that the truncated Fourier series $G_{p,m}(\cdot)$ involving $g_{-m}, \ldots, g_m$ is proportional to the convolution of the Fourier transform $G_p(\cdot)$ of the infinite series $\{g_t\}$ with Dirichlet's kernel.

On the other hand, if we consider our finite sequence as coming from the function $g(\cdot)$, i.e., $g_t = g(t \, \Delta t)$, we obtain instead

$$G_n = \int_{-\infty}^{\infty} G(f) P(f_n - f) \, df.$$

Clearly, if $G(\cdot)$ is band-limited to $|W| < f_{(N)}$, then

$$G_n = \int_{-W}^{W} G(f) P(f_n - f) \, df = \int_{-W}^{W} G_p(f) P(f_n - f) \, df.$$

Comments and Extensions to Section 3.10

[1] Let $\{g_t\}$ be an infinite sequence such that $g_t = 0$ for $t < 0$ and $t > N - 1$. The DFT of the subsequence $g_0, \ldots, g_{N-1}$ is

$$G_n \equiv \Delta t \sum_{t=0}^{N-1} g_t e^{-i2\pi nt/N} = \Delta t \sum_{t=-\infty}^{\infty} g_t e^{-i2\pi nt/N} = G_p(f_n),$$

for $n = 0, \ldots, N - 1$, where $G_p(\cdot)$ is the Fourier transform of $\{g_t\}$. Thus, as we have noted before, the DFT of the subsequence is equal to the Fourier transform of the entire sequence – at the Fourier frequencies – in the special case where $g_t = 0$ outside the subsequence. Suppose that we now consider a larger subsequence $g_0, \ldots, g_{N'-1}$ with $N' > N$. Its DFT is, say,

$$G'_n \equiv \Delta t \sum_{t=0}^{N'-1} g_t e^{-i2\pi nt/N'} = \Delta t \sum_{t=-\infty}^{\infty} g_t e^{-i2\pi nt/N'} = G_p(f'_n)$$

for $n = 0, \ldots, N' - 1$ and $f'_n \equiv n/(N' \Delta t)$. Hence, by *zero padding* our original subsequence with $N' - N$ zeros, the DFT of the extended subsequence effectively evaluates $G_p(\cdot)$ over a finer grid of frequencies $\{f'_n\}$ than that given by the Fourier frequencies $\{f_n\}$. Note that this only works because g_t is in fact zero for $t < 0$ and $t > N - 1$.

Zero padding is often used in conjunction with FFT algorithms. These can quickly compute the DFT of a finite sequence if the length of the sequence N satisfies certain restrictions (the most common restriction is that $N = 2^k$ for some positive integer k, but some FFT algorithms work for $N = N_1 \times N_2 \times \cdots \times N_q$ with each N_j being a small prime number – see Singleton, 1969). Thus, if N is not a number acceptable to an FFT algorithm but $N' > N$ is, then $N' - N$ zeros are appended to the original sequence, and the DFT of this extended sequence is computed using the FFT algorithm. This scheme works fine as long as (i) we can reasonably assume that $g_N, \ldots, g_{N'}$ are in fact zero and/or (ii) the zero padded sequence $\{G'_n\}$ is as much use or interest to us as the original sequence $\{G_n\}$. When this is not the case so that we really require the DFT of a sequence with an 'unfriendly' length, we can still take advantage of FFTs to compute the DFT by use of the following *chirp transform algorithm* (Oppenheim and Schafer, 1989, Section 9.7.2).

Let a_t for $t = 0, \ldots, N-1$ and b_t for $t = -(N-1), \ldots, N-1$ be two sequences of complex or real-valued numbers of length N and $2N - 1$, respectively, and define the following sequence of length N:

$$c_n = \sum_{t=0}^{N-1} a_t b_{n-t} \text{ for } n = 0, \ldots, N - 1. \tag{114}$$

Exercise [3.14] indicates how this convolution can be computed efficiently using zero padding along with three invocations of a standard 'powers of 2' FFT algorithm. Now, since $-2nt = (n - t)^2 - t^2 - n^2$ (a 'trick' attributed to L. I. Bluestein), we have

$$G_n = \Delta t \sum_{t=0}^{N-1} g_t e^{-i2\pi nt/N} = \Delta t \sum_{t=0}^{N-1} g_t e^{i\pi\left[(n-t)^2 - t^2 - n^2\right]/N}$$

$$= \Delta t\, e^{-i\pi n^2/N} \sum_{t=0}^{N-1} g_t e^{-i\pi t^2/N} e^{i\pi(n-t)^2/N} = \Delta t\, b_n^* \sum_{t=0}^{N-1} a_t b_{n-t},$$

where $b_t \equiv \exp\left(i\pi t^2/N\right)$ and $a_t \equiv g_t b_t^*$. The last summation above is of the same form as Equation (114) and can hence be computed efficiently using FFTs. Bluestein's substitution thus enables us to use a standard FFT algorithm to evaluate a DFT of general length. A FORTRAN implementation of the chirp transform algorithm is given in Monro and Branch (1977).

[2] The discrete time/discrete frequency concentration problem was first studied in detail by Grünbaum (1981). We formulate it by analogy to the cases of continuous time/continuous frequency and discrete time/continuous frequency. Thus we define the following measures of energy concentration:

$$\alpha^2(N, m_1, m_2) \equiv \sum_{t=m_1}^{m_2} |g_t|^2 \Big/ \sum_{t=0}^{N-1} |g_t|^2$$

and

$$\beta^2(N, m_1, m_2) \equiv \sum_{n=m_1}^{m_2} |G_n|^2 \Big/ \sum_{n=0}^{N-1} |G_n|^2,$$

where m_1 and m_2 are integers such that $0 \leq m_1 \leq m_2 \leq N-1$. Because of the obvious symmetries, we shall only study $\alpha^2(N, m_1, m_2)$. If, for simplicity, we set $\Delta t = 1$ in Equation (111a), we can use that equation and (111b) to write

$$\alpha^2(N, m_1, m_2) = \frac{\sum_{n'=1}^{N-1} \sum_{n=1}^{N-1} G_{n'} G_n^* \sum_{t=m_1}^{m_2} e^{i2\pi(n'-n)t/N}}{N \sum_{n'=1}^{N-1} |G_{n'}|^2}.$$

Now, if we let $M \equiv m_2 - m_1 + 1$, we can write

$$\sum_{t=m_1}^{m_2} e^{i2\pi(n'-n)t/N} = \sum_{t=1}^{M} e^{i2\pi(n'-n)(t+m_1-1)/N}$$

$$= e^{i2\pi(n'-n)(m_1-1)/N} \sum_{t=1}^{M} e^{i2\pi(n'-n)t/N}$$

$$= e^{i\pi(n'-n)(m_1+m_2)/N} M \mathcal{D}_M \left(\frac{n'-n}{N} \right),$$

where the last line was derived using Exercises [1.3c] and [1.3d]. If we now define

$$H_n \equiv G_n e^{i\pi n(m_1+m_2)/N},$$

we have

$$\alpha^2(N, m_1, m_2) = \frac{M \sum_{n'=1}^{N-1} \sum_{n=1}^{N-1} H_{n'} \mathcal{D}_M \left(\frac{n'-n}{N} \right) H_n^*}{N \sum_{n=1}^{N-1} |H_n|^2}.$$

By arguments similar to those used for Exercise [3.11], the sequence that maximizes $\alpha^2(N, m_1, m_2)$ must satisfy

$$\sum_{n'=1}^{N-1} \frac{M}{N} \mathcal{D}_M \left(\frac{n'-n}{N} \right) H_{n'} = \lambda_l H_n, \qquad n = 0, \ldots, N-1.$$

The highest concentration is given by the largest eigenvalue λ_0 of the corresponding matrix equation

$$A\mathbf{h} = \lambda_l \mathbf{h},$$

where A is an $N \times N$ matrix whose (n', n)th element is

$$\frac{M}{N} \mathcal{D}_M \left(\frac{n' - n}{N} \right)$$

and $\mathbf{h}$ is an N-dimensional vector with nth element H_n. As in the discrete time/continuous frequency case, there is an easy way to find the eigenvalues and eigenvectors of A by using an associated commuting tridiagonal matrix (Grünbaum, 1981). The sequences resulting from these calculations have recently been used in image processing (Wilson, 1987; Wilson and Spann, 1988).

3.11 Summary of Fourier Theory

We assume throughout this section that the functions $g(\cdot)$, $g_p(\cdot)$, $h(\cdot)$ and $h_p(\cdot)$ and sequences $\{h_t\}$ and $\{g_t\}$ are real or complex-valued. (The subscript 'p' in, for example, $g_p(\cdot)$ is added to emphasize here that the function is periodic – we do not follow this convention in subsequent chapters.)

[1] *Continuous time $g_p(\cdot)$ with period T and $\int_{-T/2}^{T/2} |g_p(t)|^2 \, dt < \infty$*
   a) Fourier representation (valid for $-\infty < t < \infty$):

$$g_p(t) \overset{\text{ms}}{=} \sum_{n=-\infty}^{\infty} G_n e^{i2\pi f_n t} \text{ with } f_n \equiv \frac{n}{T}, \qquad \text{(see (58e))}$$

   where, for $n = 0, \pm 1, \pm 2, \ldots,$

$$G_n \equiv \frac{1}{T} \int_{-T/2}^{T/2} g_p(t) e^{-i2\pi f_n t} \, dt. \qquad \text{(see (58d))}$$

   The Fourier relationship between $g_p(\cdot)$ and $\{G_n\}$ is noted as

$$g_p(\cdot) \longleftrightarrow \{G_n\}.$$

   b) Parseval's theorem:

$$\int_{-T/2}^{T/2} |g_p(t)|^2 \, dt = T \sum_{n=-\infty}^{\infty} |G_n|^2. \qquad \text{(see (59b))}$$

c) Spectral properties:
   $|G_n|$ defines an amplitude spectrum (in view of (58e)) and $|G_n|^2$ defines a discrete power spectrum; $|G_n|^2$ is the contribution to the power due to the sinusoid with frequency f_n.

d) Convolution theorem: the Fourier transform of the convolution

$$g_p * h_p(t) \equiv \frac{1}{T} \int_{-T/2}^{T/2} g_p(u) h_p(t-u) \, du \qquad (117a)$$

is

$$\frac{1}{T} \int_{-T/2}^{T/2} g_p * h_p(t) e^{-i2\pi f_n t} \, dt = G_n H_n. \qquad (117b)$$

[2] *Continuous time $g(\cdot)$ with $\int_{-\infty}^{\infty} |g(t)|^2 \, dt < \infty$*

   a) Fourier representation (valid for $-\infty < t < \infty$):

$$g(t) \stackrel{\text{ms}}{=} \int_{-\infty}^{\infty} G(f) e^{i2\pi f t} \, df, \qquad (\text{see } (65a))$$

where, for $-\infty < f < \infty$,

$$G(f) \equiv \int_{-\infty}^{\infty} g(t) e^{-i2\pi f t} \, dt. \qquad (\text{see } (65b))$$

The Fourier relationship between $g(\cdot)$ and $G(\cdot)$ is noted as

$$g(\cdot) \longleftrightarrow G(\cdot).$$

b) Parseval's theorem:

$$\int_{-\infty}^{\infty} |g(t)|^2 \, dt = \int_{-\infty}^{\infty} |G(f)|^2 \, df. \qquad (\text{see } (66))$$

c) Spectral properties:
   $|G(f)|$ defines an amplitude spectrum (in view of (65a)), and $|G(f)|^2$ defines an energy spectral density function (in view of (66)); $|G(f)|^2 \, df$ is the contribution to the energy due to sinusoids with frequencies in a small interval about f.

d) Convolution theorem: the Fourier transform of the convolution

$$g * h(t) \equiv \int_{-\infty}^{\infty} g(u) h(t-u) \, du \qquad (\text{see } (81))$$

is

$$\int_{-\infty}^{\infty} g * h(t) e^{-i2\pi f t} \, dt = G(f) H(f).$$

[3] *Discrete time* $\{g_t\}$ *sampled at intervals of* Δt *with* $\displaystyle\sum_{t=-\infty}^{\infty} |g_t|^2 < \infty$

a) Fourier representation (valid for $t = 0, \pm 1, \pm 2, \dots$):

$$g_t = \int_{-1/(2\,\Delta t)}^{1/(2\,\Delta t)} G_p(f) e^{i2\pi ft\,\Delta t}\, df, \qquad \text{(see (89b))}$$

where, for $|f| \leq 1/(2\,\Delta t) \equiv f_{(N)}$,

$$G_p(f) \equiv \Delta t \sum_{t=-\infty}^{\infty} g_t e^{-i2\pi ft\,\Delta t}. \qquad \text{(see (87))}$$

The Fourier relationship between $\{g_t\}$ and $G_p(\cdot)$ is noted as

$$\{g_t\} \longleftrightarrow G_p(\cdot).$$

b) Parseval's theorem:

$$\Delta t \sum_{t=-\infty}^{\infty} |g_t|^2 = \int_{-1/(2\,\Delta t)}^{1/(2\,\Delta t)} |G_p(f)|^2\, df. \qquad \text{(see (89c))}$$

c) Spectral properties:
$|G_p(f)|$ defines an amplitude spectrum (in view of (89b)), and $|G_p(f)|^2$ defines an energy spectral density function (in view of Parseval's theorem); $|G_p(f)|^2\, df$ is the contribution to the energy due to sinusoids with frequencies in a small interval about f.

d) Convolution theorem: the Fourier transform of the convolution

$$g * h_t \equiv \Delta t \sum_{u=-\infty}^{\infty} g_u h_{t-u} \qquad (118a)$$

is

$$\Delta t \sum_{t=-\infty}^{\infty} g * h_t e^{-i2\pi ft\,\Delta t} = G_p(f) H_p(f). \qquad (118b)$$

[4] *Segment of discrete time* $\{g_t\}$ *sampled at intervals of* Δt
a) Fourier representation (valid for $t = 0, 1, \dots, N-1$):

$$g_t = \frac{1}{N\,\Delta t} \sum_{n=0}^{N-1} G_n e^{i2\pi nt/N}, \qquad \text{(see (111a))}$$

where, for $n = 0, 1, \dots, N-1$,

$$G_n \equiv \Delta t \sum_{t=0}^{N-1} g_t e^{-i2\pi nt/N}. \qquad \text{(see (110a))}$$

G_n is associated with frequency $f_n \equiv n/(N\,\Delta t)$. The Fourier relationship between $\{g_t\}$ and $\{G_n\}$ is noted as

$$\{g_t\} \longleftrightarrow \{G_n\}.$$

Since both $\{g_t\}$ and $\{G_n\}$ can be considered as periodic sequences with period N, it is useful to arbitrarily shift the indices t and n; hence the notation

$$\{\, g_t : t = j, \ldots, N+j-1 \,\} \longleftrightarrow \{\, G_n : n = k, \ldots, N+k-1 \,\} \tag{119a}$$

means that, for $t = j, \ldots, N+j-1$,

$$g_t = \frac{1}{N\,\Delta t} \sum_{n=k}^{N+k-1} G_n e^{i2\pi nt/N},$$

where, for $n = k, \ldots, N+k-1$,

$$G_n \equiv \Delta t \sum_{t=j}^{N+j-1} g_t e^{-i2\pi nt/N}.$$

b) Parseval's theorem:

$$\Delta t \sum_{t=0}^{N-1} |g_t|^2 = \frac{1}{N\,\Delta t} \sum_{n=0}^{N-1} |G_n|^2. \qquad \text{(see (111b))}$$

c) Spectral properties:
   $|G_n|^2$ could be used to define a discrete power spectrum (but only in a very limited sense).

d) Convolution theorem: the Fourier transform of the convolution

$$g * h_t \equiv \Delta t \sum_{u=0}^{N-1} g_u h_{t-u} \tag{119b}$$

(where $h_s \equiv h_{\mathrm{mod}\,(s,N)}$ for s outside the range 0 to $N-1$) is

$$\Delta t \sum_{t=0}^{N-1} g * h_t e^{-i2\pi nt/N} = G_n H_n. \tag{119c}$$

This type of convolution is called *cyclic* (see Exercise [3.14]).

3.12 Exercises

[3.1] a) Express a_n and b_n in terms of the G_n defined in Equation (58b).

   b) Assume now that the a_n and b_n in Equation (57) are real-valued. Show that this equation can be written in the following form:

$$\tilde{g}_p(t) = \frac{c_0}{2} + \sum_{n=1}^{\infty} c_n \cos\left(\frac{2\pi nt}{T} + \phi_n\right),$$

   where $\{c_n\}$ and $\{\phi_n\}$ are sequences of real-valued constants. Express c_n and ϕ_n in terms of the G_n of Equation (58b) and vice versa.

[3.2] Use parts b) and a) of Exercise [1.3] to show that Equation (60b) can be rewritten as

$$g_{p,m}(t) = 1 + 2\Re\left(\frac{1 - \phi^m e^{imt}}{\phi^{-1} e^{-it} - 1}\right),$$

where, for a complex number $z = x + iy$, $\Re(z) \equiv x$. Conclude that

$$g_{p,m}(t) = \frac{1 - \phi^2 + f_m}{1 + \phi^2 - 2\phi \cos(t)}$$

with an appropriate definition for f_m. Use this to show that, for all t,

$$\lim_{m \to \infty} g_{p,m}(t) = g_p(t),$$

where $g_p(\cdot)$ is given by Equation (60a) and $|\phi| < 1$. (This result shows that we can in fact replace the '$\overset{\text{ms}}{=}$' in the Fourier series representation of $g_p(\cdot)$ by '$=$'; i.e., the Fourier series representation of $g_p(t)$ is in fact equal to $g_p(t)$ itself for all t.)

[3.3] a) Prove the following 'two function' version of Parseval's theorem, namely,

$$\int_{-\infty}^{\infty} g^*(t)h(t)\, dt = \int_{-\infty}^{\infty} G^*(f)H(f)\, df, \qquad (120)$$

   in which $g(\cdot)$ and $h(\cdot)$ are $L^2(-\infty, \infty)$ functions with Fourier transforms $G(\cdot)$ and $H(\cdot)$ – cf. Equation (66).

   b) Formulate a version of the above for the discrete time/continuous frequency case (cf. Equation (89c)).

[3.4] a) Generalize Heisenberg's uncertainty principle in Section 3.4 to the case where the signal $g(\cdot)$ does not necessarily have unit energy; i.e.,

$$\int_{-\infty}^{\infty} |g(t)|^2\, dt = C < \infty.$$

To simplify matters, assume that

$$\mu_g = \int_{-\infty}^{\infty} t\,|g(t)|^2\,dt = 0 \text{ and } \mu_G = \int_{-\infty}^{\infty} f\,|G(f)|^2\,df = 0.$$

b) Generalize to the case where μ_g and μ_G are not necessarily zero.

[3.5] Figure 72 indicates that width$_e$ $\{g(\cdot)\}$ for the function defined by $g(t) = \exp(-|t|)$ is 2. For comparison, compute the alternative width measures, width$_v$ $\{g(\cdot)\}$ of Equation (73a) and width$_a$ $\{g(\cdot)\}$ of Equation (85c).

[3.6] Let $g(\cdot)$ and $h(\cdot)$ be $L^2(-\infty, \infty)$ functions with Fourier transforms $G(\cdot)$ and $H(\cdot)$, and assume that $G(\cdot)H(\cdot)$ is also an $L^2(-\infty, \infty)$ function. Show that $g * h(\cdot)$ is an $L^2(-\infty, \infty)$ function and that its Fourier transform is $G(\cdot)H(\cdot)$. Hint: verify that, for each real number t,

$$h^*(t - \cdot) \longleftrightarrow H^*(\cdot)e^{-i2\pi ft}$$

and then evoke Equation (120).

[3.7]   a) Show that relationships (85a) and (85b) involving complex cross-correlation and complex autocorrelation are true.

b) Because of the duality of the time and frequency domains, convolution in the *frequency* domain is equivalent to multiplication in the *time* domain. Prove this result; i.e.,

$$g(\cdot)h(\cdot) \longleftrightarrow G * H(\cdot).$$

[3.8] Prove the following version of the convolution theorem (cf. Equations (118a) and (118b) with $\Delta t = 1$). Let $\{g_t\}$ and $\{h_t\}$ be two real-valued sequences of numbers that are square summable and hence have Fourier transforms

$$G(f) = \sum_{t=-\infty}^{\infty} g_t e^{-i2\pi ft} \text{ and } H(f) = \sum_{t=-\infty}^{\infty} h_t e^{-i2\pi ft},$$

$|f| \leq 1/2$. Show that the Fourier transform of the convolution of $\{g_t\}$ and $\{h_t\}$, namely,

$$g * h_t \equiv \sum_{u=-\infty}^{\infty} g_u h_{t-u},$$

is the function defined by $G(f)H(f)$.

[3.9] Use the following three facts to verify Equation (101c): first, $\{g_t\}$ has the Fourier integral representation

$$g_t \stackrel{\text{ms}}{=} \int_{-W}^{W} G_p(f)e^{i2\pi ft}\,df$$

(assuming $\Delta t = 1$); second, $|g_t|^2 = g_t g_t^*$; and, finally, the summation $\sum_{t=0}^{N-1} e^{i2\pi ft}$ can be reduced using Exercise [1.3d].

[3.10] Let $g(\cdot)$ be an $L^2(-\infty, \infty)$ function with a Fourier transform $G(\cdot)$ that is band-limited to $[-W, W]$. For complex z, define

$$\tilde{g}(z) \equiv \int_{-W}^{W} G(f) e^{i2\pi fz} \, df$$

(see Section 3.3). Use the Schwarz inequality in Equation (68) to prove that the kth derivative of $\tilde{g}(z)$

$$\frac{d^k \tilde{g}(z)}{dz^k} = \int_{-W}^{W} G(f) \, (i2\pi f)^k \, e^{i2\pi fz} \, df$$

satisfies, for all complex numbers z,

$$\left| \frac{d^k \tilde{g}(z)}{dz^k} \right|^2 < (2\pi W)^{2k} \frac{e^{4\pi W |y|}}{4\pi |y|} \int_{-\infty}^{\infty} |g(t)|^2 \, dt,$$

where y is the imaginary part of z.

[3.11] For real-valued g_t, the concentration of energy $\beta^2(W)$ in the frequency range $|f| \le W < 1/2$ (defined in Equation (101b)) takes the form

$$\beta^2(W) = \sum_{t'=0}^{N-1} \sum_{t=0}^{N-1} g_t \frac{\sin[2\pi W(t'-t)]}{\pi(t'-t)} g_{t'} \bigg/ \sum_{t=0}^{N-1} g_t^2$$

(cf. Equation (105a)). Rewrite the above as

$$\beta^2(W) = \mathbf{g}^T A \mathbf{g} \bigg/ \mathbf{g}^T \mathbf{g}$$

(here $\mathbf{g}$ is an N-dimensional vector, and A is an $N \times N$ matrix), and differentiate both sides with respect to $\mathbf{g}$ to show that the sequence which maximizes $\beta^2(W)$ must satisfy Equation (105b).

[3.12] As stated in the text following Equation (103b), the kth order discrete prolate spheroidal sequence is given by

$$v_{t,k}(N, W) = \frac{(-1)^k}{\epsilon_k \lambda_k(N, W)} \int_{-W}^{W} U_k(f; N, W) e^{i2\pi f[t-(N-1)/2]} \, df$$

for $t = 0, \pm 1, \pm 2, \ldots$. Use this to show that

$$U_k(f; N, W) = (-1)^k \epsilon_k \sum_{t=0}^{N-1} v_{t,k}(N, W) e^{-i2\pi f[t-(N-1)/2]}.$$

Show that $U_k(\cdot; N, W)$ is symmetric for even k, i.e.,

$$U_k(-f; N, W) = U_k(f; N, W),$$

and skew-symmetric for odd k, i.e.,

$$U_k(-f; N, W) = -U_k(f; N, W),$$

so that we can write the above more compactly as

$$U_k(f; N, W) = \epsilon_k \sum_{t=0}^{N-1} v_{t,k}(N, W) e^{i2\pi f[t-(N-1)/2]}$$

(Slepian, 1978).

[3.13] Prove that, if $g(\cdot) \longleftrightarrow G(\cdot)$, then, for all $a \neq 0$,

$$|a|^{1/2} g(at) \longleftrightarrow \frac{1}{|a|^{1/2}} G(f/a)$$

(this is Equation (70a)).

[3.14] a) Suppose that $g_0, \ldots, g_{N-1}$ and $h_0, \ldots, h_{N-1}$ are two finite sequences of length N. Prove the *cyclic convolution theorem* of Equation (119c), and describe how $g * h_0, \ldots, g * h_{N-1}$ can be computed using two DFTs and one inverse DFT. (As an aside, we note that this result has some important practical implications. If N is a sample size acceptable to an FFT algorithm, computation of $\{g * h_t\}$ directly from Equation (119b) would require on the order of N^2 floating point operations on a digital computer, whereas use of three FFTs would typically require on the order of $N \log(N)$ such operations. It is thus often more efficient (particularly for large N) to compute cyclic convolutions via FFTs rather than via the defining Equation (119b).)

b) Suppose now that $g_0, \ldots, g_{N_g-1}$ and $h_0, \ldots, h_{N_h-1}$ are two finite sequences of lengths N_g and N_h with $N_g < N_h$. Suppose we are interested in computing the following sequence of length $N_h - N_g + 1$ (a portion of a noncyclic convolution):

$$c_n = \sum_{t=0}^{N_g-1} g_t h_{n-t} \quad \text{for } n = N_g - 1, \ldots, N_h - 1.$$

Show that, if N is large enough, the c_n can be computed via cyclic convolution of the following two sequences of length N:

$$g'_t = \begin{cases} g_t, & 0 \leq t \leq N_g - 1; \\ 0, & N_g \leq t \leq N - 1; \end{cases}$$

$$h'_t = \begin{cases} h_t, & 0 \leq t \leq N_h - 1; \\ 0, & N_h \leq t \leq N - 1. \end{cases}$$

How large is 'large enough?' (This is another example of the use of zero padding.)

[3.15] a) For $a > 0$, define $I(t; a) = 1$ for $|t| \leq a$ and zero otherwise. Show that

$$I(t; a) \longleftrightarrow \frac{\sin(2\pi f a)}{\pi f}.$$

b) For $N > 0$, define the sequence $\{i_{t,N}\}$ by $i_{t,N} = 1$ for $t = 0, 1, \ldots, N-1$ and zero otherwise. Show that

$$\{i_{t,N}\} \longleftrightarrow e^{-i(N-1)\pi f} \frac{\sin(N\pi f)}{\sin(\pi f)}$$

(hint: consider Exercise [1.3d]).

[3.16] The aim of this exercise is to derive some of the properties of the dpswf $U_k(\cdot; N, W)$ and the dpss $\{v_{t,k}(N, W)\}$ that are stated in Section 3.9. Let $V_k(f; N, W) \equiv I(f; W)U_k(f; N, W)$ on $[-1/2, 1/2]$, where $I(\cdot; W)$ is as defined in part a) of the previous exercise. (Note that all functions here are periodic with period 1.) Then the integral equation (103a) with $U(\cdot) = U_k(\cdot; N, W)$ and $\lambda = \lambda_k(N, W)$ can be written

$$N\mathcal{D}_N * V_k(f; N, W) = \lambda_k(N, W)U_k(f; N, W).$$

The sequences $\{v_{t,k}(N, W)\}$ and $\{u_{t,k}(N, W)\}$ are defined by the discrete/continuous Fourier transform relationships

$$\{(-1)^k \epsilon_k \lambda_k(N, W)v_{t,k}(N, W)\} \longleftrightarrow e^{-i\pi(N-1)f}V_k(f; N, W)$$
$$\{(-1)^k \epsilon_k u_{t,k}(N, W)\} \longleftrightarrow e^{-i\pi(N-1)f}U_k(f; N, W).$$

a) By expanding the Dirichlet kernel in the convolution above, show that

$$u_{t,k}(N, W) = \begin{cases} v_{t,k}(N, W), & t = 0, 1, \ldots, N-1; \\ 0, & \text{otherwise.} \end{cases}$$

b) The properties of the solutions to integral equation (103a) imply that the $U_k(\cdot; N, W)$ functions are orthogonal on $[-W, W]$. We have normalized them to have unit energy on $[-1/2, 1/2]$. Use the fact that

$$\sum_{t=-\infty}^{\infty} u_{t,j}^* u_{t,k} = \sum_{t=-\infty}^{\infty} v_{t,j}^* u_{t,k}$$

along with the 'two function' version of Parseval's theorem (see Exercise [3.3a]) to show that the $U_k(\cdot; N, W)$ functions are orthonormal on $[-1/2, 1/2]$ and have energy $\lambda_k(N, W)$ on $[-W, W]$.

c) Show that the $\{v_{t,k}(N, W)\}$ sequences are orthonormal over $t = 0, 1, \ldots, N - 1$.

d) Show that the $\{v_{t,k}(N, W)\}$ sequences are orthogonal over $t = 0, \pm 1, \pm 2, \ldots$ with energy $1/\lambda_k(N, W)$.

e) Show that the $\{v_{t,k}(N, W)\}$ sequences satisfy the $N \times N$ linear system (104).

4

Foundations for
Stochastic Spectral Analysis

4.0 Introduction

In the previous chapter we produced representations for various deterministic functions and sequences in terms of linear combinations of sinusoids with different frequencies (for mathematical convenience we actually used complex exponentials instead of sinusoids directly). These representations allow us to easily define various energy and power spectra and to attach a physical meaning to them. For example, subject to square integrability conditions, we found that periodic functions are representable (in the mean square sense) by sums of sinusoids over a discrete set of frequency components, while nonperiodic functions are representable (also in the mean square sense) by an integral of sinusoids over a continuous range of frequencies. For periodic functions, the energy from $-\infty$ to ∞ is infinite, so we can define their spectral properties in terms of distributions of power over a discrete set of frequencies. For nonperiodic functions, the energy from $-\infty$ to ∞ is finite, so we can define their properties in terms of an energy distribution over a continuous range of frequencies.

We now want to find some way of representing a stationary process in terms of a 'sum' of sinusoids so that we can meaningfully define an appropriate spectrum for it; i.e., we want to be able to directly relate our representation for a stationary process to its spectrum in much the same way we did for deterministic functions. Now a stationary process has associated with it an ensemble of realizations that describe the possible outcomes of a random experiment. Each realization is a fixed function or sequence of time, so we could try to apply the theory for deterministic functions and sequences on a realization by realization basis and then extend it somehow to the entire ensemble. However, the functions

126

and sequences we can handle easily either have finite energy or are periodic. Because the variance of a stationary process is constant over time, a typical realization has infinite energy; moreover, with the important exception of harmonic processes, a typical realization is also not periodic. This formally rules out merely applying the deterministic theory on a realization by realization basis (by use of a limiting argument, this approach can be carried out to a certain extent for some stationary processes – see Section 4.2 for details). There is a different approach that works for stationary processes and is discussed in the next section. It is based upon the 'important exception' of the harmonic processes and leads to a spectral representation theorem for stationary processes.

4.1 Spectral Representation of Stationary Processes

In this section we motivate and state (but do not prove) the spectral representation theorem for stationary processes due to Cramér (1942). A rigorous proof is rather involved – the interested reader is referred to Priestley (1981, Section 4.11) for details. This theorem is fundamental in the spectral analysis of stationary processes since it allows us to relate the spectrum of such a process directly to a representation for the process itself. Indeed, in the words of Koopmans (1974, p. 36), 'One of the essential reasons for the central position held by stationary stochastic processes in time series analysis is the existence of a spectral representation for the process from which [the spectrum] can be directly computed.'

We motivate the spectral representation theorem for discrete parameter stationary processes by considering a special case, namely, a real-valued discrete time harmonic process

$$X_t = \sum_{l=1}^{L} D_l \cos\left(2\pi f_l t + \phi_l\right), \qquad t = 0, \pm 1, \pm 2, \ldots, \tag{127}$$

where $L \geq 1$; D_l and f_l are real-valued constants; and the ϕ_l terms are independent rv's, each having a rectangular distribution on the interval $[-\pi, \pi]$ (this is Equation (46b) with the process mean μ assumed to be zero). We assume that the frequencies f_l are ordered such that $f_l < f_{l+1}$ and that $0 < f_l < 1/2$ for all l. Except for the exclusion of the frequencies 0 and $1/2$ (which simplifies our discussion somewhat), this latter stipulation is not really a restriction due to the aliasing phenomenon (see Sections 3.8 and 4.5) – here the Nyquist frequency is $1/2$ because we are assuming that Δt is 1. Since

$$D_l \cos\left(2\pi f_l t + \phi_l\right) = \frac{D_l}{2}\left(e^{i\phi_l} e^{i2\pi f_l t} + e^{-i\phi_l} e^{-i2\pi f_l t}\right),$$

we can rewrite (127) as

$$X_t = \sum_{l=-L}^{L} C_l e^{i2\pi f_l t}, \tag{128a}$$

where

$$C_l \equiv D_l e^{i\phi_l}/2 \text{ and } C_{-l} \equiv D_l e^{-i\phi_l}/2, \qquad l = 1, \ldots, L;$$

$C_0 \equiv 0$; $f_0 \equiv 0$; and $f_{-l} \equiv -f_l$. Since the ϕ_l rv's are assumed to be independent, it follows that $C_1, C_2, \ldots, C_L$ are also independent rv's. Since $C_{-l} = C_l^*$, the rv's C_{-l} and C_l are certainly not independent, but – rather surprisingly – it can be shown that they are uncorrelated (see Exercise [4.1]). It follows that the $2L + 1$ rv's $C_{-L}, C_{-L+1}, \ldots, C_L$ are all mutually uncorrelated with means and variances given by

$$E\{C_l\} = 0 \text{ and } \operatorname{var}\{C_l\} = E\{|C_l|^2\} = D_l^2/4$$

for all l if we define $D_0 \equiv 0$ and $D_{-l} \equiv D_l$ (these results follow from Section 2.6). We thus have

$$\operatorname{var}\{X_t\} = \sum_{l=-L}^{L} \operatorname{var}\{C_l e^{i2\pi f_l t}\}$$

$$= \sum_{l=-L}^{L} E\{|C_l|^2\} = \sum_{l=-L}^{L} D_l^2/4; \tag{128b}$$

i.e., the variance of the stationary process $\{X_t\}$ can be decomposed into a sum of components $E\{|C_l|^2\}$, each of which is the expected squared amplitude of the complex exponential of frequency f_l in model (128a). We can thus define a useful variance spectrum by

$$S^{(V)}(f) \equiv \begin{cases} D_l^2/4, & \text{if } f = f_l, l = 0, \pm 1, \ldots, \pm L; \\ 0, & \text{otherwise.} \end{cases}$$

Let us now define the complex-valued stochastic process

$$Z(f) \equiv \sum_{j=0}^{l} C_j, \quad f_l < f \le f_{l+1} \text{ with } l = 0, \ldots, L, \tag{128c}$$

where $f_{L+1} \equiv 1/2$. For completeness, we define $Z(0) = 0$. With this definition, $\{Z(f)\}$ is a 'jump' process that is defined on the interval $[0, 1/2]$ and has a random complex-valued jump at each f_l. Thus

$$Z(f) = \begin{cases} 0, & \text{for } 0 \le f \le f_1; \\ C_1, & \text{for } f_1 < f \le f_2; \\ C_1 + C_2, & \text{for } f_2 < f \le f_3; \\ C_1 + C_2 + C_3, & \text{for } f_3 < f \le f_4, \end{cases}$$

and so forth. Note that the 'time' index for this process is actually frequency.

Let us define

$$dZ(f) \equiv \begin{cases} Z(f + df) - Z(f), & 0 \le f < 1/2; \\ 0, & f = 1/2; \\ dZ^*(-f), & -1/2 \le f < 0, \end{cases} \qquad (129\text{a})$$

where df is a small positive increment. For $l \ge 0$ we thus have

$$dZ(f_l) = Z(f_l + df) - Z(f_l)$$

$$= \sum_{j=0}^{l} C_j - \sum_{j=0}^{l-1} C_j = C_l,$$

and, for any $f \ne f_l$ for some l, $dZ(f) = 0$ for df sufficiently small. Since $E\{C_l\} = 0$, it follows that $E\{dZ(f)\} = 0$ for all f. It can be shown (see Exercise [4.2]) that, if f, f', df and df' are such that the intervals $[f, f + df]$ and $[f', f' + df']$ are nonintersecting subintervals of $[-1/2, 1/2]$, then the rv's $dZ(f)$ and $dZ(f')$ are uncorrelated; i.e.,

$$\text{cov}\,\{dZ(f'), dZ(f)\} = E\{dZ^*(f')\,dZ(f)\} = 0.$$

Because of this property, the process $\{Z(f)\}$ is said to have *orthogonal increments*, and the process itself is called an *orthogonal process*. Note that $\{Z(f)\}$ gives us each C_l in $\{X_t\}$ and that the expected squared magnitude of the jump in $\{Z(f)\}$ at f_l is just $E\{|C_l|^2\}$; i.e.,

$$E\{|dZ(f_l)|^2\} = E\{|C_l|^2\} = D_l^2/4.$$

Now let $g(\cdot)$ be a function that is continuous over the interval $[-1/2, 1/2]$, and let $H(\cdot)$ be a step function defined over that same interval with jumps at

$$-1/2 < a_1 < a_2 < \cdots < a_N < 1/2$$

of finite sizes b_1, b_2, ..., b_N. From the definition of the Riemann–Stieltjes integral it follows that

$$\int_{-1/2}^{1/2} g(f)\,dH(f) = \sum_{k=1}^{N} g(a_k) b_k$$

(see the Comments and Extensions to Section 2.3). If we equate $g(f)$ with $\exp{(i2\pi ft)}$ and $H(f)$ with $Z(f)$, we can rewrite Equation (128a) in the following way:

$$X_t = \int_{-1/2}^{1/2} e^{i2\pi ft}\,dZ(f). \qquad (129\text{b})$$

This is a stochastic version of the Riemann–Stieltjes integral, which is defined in a way analogous to the usual version (see Equation (34) and Exercise [4.3]). Equation (129b) is called the *spectral representation* for the stationary process (128a). It is useful mainly due to the properties of the orthogonal increments process. As an example, let us show how we can derive Equation (128b) using (129b). Since $\{X_t\}$ is a real-valued process with $E\{X_t\} = 0$,

$$
\begin{aligned}
\text{var}\,\{X_t\} &= E\{X_t^2\} = E\{|X_t|^2\} = E\{X_t^* X_t\} \\
&= E\left\{ \int_{-1/2}^{1/2} e^{-i2\pi f't}\, dZ^*(f') \int_{-1/2}^{1/2} e^{i2\pi ft}\, dZ(f) \right\} \quad \text{(130a)} \\
&= \int_{-1/2}^{1/2} \int_{-1/2}^{1/2} e^{i2\pi(f-f')t} E\{dZ^*(f')\, dZ(f)\}.
\end{aligned}
$$

Because of the properties of $\{Z(f)\}$, the only time the expectation within the double integral is nonzero occurs when $f = f' = f_l$ for $l = -L, \ldots, L$, in which case we have

$$
E\{dZ^*(f_l)\, dZ(f_l)\} = E\{|dZ(f_l)|^2\} = E\{|C_l|^2\}.
$$

Since $\exp[-i2\pi(f_l - f_l)t] = 1$, the Riemann–Stieltjes integral reduces to (128b). (The above manipulations are purely formal; see Exercise [4.3] for some justification.)

Rather surprisingly, we can develop a spectral representation for *any* discrete parameter stationary process by considering what happens to model (128a) in the limit as $L \to \infty$ in such a way that the maximum difference between adjacent frequencies f_{l-1} and f_l goes to 0 and the process variance (128b) converges to a finite number. We are now in a position to state (without proof) the *spectral representation theorem for discrete parameter stationary processes* (for more details, see Priestley, 1981, p. 251). Let $\{X_t\}$ be a real-valued discrete parameter stationary process with zero mean. There exists an orthogonal process, $\{Z(f)\}$, defined on the interval $[-1/2, 1/2]$, such that

$$
X_t = \int_{-1/2}^{1/2} e^{i2\pi ft}\, dZ(f) \quad \text{(130b)}
$$

for all integers t, where the above equality is in the mean square sense. The process $\{Z(f)\}$ has the following properties:

[1] $E\{dZ(f)\} = 0$ for all $|f| \leq 1/2$;

[2] $E\{|dZ(f)|^2\} \equiv dS^{(I)}(f)$, say, for all $|f| \leq 1/2$, where the bounded nondecreasing function $S^{(I)}(\cdot)$ is called the *integrated spectrum* of $\{X_t\}$; and

[3] for any two distinct frequencies f and f' contained in the interval $[-1/2, 1/2]$,

$$\text{cov}\{dZ(f'), dZ(f)\} = E\{dZ^*(f')\, dZ(f)\} = 0.$$

Equation (130b) is called the *spectral representation* of the process $\{X_t\}$. It says that we can represent *any* discrete parameter stationary process as an infinite sum of complex exponentials (i.e., sine and cosine functions) at frequencies f with associated random amplitudes $|dZ(f)|$ and random phases $\arg(dZ(f))$, where $\arg(z)$ is the phase angle of the complex number z. Moreover, the expected value of the square modulus of $dZ(f)$ *defines* an integrated spectrum $S^{(I)}(\cdot)$ for $\{X_t\}$.

The property of $\{Z(f)\}$ that makes the representation (130b) helpful in the proof of many results in spectral analysis is that its increments at different frequencies are uncorrelated. As an example of the use of this property, we now show a fundamental relationship between the autocovariance sequence $\{s_\tau\}$ and the integrated spectrum $S^{(I)}(\cdot)$ for a stationary process $\{X_t\}$. Since

$$X_t^* X_{t+\tau} = \int_{-1/2}^{1/2} e^{-i2\pi f't}\, dZ^*(f') \int_{-1/2}^{1/2} e^{i2\pi f(t+\tau)}\, dZ(f)$$

$$= \int_{-1/2}^{1/2} \int_{-1/2}^{1/2} e^{-i2\pi f't} e^{i2\pi f(t+\tau)}\, dZ^*(f')\, dZ(f),$$

it follows that the acvs can be written as

$$s_\tau = E\{X_t X_{t+\tau}\} = E\{X_t^* X_{t+\tau}\}$$

$$= \int_{-1/2}^{1/2} \int_{-1/2}^{1/2} e^{i2\pi(f-f')t} e^{i2\pi f\tau} E\{dZ^*(f')\, dZ(f)\}.$$

Because of the orthogonality property of $\{Z(f)\}$, the only contribution to the double integral occurs when $f = f'$, so we have

$$s_\tau = \int_{-1/2}^{1/2} e^{i2\pi f\tau} E\{|dZ(f)|^2\} = \int_{-1/2}^{1/2} e^{i2\pi f\tau}\, dS^{(I)}(f), \qquad (131)$$

which shows that the integrated spectrum determines the acvs for a stationary process.

If in fact $S^{(I)}(\cdot)$ is differentiable everywhere with a derivative denoted by $S(\cdot)$, we have

$$E\{|dZ(f)|^2\} = dS^{(I)}(f) = S(f)\, df.$$

The function $S(\cdot)$ is called the *spectral density function* (sdf). We can now rewrite Equation (131) in terms of this sdf as

$$s_\tau = \int_{-1/2}^{1/2} S(f)e^{i2\pi f\tau}\,df. \tag{132a}$$

From Section 3.7, we know that a square summable deterministic sequence $\{g_t\}$ has the Fourier representation

$$g_t = \int_{-1/2}^{1/2} G_p(f)e^{i2\pi ft}\,df, \quad \text{where } G_p(f) \equiv \sum_{t=-\infty}^{\infty} g_t e^{-i2\pi ft} \tag{132b}$$

(these are Equations (89b) and (87) with $\Delta t = 1$). Since $\{s_\tau\}$ is just a deterministic sequence, Equation (132a) indicates that it is the inverse Fourier transform of $S(\cdot)$. If we assume that $S(\cdot)$ is square integrable (and thus, from Parseval's theorem, that the acvs is square summable), we have the important fact that $S(\cdot)$ is the Fourier transform of $\{s_\tau\}$:

$$S(f) = \sum_{\tau=-\infty}^{\infty} s_\tau e^{-i2\pi f\tau} \tag{132c}$$

(strictly speaking, the above equality is in the mean square sense, but it can be regarded as a pointwise equality in almost all practical applications). Hence we have

$$\{s_\tau\} \longleftrightarrow S(\cdot). \tag{132d}$$

Note that we can regard $S(\cdot)$ as an amplitude spectrum for $\{s_\tau\}$.

The stationary process $\{X_t\}$ and the orthogonal process $\{Z(f)\}$ have a Fourier relationship, and so do the acvs $\{s_\tau\}$ and the sdf $S(\cdot)$. Let us compare these relationships with the spectral representation for a square summable sequence $\{g_t\}$ developed in Section 3.7 and repeated in (132b). For this discussion we will let $\{g_t\}$ play the role of $\{X_t\}$ even though the former is deterministic with finite energy and the latter is stochastic with infinite expected energy. If we define

$$G_p^{(I)}(f) \equiv \int_{-1/2}^{f} G_p(f')\,df' \quad \text{so that} \quad \frac{dG_p^{(I)}(f)}{df} = G_p(f),$$

then we can represent g_t using an ordinary Riemann–Stieltjes integral involving increments of $G_p^{(I)}(\cdot)$:

$$g_t = \int_{-1/2}^{1/2} e^{i2\pi ft}\,dG_p^{(I)}(f).$$

This can be compared with the spectral representation for $\{X_t\}$ in Equation (130b). The reason we cannot reduce (130b) to just a stochastic Riemann integral is that, in contrast to $G_p^{(I)}(\cdot)$, the stochastic process $\{Z(f)\}$ does *not* possess a derivative in any well-defined sense. In particular, this prevents us from establishing a relationship for $\{Z(f)\}$ in terms of $\{X_t\}$; in contrast, the derivative of $G_p^{(I)}(\cdot)$ – namely, $G_p(\cdot)$ – is the Fourier transform of $\{g_t\}$.

Let us now examine the autocorrelation of $\{g_t\}$, which is the sequence $\{g^* \star g_\tau\}$ defined by

$$g^* \star g_\tau \equiv \sum_{u=-\infty}^{\infty} g_u^* g_{u+\tau}$$

(note that we have called $g^* \star g_\tau$ an auto*correlation* rather than an auto*covariance*, in agreement with engineering practice for deterministic sequences – since $|g^* \star g_\tau| > 1$ is not precluded, this terminology conflicts with that of the statistical community). From results for square summable sequences analogous to those of Section 3.6, the autocorrelation of $\{g_t\}$ is the inverse Fourier transform of $|G_p(\cdot)|^2$ (see also Exercise [4.7]). Hence

$$g^* \star g_\tau = \int_{-1/2}^{1/2} |G_p(f)|^2 e^{i2\pi f\tau}\, df. \qquad (133)$$

If we define

$$H_p^{(I)}(f) \equiv \int_{-1/2}^{f} |G_p(f')|^2\, df' \text{ so that } \frac{dH_p^{(I)}(f)}{df} = |G_p(f)|^2,$$

we can rewrite the autocorrelation as an ordinary Riemann–Stieltjes integral:

$$g^* \star g_\tau = \int_{-1/2}^{1/2} e^{i2\pi f\tau}\, dH_p^{(I)}(f).$$

If we compare the above to Equation (131), we see that

$$E\{|dZ(f)|^2\} = dS^{(I)}(f)$$

in the stochastic model plays the role of $dH_p^{(I)}(f)$ in the deterministic case. The reason we cannot in general express (131) as an ordinary Riemann integral analogous to (133) is that the integrated spectrum $S^{(I)}(\cdot)$ does not always possess a derivative; when its derivative $S(\cdot)$ – the sdf – does exist, we then have Equation (132a), which is analogous to (133). (If we allow the use of the Dirac delta function, we can define a derivative for $S^{(I)}(\cdot)$ – see the remarks at the end of Section 4.4.)

As a final comparison, in the deterministic case, we have claimed that $|G_p(\cdot)|^2$ is an energy spectral density function so that $|G_p(f)|^2\,df$ gives the contribution to the energy in $\{g_t\}$ due to frequencies in a small interval about f. Since it follows from Equations (131) and (132a) that

$$\text{var}\,\{X_t\} = s_0 = \int_{-1/2}^{1/2} dS^{(I)}(f) = \int_{-1/2}^{1/2} S(f)\,df,$$

clearly $S(f)\,df$ is the contribution to the variance in the stationary process due to frequencies in a small interval about f. In the next section we note that the variance of a process is in fact closely related to the concept of power so that $S(\cdot)$ is often called a *power spectral density function* (psdf).

Comments and Extensions to Section 4.1

[1] Why does the integral in Equation (130b) only include the range of frequencies from $-1/2$ to $1/2$? The answer is aliasing (see Sections 3.8 and 4.5): the values of $\exp{[i2\pi ft]}$ and $\exp{[i2\pi(f \pm k)t]}$ are identical for all integers t and k, and so it is only necessary to have complex exponentials with frequencies in the range $[-1/2, 1/2]$ in representation (130b).

[2]   In the more general case when $E\{X_t\} = \mu \neq 0$ and the sampling interval Δt is not unity, the spectral representation for $\{X_t\}$ becomes

$$X_t = \mu + \int_{-f_{(N)}}^{f_{(N)}} e^{i2\pi ft\,\Delta t}\,dZ(f) \ \text{ for } \ |f| \leq f_{(N)} \equiv \frac{1}{2\,\Delta t}.$$

Equations (131) and (132a) are replaced by

$$s_\tau = \int_{-f_{(N)}}^{f_{(N)}} e^{i2\pi f\tau\,\Delta t}\,dS^{(I)}(f) = \int_{-f_{(N)}}^{f_{(N)}} S(f)e^{i2\pi f\tau\,\Delta t}\,df \qquad (134a)$$

for $\tau = 0, \pm 1, \ldots$, while Equation (132c) becomes

$$S(f) = \Delta t \sum_{\tau=-\infty}^{\infty} s_\tau e^{-i2\pi f\tau\,\Delta t} \ \text{ for } \ |f| \leq f_{(N)}. \qquad (134b)$$

[3] The *spectral representation theorem for continuous parameter stationary processes* is quite similar to the discrete parameter theorem. We merely state it and refer the reader to Priestley (1981, p. 246) for details. Let $\{X(t)\}$ be a real-valued, *stochastically continuous*, continuous parameter stationary process with zero mean. Then there exists an orthogonal process, $\{Z(f)\}$, such that, for all t, $X(t)$ can be expressed in the form

$$X(t) = \int_{-\infty}^{\infty} e^{i2\pi ft}\,dZ(f).$$

The process $\{Z(f)\}$ has the following properties:

a) $E\{dZ(f)\} = 0$ for all f;
b) $E\{|dZ(f)|^2\} = dS^{(I)}(f)$, say, for all f, where the bounded nondecreasing function $S^{(I)}(\cdot)$ is called the *integrated spectrum* of $\{X(t)\}$; and
c) for any two frequencies, f, f', with $f \neq f'$

$$\text{cov}\{dZ(f'), dZ(f)\} = E\{dZ^*(f')\,dZ(f)\} = 0.$$

Note that this theorem only holds for a subclass of continuous parameter stationary processes, namely, those that are stochastically continuous. We shall not define that concept here – it plays no importance in this book (see Priestley, 1981, p. 151, for details). It suffices to say that stochastic continuity is a rather mild regularity condition that rules out some pathological processes of little interest in practical applications. It holds if and only if the acvf $s(\cdot)$ for the process is continuous at the origin.

For the continuous parameter case, Equation (131) becomes

$$s(\tau) = \int_{-\infty}^{\infty} e^{i2\pi f\tau}\,dS^{(I)}(f).$$

If $S^{(I)}(\cdot)$ is differentiable and $s(\cdot)$ is square integrable, Equations (132a) and (132c) become

$$s(\tau) = \int_{-\infty}^{\infty} S(f)e^{i2\pi f\tau}\,df \quad \text{and} \quad S(f) = \int_{-\infty}^{\infty} s(\tau)e^{-i2\pi f\tau}\,d\tau. \quad (135)$$

[4]   With a few changes, the approach we have outlined above can also be used to motivate a spectral representation theorem for complex-valued stationary processes. In particular, we let the C_l rv's for $l \neq 0$ in Equation (128a) be *any* collection of $2L$ uncorrelated complex-valued rv's with zero means (i.e., we drop the restriction that $C_{-l} = C_l^*$), and in Equation (128c) we use the C_l rv's with $l < 0$ to construct $Z(f)$ for $f < 0$ in an obvious way – this in turn is used to define $dZ(f)$ for $f < 0$ rather than stipulating that $dZ(f) = dZ^*(-f)$ for $f < 0$ as is done in Equation (129a). The statement of the spectral representation theorem itself is the same except that 'real-valued' is replaced by 'complex-valued.'

Because we no longer have the constraint $dZ(f) = dZ^*(-f)$ for a complex-valued process, it follows that the sdf (when it exists) need not be an even function; i.e., in general, we no longer have $S(-f) = S(f)$. An example of this is the so-called *'analytic'* series for a real-valued stationary process $\{X_t\}$, which, by definition, is constructed by taking the Hilbert transform $\mathcal{HT}\{X_t\}$ of the process and then forming the complex-valued process given by $X_t + i\mathcal{HT}\{X_t\}$. The sdf of this complex-valued process is proportional to the sdf of $\{X_t\}$ for $f \geq 0$, but is identically zero for $f < 0$. For details (including the definition of the Hilbert transform), see Section 10.16 here or Section 10–3 of Papoulis (1991).

4.2 Alternative Definitions for the Spectral Density Function

In the previous section we defined the integrated spectrum and the sdf (when it exists) in terms of the spectral representation of a stationary process. This approach allows us to relate the integrated spectrum and sdf directly to the representation of the process itself.

It is often stated in the literature that by *definition* the sdf is the Fourier transform of the acvs, i.e., that Equation (132c) for discrete parameter stationary processes (or (135) for continuous parameter processes) *defines* $S(\cdot)$ (see, for example, Chapter 10 of Papoulis, 1991). It is difficult to attach much meaning to $S(\cdot)$ from this definition alone. The usual approach is to appeal to the theory of linear filters in order to establish a physical meaning for the sdf (see Section 5.6).

There is yet a third way of defining – but only heuristically – the sdf by using the Fourier theory for deterministic sequences with finite energy (Section 3.7). Here, instead of just drawing comparisons between square summable sequences and stationary processes as we did in the previous section, we treat portions of realizations of a stationary process as a square summable sequence and apply some limiting arguments. We present this definition here because it is particularly informative (see Sections 4.7 and 4.8 of Priestley, 1981). Let $\{x_t\}$ be any realization of the discrete parameter stationary process $\{X_t\}$ with zero mean. If $\{x_t\}$ is a 'typical' realization,

$$\frac{1}{N} \sum_{t=1}^{N} x_t^2 \approx \sigma^2 \equiv E\{X_t^2\}$$

for large N. It is thus unlikely that $\{x_t\}$ has finite energy since we should have

$$\lim_{N\to\infty} \sum_{t=1}^{N} x_t^2 = \infty,$$

but it is plausible that it has finite power since we should have

$$\lim_{N\to\infty} \frac{1}{N} \sum_{t=1}^{N} x_t^2 = \sigma^2$$

(this argument says that the variance of a stationary process is in fact just its power). Let us define a new sequence $\{x_{t,N}\}$ by

$$
x_{t,N} = \begin{cases} x_t, & \text{for } 0 \leq t \leq N; \\ 0, & \text{otherwise.} \end{cases}
$$

Since $\{x_{t,N}\}$ is square summable (only a finite number of its terms are nonzero), the theory of Section 3.7 says that it has the following Fourier integral representation:

$$
x_{t,N} = \int_{-1/2}^{1/2} G_{p,N}(f) e^{i2\pi ft} \, df,
$$

where

$$
G_{p,N}(f) \equiv \sum_{t=-\infty}^{\infty} x_{t,N} e^{-i2\pi ft} = \sum_{t=1}^{N} x_t e^{-i2\pi ft}. \tag{137a}
$$

By Parseval's theorem

$$
\sum_{t=1}^{N} x_{t,N}^2 = \int_{-1/2}^{1/2} |G_{p,N}(f)|^2 \, df,
$$

it follows that $|G_{p,N}(f)|^2 \, df$ is equal to the contribution to the energy of $\{x_{t,N}\}$ from components with frequencies in a small interval about f. This suggests that, even if

$$
\lim_{N \to \infty} |G_{p,N}(f)|^2
$$

were to define a reasonable function of f, it would not be finitely integrable since the energy of $\{x_t\}$ is infinite. Previously, in such cases where energy is infinite, we have found that power is well defined. This suggests that, under suitable conditions, the function of f defined via a proper interpretation of

$$
\lim_{N \to \infty} \frac{|G_{p,N}(f)|^2}{N} \tag{137b}
$$

might be well behaved enough to be finitely integrable. If this is true, an integral of a proper interpretation of the above quantity over a small interval of length df that is centered about f will equal the contribution to the power (variance) of $\{x_t\}$ from components in that frequency range.

Now the quantity in (137b) obviously depends on $\{x_t\}$, the particular realization of the stationary process we have chosen. To construct a corresponding quantity that reflects the power per frequency properties of the entire stochastic process (and not just one realization), it is natural to average the values of $|G_{p,N}(f)|^2/N$ over all realizations. If we now *redefine* $G_{p,N}(f)$ to be an rv defined for each realization of $\{X_t\}$ by Equation (137a), we are led to the quantity

$$S(f) \equiv \lim_{N \to \infty} E\left\{\frac{|G_{p,N}(f)|^2}{N}\right\}. \tag{138}$$

When it exists, $S(\cdot)$ has the following interpretation: $S(f)\,df$ is the average contribution (over all realizations) to the power from components with frequencies in a small interval about f. The power (variance) is just

$$\int_{-1/2}^{1/2} S(f)\,df.$$

The function $S(\cdot)$ is called the *power spectral density function* of the process $\{X_t\}$. This definition for $S(\cdot)$ is obviously *not* rigorous without qualifications on $\{X_t\}$ to validate the manipulations that lead to (138). In fact, this cannot be done for all stationary processes; when it can be done, the function $S(\cdot)$ is identical to the sdf that we defined via the spectral representation theorem. (In Chapter 6 we will learn that the quantity inside the expectation operation in Equation (138) is known as the *periodogram*; hence this definition of the sdf is just the limit – as the sample size N gets large – of the expected value of the periodogram.)

4.3 Basic Properties of the Spectrum

In this section we summarize some of the properties of the integrated spectrum $S^{(I)}(\cdot)$ and the sdf $S(\cdot)$. When the latter exists, the integrated spectrum is related to it by

$$S^{(I)}(f) = \int_{-1/2}^{f} S(f')\,df'.$$

If we let σ^2 represent the variance of $\{X_t\}$, then it can be argued that $S^{(I)}(\cdot)$ and $S(\cdot)$ have the following properties:

$$0 \le S^{(I)}(f) \le \sigma^2; \quad S(f) \ge 0$$
$$S^{(I)}(-1/2) = 0 \text{ and } S^{(I)}(1/2) = \sigma^2; \quad \int_{-1/2}^{1/2} S(f)\,df = \sigma^2$$
$$f < f' \text{ implies } S^{(I)}(f) \le S^{(I)}(f'); \quad S(-f) = S(f).$$

Except for the scaling factor σ^2 and (possibly) the relatively unimportant convention of right continuity, $S^{(I)}(\cdot)$ has all the properties of a probability distribution function and for that reason is sometimes called a *spectral distribution function.* Likewise, except for the same scaling factor, $S(\cdot)$ has the properties of a probability density function. We have suggested above that $S(\cdot)$ will not always exist for certain stationary processes. Similarly, there are certain probability distribution functions (such as the binomial distribution) that do not have a corresponding probability density function. However, all random variables do possess a probability distribution function, so this function is considered more fundamental than a probability density function. The relationship is similar between $S^{(I)}(\cdot)$ and $S(\cdot)$: the former exists for all discrete parameter stationary processes, whereas the latter exists only in certain cases.

These ideas can be formalized in *Wold's theorem*: a necessary and sufficient condition for the sequence $\{s_\tau\}$ to be the acvs for some discrete parameter stationary process with variance σ^2 is that there exists a function $S^{(I)}(\cdot)$ – the integrated spectrum for the process – defined on $[-1/2, 1/2]$ such that $S^{(I)}(-1/2) = 0$; $S^{(I)}(1/2) = \sigma^2$; $S^{(I)}(\cdot)$ is nondecreasing on $[-1/2, 1/2]$; and for all integers τ

$$s_\tau = \int_{-1/2}^{1/2} e^{i2\pi f\tau} \, dS^{(I)}(f).$$

For example, a white noise process has an acvs defined by $s_0 = \sigma^2$ and $s_\tau = 0$ for $|\tau| \neq 0$ (see Section 2.6). Because the function defined by

$$S^{(I)}(f) \equiv (f + 1/2)\,\sigma^2$$

satisfies the requirements of Wold's theorem including that

$$s_\tau = \int_{-1/2}^{1/2} e^{i2\pi f\tau} \, dS^{(I)}(f) = \int_{-1/2}^{1/2} \sigma^2 e^{i2\pi f\tau} \, df$$

for all τ, we can conclude that $S^{(I)}(\cdot)$ is in fact the integrated spectrum for a white noise process. Since it is differentiable with derivative

$$S(f) = \frac{dS^{(I)}(f)}{df} = \sigma^2,$$

we see that the sdf for white noise is constant over f (analogous to the electromagnetic spectrum of white light, which has equal contributions of radiation from all visible frequencies of light).

As a second example, it follows from Equation (47a) that the acvs for a harmonic process can be written as

$$s_\tau = \sum_{l=1}^{L} D_l^2 \cos\left(2\pi f_l\tau\right)/2 = \sum_{l=1}^{L} \frac{D_l^2}{2} \left(e^{i2\pi f_l\tau} + e^{-i2\pi f_l\tau}\right)/2$$

$$= \sum_{l=-L}^{L} D_l^2 e^{i2\pi f_l\tau}/4,$$

where $D_{-l} = D_l$, $D_0 \equiv 0$ and $f_{-l} \equiv -f_l$. If we define $S^{(I)}(\cdot)$ to be a step function with jumps at $\pm f_l$ of size $D_l^2/4$, we have

$$s_\tau = \int_{-1/2}^{1/2} e^{i2\pi f \tau} \, dS^{(I)}(f),$$

so Wold's theorem tells us that $S^{(I)}(\cdot)$ must be the integrated spectrum for a harmonic process. An sdf does not exist for this process since $S^{(I)}(\cdot)$ is not differentiable.

Comments and Extensions to Section 4.3

[1] All of the above results hold, by and large, for continuous parameter stationary processes $\{X(t)\}$ if we replace each '$1/2$' by '∞' throughout. The main difference is in Wold's theorem, which is known as the *Wiener–Khintchine theorem* in the continuous parameter case: a necessary and sufficient condition for $s(\cdot)$ to be the acvf of some *stochastically continuous*, continuous parameter stationary process with variance σ^2 is that there exists a function $S^{(I)}(\cdot)$ – the integrated spectrum – such that $S^{(I)}(-\infty) = 0$; $S^{(I)}(\infty) = \sigma^2$; $S^{(I)}(\cdot)$ is nondecreasing; and

$$s(\tau) = \int_{-\infty}^{\infty} e^{i2\pi f \tau} \, dS^{(I)}(f)$$

for all τ.

[2] The corresponding theory for complex-valued stationary processes closely parallels the real-valued case. The only major difference is that $S(-f) \neq S(f)$ in general; i.e., the sdf for a complex-valued stationary process is not necessarily symmetric about the origin. The stated Wold and Wiener–Khintchine theorems in fact encompass both the real-valued and complex-valued stationary processes.

[3] Throughout this book all sdf's $S(\cdot)$ for real-valued processes are *two-sided* (or *double-sided*) in the sense that they are symmetric about the origin; i.e., $S(-f) = S(f)$. In some communities *one-sided* (or *single-sided*) sdf's are the convention. In terms of our sdf's, these are defined as

$$S_{\text{single}}(f) \equiv \begin{cases} 2S(f), & f \geq 0; \\ 0, & f < 0, \end{cases}$$

so that we have, for example,

$$s_0 = \int_0^{1/2} S_{\text{single}}(f) \, df \text{ in comparison to } s_0 = \int_{-1/2}^{1/2} S(f) \, df.$$

Because sdf's are usually plotted on a decibel scale, and since a factor of 2 translates approximately into a 3 dB shift, failure to pay attention to the distinction between one-sided and two-sided sdf's can lead to mysterious discrepancies of 3 dB!

4.4 Classification of Spectra

Since $S^{(I)}(\cdot)$ is quite similar to a probability distribution function, we have the following theorem, which is an analog to the *Lebesgue decomposition theorem* for distribution functions. Any integrated spectrum, $S^{(I)}(\cdot)$, can be written in the form

$$S^{(I)}(f) = S_1^{(I)}(f) + S_2^{(I)}(f) + S_3^{(I)}(f),$$

where the $S_i^I(\cdot)$ functions are all nonnegative and nondecreasing with $S_i^{(I)}(-1/2) = 0$ and are of the following types:

[1] $S_1^{(I)}(\cdot)$ is 'absolutely continuous' – this means that its derivative exists for almost all f and is equal almost everywhere to an sdf $S(\cdot)$ such that

$$S_1^{(I)}(f) = \int_{-1/2}^{f} S(f')\, df'.$$

[2] $S_2^{(I)}(\cdot)$ is a step function with jumps of size $\{\, p_l : l = 1, 2, \dots \,\}$ at the frequencies $\{\, f_l : l = 1, 2, \dots \,\}$.

[3] $S_3^{(I)}(\cdot)$ is a continuous singular function – although, by definition, a singular function has a derivative of zero almost everywhere, such a function can in fact be continuous and, rather surprisingly, strictly increasing (Chung, 1974, Section 1.3).

Now $S_3^{(I)}(\cdot)$ is pathological and of no practical use in spectral analysis, so let us assume that it is identically equal to 0 from now on. Thus the integrated spectrum is a combination of two 'pure' forms, which we now consider in more detail for real-valued processes.

[1] $S_1^{(I)}(f) \geq 0$, $S_2^{(I)}(f) = 0$. Here the integrated spectrum consists entirely of the absolutely continuous component, and the process $\{X_t\}$ is said to have a *purely continuous spectrum*. The sdf $S(\cdot)$ exists in this case for all f, and it can be shown (from a theorem called the Riemann–Lebesgue lemma) that the acvs $\{s_\tau\}$ of the process decays to 0 as $|\tau| \to \infty$. Most of the standard models for stationary processes belong to this class. These include white noise and ARMA processes.

[2] $S_1^{(I)}(f) = 0$, $S_2^{(I)}(f) \geq 0$. Here the integrated spectrum consists entirely of a step function, and the stationary process is said to have a *purely discrete spectrum* or a *line spectrum*. The harmonic process has this type of spectrum. It can be shown that the acvs for a process with a line spectrum never damps down to 0.

A process for which both $S_1^{(I)}(\cdot)$ and $S_2^{(I)}(\cdot)$ are not identically zero has a *discrete spectrum* (rather than being purely discrete) if $S(\cdot)$ is the sdf for a white noise process. We say it has a *mixed spectrum* if $S(\cdot)$

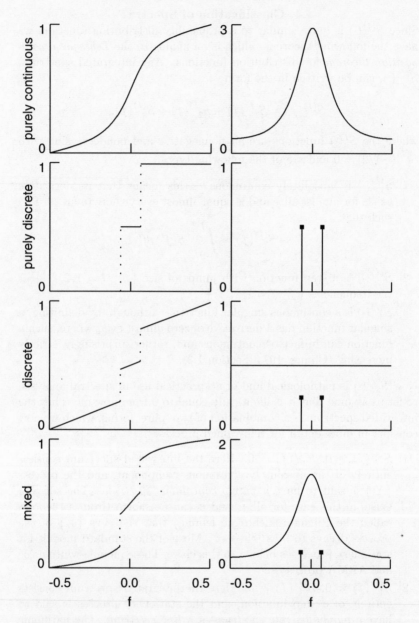

Figure 142. Examples of integrated spectra (left-hand plots) and corresponding spectral density functions (right-hand plots).

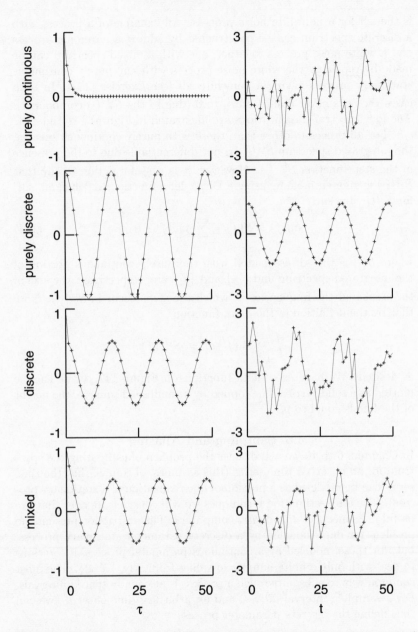

Figure 143. Autocovariance sequences from lag 0 to lag 50 (left-hand plots) and one realization of length 50 (right-hand plots) for each process with spectra shown in Figure 142.

is the sdf for a nonwhite noise process. An example of a process with a discrete spectrum can be constructed by adding a harmonic process and a white noise process together; one with a mixed spectrum can be made by replacing the white noise process with any purely continuous stationary process having a nonwhite sdf (see Exercise [2.1]). In both cases the acvs never damps down to 0 (due to the $S_2^{(I)}(\cdot)$ component). The four spectral classifications are illustrated in Figures 142 and 143.

For all processes other than those with purely continuous spectra, the integrated spectrum $S^{(I)}(\cdot)$ is not differentiable due to the presence of the step function $S_2^{(I)}(\cdot)$. However, it is a useful fiction to say that $S^{(I)}(\cdot)$ is differentiable by using a Dirac delta function to define an 'sdf' for $S^{(I)}(\cdot)$ by, say,

$$S_0(f) \equiv S(f) + \sum_l p_l \delta(f - f_l), \qquad (144)$$

where $S(\cdot)$ is the sdf associated with the purely continuous portion of the integrated spectrum and $\{p_l\}$ and $\{f_l\}$ are, respectively, the size of the steps and the frequencies at which the steps occur in $S_2^{(I)}(\cdot)$. Note that by the definition of the delta function

$$\int_{-1/2}^{f} S_0(f') \, df' = S^{(I)}(f)$$

as desired. We have used delta functions in Figure 142, where each is plotted as a solid vertical line topped by a small solid square – the height of the line is equal to p_l.

4.5 Sampling and Aliasing

In Chapters 6 to 10 we will discuss the problem of estimating the spectrum for an observed time series (this assumes, of course, that the time series can be modeled as a portion of one realization of a stationary process). The only estimation techniques we will consider are ones that are useful in connection with digital computers. This restriction presents no problem for data modeled by a discrete parameter stationary process, but, for those modeled by a continuous parameter process, it forces us to deal with only a finite number of values from, say, $\{X(t)\}$. The most common way to subsample such a process is at equally spaced intervals. For a sampling interval $\Delta t > 0$ and an arbitrary time offset t_0, we can thus define the discrete parameter process

$$X_t \equiv X(t_0 + t \, \Delta t), \qquad t = 0, \pm 1, \pm 2, \ldots.$$

If $\{X(t)\}$ is a stationary process with, say, sdf $S_{X(t)}(\cdot)$ and acvf $s(\tau)$, then $\{X_t\}$ is necessarily also a stationary process with, say, sdf $S_{X_t}(\cdot)$ and acvs $\{s_\tau\}$. Now,

$$s_\tau = \operatorname{cov}\{X_t, X_{t+\tau}\} = \operatorname{cov}\{X(t_0 + t \, \Delta t), X(t_0 + [t + \tau] \, \Delta t)\} = s(\tau \, \Delta t),$$

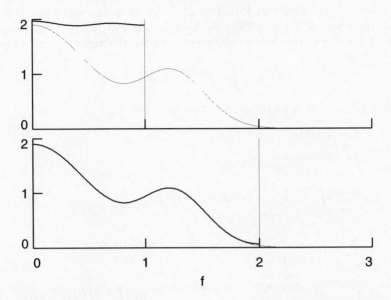

Figure 145. Spectral density functions for a continuous parameter process $\{X(t)\}$ (thin solid curves on the two plots) and for two discrete parameter processes sampled from $\{X(t)\}$ (thick solid curves). The sampling interval Δt associated with the top plot is $1/2$ (yielding a Nyquist frequency of 1, indicated by the thin vertical line), while Δt is $1/4$ for the bottom plot (the Nyquist frequency is 2). Because there is substantial power in the sdf of $\{X(t)\}$ for frequencies between 1 and 2 but little beyond $f = 2$, the sdf for the $\Delta t = 1/2$ sampled process in the upper plot does not agree well with the sdf for the continuous parameter process, whereas the bottom plot indicates good agreement when $\Delta t = 1/4$ – in fact, the agreement is so good that the thin curve for the sdf of $\{X(t)\}$ is only visible quite close to the Nyquist frequency 2.

so $\{s_\tau\}$ is sampled at intervals of Δt from $s(\cdot)$. Since $s(\cdot) \longleftrightarrow S_{X(t)}(\cdot)$ and $\{s_\tau\} \longleftrightarrow S_{X_t}(\cdot)$, the relationship between $S_{X(t)}(\cdot)$ and $S_{X_t}(\cdot)$ follows from Equation (98a):

$$S_{X_t}(f) = \sum_{k=-\infty}^{\infty} S_{X(t)}(f + k/\Delta t), \qquad |f| \le \frac{1}{2\,\Delta t} \equiv f_{(N)}. \qquad (145)$$

If $S_{X(t)}(\cdot)$ is essentially zero for $|f| > f_{(N)}$, we can expect good correspondence between $S_{X_t}(\cdot)$ and $S_{X(t)}(\cdot)$ for $|f| \le f_{(N)}$; if $S_{X(t)}(\cdot)$ is large for some $|f| > f_{(N)}$, the correspondence can be quite poor – an estimate of $S_{X_t}(\cdot)$ will then not tell us much about the sdf $S_{X(t)}(\cdot)$. An example of these two possibilities is shown in Figure 145 for the sdf

$$S_{X(t)}(f) = 1.9e^{-2f^2} + e^{-6(|f|-1.25)^2}.$$

We can also arrive at Equation (145) by using the spectral representations for $\{X(t)\}$ and $\{X_t\}$ in the following way. For convenience, let $E\{X(t)\} = 0$. By the spectral representation theorem

$$X(t) = \int_{-\infty}^{\infty} e^{i2\pi ft} \, dZ_{X(t)}(f),$$

where $\{Z_{X(t)}(f)\}$ is an orthogonal process. Now

$$
\begin{aligned}
X_t &= X(t_0 + t\,\Delta t) \\
&= \int_{-\infty}^{\infty} e^{i2\pi f(t_0 + t\,\Delta t)} \, dZ_{X(t)}(f) \\
&= \sum_{k=-\infty}^{\infty} \int_{(2k-1)/(2\,\Delta t)}^{(2k+1)/(2\,\Delta t)} e^{i2\pi ft\,\Delta t} e^{i2\pi f t_0} \, dZ_{X(t)}(f) \\
&= \sum_{k=-\infty}^{\infty} \int_{-1/(2\,\Delta t)}^{1/(2\,\Delta t)} e^{i2\pi(f+k/\Delta t)t\,\Delta t} e^{i2\pi(f+k/\Delta t)t_0} \, dZ_{X(t)}(f + k/\Delta t) \\
&= \int_{-f_{(N)}}^{f_{(N)}} e^{i2\pi ft\,\Delta t} \sum_{k=-\infty}^{\infty} e^{i2\pi(f+k/\Delta t)t_0} \, dZ_{X(t)}(f + k/\Delta t) \\
&= \int_{-f_{(N)}}^{f_{(N)}} e^{i2\pi ft\,\Delta t} \, dZ(f),
\end{aligned}
$$

where

$$dZ(f) \equiv \sum_{k=-\infty}^{\infty} e^{i2\pi(f+k/\Delta t)t_0} \, dZ_{X(t)}(f + k/\Delta t).$$

Note that, by the properties of $\{Z_{X(t)}(f)\}$, we have $E\{dZ(f)\} = 0$ for all $|f| \le f_{(N)}$. Suppose now that f' and f satisfy $-f_{(N)} \le f' < f \le f_{(N)}$. Then

$$
\begin{aligned}
\operatorname{cov}\{dZ(f'), dZ(f)\} &= \sum_{k=-\infty}^{\infty} \sum_{l=-\infty}^{\infty} e^{-i2\pi(f'+k/\Delta t)t_0} e^{i2\pi(f+l/\Delta t)t_0} \\
&\qquad \times E\{dZ_{X(t)}^*(f' + l/\Delta t) \, dZ_{X(t)}(f + k/\Delta t)\} \\
&= 0.
\end{aligned}
$$

We can thus take $\{dZ(f)\}$ to be the increments of the orthogonal process $\{Z(f)\}$ in the spectral representation for $\{X_t\}$ in Equation (130b). The integrated spectrum $S_{X_t}^{(I)}(\cdot)$ for $\{X_t\}$ is thus specified via

$$
\begin{aligned}
dS_{X_t}^{(I)}(f) &= E\{|dZ(f)|^2\} = \sum_{k=-\infty}^{\infty} E\{|dZ_{X(t)}(f + k/\Delta t)|^2\} \\
&= \sum_{k=-\infty}^{\infty} dS_{X(t)}^{(I)}(f + k/\Delta t), \qquad |f| \le f_{(N)},
\end{aligned}
$$

where $S_{X(t)}^{(I)}(\cdot)$ is the integrated spectrum for $\{X(t)\}$. The above reduces to Equation (145) when $S_{X(t)}^{(I)}(\cdot)$ is differentiable.

Comments and Extensions to Section 4.5

[1]   Although a sampling interval of $\Delta t = 1/4$ seems natural for the example considered in Figure 145, there are valid reasons for using both a larger and smaller value. For example, use of $\Delta t = 1/2$ yields a discrete parameter process $\{X_t\}$ that – as can be be seen from the thick solid curve in the top plot – is quite close to white noise. Sampling at this rate thus yields approximately uncorrelated variates, which might be helpful in applying certain statistical procedures. On the other hand, Kay (1981a) shows that, to approximately minimize the mean square error of the biased estimator of the acvs (see Section 6.2), we should oversample at *twice* the natural Nyquist rate, which for the example of Figure 145 would yield $\Delta t = 1/8$ approximately. Another rationale for oversampling is to eliminate potential spectral estimation problems near the Nyquist frequency – see Hardin (1986) for details.

4.6   Comparison of SDFs and ACVSs as Characterizations

One of the reasons spectral analysis is so widely used in the physical sciences is that the sdf for a purely continuous stationary process gives us an easy way of summarizing – and visualizing – the important second-order properties of the process. The spectral representation theorem provides the necessary theoretical basis for this statement. It is sometimes argued that, since the sdf and acvs contain the same amount of 'information' (in the sense that, if we know one of them, we can calculate the other), the two characterizations are equally informative. This is certainly true in some cases. A white noise process is equally well-characterized by either its sdf or acvs. In other cases, however, the sdf is clearly the preferred characterization, as we demonstrate with the following simple example.

Figure 148 shows two plots – the sdf for the AR(4) process defined in Equation (46a) and its corresponding acvs (for lags 0 to 50). If we examine the sdf, we know from the spectral representation theorem that realizations from a process with this sdf should show a tendency to oscillate close to two different frequencies, namely, 0.11 and 0.14 cycles per unit time (see the bottom plot of Figure 45). The power levels of these oscillations are within 3 dB of each other (a 3 dB difference corresponds approximately to a factor of 2). Note that it is very difficult to draw these same conclusions from an examination of the acvs plot alone.

Figure 149 shows two similar plots for a second stationary process. This has essentially the same characteristics as the one in Figure 148, with the exception of a small tendency to oscillate at a frequency close

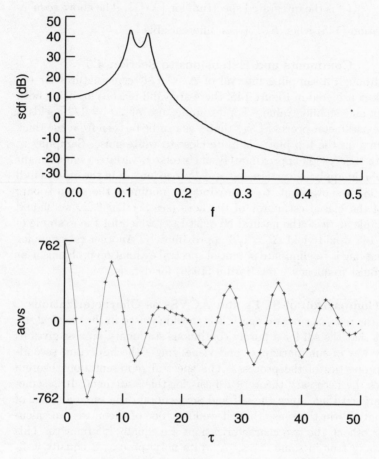

Figure 148. A spectral density function (top plot) and its corresponding autocovariance sequence (bottom plot).

to 0.35. The magnitude of this component is about 43 dB less than the two dominant components. Note, however, that the acvs for this process is virtually identical to that in Figure 148 – the effect of the additional component does not show up in any appreciable way in the acvs!

The sdf usually proves to be the more sensitive and interpretable diagnostic or exploratory tool. Also, since physical time series are usually recorded using mechanical or electronic devices, one must know the way in which the measuring equipment affects the recorded data. This knowledge is vital for understanding the limitations of the recorded data. The easiest way to express these limitations is in the frequency domain by means of the transfer (or frequency response) function for the measuring instrument (see Section 5.1). Thus the simplest way to

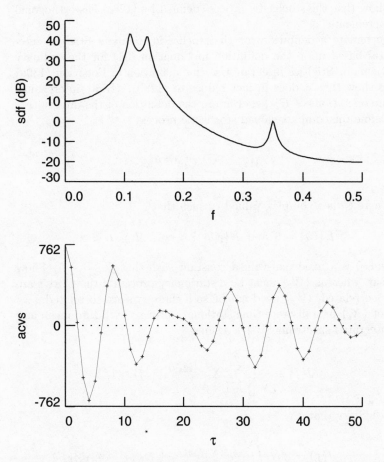

Figure 149. A second spectral density function (top plot) and its corresponding autocovariance sequence (bottom plot).

characterize recorded data with significant measurement noise is usually in the frequency domain.

4.7 Exercises

[4.1] Let ϕ be an rv with a uniform (i.e., rectangular) distribution over the interval $[-\pi, \pi]$. Consider the complex-valued rv's

$$C \equiv De^{i\phi}/2 \text{ and } C^* \equiv De^{-i\phi}/2,$$

where D is a real-valued constant. Show that C and C^* are uncorrelated. (The result of this exercise can be used to support the claim that the $2L+1$ rv's in the summation in Equation (128a) are uncorrelated.)

[4.2] Show that the stochastic process defined by (128c) has orthogonal increments.

[4.3] Formulate a definition for the stochastic Riemann–Stieltjes integral based upon the definition in Equation (34) for the ordinary Riemann–Stieltjes integral. Use this definition in Equation (130a) to show that it does in fact reduce to (128b). (Hint: interchange the operations of $E\{\cdot\}$ and lim in the definition of the integral.)

[4.4] Define the complex-valued stochastic process

$$W_t = \sum_{l=-L}^{L} C_l e^{i2\pi f_l t},$$

where C_l is a complex-valued rv such that

$$E\{C_l\} = 0 \text{ and } E\{|C_l|^2\} < \infty, \quad |l| \leq L < \infty,$$

and f_l is a fixed real-valued constant such that $f_{-l} = -f_l$. Show that, whereas $\{W_t\}$ must be a stationary process if the C_l rv's are uncorrelated, $\{W_t\}$ need not be so if these rv's are correlated.

[4.5] Let $\{X_t\}$ be a discrete time stationary process with zero mean and integrated spectrum $S^{(I)}(\cdot)$. Define

$$J(f) \equiv \frac{1}{\sqrt{N}} \sum_{t=1}^{N} X_t e^{-i2\pi f t}, \quad |f| \leq 1/2.$$

a) Show that

$$J(f) = \sqrt{N} \int_{-1/2}^{1/2} e^{i(N+1)\pi(f'-f)} \mathcal{D}_N(f'-f) \, dZ(f'),$$

where $\{Z(f)\}$ is an orthogonal process and $\mathcal{D}_N(\cdot)$ is Dirichlet's kernel (see Exercise [1.3c]).

b) Show that, if $S^{(I)}(\cdot)$ has derivative $S(\cdot)$, then

$$E\{|J(f)|^2\} = N \int_{-1/2}^{1/2} \mathcal{D}_N^2(f-f')S(f') \, df'.$$

[4.6] Consider the following sequence:

$$s_\tau = \begin{cases} 1, & \text{if } |\tau| \leq K; \\ 0, & \text{if } |\tau| > K, \end{cases}$$

where $K \geq 1$ is an integer. Is $\{s_\tau\}$ the acvs for some discrete parameter stationary process $\{X_t\}$ with sdf $S(\cdot)$?

[4.7] We know from Section 2.6 that the acvs of a qth order moving average process is

$$s_\tau = \begin{cases} \sigma_\epsilon^2 \sum_{j=0}^{q-|\tau|} \theta_{j,q}\theta_{j+|\tau|,q}, & |\tau| \le q; \\ 0, & |\tau| > q. \end{cases}$$

a) Show that, if we define $\theta_{j,q} = 0$ for $j < 0$ and $j > q$, we can rewrite the above as

$$s_\tau = \sigma_\epsilon^2 \sum_{j=-\infty}^{\infty} \theta_{j,q}\theta_{j+\tau,q}, \qquad \tau = 0, \pm 1, \pm 2, \dots.$$

b) Argue that an sdf exists for this process and that it is given by

$$S(f) = \sigma_\epsilon^2 \sum_{\tau=-q}^{q} \sum_{j=-\infty}^{\infty} \theta_{j,q}\theta_{j+\tau,q} e^{-i2\pi f\tau}.$$

c) State and prove a version of the autocorrelation theorem appropriate for square summable sequences (see Equation (85b)). Show how this theorem can be used to rewrite the above as

$$S(f) = \sigma_\epsilon^2 \left| \sum_{j=0}^{q} \theta_{j,q} e^{-i2\pi fj} \right|^2.$$

[4.8] Consider a *continuous* parameter complex-valued stationary process $\{X(t)\}$ with the following 'Gaussian-shaped' acvf:

$$s(\tau) = Ae^{-B\tau^2 + i2\pi f_0\tau}, \qquad |\tau| < \infty,$$

where $A > 0$, $B > 0$ and f_0 are all real-valued constants. (For appropriate A, B and f_0, this process has been found to be an adequate model for certain unwanted echoes in radar – 'clutter' noise – and for Doppler measurements of blood flow; see, for example, Jacovitti and Scarano, 1987.) Show that the corresponding sdf for this process is given by

$$S(f) = A\left(\frac{\pi}{B}\right)^{1/2} e^{-\pi^2(f-f_0)^2/B}, \qquad |f| < \infty.$$

(This sdf is a rescaled version of the Gaussian (or normal) probability density function with mean f_0 and is thus sometimes called a *Gaussian sdf*. Note that the term 'Gaussian' here refers to the shape of the sdf and *not* to the probability distribution of the rv $X(t)$, which need not have a Gaussian distribution.)

[4.9] Consider a *continuous* parameter real-valued stationary process
$\{X(t)\}$ with the following sdf:

$$S(f) = \frac{\sigma^4}{\sigma^4 + (\pi f)^2} \ \text{ for } -\infty < f < \infty,$$

where $\sigma^2 > 0$ is the process variance (this is known in the literature
as a *Lorenzian* sdf). Show that the acvf for $\{X(t)\}$ is given by

$$s(\tau) = \sigma^2 e^{-2\sigma^2 |\tau|} \ \text{ for } -\infty < \tau < \infty.$$

(We note in passing that, if we sample $\{X(t)\}$ at a sampling interval
of Δt, the resulting discrete parameter stationary process $\{X_t\}$ is
a first-order autoregressive process (AR(1)) with variance σ^2 and
AR coefficient $\phi_{1,1} = \exp\left(-2\sigma^2 \, \Delta t\right)$ – see Chapter 9 for details.)

5

Linear Time-Invariant Filters

5.0 Introduction

In Section 3.6 we discussed the convolution of two $L^2(-\infty, \infty)$ functions $g(\cdot)$ and $h(\cdot)$ and found that the Fourier transform of their convolution,

$$g * h(t) \equiv \int_{-\infty}^{\infty} g(u)h(t - u)\, du, \qquad (153)$$

was simply the product of $G(\cdot)$ and $H(\cdot)$, our notation for the Fourier transforms of $g(\cdot)$ and $h(\cdot)$. In Section 3.2 we argued that the quantity $|H(f)|^2$ defines an energy spectral density function (sdf) for $h(\cdot)$. Let us now regard the convolution (153) as a manipulation of a signal $h(\cdot)$ that produces a new signal $g * h(\cdot)$. The original signal $h(\cdot)$ has a distribution of energy with respect to frequency given by its energy sdf, and the new signal has a distribution of energy given by

$$|G(f)H(f)|^2 = |G(f)|^2|H(f)|^2.$$

Thus the energy sdf of the new signal $g * h(\cdot)$ is multiplicatively related to the energy sdf of the original signal $h(\cdot)$ via the Fourier transform of the manipulator $g(\cdot)$.

In this chapter we investigate and extend these ideas through the theory of linear time-invariant (LTI) filters. Our goal is to formalize ways of relating the spectra associated with inputs to an LTI filter to the spectra associated with outputs from the filter. We shall see that the feature that makes LTI filters so easy to work with is our ability to represent various signals and stationary processes as linear combinations of complex exponentials.

LTI filters play a key role in the spectral analysis of stationary processes. As we shall see, an LTI filter is nothing more than a linear time-invariant transformation of some function of time. If the function of time is a realization of a stationary process, an LTI filter transforms the process into a new process that, under rather mild conditions, is also stationary. An important feature of LTI filters is that, given the integrated spectrum of the original process, there is an easy way to determine the integrated spectrum of the new process. The theory of LTI filters thus gives us a powerful means for determining the sdf of a wide class of stationary processes.

5.1 Basic Theory of LTI Analog Filters

Let us define a continuous parameter *filter* L as a mapping, or association, between an input function $x(\cdot)$ and an output function $y(\cdot)$. Symbolically we write

$$L\{x(\cdot)\} = y(\cdot). \tag{154}$$

Since we regard both $x(\cdot)$ and $y(\cdot)$ as functions of time t, where t is a real-valued parameter such that $-\infty < t < \infty$, the qualifier 'continuous parameter' is appropriate – in the engineering literature a continuous parameter filter is often called an *analog* filter. In mathematics L is known as a *transformation* or *operator*. It is important to realize that a filter is *not* just an ordinary function: for example, a real-valued function that is defined on the real line associates a *point* on the real line with another point on the real line, whereas a filter associates a *function* from some – so far unidentified – abstract space of functions with another function in that same space.

For the remainder of this section we need the following special notation. If α is a real or complex-valued scalar and $x(\cdot)$ is a function, the notation $\alpha x(\cdot)$ refers to the function defined by $\alpha x(t)$ for $-\infty < t < \infty$. If $x_1(\cdot)$ and $x_2(\cdot)$ are two functions, then $x_1(\cdot) + x_2(\cdot)$ denotes the function defined by $x_1(t) + x_2(t)$. Finally, if τ is a real-valued scalar and $x(\cdot)$ is a function, then $x(\cdot; \tau)$ denotes the function whose value at time t is given by $x(t + \tau)$; i.e.,

$$x(t; \tau) = x(t + \tau), \quad -\infty < t < \infty.$$

Thus the filter defined by

$$L\{x(\cdot)\} = x(\cdot; \tau)$$

is a shift filter which takes as input a certain function $x(\cdot)$ and returns a function that is defined by shifting the original function by τ units to the left. For example, if the input to the shift filter is the function defined by $x(t) = \sin(t)$ and $\tau = \pi/2$, the output is the function defined by $x(t; \pi/2) = \cos(t)$.

An analog filter L is called a *linear time-invariant* (LTI) analog filter if it has the following three properties:

[1] Scale preservation:

$$L\{\alpha x(\cdot)\} = \alpha L\{x(\cdot)\};$$

i.e., multiplication of the input by the factor α results in multiplication of the output by α also.

[2] Superposition:

$$L\{x_1(\cdot) + x_2(\cdot)\} = L\{x_1(\cdot)\} + L\{x_2(\cdot)\};$$

i.e., if we define a new function by adding $x_1(\cdot)$ and $x_2(\cdot)$ and if we use it as input to the LTI filter L, the output from L is simply that function defined by adding together the outputs of L when $x_1(\cdot)$ and $x_2(\cdot)$ are separately used as input to L.

[3] Time invariance:

$$\text{if } L\{x(\cdot)\} = y(\cdot), \text{ then } L\{x(\cdot;\tau)\} = y(\cdot;\tau); \qquad (155a)$$

i.e., if two inputs to the LTI filter are the same except for a shift in time, the outputs will also be the same except for the same shift in time.

Properties [1] and [2] together express the linearity of L:

$$L\{\alpha x_1(\cdot) + \beta x_2(\cdot)\} = \alpha L\{x_1(\cdot)\} + \beta L\{x_2(\cdot)\}.$$

By induction, it follows that

$$L\left\{\sum_{j=1}^{N} \alpha_j x_j(\cdot)\right\} = \sum_{j=1}^{N} \alpha_j L\{x_j(\cdot)\}. \qquad (155b)$$

With suitable conditions the above holds 'in the limit' so that Equation (155b) is valid when the finite summation is replaced by an infinite summation.

Now suppose we take an $L^2(-\infty, \infty)$ deterministic function (i.e., one that is square integrable) or a realization of a stationary process and use it as input to some LTI filter. As we shall see, under mild conditions, the output from this LTI filter is also, respectively, an $L^2(-\infty, \infty)$ function or a realization of a different stationary process (defined on a realization by realization basis by the LTI filter). It is interesting that Equations (155a) and (155b) are all we need to derive the relationship between the spectra of the input and output to the LTI filter. Here are the details (this material is based on Koopmans, 1974).

Let the input into the LTI filter be the complex exponential

$$\mathcal{E}_f(t) \equiv e^{i2\pi ft}, \quad -\infty < t < \infty,$$

where f is some fixed frequency, and let $y_f(\cdot)$ denote the output function; i.e.,

$$y_f(\cdot) = L\{\mathcal{E}_f(\cdot)\}.$$

The rationale for this approach is simple. If we regard L as a 'black box' which accepts an input and transforms it somehow, and if we want to learn something about this 'black box,' it makes sense to feed it simple test functions such as complex exponentials to learn how it reacts. The complex exponentials are of particular interest to us because all the representations we have examined for deterministic functions and stationary processes involve linear combinations of one kind or another of complex exponentials. Now, by the properties [1] and [3] of an LTI filter, we have for all τ

$$y_f(\cdot;\tau) \overset{[3]}{=} L\{\mathcal{E}_f(\cdot;\tau)\} = L\{e^{i2\pi f\tau}\mathcal{E}_f(\cdot)\} \overset{[1]}{=} e^{i2\pi f\tau}L\{\mathcal{E}_f(\cdot)\} = e^{i2\pi f\tau}y_f(\cdot).$$

This implies that

$$y_f(t;\tau) = y_f(t+\tau) = e^{i2\pi f\tau}y_f(t) \text{ for all } t \text{ and } \tau.$$

In particular, for $t = 0$ we obtain

$$y_f(\tau) = e^{i2\pi f\tau}y_f(0).$$

Since τ can assume any real value, the above implies that

$$y_f(t) = e^{i2\pi ft}y_f(0), \text{ for all } t; \text{ i.e., } y_f(\cdot) = y_f(0)\mathcal{E}_f(\cdot).$$

Thus, when the function $\mathcal{E}_f(\cdot)$ is used as input to the LTI filter L, the output is the same function multiplied by some constant, $y_f(0)$, which is independent of time but will depend in general on the frequency f. To keep track of this frequency dependence, define

$$G(f) \equiv y_f(0).$$

We thus have shown that

$$L\{\mathcal{E}_f(\cdot)\} = G(f)\mathcal{E}_f(\cdot). \qquad \qquad (156)$$

In mathematical terms, the complex exponentials (regarded as functions of t with f fixed) would be called the *eigenfunctions* for the LTI filter L, and each $G(f)$ would be called an associated *eigenvalue*. The above relationship is of fundamental importance: if the input to an LTI filter is a complex exponential, the output is also a complex exponential with the exact same frequency multiplied by $G(f)$. Why is this important? Suppose that $x(\cdot)$ can be represented by

$$x(t) = \sum_f \alpha_f e^{i2\pi ft}; \text{ i.e., } x(\cdot) = \sum_f \alpha_f \mathcal{E}_f(\cdot). \tag{157a}$$

Then Equations (155b) and (156) tell us that the output from the LTI filter is just

$$y(\cdot) = L\{x(\cdot)\} = \sum_f \alpha_f G(f)\mathcal{E}_f(\cdot);$$

i.e., $y(\cdot)$ can be represented by

$$y(t) = \sum_f \alpha_f G(f) e^{i2\pi ft}.$$

In particular, the spectral representation theorem for continuous parameter stationary processes says that, if $x(\cdot)$ is a realization of a stationary process $\{X(t)\}$ with zero mean, then

$$x(t) = \int_{-\infty}^{\infty} e^{i2\pi ft} \, dZ_x(f)$$

for all t, where $Z_x(\cdot)$ is the corresponding realization of the orthogonal process $\{Z_X(f)\}$. As a function of t, we can regard the above equation as a special case of Equation (157a) if we equate α_f with the increments $dZ_x(f)$ (this step requires some justification because we have passed from a finite to an infinite summation). Hence we have

$$y(t) = \int_{-\infty}^{\infty} e^{i2\pi ft} G(f) \, dZ_x(f) \text{ and } Y(t) = \int_{-\infty}^{\infty} e^{i2\pi ft} G(f) \, dZ_X(f).$$

It follows that, under a mild condition, $\{Y(t)\}$ is a stationary process with spectral representation given by

$$Y(t) = \int_{-\infty}^{\infty} e^{i2\pi ft} \, dZ_Y(f), \text{ where } dZ_Y(f) \equiv G(f) \, dZ_X(f).$$

If we denote the integrated spectra for $\{X(t)\}$ and $\{Y(t)\}$ by $S_X^{(I)}(\cdot)$ and $S_Y^{(I)}(\cdot)$, respectively, we have

$$\begin{aligned} dS_Y^{(I)}(f) = E|dZ_Y(f)|^2 &= |G(f)|^2 E|dZ_X(f)|^2 \\ &= |G(f)|^2 \, dS_X^{(I)}(f). \end{aligned} \tag{157b}$$

When sdf's exist for all f,

$$dS_Y^{(I)}(f) = S_Y(f)\,df \text{ and } dS_X^{(I)}(f) = S_X(f)\,df,$$

so Equation (157b) reduces to

$$S_Y(f) = |G(f)|^2 S_X(f).$$

The function $G(\cdot)$ is called the *transfer function* (or *frequency response function*) of the LTI filter L. The transfer function relates the integrated spectra of the input to – and the output from – an LTI filter in a very simple fashion. In particular, the relationship is independent of time, and power is not transferred from one frequency to another. These facts show the importance of LTI filters within spectral analysis, and, conversely, the usefulness of spectral analysis when LTI transformations are applied to a time series.

In general, $G(\cdot)$ is a complex-valued function, so we can write

$$G(f) = |G(f)|e^{i\theta(f)},$$

where $|G(\cdot)|$ and $\theta(\cdot)$ are called, respectively, the *gain function* and the *phase function* of the LTI filter. Note that $\theta(f) = \arg(G(f))$. The quantities

$$-\frac{1}{2\pi} \cdot \frac{d\theta(f)}{df} \text{ and } -\theta(f)$$

define the *group delay* and the *phase shift function*.

Equation (156) gives us a simple rule for computing the transfer function of an LTI filter: if we apply the function $\mathcal{E}_f(\cdot)$ as input to an LTI filter, the coefficient of $\exp(i2\pi ft)$ in the resulting output from the filter *defines* $G(f)$, the transfer function at frequency f.

The results of this section that pertain to stationary processes can be summarized by the following theorem, which we call the *LTI analog filtering theorem*: if $\{X(t)\}$ is a continuous parameter stationary process with zero mean and integrated spectrum $S_X^{(I)}(\cdot)$ and if L is an LTI analog filter with transfer function $G(\cdot)$ such that

$$\int_{-\infty}^{\infty} |G(f)|^2\, dS_X^{(I)}(f) < \infty,$$

then $\{Y(t)\} \equiv L\{\{X(t)\}\}$ is a continuous parameter stationary process with zero mean and integrated spectrum $S_Y^{(I)}(\cdot)$ such that

$$dS_Y^{(I)}(f) = |G(f)|^2\, dS_X^{(I)}(f).$$

The proof of this theorem is Exercise [5.1].

In the remainder of this book we prefer to use the informal notation

$$L\{x(t)\} = y(t) \text{ instead of } L\{x(\cdot)\} = y(\cdot)$$

to define an LTI analog filter. This is done merely for notational convenience – it allows us to define functions implicitly on a point by point basis without having to come up with an explicit notation for them (as we did for $\mathcal{E}_f(\cdot)$). In all cases our informal notation means that the LTI filter L maps the function defined on a point by point basis by $x(t)$ to the function defined on a point by point basis by $y(t)$. For example, Equation (156) in this informal notation is

$$L\{e^{i2\pi ft}\} = G(f)e^{i2\pi ft}.$$

Comments and Extensions to Section 5.1

[1] A general class of linear transformations of $x(\cdot)$ is given by

$$y(t) = \int_{-\infty}^{\infty} K(t,t')x(t')\,dt'$$

for some $K(\cdot,\cdot)$ (Bracewell, 1978, Chapter 9; Champeney, 1987). Under assumption (155a) of time invariance, we have

$$y(t-\tau) = \int_{-\infty}^{\infty} K(t,t')x(t'-\tau)\,dt'.$$

If we make the change of variable $t'' = t' - \tau$, we get

$$y(t-\tau) = \int_{-\infty}^{\infty} K(t,t''+\tau)x(t'')\,dt'',$$

and if we replace $t - \tau$ by t and relabel t'' as t', we have

$$y(t) = \int_{-\infty}^{\infty} K(t+\tau,t'+\tau)x(t')\,dt'.$$

A comparison of the two expressions above for $y(t)$ shows us that time invariance implies $K(t,t') = K(t+\tau,t'+\tau)$ for all t and τ. If we let $\tau = -t'$, we have $K(t,t') = K(t-t',0)$ for all t and t' so that we can write, say, $K(t,t') = g(t-t')$, a function purely of $t-t'$. Hence $y(\cdot)$ can be expressed as a convolution:

$$y(t) = \int_{-\infty}^{\infty} g(t-t')x(t')\,dt'.$$

Linearity plus time invariance thus *implies* the convolution relationship. However, to be able to include the trivial case $y(t) = x(t)$, we must allow $g(\cdot) = \delta(\cdot)$, the Dirac delta function. If we want to exclude generalized functions like the delta function from the convolution expression, then 'linearity plus time invariance' gives a set of filters that are larger than those that can be expressed as a convolution.

[2]   The theory of LTI filters also justifies the form in which we have chosen to represent signals and stationary processes, namely, as linear combinations of complex exponentials. For example, suppose $g_p(\cdot)$ is a member of the $L^2(-T/2, T/2)$ class of functions, i.e., the class of all complex-valued continuous time functions such that

$$\int_{-T/2}^{T/2} |g_p(t)|^2 \, dt < \infty.$$

From the results of Section 3.1, we can represent $g_p(\cdot)$ over the interval $[-T/2, T/2]$ as

$$g_p(t) = \sum_{n=-\infty}^{\infty} G_n e^{i2\pi f_n t},$$

where

$$f_n \equiv \frac{n}{T} \text{ and } G_n \equiv \frac{1}{T} \int_{-T/2}^{T/2} g_p(t) e^{-i2\pi f_n t} \, dt.$$

Since we can easily extend $g_p(\cdot)$ to the whole real axis as a periodic function with period T, Parseval's theorem

$$\int_{-T/2}^{T/2} |g_p(t)|^2 \, dt = T \sum_{n=-\infty}^{\infty} |G_n|^2$$

allows us to use $|G_n|^2$ to define a discrete power spectrum.

Let us now define

$$\phi_n(t) \equiv e^{i2\pi f_n t} / \sqrt{T}, \quad n = 0, \pm 1, \pm 2, \dots. \tag{160a}$$

We say that the collection of functions $\phi_n(\cdot)$ forms an *orthonormal basis* for $L^2(-T/2, T/2)$ because, first,

$$\int_{-T/2}^{T/2} \phi_m(t) \phi_n^*(t) \, dt = \begin{cases} 0, & \text{if } m \neq n; \\ 1, & \text{if } m = n, \end{cases}$$

and, second, any function $g_p(\cdot)$ in $L^2(-T/2, T/2)$ can be written as

$$g_p(t) = \sum_{n=-\infty}^{\infty} \mathcal{G}_n \phi_n(t), \text{ where } \mathcal{G}_n \equiv \int_{-T/2}^{T/2} g_p(t) \phi_n^*(t) \, dt. \tag{160b}$$

In this new notation Parseval's theorem becomes

$$\int_{-T/2}^{T/2} |g_p(t)|^2 \, dt = \sum_{n=-\infty}^{\infty} |\mathcal{G}_n|^2, \qquad (161)$$

which allows us to use $|\mathcal{G}_n|^2$ (divided by T) to define a discrete power spectrum.

The orthonormal basis for $L^2(-T/2, T/2)$ that we have just defined is not unique. There are many other orthonormal bases that we could define such that both (160b) and (161) would still hold and $|\mathcal{G}_n|^2$ would thus define a discrete power spectrum with respect to this new basis. The theory of LTI filters tells us there is something special about the orthonormal basis defined by (160a). Thus, if we define an LTI filter L as

$$L\{g_p(t)\} = g_p * h_p(t) \equiv \frac{1}{T} \int_{-T/2}^{T/2} g_p(u) h_p(t-u) \, du,$$

we know that the discrete power spectrum for the output $g_p * h_p(\cdot)$ is

$$\frac{1}{T^2} |\mathcal{G}_n|^2 |\mathcal{H}_n|^2, \quad \text{where} \quad \mathcal{H}_n \equiv \int_{-T/2}^{T/2} h_p(t) \phi_n^*(t) \, dt$$

(assuming that $h_p(\cdot)$ is a periodic $L^2(-T/2, T/2)$ function). Such a simple relationship is unique to the basis defined by (160a); for any other nontrivially different basis, LTI filtering results in a more complicated relationship between the discrete power spectra of the input and the output.

For example, a nonsinusoidal orthonormal basis that has proved useful in practical applications can be formed from the *Walsh functions* (Beauchamp, 1984). These functions assume only the values ± 1. The first nine of one version of these functions – shifted to the interval $[-T/2, T/2]$ – are shown in Figure 162. They can be labeled from top to bottom as $W_n(\cdot)$ for $n = 0, 1, \ldots, 8$ – see Beauchamp (1984) for details on how to generate these and higher order Walsh functions. For Walsh functions, the concept corresponding to frequency is *sequency*, which is defined as half the number of zero crossings over one period T. This is just the number of transitions from ± 1 to ∓ 1 with the convention that the endpoints count as one transition if

$$\lim_{\substack{\delta \to 0 \\ \delta > 0}} W_n\left(-(T - \delta)/2\right) \neq W_n\left(T/2\right).$$

With this definition the sequency of $W_n(\cdot)$ is just $\lfloor (n+1)/2 \rfloor$, so each nonzero sequency is associated with exactly two Walsh functions. Figure 162 shows all the Walsh functions for sequencies 0 to 4. By way

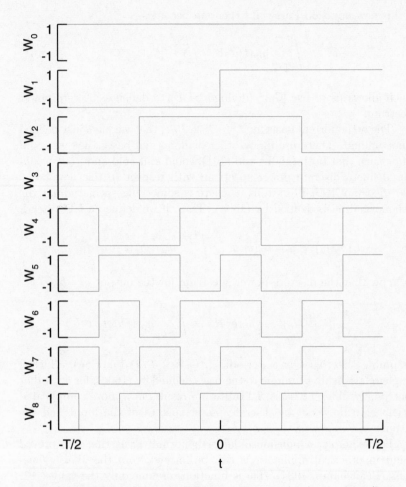

Figure 162. Walsh functions $W_n(\cdot)$ from $n = 0$ (top) to $n = 8$ (bottom). This figure was adapted from Figures 1.4 and 1.7 of Beauchamp (1984).

of comparison, note that each nonzero frequency f_m is associated with exactly two orthogonal sinusoids, namely, $\cos(2\pi f_m t)$ and $\sin(2\pi f_m t)$, and that both these sinusoids have sequency m.

If we now redefine

$$\phi_n(t) \equiv \begin{cases} W_{2n}(t)/\sqrt{T}, & \text{if } n \geq 0; \\ W_{2|n|-1}(t)/\sqrt{T}, & \text{if } n < 0, \end{cases}$$

it can be shown that Equations (160b) and (161) still hold. We can thus define a Walsh discrete power spectrum via $|\mathcal{G}_n|^2$. Since $\phi_n(t)$ and $\phi_{-n}(t)$ have sequency n, the $\pm n$ components of this spectrum tell us the contribution to the power from Walsh functions of sequency $|n|$.

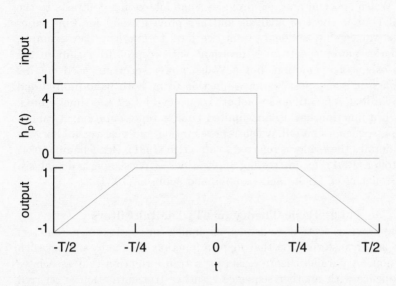

Figure 163. Filtering of $W_2(\cdot)$. When $W_2(\cdot)$ (top plot) – which has sequency 1 – is convolved with a periodic rectangular smoothing kernel $h_p(\cdot)$ (middle plot), the result is a periodic function (bottom plot) that cannot be expressed in terms of Walsh functions of sequency 1.

Now suppose that $g_p(\cdot) = W_2(\cdot)$. Since our periodic function is just the second Walsh function, its Walsh discrete power spectrum is concentrated entirely at sequency 1. If we pass this function through the LTI filter defined by, say,

$$L\{g_p(t)\} = \frac{1}{T} \int_{-T/2}^{T/2} g_p(u) h_p(t - u) \, du \equiv g_p * h_p(t)$$

(see Section 5.3 for details on this type of filter), where

$$h_p(t) \equiv \begin{cases} 4, & \text{if } |t| \leq T/8; \\ 0, & \text{if } T/8 < |t| \leq T/2; \\ h_p(t \pm T), & \text{if } |t| > T/2 \end{cases}$$

(see Figure 163), it is obvious that the Walsh discrete power spectrum for the output from this filter is no longer concentrated just at sequency 1; i.e., it cannot be expressed as just a linear combination of $W_1(\cdot)$ and $W_2(\cdot)$ (see Figure 162). In fact it can be shown that the power is now distributed over an infinite range of sequencies. By contrast, if we pass a band-limited function through an LTI filter, then the output must also be band-limited – the (Fourier) discrete power spectrum for the output from any LTI filter can have nonzero contributions only at those frequencies that are nonzero in the discrete power spectrum of the input.

Walsh spectral analysis provides some interesting contrasts to the usual Fourier spectral analysis and has proven useful for time series whose values shift suddenly from one level to another. For example, a Fourier power spectrum is invariant with respect to a shift in the time origin of a function, but a Walsh power spectrum need not be; although it is not possible for a function to be both band-limited and time-limited, a function can be both sequency-limited and time-limited; and, if a function has a band-limited Fourier representation, it cannot have a sequency-limited Walsh representation and vice versa. For further details, the reader is referred to Morettin (1981), Beauchamp (1984) or Stoffer (1991); for interesting applications of Walsh spectral analysis, see Stoffer *et al.* (1988) and Lanning and Johnson (1983).

5.2 Basic Theory of LTI Digital Filters

In the previous section we defined an analog (or continuous parameter) filter as a transformation that maps a function of time to another such function. A parallel theory exists for a transformation that associates a sequence with another sequence – such a transformation is referred to as a discrete parameter filter or *digital* filter. The theory of linear time-invariant digital filters closely parallels that of LTI analog filters, so we only sketch the key points for sequences in this section.

A digital filter L that transforms an input sequence $\{x_t\}$ into an output sequence $\{y_t\}$ is called a linear time-invariant digital filter if it has the following three properties:

[1] Scale preservation:

$$L\{\{\alpha x_t\}\} = \alpha L\{\{x_t\}\}.$$

[2] Superposition:

$$L\{\{x_{t,1} + x_{t,2}\}\} = L\{\{x_{t,1}\}\} + L\{\{x_{t,2}\}\}.$$

[3] Time invariance:

$$\text{if } L\{\{x_t\}\} = \{y_t\}, \text{ then } L\{\{x_{t+\tau}\}\} = \{y_{t+\tau}\},$$

where τ is integer-valued and the notation $\{x_{t+\tau}\}$ refers to the sequence whose tth element is $x_{t+\tau}$.

By using $\{\exp(i2\pi ft)\}$ – a sequence with tth element $\exp(i2\pi ft)$ – as the input to an LTI digital filter, it follows from the same arguments as before that

$$L\{\{e^{i2\pi ft}\}\} = G(f)\{e^{i2\pi ft}\},$$

where $G(\cdot)$ defines the transfer function (cf. Equation (156)). The result corresponding to the LTI analog filtering theorem is the following

LTI digital filtering theorem: if $\{X_t\}$ is a discrete parameter stationary process with zero mean and integrated spectrum $S_X^{(I)}(\cdot)$ and if L is an LTI digital filter with transfer function $G(\cdot)$ such that

$$\int_{-1/2}^{1/2} |G(f)|^2 \, dS_X^{(I)}(f) < \infty,$$

then $\{Y_t\} \equiv L\{\{X_t\}\}$ is a discrete parameter stationary process with zero mean and integrated spectrum $S_Y^{(I)}(\cdot)$ such that

$$dS_Y^{(I)}(f) = |G(f)|^2 \, dS_X^{(I)}(f).$$

As before, we will find it convenient to use the succinct – but formally incorrect – notation

$$L\{x_t\} = y_t \ \text{ as shorthand for } \ L\{\{x_t\}\} = \{y_t\}.$$

5.3 Convolution as an LTI Filter

We consider in this section some details about an LTI analog filter L of the following form:

$$L\{X(t)\} = \int_{-\infty}^{\infty} g(u)X(t-u)\,du \equiv Y(t) \tag{165}$$

(that this indeed satisfies the properties of an LTI filter is the subject of Exercise [5.2a]). Here the input to the LTI filter is a stationary process $\{X(t)\}$ with zero mean that has a purely continuous spectrum with associated sdf $S_X(\cdot)$. The output is the stochastic process $\{Y(t)\}$ that results from convolving $\{X(t)\}$ with the real-valued deterministic function $g(\cdot)$. The process $\{Y(t)\}$ is thus formed from an infinite linear combination of members of the process $\{X(t)\}$. The characteristics of the LTI filter are entirely determined by $g(\cdot)$, which – in the analog (continuous parameter) case – is called the *impulse response function* for the following reason. Suppose we let the input to the LTI analog filter in Equation (165) be $\delta(\cdot)$, the Dirac delta function with an infinite spike at the origin. By the properties of that function, we have

$$L\{\delta(t)\} = \int_{-\infty}^{\infty} g(u)\delta(t-u)\,du = g(t).$$

Hence, the input of an 'impulse' (the delta function) into the LTI filter yields an output that defines the impulse response function.

To find the transfer function for L in Equation (165), we apply the function of t defined by $\exp(i2\pi ft)$ as input to the LTI filter to get

$$L\{e^{i2\pi ft}\} = \int_{-\infty}^{\infty} g(u)e^{i2\pi f(t-u)}\,du = e^{i2\pi ft}\int_{-\infty}^{\infty} g(u)e^{-i2\pi fu}\,du.$$

Hence the transfer function

$$G(f) \equiv \int_{-\infty}^{\infty} g(u)e^{-i2\pi fu}\,du$$

is just the Fourier transform of the impulse response function $g(\cdot)$. Since $S_Y(\cdot)$ and $S_X(\cdot)$ are related by

$$S_Y(f) = |G(f)|^2 S_X(f),$$

we must have

$$\int_{-\infty}^{\infty} |G(f)|^2 S_X(f)\,df = \int_{-\infty}^{\infty} S_Y(f)\,df < \infty$$

for $\{Y(t)\}$ to be a stationary process. In particular, if $S_X(\cdot)$ is a bounded function, the requirement for stationarity becomes

$$\int_{-\infty}^{\infty} |G(f)|^2\,df < \infty.$$

In this case, Parseval's relationship for deterministic nonperiodic functions with finite energy says that, because $g(\cdot)$ is real-valued, the above is equivalent to

$$\int_{-\infty}^{\infty} g^2(u)\,du < \infty,$$

a condition which in turn guarantees the existence of $G(\cdot)$.

The corresponding results for the discrete parameter case are quite similar. *Any* LTI digital filter L can be expressed in the form

$$L\{X_t\} = \sum_{u=-\infty}^{\infty} g_u X_{t-u} \equiv Y_t,$$

where $\{g_u\}$ is a real-valued deterministic sequence called the *impulse response sequence* – the above is the discrete parameter version of Equation (165). If $\{X_t\}$ is a stationary process with a bounded sdf $S_X(\cdot)$ defined on the interval $[-1/2, 1/2]$ and if $\{g_u\}$ is square summable, it follows that $\{Y_t\}$ is a stationary process with an sdf given by

$$S_Y(f) = |G(f)|^2 S_X(f), \quad |f| \le 1/2,$$

where the transfer function is now obtained by using the sequence $\{\exp{(i2\pi ft)}\}$ as input to the LTI digital filter:

$$L\{e^{i2\pi ft}\} = \sum_{u=-\infty}^{\infty} g_u e^{i2\pi f(t-u)} = e^{i2\pi ft}G(f)$$

with

$$G(f) \equiv \sum_{u=-\infty}^{\infty} g_u e^{-i2\pi fu} \text{ for } |f| \le 1/2.$$

5.4 Determination of SDFs by LTI Digital Filtering

Let us now give some examples of how the above theory can be used to determine the sdf's of discrete parameter stationary processes.

• *Moving average process*

An MA(q) process with zero mean by definition satisfies the difference equation

$$X_t = \epsilon_t - \theta_{1,q}\epsilon_{t-1} - \cdots - \theta_{q,q}\epsilon_{t-q},$$

where each $\theta_{j,q}$ is a constant ($\theta_{q,q} \ne 0$) and $\{\epsilon_t\}$ is a white noise process with zero mean and variance σ_ϵ^2 (this is Equation (43a) with $\mu = 0$ and $\theta_{0,q}$ replaced by its defining value of -1). Define the filter L on sequences $\{y_t\}$ by

$$L\{y_t\} = y_t - \theta_{1,q}y_{t-1} - \cdots - \theta_{q,q}y_{t-q},$$

and note that $X_t = L\{\epsilon_t\}$. It is easy to argue that this filter satisfies the properties of an LTI digital filter. To determine its transfer function, we input the sequence defined by $\exp{(i2\pi ft)}$:

$$
\begin{aligned}
L\{e^{i2\pi ft}\} &= e^{i2\pi ft} - \theta_{1,q}e^{i2\pi f(t-1)} - \cdots - \theta_{q,q}e^{i2\pi f(t-q)} \\
&= e^{i2\pi ft}\left(1 - \theta_{1,q}e^{-i2\pi f} - \cdots - \theta_{q,q}e^{-i2\pi fq}\right),
\end{aligned}
$$

so the transfer function is given by

$$G(f) = 1 - \theta_{1,q}e^{-i2\pi f} - \cdots - \theta_{q,q}e^{-i2\pi fq}.$$

From the relationships

$$S_X(f) = |G(f)|^2 S_\epsilon(f) \text{ and } S_\epsilon(f) = \sigma_\epsilon^2,$$

we have

$$S_X(f) = \sigma_\epsilon^2 \left|1 - \theta_{1,q}e^{-i2\pi f} - \cdots - \theta_{q,q}e^{-i2\pi fq}\right|^2$$

for an MA(q) process.

● *Autoregressive process*

A stationary AR(p) process with zero mean satisfies the equation

$$X_t = \phi_{1,p}X_{t-1} + \cdots + \phi_{p,p}X_{t-p} + \epsilon_t, \qquad (168a)$$

where each $\phi_{j,p}$ is a constant ($\phi_{p,p} \neq 0$) and $\{\epsilon_t\}$ is as in the previous example (see Section 2.6). Define the filter L on sequences $\{y_t\}$ by

$$L\{y_t\} = y_t - \phi_{1,p}y_{t-1} - \cdots - \phi_{p,p}y_{t-p},$$

and note that $L\{X_t\} = \epsilon_t$. The transfer function of the filter is

$$G(f) = 1 - \phi_{1,p}e^{-i2\pi f} - \cdots - \phi_{p,p}e^{-i2\pi fp}.$$

From the relationship

$$S_\epsilon(f) = |G(f)|^2 S_X(f),$$

we conclude that

$$S_X(f) = \frac{\sigma_\epsilon^2}{\left|1 - \phi_{1,p}e^{-i2\pi f} - \cdots - \phi_{p,p}e^{-i2\pi fp}\right|^2}. \qquad (168b)$$

In Section 2.6 we mentioned that the $\phi_{j,p}$ coefficients must satisfy certain conditions for $\{X_t\}$ to be stationary. It can be shown (Grenander and Rosenblatt, 1984, pp. 36–38) that a necessary and sufficient condition for the existence of a stationary solution to $\{X_t\}$ of Equation (168a) is that $G(\cdot)$ never vanish; i.e., $G(f) \neq 0$ for any f. In this case, the stationary solution to (168a) is unique and is given by

$$X_t = \int_{-1/2}^{1/2} \frac{e^{i2\pi ft}}{G(f)} \, dZ_\epsilon(f),$$

where $\{Z_\epsilon(f)\}$ is the orthogonal process in the spectral representation for $\{\epsilon_t\}$.

● *Autoregressive moving average process*

An ARMA(p, q) process with zero mean by definition satisfies the difference equation

$$X_t = \phi_{1,p}X_{t-1} + \cdots + \phi_{p,p}X_{t-p}$$
$$+ \epsilon_t - \theta_{1,q}\epsilon_{t-1} - \cdots - \theta_{q,q}\epsilon_{t-q},$$

where $\phi_{j,p}$ and $\theta_{k,q}$ are constants ($\phi_{p,p} \neq 0$ and $\theta_{q,q} \neq 0$) and $\{\epsilon_t\}$ is as in the previous two examples. Exercise [5.4] is to show that its sdf is given by

$$S_X(f) = \sigma_\epsilon^2 \frac{\left|1 - \theta_{1,q}e^{-i2\pi f} - \cdots - \theta_{q,q}e^{-i2\pi fq}\right|^2}{\left|1 - \phi_{1,p}e^{-i2\pi f} - \cdots - \phi_{p,p}e^{-i2\pi fp}\right|^2}. \qquad (168c)$$

5.5 Some Filter Terminology

So far our discussion of filters has been rather theoretical and of limited practical use. The remainder of this chapter concentrates on defining some terminology used in filtering theory, developing an interpretation of the spectrum via filters and considering some applications. In particular we shall see that the discrete prolate spheroidal sequences (dpss's) introduced in Chapter 3 arise naturally in filter design. The literature on filtering is vast, and hence the discussion given here is not intended to do much more than introduce some ideas of importance to spectral analysis (useful references for filtering theory are Hamming, 1983, and Rabiner and Gold, 1975).

First let us define the concepts of cascaded filters, low-pass filters, high-pass filters and band-pass filters. A *cascaded filter* is defined as an arrangement of a set of n filters such that the output from the first filter is the input to the second filter and so forth. If all n filters are LTI filters and if the input to the system is considered a realization of a stationary process, it is easy to describe the spectral relationship between the input to the first filter and the output from the nth filter. Let $S_j^{(I)}(\cdot)$ be the integrated spectrum of the input to the jth filter, and let $S_{j+1}^{(I)}(\cdot)$ be the integrated spectrum of the output from the jth filter. (We assume that all these spectra exist.) Let $G_j(\cdot)$ be the transfer function of the jth filter. From Equation (157b) we have

$$dS_{j+1}^{(I)}(f) = |G_j(f)|^2 \, dS_j^{(I)}(f) \text{ for } j = 1, 2, \ldots, n.$$

The relationship between the integrated spectra of the input and the output to the cascaded filter is thus given by

$$dS_{n+1}^{(I)}(f) = |G_n(f)|^2 |G_{n-1}(f)|^2 \ldots |G_1(f)|^2 \, dS_1^{(I)}(f).$$

Note that the integrated spectrum of the output does not depend on the order in which the filters occur in the cascade (as long as the integrated spectra exist at each step in the cascade).

If the transfer function of an LTI filter has the form

$$|G(f)|^2 = \begin{cases} 1, & \text{if } |f| \leq f_0; \\ 0, & \text{otherwise,} \end{cases}$$

we call the filter an *ideal low-pass filter*, since it passes all components with frequencies lower than f_0, but completely suppresses all components with higher frequencies. Conversely, if the transfer function of an LTI filter has the form

$$|G(f)|^2 = \begin{cases} 1, & \text{if } |f| \geq f_0; \\ 0, & \text{otherwise,} \end{cases}$$

we call the filter an *ideal high-pass filter*. Finally, if the transfer function is of the form

$$|G(f)|^2 = \begin{cases} 1, & \text{if } 0 < f_1 \leq |f| \leq f_2; \\ 0, & \text{otherwise,} \end{cases}$$

we call the associated filter an *ideal band-pass filter*. The effect of this filter is to allow all frequencies in the band $[f_1, f_2]$ to pass through unattenuated, but all other frequency components are completely eliminated. We can create a band-pass filter by cascading appropriate low-pass and high-pass filters together. Such ideal filters are not realizable as finite-length LTI filters (see Section 5.8).

The above terminology applies to both analog and digital filters. Consider now an LTI digital filter L with impulse response sequence $\{g_u\}$:

$$L\{X_t\} = \sum_{u=-\infty}^{\infty} g_u X_{t-u}.$$

This filter is said to be *causal* if $g_u = 0$ for all $u < 0$ and *acausal* otherwise. If $g_u = 0$ outside a finite range for u, it is called a *finite impulse response* (FIR) filter; otherwise, it is called an *infinite impulse response* (IIR) filter.

5.6 Interpretation of Spectrum via Band-Pass Filtering

We can use ideal band-pass filters to motivate a physical interpretation for the integrated spectrum. Consider the real-valued continuous parameter stationary process $\{X(t)\}$ with integrated spectrum $S_X^{(I)}(\cdot)$. Its variance (a measure of the power for the process) is given by

$$\sigma_X^2 \equiv \int_{-\infty}^{\infty} dS_X^{(I)}(f) = \int_{-\infty}^{\infty} S_X(f)\, df,$$

where the second equality holds in the case where $S_X^{(I)}(\cdot)$ is differentiable with derivative $S_X(\cdot)$ – the sdf of $\{X(t)\}$. Suppose we use this process as input to an ideal band-pass filter with a transfer function that satisfies

$$|G(f)|^2 = \begin{cases} 1, & \text{if } f' \leq |f| \leq f' + df'; \\ 0, & \text{otherwise,} \end{cases}$$

where $f' > 0$ and df' is a small positive increment in frequency. Let $\{Y(t)\}$ represent the output from this filter, and let $S_Y^{(I)}(f)$ and $S_Y(f)$ be, respectively, its integrated spectrum and sdf (when the latter exists). Its variance is given by

$$\sigma_Y^2 \equiv \int_{-\infty}^{\infty} dS_Y^{(I)}(f) = \int_{-\infty}^{\infty} S_Y(f)\, df.$$

By the LTI analog filtering theorem, the relationship between the spectrum of the output $S_Y^{(I)}(\cdot)$ and the spectrum of the input $S_X^{(I)}(\cdot)$ is

$$dS_Y^{(I)}(f) = |G(f)|^2 \, dS_X^{(I)}(f),$$

or, in the case where the sdf's exist,

$$S_Y(f) = |G(f)|^2 S_X(f).$$

Now

$$
\begin{aligned}
\sigma_Y^2 &= \int_{-\infty}^{\infty} |G(f)|^2 \, dS_X^{(I)}(f) = \int_{-f'-df'}^{-f'} dS_X^{(I)}(f) + \int_{f'}^{f'+df'} dS_X^{(I)}(f) \\
&= S_X^{(I)}(-f') - S_X^{(I)}(-f'-df') + S_X^{(I)}(f'+df') - S_X^{(I)}(f') \\
&= 2\left[S_X^{(I)}(f'+df') - S_X^{(I)}(f') \right],
\end{aligned}
$$

because of the two-sided nature of $S_X^{(I)}(\cdot)$. Hence the incremental difference in the integrated spectrum at the frequency f' is simply the variance associated with the output of a narrow band-pass filter bordering that frequency. When sdf's exist, the above becomes

$$\sigma_Y^2 \approx 2S_X(f') \, df'$$

(again the factor of 2 is needed due to the two-sided nature of sdf's). Since the variance of the input to the filter is just $\int_{-\infty}^{\infty} S_X(f') \, df'$, the above tells us that $2S_X(f') \times df'$ represents approximately the contribution to the variance from frequencies close to $\pm f'$.

5.7 An Example of LTI Digital Filtering

As a simple example of an LTI digital filter, let us consider the following impulse response sequence:

$$
g_u^{(1)} = \begin{cases} 1/2, & u = 0; \\ 1/4, & u = \pm 1; \\ 0, & \text{otherwise.} \end{cases}
$$

This sequence defines an acausal FIR filter with a transfer function given by

$$G^{(1)}(f) \equiv g_{-1}^{(1)} e^{i2\pi f} + g_0^{(1)} + g_1^{(1)} e^{-i2\pi f} = \cos^2(\pi f)$$

for $|f| \le 1/2$. The square modulus of this function is shown by the thick solid curve in the upper plot of Figure 172. Its shape indicates that the filter poorly approximates an ideal low-pass filter: high-frequency components are attenuated – i.e., reduced in magnitude – in comparison

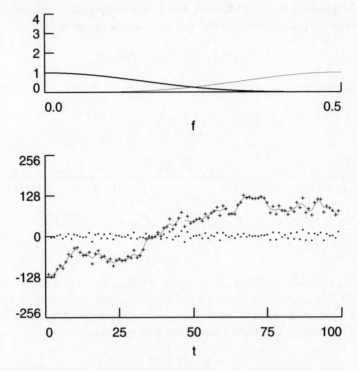

Figure 172. Example of LTI digital filter. The top plot shows $|G^{(1)}(f)|^2$ versus f (thick solid curve) and the corresponding quantity for the 'residual filter' (thin solid curve), while the bottom plot shows the unfiltered data (the pluses), the filtered data (solid curve) and the residuals (the dots).

to low-frequency components. The results of applying this filter to some data related to the rotation of the earth are shown in the lower plot of Figure 172. Here the pluses represent the original unfiltered data x_t, and the solid curve represents the filtered data, namely,

$$y_t^{(1)} \equiv g_{-1}^{(1)} x_{t+1} + g_0^{(1)} x_t + g_1^{(1)} x_{t-1}.$$

Note that the filtered data have the same shape as the 'backbone' of the original data (the low-frequency components) and have less 'local variability' (the high-frequency components) than the original data. This is consistent with what we would expect from the shape of the transfer function for the filter.

We now examine the difference between the original data and the filtered data, say,

$$r_t^{(1)} \equiv x_t - y_t^{(1)}.$$

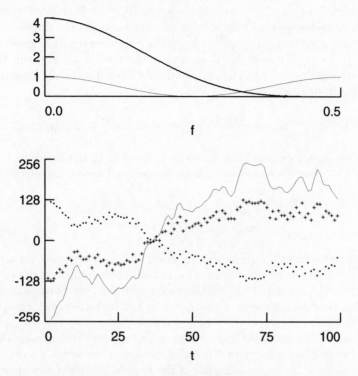

Figure 173. Second example of digital filtering.

This residual series (the dots in the lower plot of Figure 172) can also be expressed as a filtering of our original data since we have

$$r_t^{(1)} = -g_{-1}^{(1)} x_{t+1} + (1 - g_0^{(1)}) x_t - g_1^{(1)} x_{t-1}$$
$$= -(x_{t+1} - 2x_t + x_{t-1})/4.$$

The transfer function for the filter defined by the impulse response sequence $h_0^{(1)} \equiv 1 - g_0^{(1)}$ and $h_u^{(1)} \equiv -g_u^{(1)}$ for $u \neq 0$ is just $\sin^2(\pi f)$ for $|f| \leq 1/2$, and its square modulus is shown by the thin solid curve in the upper plot of Figure 172. The shape of this transfer function is that of a nonideal high-pass filter: the filter attenuates the low-frequency components and leaves the high-frequency components relatively unaltered.

In the above example we were able to construct a useful high-pass filter by subtracting the output of a low-pass filter from its input. This is a useful procedure because it allows us to decompose a time series into two parts, one containing primarily its low-frequency components, and the other its high-frequency components. This decomposition is also possible with other low-pass filters if certain criteria are satisfied. One important property concerns the normalization used for the impulse response sequence of the filter. Note that the *shape* of the square modulus of a transfer function is unchanged if we multiply each member of the impulse response sequence by a constant. For example, the impulse response sequence defined by $g_u^{(2)} = 2g_u^{(1)}$ has a transfer function given by $2G^{(1)}(\cdot)$. Its square modulus is shown by the thick solid curve in the upper plot of Figure 173. The corresponding filtered and residual series are defined, respectively, by

$$y_t^{(2)} = g_{-1}^{(2)}x_{t+1} + g_0^{(2)}x_t + g_1^{(2)}x_{t-1} \text{ and } r_t^{(2)} = x_t - y_t^{(2)}.$$

They are shown by the solid curve and the dots in the lower plot of Figure 173. The filter associated with the residual series, namely,

$$\begin{aligned} r_t^{(2)} &\equiv -g_{-1}^{(2)}x_{t+1} + (1 - g_0^{(2)})x_t - g_1^{(2)}x_{t-1} \\ &= -(x_{t+1} + x_{t-1})/2, \end{aligned}$$

has a transfer function whose square modulus is given by the thin solid curve in the upper plot of Figure 173. Note that its frequency response characteristics are not at all like those of a high-pass filter and that the resulting residual series is not composed of just high-frequency components.

If we want a low-pass filter whose output traces the backbone of a given time series rather than a rescaled version of the series, the above example shows that normalization of the impulse response sequence is important. The proper normalization can be specified by insisting that, if the time series is locally smooth around x_t, the output from the filter should be the value x_t itself. We define locally smooth as meaning locally linear. Thus, if

$$x_t = \alpha + \beta t$$

for constants α and β, and if the impulse response sequence $\{g_u\}$ is nonzero only for $|u| \leq K$, we want

$$\sum_{u=-K}^{K} g_u x_{t-u} = x_t; \text{ i.e., } \sum_{u=-K}^{K} g_u[\alpha + \beta(t - u)] = \alpha + \beta t.$$

If we assume that the impulse response sequence is symmetric about $u = 0$, i.e., $g_{-u} = g_u$, the above is satisfied for all α, β and t if

$$\sum_{u=-K}^{K} g_u = 1.$$

Note that in our examples this normalization holds for the impulse response sequence $\{g_u^{(1)}\}$ but not for $\{g_u^{(2)}\}$.

If an impulse response sequence is symmetric about $u = 0$ and sums to 1, its associated transfer function must be real-valued and is in fact given by

$$G(f) = g_0 + 2 \sum_{u=1}^{K} g_u \cos (2\pi f u).$$

Now

$$G(0) = g_0 + 2 \sum_{u=1}^{K} g_u = 1;$$

hence $|G(0)|^2 = 1$ and $10 \log_{10}(|G(0)|^2) = 0$; i.e., the response is 0 dB down at $f = 0$ – another way of stating the requirement $\sum g_u = 1$. If $\{g_u\}$ defines a low-pass filter, $G(\cdot)$ should be of the following form: first, it should be close to 1 for all f less than, say, f_L in magnitude (f_L is called the (nominal) *cutoff frequency* of the filter); and, second, it should be close to 0 for f between, say, f_H and $1/2$ in magnitude (where $f_H > f_L$ but the two are as close as possible). If we now form the residual series

$$r_t = x_t - \sum_{u=-K}^{K} g_u x_{t-u} = \sum_{u=-K}^{K} h_u x_{t-u},$$

where $h_0 = 1 - g_0$ and $h_u = -g_u$ for $|u| \geq 1$, the filter associated with $\{h_u\}$ has a transfer function given by

$$H(f) = h_0 + 2 \sum_{u=1}^{K} h_u \cos (2\pi f u) = 1 - g_0 - 2 \sum_{u=1}^{K} g_u \cos (2\pi f u) = 1 - G(f).$$

Thus, under the assumptions we have made about $\{g_u\}$, the residual filter $\{h_u\}$ should resemble a high-pass filter: first, it should be close to zero for all f less than f_L in magnitude; and, second, it should be close to one for f between f_H and $1/2$ in magnitude.

To summarize our discussion, two conditions on a low-pass FIR filter are sufficient for its associated residual filter to be a reasonable high-pass filter: first, the impulse response sequence of the filter sums to 1; and, second, the sequence is symmetric about $u = 0$. The first condition is essential (as the filter $\{g_u^{(2)}\}$ shows); the second condition was needed for our mathematical development, but it can be circumvented – it is possible to construct examples of asymmetric low-pass filters whose associated residual filter is a reasonable high-pass filter.

Comments and Extensions to Section 5.7

[1] Our example can also be used to point out the strong resemblance between detrending a time series and subjecting it to a high-pass filter. Trend can be loosely defined as a 'long term change in the mean' (Chatfield, 1984, p. 13; Granger, 1966). We might argue that the solid curve in the lower plot of Figure 172 represents the trend of the time series, while the dots represent the detrended series. Alternatively, we might fit a least squares line to the series and argue that the fitted regression line is the trend, while the residuals from the regression fit are the detrended data. (As indicated by the results of Exercise [5.11], fitting a regression line is *not* an example of LTI filtering, so unfortunately we cannot use the notion of a transfer function to help us assess the effect of this type of detrending.) This and other forms of detrending act like high-pass filters – detrending effectively removes low-frequency components in a time series.

5.8 Least Squares Filter Design

In this and the next section we consider some simple, but effective, approaches to approximating an ideal low-pass digital filter (with obvious modifications, these same methods can be used to approximate ideal high-pass and band-pass filters also). The transfer function for an ideal low-pass filter of bandwidth $2W < 1$ is given by

$$G_I(f) \equiv \begin{cases} 1, & |f| \leq W; \\ 0, & W < |f| \leq 1/2. \end{cases} \tag{176}$$

Since $G_I(\cdot)$ is square integrable over the region $[-1/2, 1/2]$, we can periodically extend it outside that region and then appeal to the theory of Section 3.7 to consider it as the Fourier transform of a sequence defined by

$$g_{u,I} \equiv \int_{-1/2}^{1/2} G_I(f) e^{i2\pi fu}\, df = \int_{-W}^{W} e^{i2\pi fu}\, df = \begin{cases} 2W, & u = 0; \\ \dfrac{\sin(2\pi Wu)}{\pi u}, & u \neq 0. \end{cases}$$

From this expression we see that the ideal low-pass digital filter defined by (176) is a symmetric (in the sense that $g_{-u,I} = g_{u,I}$) acausal IIR filter – this limits its usefulness in practical applications and motivates us to look at various approximations. For simplicity, we only consider approximations from within the class of symmetric acausal FIR filters.

Our first approach is called *least squares filter design* and is based upon the following result: if K is a fixed positive odd integer, then among all functions of the form

$$H_K(f) \equiv \sum_{u=-\lfloor K/2 \rfloor}^{\lfloor K/2 \rfloor} h_u e^{-i2\pi fu},$$

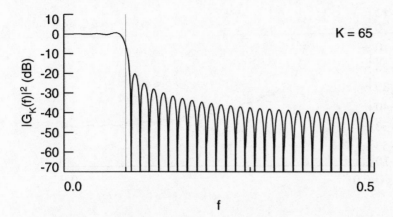

Figure 177. Squared modulus of $G_K(\cdot)$ – the Kth order least squares approximation to an ideal low-pass filter – for $W = 0.1$ and $K = 65$. The thin vertical line marks $W = 0.1$.

the one that minimizes

$$\int_{-1/2}^{1/2} |G_I(f) - H_K(f)|^2 \, df$$

is obtained by letting $h_u = g_{u,I}$ for $|u| \leq \lfloor K/2 \rfloor$ (see the discussion surrounding Equation (61); also note that $\lfloor K/2 \rfloor = (K-1)/2$ because K is odd). Thus the acausal FIR filter defined by

$$g_{u,K} \equiv \begin{cases} g_{u,I}, & |u| \leq \lfloor K/2 \rfloor; \\ 0, & \text{otherwise,} \end{cases}$$

has a transfer function

$$G_K(f) \equiv \sum_{u=-\lfloor K/2 \rfloor}^{\lfloor K/2 \rfloor} g_{u,K} e^{-i2\pi f u}$$

that is the best Kth order approximation – in the least squares sense – to the transfer function of an ideal low-pass digital filter.

As an example, Figure 177 shows $|G_K(f)|^2$ versus f for $W = 0.1$ and $K = 65$. This figure points out a potential problem with least squares filters, namely, ripples in the region where $G_I(\cdot)$ has a discontinuity – in this example, this occurs at $W = 0.1$ (indicated by the thin vertical line). These ripples are due to the Gibbs phenomenon and constitute a form of leakage in the region where $G_I(f) = 0$. We introduced these ideas in Section 3.7 and discussed there the use of convergence

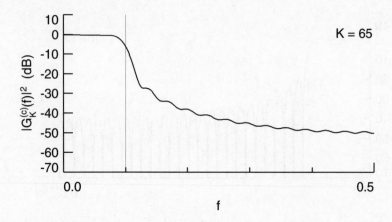

Figure 178. Squared modulus of $G_K^{(c)}(\cdot)$ with triangular convergence factors – compare this with $G_K(\cdot)$ in Figure 177.

factors $\{c_u\}$ to reduce the ripples. In the present context these factors would define a new impulse response sequence given by, say,

$$g_{u,K}^{(c)} \equiv c_u g_{u,K} = \begin{cases} c_u g_{u,I}, & |u| \leq \lfloor K/2 \rfloor; \\ 0, & \text{otherwise,} \end{cases}$$

with a corresponding transfer function given by

$$G_K^{(c)}(f) \equiv \sum_{u=-\lfloor K/2 \rfloor}^{\lfloor K/2 \rfloor} g_{u,K}^{(c)} e^{-i2\pi f u}.$$

For example, the use of Cesàro sums yields triangular convergence factors:

$$c_u = \begin{cases} \gamma \left(1 - \dfrac{2|u|}{K+1} \right), & |u| \leq \lfloor K/2 \rfloor; \\ 0, & \text{otherwise;} \end{cases}$$

here γ is a gain factor that allows us to, say, force the normalization $\sum g_{u,K}^{(c)} = G_K^{(c)}(0) = 1$ (see the discussion in the previous section). Figure 178 shows $|G_K^{(c)}(f)|^2$ versus f for the same values of W and K as in Figure 177. Note that the ripples are substantially reduced in $|G_K^{(c)}(\cdot)|^2$ but also that the discontinuity at $W = 0.1$ is now less accurately rendered than in $|G_K(\cdot)|^2$.

We also note the following interesting interpretation of convergence factors (Bloomfield, 1976). Since we can write

$$G_K^{(c)}(f) = \int_{-1/2}^{1/2} G_I(f') C_K(f-f') \, df' \text{ for } C_K(f) \equiv \sum_{u=-\lfloor K/2 \rfloor}^{\lfloor K/2 \rfloor} c_u e^{-i2\pi f u},$$

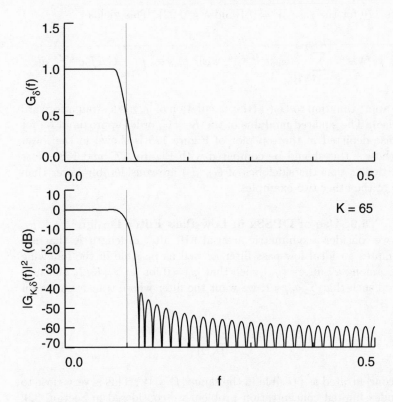

Figure 179. $G_\delta(\cdot)$ with $W = 0.1$ and $\delta = 0.02$ (top plot) and the squared modulus of its approximation $G_{K,\delta}(\cdot)$ with $K = 65$ (bottom plot). The frequency W is indicated by the thin vertical lines.

$G_K^{(c)}(\cdot)$ is a 'smoothed' version of $G_I(\cdot)$ (see Section 3.6). Because $G_I(\cdot)$ is already smooth everywhere except at the discontinuities at $\pm W$, the main effect of convolving it with the smoothing kernel $C_K(\cdot)$ is to eliminate these discontinuities, but we could just as well do this directly. For example, instead of seeking least squares approximations to $G_I(\cdot)$ itself, we could look at such approximations to the following smoothed version of $G_I(\cdot)$:

$$G_\delta(f) \equiv \begin{cases} 1, & |f| \leq W - \delta; \\ \dfrac{1}{2}\left[1 + \cos\left(\pi\dfrac{|f| - W + \delta}{2\delta}\right)\right], & W - \delta < |f| \leq W + \delta; \\ 0, & W + \delta < |f| \leq 1/2 \end{cases}$$

under the constraints $0 < \delta < W$ and $W + \delta < 1/2$ (see the top plot of

Figure 179 for the case $W = 0.1$ and $\delta = 0.02$). This yields

$$G_{K,\delta}(f) \equiv \sum_{u=-\lfloor K/2 \rfloor}^{\lfloor K/2 \rfloor} g_{u,\delta}e^{-i2\pi fu} \quad \text{with} \quad g_{u,\delta} \equiv \int_{-1/2}^{1/2} G_{\delta}(f)e^{i2\pi fu}\,df$$

as an approximation to $G_{\delta}(\cdot)$ (the calculation of $g_{u,\delta}$ is a straightforward exercise). The squared modulus of the $K = 65$ order approximation for the case depicted in the top plot of Figure 179 is shown in the lower plot there – this should be compared with Figures 177 and 178. Note in particular that the sidelobes of $G_{K,\delta}(\cdot)$ are considerably lower than those of the other two examples.

5.9 Use of DPSSs in Low-Pass Filter Design

Here we consider a symmetric acausal FIR filter of length K that approximates an ideal low-pass filter as well as possible in the following sense: among all filters $\{g_u\}$ such that $g_u = 0$ for $|u| > \lfloor K/2 \rfloor$ and normalized such that $\sum g_u = 1$, we want the filter whose transfer function

$$G_K(f) \equiv \sum_{u=-\lfloor K/2 \rfloor}^{\lfloor K/2 \rfloor} g_u e^{-i2\pi fu}$$

is as concentrated as possible in the range $|f| \leq W$. This is very close to the index-limited concentration problem we considered in Section 3.9. We again use the measure of concentration

$$\beta^2(W) \equiv \int_{-W}^{W} |G_K(f)|^2\,df \bigg/ \int_{-1/2}^{1/2} |G_K(f)|^2\,df$$

$$= \sum_{u'=-\lfloor K/2 \rfloor}^{\lfloor K/2 \rfloor} \sum_{u=-\lfloor K/2 \rfloor}^{\lfloor K/2 \rfloor} g_u^* \frac{\sin\left(2\pi W(u'-u)\right)}{\pi(u'-u)} g_{u'} \bigg/ \sum_{u=-\lfloor K/2 \rfloor}^{\lfloor K/2 \rfloor} |g_u|^2$$

(see Equation (105a)). Following the approach of Section 3.9, the solution to this concentration problem is any eigenvector associated with the largest eigenvalue of the following set of matrix eigenvalue equations:

$$\sum_{u'=-\lfloor K/2 \rfloor}^{\lfloor K/2 \rfloor} \frac{\sin\left(2\pi W(u'-u)\right)}{\pi(u'-u)} g_{u'} = \lambda g_u, \quad u = -\lfloor K/2 \rfloor, \ldots, \lfloor K/2 \rfloor.$$

$$(180)$$

Let $\lambda_0(W)$ denote this eigenvalue, and let

$$\tilde{\mathbf{v}}_0(W) \equiv \left[\tilde{v}_{-\lfloor K/2 \rfloor,0}(W), \ldots, \tilde{v}_{0,0}(W), \ldots, \tilde{v}_{\lfloor K/2 \rfloor,0}(W)\right]^T$$

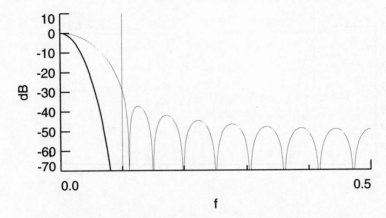

Figure 181. Squared modulus of transfer functions corresponding to $\tilde{\mathbf{v}}_0(W)$ with $W = 0.1$ and $K = 17$ (thin solid curve) and 65 (thick curve). The thin vertical line marks the frequency W.

be the corresponding eigenvector normalized so that its elements sum to unity. The elements of this eigenvector are renormalized and reindexed portions of the zeroth-order *discrete prolate spheroidal sequence* (dpss), which we denoted as $\{v_{t,0}(N, W)\}$ in Section 3.9. If we let A be the $K \times K$ matrix whose (u', u)th element is $\sin[2\pi W(u' - u)]/[\pi(u' - u)]$ (defined to be $2W$ when $u' = u$), then the set of K equations in (180) can be written as $A\tilde{\mathbf{v}}_0(W) = \lambda_0(W)\tilde{\mathbf{v}}_0(W)$, and the concentration measure is

$$\beta^2(W) = \frac{\tilde{\mathbf{v}}_0^T(W)A\tilde{\mathbf{v}}_0(W)}{\tilde{\mathbf{v}}_0^T(W)\tilde{\mathbf{v}}_0(W)} = \frac{\lambda_0(W)\tilde{\mathbf{v}}_0^T(W)\tilde{\mathbf{v}}_0(W)}{\tilde{\mathbf{v}}_0^T(W)\tilde{\mathbf{v}}_0(W)} = \lambda_0(W).$$

The solid curves in Figure 181 show the squared magnitude of the resulting transfer function $G_K(\cdot)$ for $W = 0.1$ and the cases $K = 17$ (thin curve) and 65 (thick). The thin vertical line indicates the location of the cutoff frequency $W = 0.1$. This plot shows that each $G_K(\cdot)$ captures one important aspect of the ideal low-pass digital filter – a small amount of energy in the frequency range $W < |f| \le 1/2$. However, both fail miserably in another aspect – the squared modulus of the transfer function of the ideal filter is not anywhere near constant for $0 \le |f| \le W$.

The $K \times K$ matrix A from which the eigenvectors are derived is a symmetric Toeplitz matrix (see Section 2.4). Makhoul (1981a) has shown that the transfer function for an eigenvector of a symmetric Toeplitz matrix will typically be zero at $K - 1$ different frequencies f such that $-1/2 < f \le 1/2$ if the eigenvector corresponds to either the maximum or minimum eigenvalue. Thus we expect the transfer function of our FIR filter to have zero response (a notch) at certain frequencies.

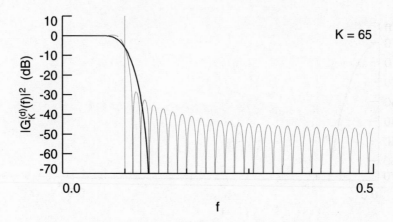

Figure 182. Squared modulus of $G_K^{(d)}(\cdot)$ using dpss convergence factors for $K = 65$ and the cases $\delta = 0.04$ (thick solid curve) and $\delta = 0.01$ (thin). The thin vertical line marks the position of the cutoff frequency $W = 0.1$.

For example, 8 notches are apparent in the thin solid curve in Figure 181 for $K = 17$, and by symmetry there must be 8 additional notches in the frequency range $[-1/2, 0]$ for a total of $K - 1 = 16$ notches.

A more reasonable use for dpss's in low-pass filter design is as convergence factors (Rabiner and Gold, 1975, Sections 3.8 and 3.11). Let us define

$$d_u \equiv \begin{cases} \gamma \tilde{v}_{u,0}(\delta), & |u| \leq \lfloor K/2 \rfloor; \\ 0, & \text{otherwise}, \end{cases}$$

where, as before, γ is a gain factor; however, note that the cutoff frequency for the dpss is now δ instead of W. We can now define a FIR filter with impulse response sequence

$$g_{u,K}^{(d)} \equiv d_u g_{u,I},$$

where, as in the previous section, $\{g_{u,I}\}$ is the impulse response sequence of an ideal low-pass filter of bandwidth $2W$. The transfer function corresponding to $\{g_{u,K}^{(d)}\}$ is given by

$$G_K^{(d)}(f) = \int_{-1/2}^{1/2} G_I(f') D_K(f-f')\, df' \text{ for } D_K(f) \equiv \sum_{u=-\lfloor K/2 \rfloor}^{\lfloor K/2 \rfloor} d_u e^{-i2\pi fu},$$

where $G_I(\cdot)$ is the transfer function given in Equation (176) for the ideal low-pass digital filter. Because $G_I(\cdot)$ is convolved with $D_K(\cdot)$ to form $G_K^{(d)}(\cdot)$, the discontinuity in $G_I(\cdot)$ is smeared out into a transition band. The width of this transition band is proportional to the width of the central lobe of $D_K(\cdot)$, while the sidelobes of the latter contribute to leakage outside the passband. For a given filter length K, by adjusting the bandwidth parameter δ for $\{\tilde{v}_{u,0}(\delta)\}$, we can trade off between a small central lobe (its width decreases as δ decreases) and small sidelobes (their heights decrease as δ increases). The dpss is a natural choice for the convergence factors because the squared modulus of its transfer function is a good approximation to a Dirac delta function.

As an example, Figure 182 shows the squared modulus of $G_K^{(d)}(\cdot)$ for $K = 65$, $W = 0.1$ and the cases $\delta = 0.04$ (thick solid curve) and $\delta = 0.01$ (thin). In each case the gain factor γ was set so that $G_K^{(d)}(0) = 1$. As expected, the transfer function for $\delta = 0.04$ has a wider transition region about $W = 0.1$ than does the one for $\delta = 0.01$, whereas the sidelobe level is much higher in the latter case than in the former. Of course, if we allow ourselves the liberty of increasing the filter length K beyond 65, the tradeoff between transition region width and sidelobe level becomes easier to manage.

5.10 Exercises

[5.1] Let $\{X(t)\}$ be a stochastically continuous, continuous parameter stationary process with zero mean and spectral representation

$$X(t) = \int_{-\infty}^{\infty} e^{i2\pi f t}\, dZ_X(f),$$

where $\{Z_X(\cdot)\}$ is an orthogonal process with

$$E\{|dZ_X(f)|^2\} = dS_X^{(I)}(f).$$

If $G(\cdot)$ is a complex-valued function such that

$$\int_{-\infty}^{\infty} |G(f)|^2\, dS_X^{(I)}(f) < \infty, \qquad (183)$$

show that

$$Y(t) \equiv \int_{-\infty}^{\infty} e^{i2\pi f t} G(f)\, dZ_X(f)$$

is a stationary process. Why is condition (183) necessary?

[5.2] a) Show that Equation (165) defines an LTI filter.

      b) Show that

$$L\{X(t)\} \equiv \frac{d^p X(t)}{dt^p}$$

is an LTI filter, and determine its transfer function. This is an example of an LTI analog filter that must make use of Dirac delta functions if it is to be expressed as a convolution.

[5.3] Let $\{X_t\}$ be a discrete parameter stationary process with sdf given by $S_X(\cdot)$ defined over the interval $[-1/2, 1/2]$. Let the first difference process $\{Y_t\}$ be given by

$$Y_t \equiv X_t - X_{t-1}.$$

Show that the sdf of $\{Y_t\}$ is given by

$$S_Y(f) = 4\sin^2(\pi f)\, S_X(f).$$

Does a first difference filter resemble a low-pass or high-pass filter?

[5.4] Let $\{X_t\}$ be a discrete parameter stationary ARMA(p,q) process; i.e., $\{X_t\}$ satisfies the equation

$$X_t = \sum_{k=1}^{p} \phi_{k,p} X_{t-k} + \epsilon_t - \sum_{k=1}^{q} \theta_{k,q} \epsilon_{t-k},$$

where $\{\epsilon_t\}$ is a white noise process with zero mean and variance σ_ϵ^2. Show that the sdf of this process is given by Equation (168c).

[5.5] Let $\{X_t\}$ be a discrete parameter stationary process with an sdf $S_X(\cdot)$ defined over the interval $[-1/2, 1/2]$. Let

$$Y_t \equiv X_t - \frac{1}{2K+1} \sum_{j=-K}^{K} X_{t+j},$$

where $K > 0$ is an integer. Find the sdf for $\{Y_t\}$. Sketch this sdf for the case where $\{X_t\}$ is a white noise process for $K = 4$ and 16.

[5.6] Let $\{X_t\}$ be a discrete parameter stationary process with an sdf $S_X(\cdot)$ defined over the interval $[-1/2, 1/2]$.

a) Let

$$\bar{X}_{t,K} \equiv \frac{1}{K} \sum_{k=0}^{K-1} X_{t-k} \quad \text{and} \quad Y_{t,K} \equiv \bar{X}_{t,K} - \bar{X}_{t-K,K}.$$

Show that $\{Y_{t,K}\}$ is a zero mean stationary process with sdf

$$S_{Y,K}(f) = |G_K(f)|^2 S_X(f) \quad \text{with} \quad |G_K(f)|^2 = \frac{4\sin^4(K\pi f)}{K^2 \sin^2(\pi f)}.$$

b) Plot $|G_K(\cdot)|^2$ for $K = 1$, 2, 4, 8 and 16 to see that $|G_K(\cdot)|^2$ is an approximation to the squared magnitude of a transfer function for a band-pass filter with a passband defined by $|f| \in [1/(4K), 1/(2K)]$. (This sequence of filters is the basis for a crude method of estimating the sdf called *pilot analysis*.

This method was useful as a simple way of computing a rough estimate of the sdf in days before modern computers and the popularization of the fast Fourier transform. For some details, see Section 7.3.2 of Jenkins and Watts, 1968.)

[5.7] a) Let $\{X_t\}$ be a real-valued zero mean stationary process with acvs $\{s_{\tau,X}\}$ and sdf $S_X(\cdot)$ such that $\{s_{\tau,X}\}$ and $S_X(\cdot)$ are a Fourier transform pair. Define the complex-valued process $\{Z_t\}$ by

$$Z_t \equiv X_t e^{-i2\pi f_0 t},$$

where f_0 is a fixed frequency such that $0 < f_0 \leq 1/2$. Show that $\{Z_t\}$ is a stationary process with acvs and sdf given, respectively, by

$$s_{\tau,Z} = s_{\tau,X} e^{-i2\pi f_0 \tau} \text{ and } S_Z(f) = S_X(f_0 + f).$$

Does the transformation from $\{X_t\}$ to $\{Z_t\}$ constitute an LTI filter? (This result is an example of a technique called *complex demodulation*; see Bloomfield, 1976, and Hasan, 1983.)

 b) Suppose $f_0 = 1/2$ so that $\exp(-i2\pi f_0 t) = (-1)^t$. Show that we can effectively turn a low-pass filter into a high-pass filter by the following three operations: (i) form $Z_t = (-1)^t X_t$; (ii) pass $\{Z_t\}$ through a low-pass filter to produce, say, $\{Z_t'\}$; and (iii) form $X_t' \equiv (-1)^t Z_t'$, where $\{X_t'\}$ can be regarded as a high-pass filtered version of $\{X_t\}$.

 c) Let $\{X_t\}$ and f_0 be as defined in part a). Define the real-valued process $\{Y_t\}$ by

$$Y_t \equiv X_t \cos(2\pi f_0 t).$$

Is $\{Y_t\}$ a stationary process?

[5.8] Let $\{X(t)\}$ be a continuous parameter stationary process with sdf $S_{X(t)}(\cdot)$ defined over $(-\infty, \infty)$. For $\Delta t > 0$ let

$$\bar{X}_{\Delta t}(t) \equiv \frac{1}{\Delta t} \int_{t-\Delta t}^{t} X(u)\, du$$

represent the average value of the process over the interval $[t-\Delta t, t]$, and let

$$\bar{Y}_{\Delta t}(t) \equiv \bar{X}_{\Delta t}(t) - \bar{X}_{\Delta t}(t - \Delta t)$$

be the difference between average values that are Δt time units apart. Finally, let

$$Y_t \equiv \bar{Y}_{\Delta t}(t\, \Delta t), \qquad t = 0, \pm 1, \pm 2, \ldots,$$

define a discrete parameter process formed by taking samples Δt time units apart from $\{Y_{\Delta t}(t)\}$. Show that, for $|f| \leq 1/(2\,\Delta t)$, the sdf for the stationary process $\{Y_t\}$ is given by

$$S_Y(f) \equiv 4\sin^4(\pi f\,\Delta t) \sum_{k=-\infty}^{\infty} \frac{S_{X(t)}(f + k/\Delta t)}{[\pi(f + k/\Delta t)\Delta t]^2}.$$

(This way of processing $\{X(t)\}$ arises in the study of the timekeeping properties of atomic clocks; see Barnes *et al.*, 1971.)

[5.9] If the action of L on complex exponentials is described by Equation (156), and if $G(-f) = G^*(f)$ for all f, what does L do to sines and cosines?

[5.10] Suppose that $g_p(\cdot)$ and $h_p(\cdot)$ are two periodic functions with period T such that

$$\int_{-T/2}^{T/2} |g_p(t)|^2\,dt < \infty \ \text{ and } \ \int_{-T/2}^{T/2} |h_p(t)|^2\,dt < \infty.$$

Let $\{G_n\}$ and $\{H_n\}$ be their Fourier coefficients as defined by Equation (58d). If we regard $h_p(\cdot)$ as a signal and $g_p(\cdot)/T$ as a filter, the convolution of Equation (117a), namely,

$$g_p * h_p(t) \equiv \frac{1}{T}\int_{-T/2}^{T/2} g_p(u)h_p(t - u)\,du,$$

can be regarded as a filtered version of the signal. What plays the role of the transfer function here?

[5.11] Show that
$$L\{X(t)\} \equiv \alpha + \beta X(t)$$

is *not* an LTI filter (here α and β are arbitrary nonzero constants).

6

Nonparametric Spectral Estimation

6.0 Introduction

Here we study nonparametric estimators of the spectral density function (sdf) for stationary processes with purely continuous spectra. The spectral properties for these processes are more readily seen from their sdf's than from their integrated spectra. Of course in this case the sdf and the integrated spectrum contain equivalent information, but the former is usually easier to interpret – just as probability density functions (rather than cumulative probability distribution functions) give a better visual indication of the distribution of probabilities for a random variable. Our discussion will thus concentrate on estimation of the sdf.

We shall base our sdf estimators on the observed values of a time series. We consider a time series as a portion of sample size N of one realization of a contiguous portion $X_1, X_2, \ldots, X_N$ of a discrete parameter stationary process $\{X_t\}$ with sdf $S(\cdot)$ (see Section 4.5 for a discussion on sampling from a continuous parameter process). The motivation for nonparametric sdf estimators is the relationship between the sdf and the autocovariance sequence (acvs) given by Equation (134b), namely,

$$S(f) = \Delta t \sum_{\tau=-\infty}^{\infty} s_\tau e^{-i2\pi f \tau \, \Delta t} \text{ for } |f| \leq f_{(N)} \equiv \frac{1}{2\,\Delta t}.$$

The key idea is to estimate $\{s_\tau\}$ in some fashion and then apply the above formula to estimate the sdf. For a real-valued stationary process,

$$s_\tau = \text{cov}\{X_t, X_{t+\tau}\} = E\{(X_t - \mu)(X_{t+\tau} - \mu)\},$$

where μ is the mean of the process. In the next two sections we consider estimation of μ and $\{s_\tau\}$.

187

6.1 Estimation of the Mean

Let $\{X_t\}$ be a discrete parameter (second-order) stationary process with mean μ and acvs $\{s_\tau\}$. We want to estimate these time domain quantities from a sample of N observations of our process, which we consider as observed values of the rv's $X_1, X_2, \ldots, X_N$.

A natural estimator of μ is just the sample mean, given by the rv

$$\bar{X} \equiv \frac{1}{N} \sum_{t=1}^{N} X_t.$$

Since

$$E\{\bar{X}\} = \frac{1}{N} \sum_{t=1}^{N} E\{X_t\} = \mu,$$

this estimator is unbiased. Moreover, under certain mild conditions (see Section 6.1 of Fuller, 1976), $\bar{X}$ is a consistent estimator of μ, by which we mean that, for every $\epsilon > 0$,

$$\lim_{N \to \infty} \mathbf{P}\left[|\bar{X} - \mu| > \epsilon\right] = 0,$$

where $\mathbf{P}[\cdot]$ is the probability measure associated with the stationary process. Consistency follows directly from Chebyshev's inequality

$$\mathbf{P}\left[|\bar{X} - \mu| > \epsilon\right] \leq \frac{E\{(\bar{X} - \mu)^2\}}{\epsilon^2} = \frac{\text{var}\{\bar{X}\}}{\epsilon^2}$$

as soon as we can show that $\text{var}\{\bar{X}\} \to 0$ as $N \to \infty$. It is beyond the scope of this book to prove this in general, but we do note that, for the important special case of an absolutely summable acvs,

$$\text{var}\{\bar{X}\} \approx \frac{S(0)}{N \, \Delta t} \tag{188}$$

for large N, where $S(0)$ is a finite number (if $S(0) = 0$, this approximation is not particularly useful). It follows from this that the variance of the sample mean does decrease to 0 as N goes to ∞. A proof of this fact is outlined in the Comments and Extensions below.

Comments and Extensions to Section 6.1

[1]   We wish to show that approximation (188) is valid for large N for a stationary process with an acvs that is absolutely summable; i.e.,

$$\sum_{\tau=-\infty}^{\infty} |s_\tau| < \infty.$$

First, we note that

$$\text{var}\{\bar{X}\} = E\{(\bar{X} - \mu)^2\} = \frac{1}{N^2} \sum_{t=1}^{N} \sum_{u=1}^{N} E\{(X_t - \mu)(X_u - \mu)\}$$

$$= \frac{1}{N^2} \sum_{t=1}^{N} \sum_{u=1}^{N} s_{u-t} = \frac{1}{N^2} \sum_{\tau=-(N-1)}^{N-1} (N - |\tau|) s_\tau$$

$$= \frac{1}{N} \sum_{\tau=-(N-1)}^{N-1} \left(1 - \frac{|\tau|}{N}\right) s_\tau. \tag{189a}$$

As $N \to \infty$, we can reduce the above to a compact form by appealing to the Cesàro summability theorem (see Section 3.7). In the present context, this theorem states that, if $\sum_{\tau=-(N-1)}^{N-1} s_\tau$ converges to a limit as $N \to \infty$ (this must be true since we are assuming that the acvs is absolutely summable), then $\sum_{\tau=-(N-1)}^{N-1} (1 - |\tau|/N) s_\tau$ converges to the same limit. We can thus conclude that

$$\lim_{N \to \infty} N \text{ var}\{\bar{X}\} = \lim_{N \to \infty} \sum_{\tau=-(N-1)}^{N-1} \left(1 - \frac{|\tau|}{N}\right) s_\tau \tag{189b}$$

$$= \lim_{N \to \infty} \sum_{\tau=-(N-1)}^{N-1} s_\tau = \sum_{\tau=-\infty}^{\infty} s_\tau.$$

The assumption of absolute summability can be shown to imply that $\{X_t\}$ has a purely continuous spectrum with sdf given by

$$S(f) = \Delta t \sum_{\tau=-\infty}^{\infty} s_\tau e^{-i2\pi f \tau \, \Delta t} \text{ and hence } S(0) = \Delta t \sum_{\tau=-\infty}^{\infty} s_\tau.$$

Equation (189b) now becomes

$$\lim_{N \to \infty} N \text{ var}\{\bar{X}\} = \frac{S(0)}{\Delta t}; \text{ i.e., } \text{var}\{\bar{X}\} \approx \frac{S(0)}{N \, \Delta t}$$

for large N, which completes the proof.

[2]   The reader should be aware of the fact that there are stationary processes for which the sample mean is *not* a consistent estimator of the true mean of the process (see Exercise [2.9] for a simple example). Theorems that state conditions under which the sample mean converges in some sense to the true mean come under the classification of *ergodic theorems* in the literature (see Section 16 of Yaglom, 1987, for a nice discussion of these issues). We do not give any attention to ergodicity

in this book – our experience is that it is of limited use in practical applications.

[3]   If we denote the variance-covariance matrix of the vector $\mathbf{X} \equiv [X_1, X_2, \ldots, X_N]^T$ by Γ_N (see Section 2.4), and if we assume that Γ_N^{-1} exists, the best (in the sense of having minimum variance) linear unbiased estimator of μ is given by

$$\hat{\mu} \equiv \frac{\mathbf{O}_N^T \Gamma_N^{-1} \mathbf{X}}{\mathbf{O}_N^T \Gamma_N^{-1} \mathbf{O}_N}, \quad \text{where } \mathbf{O}_N \equiv [\, \underbrace{1, 1, \ldots, 1}_{N \text{ of these}} \,]^T.$$

Note that this estimator requires knowledge of the acvs out to lag $N-1$, which is rarely the case in situations where spectral analysis is of interest. If the sdf $S(\cdot)$ is piecewise continuous with no discontinuity at $f = 0$ and $0 < S(f) < \infty$ for all f, then it can be argued that

$$e(\bar{X}, \hat{\mu}) \equiv \lim_{N \to \infty} \frac{\text{var}\{\hat{\mu}\}}{\text{var}\{\bar{X}\}} = 1. \tag{190a}$$

If the above is true, we say that $\bar{X}$ is an *asymptotically efficient* estimator of μ. Samarov and Taqqu (1988) discuss some interesting cases in which $\bar{X}$ is not asymptotically efficient; i.e., $e(\bar{X}, \hat{\mu}) < 1$. They consider the class of *fractional difference processes*, for which the sdf is given by

$$S(f) = C \left| \sin\left(\pi f \, \Delta t\right) \right|^\alpha, \quad \text{where } C > 0 \text{ and } \alpha > -1.$$

Note that, for small $|f|$, we have $S(f) \propto |f|^\alpha$ approximately, so $S(f) \to 0$ as $f \to 0$ if $\alpha > 0$ while $S(f) \to \infty$ as $f \to 0$ if $-1 < \alpha < 0$. Samarov and Taqqu show that $0.98 < e(\bar{X}, \hat{\mu}) < 1$ when $-1 < \alpha < 0$; $e(\bar{X}, \hat{\mu})$ decreases from 1 to 0 as α increases from 0 to 1; and $e(\bar{X}, \hat{\mu}) = 0$ for all $\alpha \geq 1$. Hence the sample mean can be a poor estimator of μ when $S(\cdot)$ decreases to 0 rapidly as $f \to 0$.

6.2   Estimation of the Autocovariance Sequence

Because

$$s_\tau \equiv E\left\{ (X_t - \mu)(X_{t+\tau} - \mu) \right\},$$

a natural estimator for the acvs is

$$\hat{s}_\tau^{(u)} \equiv \frac{1}{N - |\tau|} \sum_{t=1}^{N-|\tau|} (X_t - \bar{X})(X_{t+|\tau|} - \bar{X}), \tag{190b}$$

$\tau = 0, \pm 1, \ldots, \pm(N - 1)$. Note that $\hat{s}_{-\tau}^{(u)} = \hat{s}_\tau^{(u)}$, so this estimator mimics the symmetry property of $\{s_\tau\}$. For a sample of N observations there are no observations more than $N - 1$ lag units apart, so we cannot estimate s_τ for $|\tau| \geq N$ by an estimator of the above form. If we momentarily replace $\bar{X}$ by μ in Equation (190b), we see that

$$E\{\hat{s}_\tau^{(u)}\} = \frac{1}{N - |\tau|} \sum_{t=1}^{N-|\tau|} E\{(X_t - \mu)(X_{t+|\tau|} - \mu)\}$$

$$= \frac{1}{N - |\tau|} \sum_{t=1}^{N-|\tau|} s_\tau = s_\tau$$

for all $|\tau| \leq N - 1$. Thus $\hat{s}_\tau^{(u)}$ is an unbiased estimator of s_τ when μ is known. If μ is estimated by $\bar{X}$, it is a straightforward (but messy) exercise to derive $E\{\hat{s}_\tau^{(u)}\}$ exactly (Anderson, 1971, pp. 448–9). Such calculations can be used to argue that, when the process mean must be estimated from the sample mean, $\hat{s}_\tau^{(u)}$ is typically a biased estimator of s_τ (see Exercise [6.1]).

There is a second estimator of s_τ that is usually preferred to $\hat{s}_\tau^{(u)}$, namely,

$$\hat{s}_\tau^{(p)} \equiv \frac{1}{N} \sum_{t=1}^{N-|\tau|} (X_t - \bar{X})(X_{t+|\tau|} - \bar{X}) \tag{191a}$$

(the rationale for the superscript (p) is given in the next section). The only difference between the two estimators is the multiplicative factor in front of the summation, i.e., $1/(N - |\tau|)$ for $\hat{s}_\tau^{(u)}$, and $1/N$ for $\hat{s}_\tau^{(p)}$. If we again momentarily replace $\bar{X}$ by μ, it follows that

$$E\{\hat{s}_\tau^{(p)}\} = \frac{1}{N} \sum_{t=1}^{N-|\tau|} s_\tau = \left(1 - \frac{|\tau|}{N}\right) s_\tau, \tag{191b}$$

so that $\hat{s}_\tau^{(p)}$ is a biased estimator, and the magnitude of its bias increases as $|\tau|$ increases. In the literature $\hat{s}_\tau^{(u)}$ and $\hat{s}_\tau^{(p)}$ are frequently called, respectively, the 'unbiased' and 'biased' estimators of s_τ (even though, when $\bar{X}$ is used, both are typically biased, and, in fact, the magnitude of the bias in $\hat{s}_\tau^{(u)}$ can be *greater* than that in $\hat{s}_\tau^{(p)}$ – see Exercise [6.1] and the AR(2) example in Figure 192).

Why should we prefer the biased estimator $\hat{s}_\tau^{(p)}$ to the unbiased estimator $\hat{s}_\tau^{(u)}$? Here are some thoughts:

[1] For many stationary processes of practical interest, the mean square error (mse) of $\hat{s}_\tau^{(p)}$ is smaller than that of $\hat{s}_\tau^{(u)}$; i.e.,

$$\text{mse}\,\{\hat{s}_\tau^{(p)}\} \equiv E\{(\hat{s}_\tau^{(p)} - s_\tau)^2\} < E\{(\hat{s}_\tau^{(u)} - s_\tau)^2\} \equiv \text{mse}\,\{\hat{s}_\tau^{(u)}\}. \tag{191c}$$

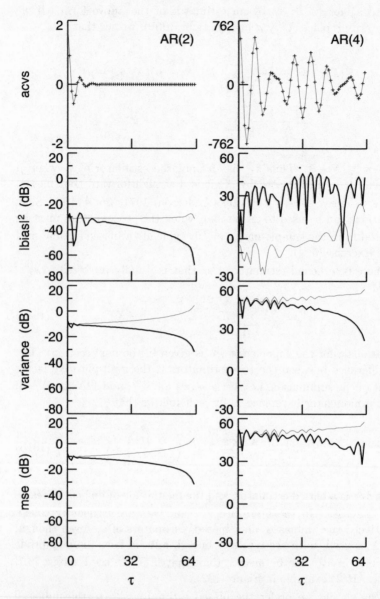

Figure 192. Comparison of bias, variance and mean square error of the unbiased and biased estimators of the acvs for the AR(2) process of Equation (45) (left column) and the AR(4) process of Equation (46a) (right). The true acvs's are shown (to lag 63) in the first row of plots. In the other three rows, the thin and thick curves refer to, respectively, the unbiased and biased estimators. The sample size is 64; the process means are assumed unknown; and both processes are Gaussian.

Here is the basis for this statement. Since

$$\text{mse} = \text{variance} + (\text{bias})^2,$$

where the bias for an estimator $\hat{\theta}$ of the parameter θ is defined as the quantity $E\{\hat{\theta}\} - \theta$, inequality (191c) evidently says that the variability in $\hat{s}_\tau^{(u)}$ is more harmful than the bias in $\hat{s}_\tau^{(p)}$. This effect is particularly marked for $|\tau|$ close to N, where the variability of $\hat{s}_\tau^{(u)}$ is inherently large. Consider, for example, $\hat{s}_{N-1}^{(u)}$ and $\hat{s}_{N-1}^{(p)}$ – these two quantities are, respectively,

$$(X_1 - \bar{X})(X_N - \bar{X}) \text{ and } \frac{1}{N}(X_1 - \bar{X})(X_N - \bar{X}), \qquad (193)$$

from which we can see that the variance of $\hat{s}_{N-1}^{(u)}$ is N^2 times the variance of $\hat{s}_{N-1}^{(p)}$. As concrete examples, Figure 192 shows the true acvs for the AR(2) process defined by Equation (45) (upper left-hand plot) and for the AR(4) process defined by (46a) (upper right-hand plot) for lags 0 to 63. Under the assumptions that these processes are Gaussian, that the sample mean is used to estimate the process mean and that the sample size N is 64, plots of the squared bias, variance and mse versus the lag τ for $\{\hat{s}_\tau^{(u)}\}$ (thin curves) and $\{\hat{s}_\tau^{(p)}\}$ (thick curves) are shown underneath their corresponding acvs plots. Except for the biased estimator in the AR(4) case at large lags, the curves for the variance and the mse are practically the same – this supports the assertion that the variance is the dominant contributor to the mse. For both cases, the mse of $\hat{s}_\tau^{(p)}$ is strictly smaller than that of $\hat{s}_\tau^{(u)}$ for all nonzero τ (note also that, except at a few small lags, the squared bias of $\{\hat{s}_\tau^{(p)}\}$ is smaller than that of $\{\hat{s}_\tau^{(u)}\}$ in the AR(2) case).

[2] If $\{X_t\}$ has a purely continuous spectrum, it can be shown that

$$s_\tau \to 0 \text{ as } |\tau| \to \infty.$$

It makes some sense – but not very much – to prefer an estimator of the sequence $\{s_\tau\}$ that decreases to 0 nicely as $|\tau|$ approaches $N - 1$. Consider, again, (193), which shows that $\hat{s}_{N-1}^{(p)}$ is much closer to 0 than $\hat{s}_{N-1}^{(u)}$ is.

[3] We have shown previously that the acvs of a stationary process must be positive semidefinite. Whereas the sequence $\{\hat{s}_\tau^{(p)}\}$ is always positive semidefinite, the sequence $\{\hat{s}_\tau^{(u)}\}$ need not be so (here we assume that both $\hat{s}_\tau^{(p)}$ and $\hat{s}_\tau^{(u)}$ are defined to be zero for all $|\tau| \geq N$). The following simple proof of the first claim is given in McLeod

and Jiménez (1984, 1985). (The second claim is the subject of Exercise [6.3].) Let

$$d_k \equiv x_k - \bar{x},$$

where the right-hand side is any observed value of $X_k - \bar{X}$. Let $\{\epsilon_t\}$ be a zero mean white noise process with variance $\sigma_\epsilon^2 = 1/N$, and construct the following Nth order moving average process:

$$Y_t = \sum_{k=1}^{N} d_k \epsilon_{t-k}.$$

Then the *theoretical* (and hence necessarily positive semidefinite) acvs of $\{Y_t\}$,

$$s_{\tau,Y} = \sigma_\epsilon^2 \sum_{t=1}^{N-|\tau|} d_t d_{t+|\tau|} = \frac{1}{N} \sum_{t=1}^{N-|\tau|} (x_t - \bar{x})(x_{t+|\tau|} - \bar{x}),$$

equals the *estimated* acvs of $\{X_t\}$ using $\{\hat{s}_\tau^{(p)}\}$ (cf. Equation (43b)). Because this result holds for all realizations, we can conclude that the estimated acvs $\{\hat{s}_\tau^{(p)}\}$ is positive semidefinite. One practical consequence of the possible failure of $\{\hat{s}_\tau^{(u)}\}$ to be positive semidefinite is 'curious behavior of the estimates of the spectrum' (Jenkins and Watts, 1968, p. 183); for example, use of $\{\hat{s}_\tau^{(u)}\}$ can lead to estimates of the sdf – a nonnegative quantity, recall – that are negative for some frequencies.

Figure 195 illustrates points [2] and [3] above. The upper plot shows a time series proportional to the number of cars crossing the Golden Gate Bridge each month versus a running count of months. The lower plot shows the normalized estimates $\hat{s}_\tau^{(u)}/\hat{s}_0^{(u)}$ (thin curve) and $\hat{s}_\tau^{(p)}/\hat{s}_0^{(p)}$ (thick curve) versus the lag τ. Note, first, that $\{\hat{s}_\tau^{(p)}\}$ damps to zero gracefully while $\{\hat{s}_\tau^{(u)}\}$ does not and, second, that $|\hat{s}_\tau^{(u)}| > \hat{s}_0^{(u)}$ for some large values of τ – this cannot happen for a valid theoretical acvs.

For these reasons, statisticians prefer $\hat{s}_\tau^{(p)}$ over $\hat{s}_\tau^{(u)}$ as the estimator of s_τ. Hereinafter we shall use it as our standard 'raw' estimator of the acvs.

Comments and Extensions to Section 6.2

[1] An examination of Equation (191a) reveals that $\hat{s}_\tau^{(p)}$ looks similar to a convolution of the finite sequence $X_1 - \bar{X}, X_2 - \bar{X}, \ldots, X_N - \bar{X}$, with the time reverse of itself (if proper care is taken at the 'end points') – see Equation (119b). This fact can be exploited to develop a procedure to evaluate $\hat{s}_\tau^{(p)}$ for $\tau = 0, \ldots, N-1$ that requires fewer numerical

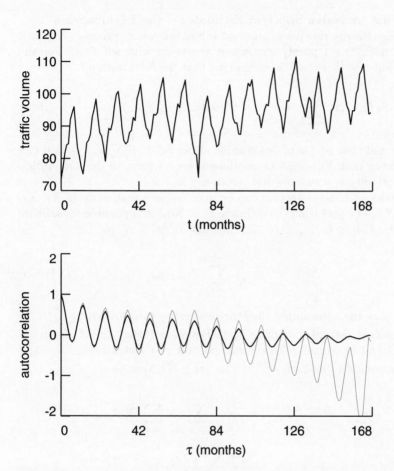

Figure 195. Comparison of unbiased and biased estimates of acvs. The top plot shows a time series related to the monthly traffic volume across the Golden Gate Bridge, while the bottom plot shows two estimates of its autocorrelation sequence – the thin and thick curves are based on, respectively, the unbiased and biased estimates of the acvs.

operations than use of (191a) directly. The details are the subject of Exercise [6.4].

[2]   We have shown that the sequence $\{\hat{s}_\tau^{(p)}\}$ is necessarily positive semidefinite. Since the $\hat{s}_\tau^{(p)}$ terms are a sequence of rv's, this statement means that each realization of these rv's (a sequence of numbers) is positive semidefinite. Newton (1988, p. 165) gives a stronger result: a realization of the sequence $\{\hat{s}_\tau^{(p)}\}$ as defined by Equation (191a) is positive definite if and only if the corresponding realizations of $X_1, \ldots, X_N$ are not all identical (see also McLeod and Jiménez, 1985).

6.3 A Naive Spectral Estimator – the Periodogram

Suppose the discrete parameter real-valued stationary process $\{X_t\}$ with zero mean has a purely continuous spectrum with sdf $S(\cdot)$. For the remainder of this chapter, we assume that the relationship

$$S(f) = \Delta t \sum_{\tau=-\infty}^{\infty} s_\tau e^{-i2\pi f \tau \, \Delta t} \quad \text{with } |f| \le f_{(N)} \equiv \frac{1}{2\,\Delta t} \tag{196a}$$

holds and that $S(\cdot)$ is in fact continuous for all f. If $\sum |s_\tau| < \infty$, it can be shown that $S(\cdot)$ must be continuous everywhere, so the assumption of continuity is restrictive but not overly so.

Given a time series that can be regarded as a realization of X_1, X_2, ..., X_N, our problem is to estimate $S(\cdot)$. Now it is possible to estimate s_τ for $\tau = 0, \pm 1, \ldots, \pm(N-1)$, but *not* for $|\tau| \ge N$, by

$$\hat{s}_\tau^{(p)} = \frac{1}{N} \sum_{t=1}^{N-|\tau|} X_t X_{t+|\tau|} \tag{196b}$$

(this uses the assumption that the process mean is known to be 0). It thus seems natural to replace s_τ in Equation (196a) by $\hat{s}_\tau^{(p)}$ for $|\tau| \le N-1$ and to truncate the summation over τ at the points $\pm(N-1)$ – this amounts to *defining* $\hat{s}_\tau^{(p)} = 0$ for $|\tau| \ge N$. Now

$$\Delta t \sum_{\tau=-(N-1)}^{N-1} \hat{s}_\tau^{(p)} e^{-i2\pi f \tau \, \Delta t} = \frac{\Delta t}{N} \sum_{\tau=-(N-1)}^{N-1} \sum_{t=1}^{N-|\tau|} X_t X_{t+|\tau|} e^{-i2\pi f \tau \, \Delta t}$$

$$= \frac{\Delta t}{N} \sum_{j=1}^{N} \sum_{k=1}^{N} X_j X_k e^{-i2\pi f(k-j)\,\Delta t}$$

$$= \frac{\Delta t}{N} \left| \sum_{t=1}^{N} X_t e^{-i2\pi f t \, \Delta t} \right|^2 \equiv \hat{S}^{(p)}(f) \tag{196c}$$

after a change of variables in the double summation, which we can justify as follows (see also item [2] in the Comments and Extensions). Consider the $N \times N$ matrix whose (j, k)th element is $X_j X_k \exp\left[-i2\pi f(k-j)\,\Delta t\right]$. The double summation in the second line of (196c) is just the summation of all the elements of this matrix. The inner summation adds up the elements of the jth row, while the outer summation ranges over all rows. The double summation in the first line of (196c) again sums up all the elements of the matrix, but now by diagonals defined by $\tau = k - j$: here the inner summation adds up the elements of the τth diagonal (note that there are $N - |\tau|$ elements in that diagonal), while the outer summation ranges over all $2N - 1$ diagonals.

The function $\hat{S}^{(p)}(\cdot)$ defined above is known as the *periodogram* (even though it is a function of frequency and not period). Like $S(\cdot)$, it is defined over the interval $[-f_{(N)}, f_{(N)}]$. If we restrict our attention momentarily to just the frequency $f = k/(N\,\Delta t)$, where k is an integer such that $|k| \leq \lfloor N/2 \rfloor$, we see that the periodogram at that frequency is (almost) the squared modulus of the kth component of the discrete Fourier transform of the sequence $X_1, \ldots, X_N$ (see Equation (110a); the qualifier 'almost' is due to the presence of the scaling factor $\Delta t/N$ instead of $(\Delta t)^2$). We note that $\{\hat{s}_\tau^{(p)}\}$ and $\hat{S}^{(p)}(\cdot)$ are a Fourier transform pair (this explains the superscript (p) on $\hat{s}_\tau^{(p)}$):

$$\{\hat{s}_\tau^{(p)}\} \longleftrightarrow \hat{S}^{(p)}(\cdot)$$

(see Equation (87) and recall that $\hat{s}_\tau^{(p)} \equiv 0$ for $|\tau| \geq N$); hence we have, for continuous frequency,

$$\hat{s}_\tau^{(p)} = \int_{-f_{(N)}}^{f_{(N)}} \hat{S}^{(p)}(f) e^{i2\pi f \tau\,\Delta t}\, df, \quad \tau = 0, \pm 1, \ldots, \pm(N-1)$$

(see Equation (89b)). From Equation (119a) we also have

$$\{\, \hat{s}_\tau^{(p)} : \tau = -(N-1), \ldots, N \,\} \longleftrightarrow \{\, \hat{S}^{(p)}(\tilde{f}_k) : k = -(N-1), \ldots, N \,\}$$

with $\tilde{f}_k \equiv k/(2N\,\Delta t)$. Hence we have, for discrete frequency,

$$\hat{s}_\tau^{(p)} = \frac{1}{2N\,\Delta t} \sum_{k=-(N-1)}^{N} \hat{S}^{(p)}(\tilde{f}_k) e^{i\pi k\tau/N}, \quad \tau = 0, \pm 1, \ldots, \pm(N-1), N.$$

Let us now investigate the sampling properties of $\hat{S}^{(p)}(\cdot)$. We will be chiefly concerned at first with

$$E\{\hat{S}^{(p)}(f)\}, \quad \operatorname{var}\{\hat{S}^{(p)}(f)\}, \quad \text{and} \quad \operatorname{cov}\{\hat{S}^{(p)}(f'), \hat{S}^{(p)}(f)\}, \quad f' \neq f.$$

If $\hat{S}^{(p)}(\cdot)$ were an ideal estimator of $S(\cdot)$, we would have

[1] $E\{\hat{S}^{(p)}(f)\} \approx S(f)$ for all f,
[2] $\operatorname{var}\{\hat{S}^{(p)}(f)\} \to 0$ as $N \to \infty$ and
[3] $\operatorname{cov}\{\hat{S}^{(p)}(f'), \hat{S}^{(p)}(f)\} \approx 0$ for $f' \neq f$.

We shall find, however, that, while [1] is a good approximation for some processes and some values of f, it can be very poor in other cases; [2] is blatantly false when $S(f) > 0$; and [3] holds if f' and f are certain distinct frequencies, namely, the Fourier frequencies $f_k \equiv k/(N\,\Delta t)$.

We start by examining $E\{\hat{S}^{(p)}(f)\}$ and determining how close it is to $S(f)$, the quantity it is supposed to be estimating. From Equations (196c) and (191b) it follows that

$$E\{\hat{S}^{(p)}(f)\} = \Delta t \sum_{\tau=-(N-1)}^{N-1} E\{\hat{s}_\tau^{(p)}\} e^{-i2\pi f\tau\,\Delta t}$$

$$= \Delta t \sum_{\tau=-(N-1)}^{N-1} \left(1 - \frac{|\tau|}{N}\right) s_\tau e^{-i2\pi f\tau\,\Delta t}. \qquad (198a)$$

This is a convenient expression for computing $E\{\hat{S}^{(p)}(f)\}$ for various stationary processes with known acvs, but we can obtain more insight from the following (this can be justified using Exercise [4.5b]):

$$E\{\hat{S}^{(p)}(f)\} = N\,\Delta t \int_{-f_{(N)}}^{f_{(N)}} \mathcal{D}_N^2\left([f-f']\,\Delta t\right) S(f')\,df',$$

where $\mathcal{D}_N(\cdot)$ is Dirichlet's kernel, which we met in Exercise [1.3c] and again in Section 3.7. If we define

$$\mathcal{F}(f) \equiv N\,\Delta t\,\mathcal{D}_N^2(f\,\Delta t) = \frac{\Delta t \sin^2(N\pi f\,\Delta t)}{N \sin^2(\pi f\,\Delta t)}, \qquad (198b)$$

we have

$$E\{\hat{S}^{(p)}(f)\} = \int_{-f_{(N)}}^{f_{(N)}} \mathcal{F}(f-f')S(f')\,df'. \qquad (198c)$$

The function $\mathcal{F}(\cdot)$ is known in the literature as *Fejér's kernel*. Note that the true sdf is thus convolved with Fejér's kernel to give $E\{\hat{S}^{(p)}(\cdot)\}$.

To understand the implications of Equation (198c), we need to investigate the properties of Fejér's kernel.

[1] For all integers $N \geq 1$, $\mathcal{F}(f) \to N\,\Delta t$ as $f \to 0$ (an application of l'Hospital's rule);

[2] for $N > 1$, $f \in [-f_{(N)}, f_{(N)}]$ and $f \neq 0$, $\mathcal{F}(f) < \mathcal{F}(0)$ (a calculus problem);

[3] for $f \in [-f_{(N)}, f_{(N)}]$ and $f \neq 0$, $\mathcal{F}(f) \to 0$ as $N \to \infty$ (the N in the denominator of (198b) dominates);

[4] for any integer $k \neq 0$ such that $f_k \equiv k/(N\,\Delta t)$ satisfies $|f_k| \leq f_{(N)}$, $\mathcal{F}(f_k) = 0$ (because $\sin(k\pi) = 0$); and

[5] it follows from Exercise [1.3c] that

$$\left| \sum_{t=1}^{N} e^{i2\pi ft\,\Delta t} \right|^2 = N^2 \mathcal{D}_N^2(f\,\Delta t),$$

so we can write

$$\mathcal{F}(f) = \frac{\Delta t}{N} \sum_{t=1}^{N} \sum_{u=1}^{N} e^{i2\pi f(t-u)\,\Delta t}.$$

It follows from a term by term integration of the right-hand side that

$$\int_{-f_{(N)}}^{f_{(N)}} \mathcal{F}(f)\,df = 1. \tag{199}$$

The quantity $10\log_{10}(\mathcal{F}(f))$ is plotted versus f in Figure 200 for $\Delta t = 1$ and $N = 4$, 16 and 64. We see that $\mathcal{F}(\cdot)$ is symmetric about the origin and consists of a broad central peak (also called the central lobe) and $N - 2$ sidelobes which decrease in magnitude as $|f|$ increases. The squared magnitude of the first sidelobes is only about 13 dB below that of the central lobe.

From properties [1], [3] and [5], it follows that, as $N \to \infty$, Fejér's kernel acts as a Dirac delta function with an infinite spike at $f = 0$. Since $S(\cdot)$ is assumed to be continuous, we have from Equation (198c)

$$\lim_{N \to \infty} E\{\hat{S}^{(p)}(f)\} = S(f);$$

i.e., for all f, $\hat{S}^{(p)}(f)$ is an asymptotically unbiased estimator of $S(f)$.

Now, the fact that $\hat{S}^{(p)}(f)$ is asymptotically unbiased does not mean that its bias is necessarily small for any particular N. A quote from Thomson (1982) is appropriate here:

> ... for processes with spectra typical of those encountered in engineering, the sample size must be extraordinarily large for the periodogram to be reasonably unbiased. While it is not clear what sample size, if any, gives reasonably valid results, in my experience periodogram estimates computed from 1.2 million data points on the WT4 waveguide project were too badly biased to be useful. The best that could be said for them is that they were so obviously incorrect as not to be dangerously misleading. In other applications where less is known about the process, such errors may not be so obvious.

There are some stationary processes, however, for which the bias in the periodogram is negligible for small N. For example, the following theorem gives us a rate of decrease for the bias under a certain regularity condition (Brillinger, 1981a, p. 123): if $\{X_t\}$ is a stationary process with acvs $\{s_\tau\}$ such that

$$\sum_{\tau=-\infty}^{\infty} |\tau s_\tau| < \infty, \quad \text{then} \quad E\{\hat{S}^{(p)}(f)\} = S(f) + O\left(\frac{1}{N}\right).$$

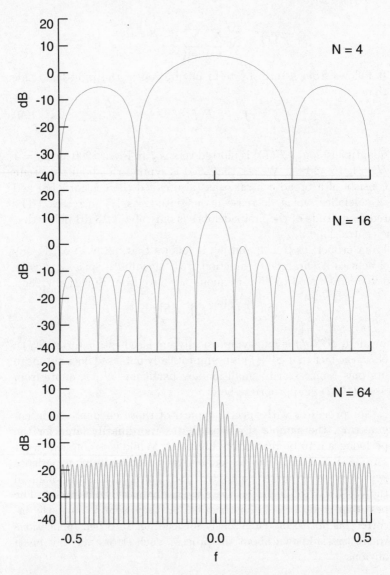

Figure 200. Fejér's kernel for sample sizes $N = 4$, 16 and 64.

The regularity condition essentially implies that the acvs dies down to zero quickly; alternatively, it implies that $S(\cdot)$ is a smooth function (for example, the condition holds if $S(\cdot)$ has a continuous first derivative). When this is true, the bias in the periodogram decreases at the rate of $1/N$ (but note that even here nothing is said about the *absolute* magnitude of the bias). With a stronger regularity condition (say,

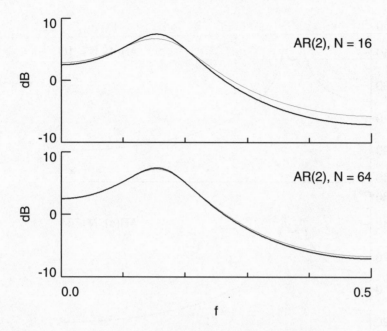

Figure 201. Bias properties of periodogram for process with low dynamic range. The thick curve in each plot is the true sdf $S(\cdot)$ for the AR(2) process described by Equation (45). The thin curves are $E\{\hat{S}^{(p)}(\cdot)\}$ for sample sizes $N = 16$ and 64 plotted versus frequency. The vertical axis is in decibels – either $10 \log_{10}(S(f))$ or $10 \log_{10}(\hat{S}^{(p)}(f))$.

$\sum_{\tau=-\infty}^{\infty} |\tau^2 s_\tau| < \infty$), the rate of decrease is even faster (see Brillinger, 1981b, for details).

Let us now look at the bias in the periodogram for three specific stationary processes (each with $\Delta t = 1$ so that $f_{(N)} = 1/2$). The first process $\{X_{t,1}\}$ is white noise with a variance of 1; the next is the second-order autoregressive (AR) process $\{X_{t,2}\}$ defined by Equation (45); and the third is the fourth-order AR process $\{X_{t,4}\}$ defined by Equation (46a). The shape of the sdf for $\{X_{t,2}\}$ is shown by the thick curves in Figure 201, while that of $\{X_{t,4}\}$ is shown by the thick curves in Figures 202 and 203. A useful crude characterization of these three processes is in terms of their *dynamic range*, which we define by the ratio

$$10 \log_{10}\left(\frac{\max_f S(f)}{\min_f S(f)}\right).$$

The dynamic range of a white noise process is 0; $\{X_{t,2}\}$ and $\{X_{t,4}\}$ have dynamic ranges of about 14 dB and 65 dB, respectively (the latter is typical of the spectra for many geophysical processes).

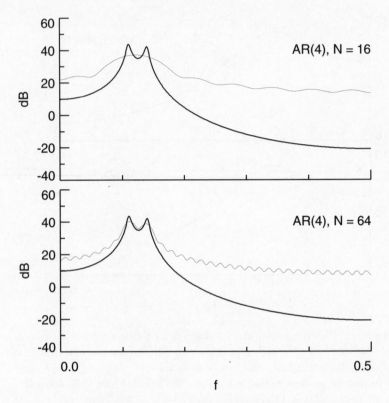

Figure 202. Bias properties of periodogram for process with high dynamic range, part 1.

The sdf for the white noise process $\{X_{t,1}\}$ is just 1 for all $f \in [-1/2, 1/2]$. It follows from Equations (198c) and (199) that

$$E\{\hat{S}^{(p)}(f)\} = \int_{-1/2}^{1/2} \mathcal{F}(f - f') \, df' = 1,$$

so the periodogram is unbiased in the case of white noise.

Figure 201 shows $E\{\hat{S}^{(p)}(f)\}$ as a function of f for $N = 16$ and 64 for $\{X_{t,2}\}$. For both plots in the figure the thin curve shows $E\{\hat{S}^{(p)}(f)\}$, while the thick curve is the true sdf. For $N = 16$ we see that $E\{\hat{S}^{(p)}(f)\}$ is within 2 dB of $S(f)$ for all f, while the two functions nearly coincide for $N = 64$. Thus the periodogram is already largely unbiased by $N = 64$ for this process of rather small dynamic range.

The situation is quite different for $\{X_{t,4}\}$, the process with the largest dynamic range of the three. For $N = 64$ (bottom plot of Figure 202) there are biases of more than 30 dB (i.e., three orders of magnitude) for f close to $1/2$. Even for $N = 1024$ (bottom plot of Figure 203),

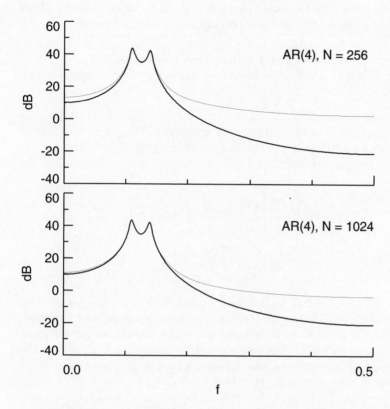

Figure 203. Bias properties of periodogram for process with high dynamic range, part 2.

$E\{\hat{S}^{(p)}(f)\}$ and $S(f)$ differ by almost 20 dB (two orders of magnitude) for values of f for which $S(f)$ is small.

The bias in the periodogram for processes with high dynamic range can be attributed to the sidelobes of Fejér's kernel. Equation (198c) shows that $E\{\hat{S}^{(p)}(f)\}$ is the value of a convolution at f of that kernel with the true sdf of the process. If Fejér's kernel consisted of just a central lobe, the convolution would involve just frequencies close to $S(f)$; because of the sidelobes, the convolution also involves distant frequencies. For a process such as $\{X_{t,4}\}$ with a high dynamic range, $E\{\hat{S}^{(p)}(f)\}$ is badly biased at those frequencies for which $S(f)$ is small compared with the rest of the sdf. This transfer of power from one region of the sdf to another via the kernel of the convolution is known as *leakage*.

There are two common techniques for lessening the bias in the periodogram – tapering and prewhitening. The former modifies the kernel in the convolution in Equation (198c); the latter preprocesses a time series to reduce the dynamic range of the sdf to be estimated. These techniques are defined and discussed in the next two sections.

Comments and Extensions to Section 6.3

[1] If $E\{X_t\} = \mu \neq 0$ as we have assumed so far, we can replace X_t by either

$$X_t - \mu \ \text{ or } \ X_t - \bar{X}$$

accordingly as μ is either known or unknown. Two interesting points about this replacement are the subject of Exercise [6.5]. First, for *any* value C independent of t,

$$\tilde{S}^{(p)}(f_k) \equiv \frac{\Delta t}{N} \left| \sum_{t=1}^{N} (X_t - C) e^{-i2\pi f_k t \Delta t} \right|^2$$

$$= \frac{\Delta t}{N} \left| \sum_{t=1}^{N} X_t e^{-i2\pi f_k t \Delta t} \right|^2 \equiv \hat{S}^{(p)}(f_k),$$

where $f_k \equiv k/(N \Delta t)$ for integers k such that $0 < f_k \leq f_{(N)}$; i.e., the frequency f_k is one of the nonzero Fourier frequencies. Second, $\tilde{S}^{(p)}(0) = 0$ when $C = \bar{X}$, the sample mean. Since the periodogram is a continuous function of frequency, this latter fact implies that the periodogram of a centered time series can be a badly biased estimate of $S(f)$ for f close to 0 (say, $|f| < f_1 = 1/(N \Delta t)$).

[2] Note that Equation (196c) follows from a variation of the autocorrelation theorem (see Equation (85b)):

$$\sum_{\tau=-(N-1)}^{N-1} \hat{s}_\tau^{(p)} e^{-i2\pi f \tau \Delta t} = \frac{1}{N} \sum_{\tau=-(N-1)}^{N-1} \left(\sum_{t=1}^{N-|\tau|} X_t X_{t+|\tau|} \right) e^{-i2\pi f \tau \Delta t}$$

is essentially the Fourier transform of a convolution of a sequence with its time reverse, and

$$\left| \sum_{t=1}^{N} X_t e^{-i2\pi f t \Delta t} \right|^2$$

is essentially the product of the Fourier transform of the sequence and its complex conjugate. Note also that Equations (198a) and (198c) can be related via an appropriate convolution theorem: (198a) is essentially a Fourier transform of a *product* in the lag domain, which becomes the *convolution* (198c) in the frequency domain.

[3] In older works on spectral analysis, the spectrum at frequency f of a segment $X_1, X_2, \ldots, X_N$ of a stationary process is sometimes *defined* – at least implicitly – to be $E\{\hat{S}^{(p)}(f)\}$. This matches our definition

in the case of white noise, but the plots in Figures 202 and 203 show that the two definitions can be quite different in general. In fact, if we translate the definition (138) for the sdf discussed in Section 4.2 into the notation of this chapter, we have

$$S(f) = \lim_{N \to \infty} E\{\hat{S}^{(p)}(f)\}.$$

Thus an operational difficulty with using $E\{\hat{S}^{(p)}(f)\}$ as the definition for the spectrum is that the spectrum can depend critically on the sample size. This is undesirable, if for no other reason than the difficulty it introduces in meaningfully comparing the spectra from data collected in different experiments concerning the same physical phenomenon.

[4]   We have chosen to label our sample of size N as $X_1, X_2, \ldots, X_N$; a second common convention is $X_0, X_1, \ldots, X_{N-1}$. The periodogram is the same for both conventions: if we let $X_t' \equiv X_{t+1}$, then

$$\left| \sum_{t=0}^{N-1} X_t' e^{-i2\pi f t\, \Delta t} \right|^2 = \left| \sum_{t=1}^{N} X_t e^{-i2\pi f t\, \Delta t} \right|^2,$$

even though the complex-valued quantities

$$\sum_{t=0}^{N-1} X_t' e^{-i2\pi f t\, \Delta t} \text{ and } \sum_{t=1}^{N} X_t e^{-i2\pi f t\, \Delta t}$$

differ in phase.

[5]   Suppose we let Γ_N denote the variance-covariance matrix of the vector $[X_1, X_2, \ldots, X_N]^T$ whose elements are a segment of size N from a stationary process with sdf $S(\cdot)$ (see Section 2.4). The dynamic range of $S(\cdot)$ has an interesting relationship to an accepted measure of the ill-conditioning of the matrix Γ_N (this is important to know if we wish to compute its inverse). This measure d_N is given by the ratio of the largest eigenvalue of Γ_N, say λ_N, to its smallest eigenvalue, say ν_N:

$$d_N \equiv \frac{\lambda_N}{\nu_N}.$$

Grenander and Szegő (1984) have shown that

$$\lim_{N \to \infty} d_N = \frac{\max_f S(f)}{\min_f S(f)}.$$

Hence the dynamic range of $S(\cdot)$ is approximately equal to $10 \log_{10}(d_N)$, and, conversely, an approximation to d_N can be obtained from the dynamic range.

6.4 Bias Reduction – Tapering

We said in the previous section that much of the bias in the periodogram can be attributed to the sidelobes of Fejér's kernel $\mathcal{F}(\cdot)$. If these sidelobes were substantially smaller, we would have considerably less bias in $\hat{S}^{(p)}(\cdot)$ due to the leakage of power from one portion of the sdf to another. Tapering is a technique that effectively reduces the sidelobes associated with Fejér's kernel (with some tradeoffs). It was introduced in the context of spectral analysis by Blackman and Tukey (1958, p. 93).

Let X_1, X_2, ..., X_N be a portion of length N of a zero mean stationary process with sdf $S(\cdot)$. We form the product $h_t X_t$ for each value of t, where $\{h_t\}$ is a suitable sequence of real-valued constants called a *data taper* (other names in the literature are *data window*, *linear taper*, *linear window*, *fader* and *shading sequence*). Let

$$J(f) \equiv (\Delta t)^{1/2} \sum_{t=1}^{N} h_t X_t e^{-i2\pi f t \, \Delta t}.$$

By the spectral representation theorem

$$X_t = \int_{-f_{(N)}}^{f_{(N)}} e^{i2\pi f' t \, \Delta t} \, dZ(f'),$$

where $\{Z(\cdot)\}$ is an orthogonal process. Thus

$$\begin{aligned}
J(f) &= (\Delta t)^{1/2} \sum_{t=1}^{N} h_t \left(\int_{-f_{(N)}}^{f_{(N)}} e^{i2\pi f' t \, \Delta t} \, dZ(f') \right) e^{-i2\pi f t \, \Delta t} \\
&= \int_{-f_{(N)}}^{f_{(N)}} (\Delta t)^{1/2} \sum_{t=1}^{N} h_t e^{-i2\pi (f-f') t \, \Delta t} \, dZ(f') \\
&= \frac{1}{(\Delta t)^{1/2}} \int_{-f_{(N)}}^{f_{(N)}} H(f - f') \, dZ(f'),
\end{aligned}$$

(206a)

where $\{h_t\} \longleftrightarrow H(\cdot)$ under the assumption that $\{h_t\}$ is an infinite sequence with $h_t = 0$ for $t < 1$ and $t > N$; i.e.,

$$H(f) \equiv \Delta t \sum_{t=1}^{N} h_t e^{-i2\pi f t \, \Delta t}.$$

(206b)

Let us define

$$\hat{S}^{(d)}(f) \equiv |J(f)|^2 = \Delta t \left| \sum_{t=1}^{N} h_t X_t e^{-i2\pi f t \, \Delta t} \right|^2.$$

(206c)

Because $\{Z(\cdot)\}$ has orthogonal increments, we have (by an argument identical to that used in Exercise [4.5b])

$$E\{\hat{S}^{(d)}(f)\} = \int_{-f_{(N)}}^{f_{(N)}} \mathcal{H}(f - f')S(f')\, df', \qquad (207a)$$

where

$$\mathcal{H}(f) \equiv \frac{1}{\Delta t}\,|H(f)|^2 = \Delta t \left| \sum_{t=1}^{N} h_t e^{-i2\pi ft\,\Delta t} \right|^2 .$$

For computational purposes, we note that

$$E\{\hat{S}^{(d)}(f)\} = \Delta t \sum_{\tau=-(N-1)}^{N-1} \left(s_\tau \sum_{t=1}^{N-|\tau|} h_t h_{t+|\tau|} \right) e^{-i2\pi f\tau\,\Delta t} \qquad (207b)$$

(see Exercise [6.8]).

Spectral estimators that take the form of Equation (206c) are referred to as *direct spectral estimators*, hence the superscript (d); they are also called *modified periodograms*. If $h_t = 1/\sqrt{N}$ for $1 \le t \le N$ (the so-called 'rectangular' or 'default' taper), we have

$$\hat{S}^{(d)}(f) = \hat{S}^{(p)}(f) \text{ and } \mathcal{H}(f) = \mathcal{F}(f).$$

That is, $\hat{S}^{(d)}(\cdot)$ reverts to the periodogram; $\mathcal{H}(\cdot)$ becomes Fejér's kernel; and Equation (207b) reduces to Equation (198a). The function $\mathcal{H}(\cdot)$ is often called the *spectral window* of the direct spectral estimator $\hat{S}^{(d)}(\cdot)$. We note that

$$\hat{S}^{(d)}(f) = \Delta t \sum_{\tau=-(N-1)}^{N-1} \hat{s}_\tau^{(d)} e^{-i2\pi f\tau\,\Delta t}, \qquad (207c)$$

where

$$\hat{s}_\tau^{(d)} \equiv \begin{cases} \sum_{t=1}^{N-|\tau|} h_t X_t h_{t+|\tau|} X_{t+|\tau|}, & |\tau| \le N-1; \\ 0, & |\tau| \ge N \end{cases} \qquad (207d)$$

(see Exercise [6.9]). Since $\{\hat{s}_\tau^{(d)}\} \longleftrightarrow \hat{S}^{(d)}(\cdot)$, we see that the sequence $\{\hat{s}_\tau^{(d)}\}$ is the estimator of the acvs corresponding to $\hat{S}^{(d)}(\cdot)$, the sdf estimator.

Integration of both sides of (207b) yields

$$\int_{-f_{(N)}}^{f_{(N)}} E\{\hat{S}^{(d)}(f)\}\, df$$

$$= \Delta t \sum_{\tau=-(N-1)}^{N-1} \left(s_\tau \sum_{t=1}^{N-|\tau|} h_t h_{t+|\tau|} \right) \int_{-f_{(N)}}^{f_{(N)}} e^{-i2\pi f\tau\,\Delta t}\, df$$

$$= s_0 \sum_{t=1}^{N} h_t^2$$

since

$$\int_{-f_{(N)}}^{f_{(N)}} e^{-i2\pi f \tau \, \Delta t} \, df = \begin{cases} 2f_{(N)} = 1/\Delta t, & \tau = 0; \\ 0, & \text{otherwise.} \end{cases}$$

If we normalize $\{h_t\}$ such that

$$\sum_{t=1}^{N} h_t^2 = 1 \qquad\qquad (208a)$$

(note that this holds for the rectangular data taper), we have the desirable property

$$\int_{-f_{(N)}}^{f_{(N)}} E\{\hat{S}^{(d)}(f)\} \, df = s_0 = \int_{-f_{(N)}}^{f_{(N)}} S(f) \, df. \qquad (208b)$$

Because $\{h_t\} \longleftrightarrow H(\cdot)$ and since $\mathcal{H}(\cdot)$ is proportional to the square modulus of $H(\cdot)$, it follows from Parseval's theorem (Equation (89c)) that

$$\sum_{t=1}^{N} h_t^2 = \int_{-f_{(N)}}^{f_{(N)}} \mathcal{H}(f) \, df.$$

The normalization (208a) is thus equivalent to

$$\int_{-f_{(N)}}^{f_{(N)}} \mathcal{H}(f) \, df = 1. \qquad\qquad (208c)$$

For a white noise process with variance σ^2 and sdf $S(f) = \sigma^2 \, \Delta t$, it follows from (207a) that

$$E\{\hat{S}^{(d)}(f)\} = \sigma^2 \, \Delta t \int_{-f_{(N)}}^{f_{(N)}} \mathcal{H}(f - f') \, df'$$

$$= \sigma^2 \, \Delta t \int_{-f_{(N)}}^{f_{(N)}} \mathcal{H}(f) \, df = \sigma^2 \, \Delta t$$

(the second line above follows from the first since $\mathcal{H}(\cdot)$ is periodic with period $2f_{(N)}$). Thus direct spectral estimators that satisfy (208a) are unbiased in the case of white noise – a fact that we have already noted for the special case of a rectangular data taper.

The key idea behind tapering is to select $\{h_t\}$ so that $\mathcal{H}(\cdot)$ has much smaller sidelobes than $\mathcal{F}(\cdot)$. A discontinuity in a sequence of numbers means that there will be ripples in its Fourier transform (see Section 3.7). Since $h_t = 0$ for $t < 1$ and $t > N$, we can have small sidelobes in $\mathcal{H}(\cdot)$ by making its associated h_t terms go to 0 in a smoother fashion than the abrupt change in the rectangular data taper. That this indeed can be

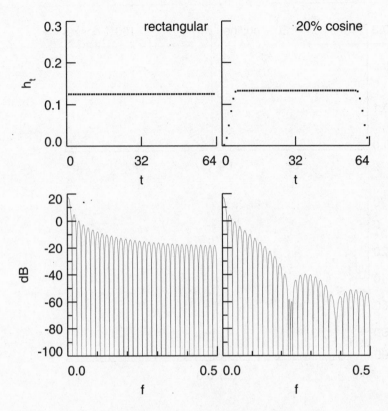

Figure 209. Two different data tapers (top plots) and, under them, the associated spectral windows $\mathcal{H}(\cdot)$ for sample size $N = 64$ and $\Delta t = 1$. For the top plots, the horizontal axes are the index t, while the vertical axes are h_t, the value of a taper at index t. For the bottom plots, the horizontal axes are frequency (cycles per unit time), while the vertical axes are $\mathcal{H}(\cdot)$ on a decibel scale.

done is shown in Figures 209 and 210. The top plots of these figures show four different tapers for sample size $N = 64$. For the sake of comparison, the first taper (upper left plot of Figure 209) is just the rectangular data taper. The second taper (upper right plot) is the so-called 20% cosine taper. The $p \times 100\%$ cosine taper is defined by

$$
h_t = \begin{cases} \dfrac{C}{2}\left[1 - \cos\left(\dfrac{2\pi t}{\lfloor pN\rfloor + 1}\right)\right], & 1 \le t \le \dfrac{\lfloor pN\rfloor}{2}; \\ C, & \dfrac{\lfloor pN\rfloor}{2} < t < N + 1 - \dfrac{\lfloor pN\rfloor}{2}; \\ \dfrac{C}{2}\left[1 - \cos\left(\dfrac{2\pi(N+1-t)}{\lfloor pN\rfloor + 1}\right)\right], & N + 1 - \dfrac{\lfloor pN\rfloor}{2} \le t \le N, \end{cases}
$$

$$(209)$$

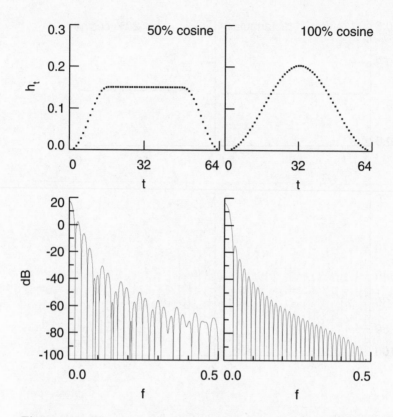

Figure 210. Two more data tapers and associated spectral windows.

where C is a normalizing constant that forces $\sum_{t=1}^{N} h_t^2 = 1$. Thus the 20% cosine taper is like the rectangular taper, except the first 10% and last 10% of the points have been replaced by portions of a raised cosine. The third taper (upper left plot of Figure 210) is a 50% cosine taper. The fourth taper (upper right plot) is a 100% cosine taper – sometimes called a *Hanning taper* (see Exercise [6.10]). Note that the effect of the latter three tapers is to 'force' the tapered series $\{h_t X_t\}$ toward zero in a smooth fashion near $t = 1$ and $t = N$.

The lower plots of Figures 209 and 210 show the spectral windows corresponding to the tapers in the upper plots. The spectral window for the default rectangular taper is just Fejér's kernel again. Note that the other spectral windows all have significantly smaller sidelobes. However, notice that the widths of the main lobes of these spectral windows are all larger than that of Fejér's kernel. We have suppressed the sidelobes in the windows at the expense of wider main lobes.

A particularly interesting class of data tapers to consider are those formed from zeroth-order discrete prolate spheroidal sequences (dpss's)

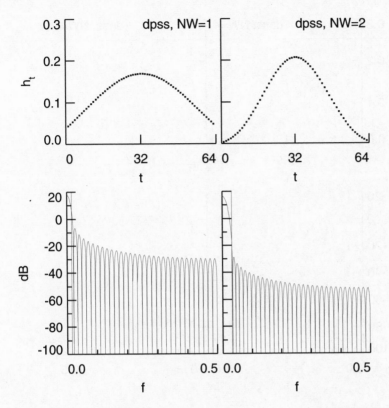

Figure 211. Two different zeroth-order dpss data tapers and associated spectral windows $\mathcal{H}(\cdot)$ ($NW = 1$ for left-hand plots and $NW = 2$ for right-hand plots). The thin vertical lines near the vertical axes in the bottom plots indicate the location of $f = W$.

– these were introduced in Section 3.9 (see the discussion following Equation (105b)) and also play a prominent role in Chapter 7. Recall that the zeroth-order dpss with parameters N and W is a sequence of length N that has the largest possible concentration of its energy in the frequency interval $[-W, W]$ as measured by the ratio

$$\beta^2(W) \equiv \int_{-W}^{W} |H(f)|^2 \, df \bigg/ \int_{-f_{(N)}}^{f_{(N)}} |H(f)|^2 \, df$$

(this is Equation (101b) under the assumption that Δt is not necessarily 1). Now the spectral window $\mathcal{H}(\cdot)$ is proportional to $|H(\cdot)|^2$ and has most of its energy concentrated in its central lobe for a zeroth-order dpss. This implies that the sidelobes of $\mathcal{H}(\cdot)$ should be small. There is, however, a natural tradeoff between the width of the central lobe

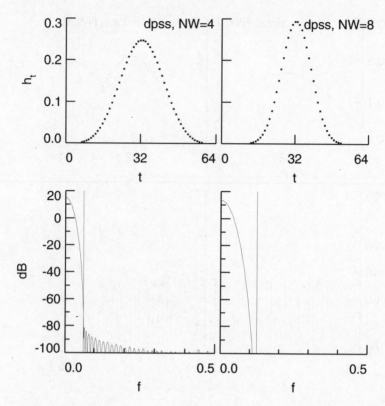

Figure 212. Two more zeroth-order dpss data tapers and associated spectral windows $\mathcal{H}(\cdot)$ ($NW = 4$ for left-hand plots and $NW = 8$ for right-hand plots).

and the energy in the sidelobes – as the former increases, the latter decreases. This is illustrated in Figures 211 and 212. The upper plots for these figures show the zeroth-order dpss tapers for $N = 64$ and $NW = 1, 2, 4$ and 8 (here we assume $\Delta t = 1$). The lower plots show the corresponding spectral windows. Note that the energy in the side-lobes decreases markedly as W (proportional to the width of the central lobe) increases. It is particularly interesting to compare the spectral window for the $NW = 1$ dpss taper (lower left-hand plot on Figure 211) with the spectral window for the periodogram (lower left-hand plot on Figure 209). The central lobe widths of these two windows are nearly the same, yet the sidelobes for the window associated with the dpss taper are about 10 dB below those of Fejér's kernel (the window for the rectangular taper).

Figures 213 and 214 illustrate how the tradeoff between the central lobe width and the size of the sidelobes affects the first moment of direct

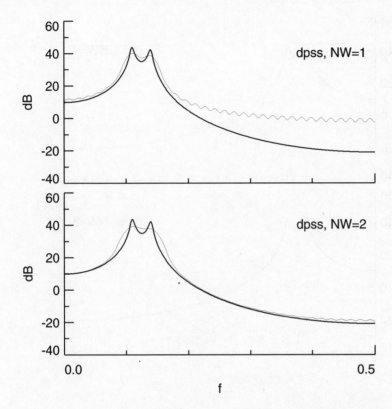

Figure 213. Comparison of true sdf and $E\{\hat{S}^{(d)}(\cdot)\}$, part a. The thick curves are the true sdf of the AR(4) process of Equation (46a). The thin curves are $E\{\hat{S}^{(d)}(\cdot)\}$ for two different direct spectral estimators that use different zeroth-order dpss data tapers. The periodogram for this case is the lower plot of Figure 202. The horizontal axis is frequency (in cycles/Δt with Δt assumed to be 1), while the vertical axis is $10 \log_{10}(E\{\hat{S}^{(d)}(\cdot)\})$. The sample size is $N = 64$.

spectral estimators. The thin curve in each plot shows $E\{\hat{S}^{(d)}(f)\}$ as a function of f for the AR(4) process $\{X_{t,4}\}$ with $N = 64$. The thick curves are the true sdf. The four estimators in the four plots use the dpss data tapers illustrated in Figures 211 and 212. The expected value of the periodogram for this case is the bottom plot of Figure 202. If we compare the expected value of the periodogram with that of the $NW = 1$ dpss direct spectral estimator (the lower plot of Figure 202 and upper plot of Figure 213, respectively), we see that the latter is uniformly closer to the true sdf than the former, except in the region of the twin peaks. Note, in particular, that the bias in the dpss estimator is down about 10 dB in the high frequency portions of the sdf. We can

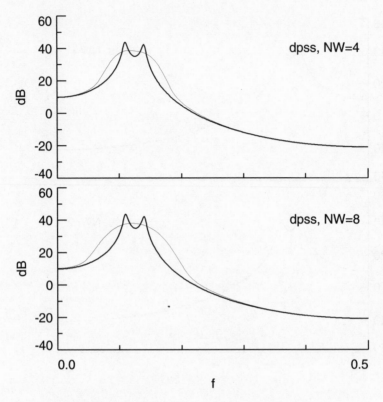

Figure 214. Comparison of true sdf and $E\{\hat{S}^{(d)}(\cdot)\}$, part b.

further reduce the bias in that region by increasing NW, but note the tradeoff apparent in Figures 213 and 214: as the width of the central lobe of the spectral window increases (i.e., as W increases), the two peaks in the true sdf are no longer apparent in $E\{\hat{S}^{(d)}(\cdot)\}$. Since this expectation is in fact the convolution of the true sdf with the spectral window (Equation (207a)), we are in effect distorting the sdf when we use a spectral window with a wide central lobe. This example shows that there is an inherent tradeoff between reducing bias due to leakage (i.e., due to the sidelobes) and introducing bias due to the width of the central lobe. We consider this tradeoff in more detail in Chapter 7, when we introduce the concepts of broad-band bias and local bias.

In summary, tapering is a useful method for reducing the bias due to leakage in direct spectral estimators. This is particularly important for spectra with large dynamic range (and of less importance for those with small dynamic range). When spectral analysis is being used as a tool for an exploratory data analysis (i.e., little is known about the true sdf of the process generating the data), it is important to use spectral estimators with good bias characteristics to make sure that it is the data and *not* the spectral window that is determining the shape of the estimated sdf. Note that so far we have considered the effects of tapering only on the first moment properties of direct spectral estimators. Other advantages and disadvantages of tapering are presented in Section 6.6, where we discuss second moment properties. (For a good overall discussion of tapering, see Brillinger, 1981b.)

Comments and Extensions to Section 6.4

[1]   As we shall see in Chapter 7, tapering plays a fundamental role in modern nonparametric spectral estimation. Some researchers, however, regard it as an undesirable (and suspicious) operation. For example, Yuen (1979) states the following (edited slightly to follow the notation and nomenclature of this chapter):

> ... the idea [of tapering] is statistically quite unsound. ... the most serious criticism is this: [tapering] effectively 'throws data away' because some of the values of X_t are multiplied into small weighting factors. ... Why is this statistically bad? In the theory of estimation it is the basic idiom of the subject that the weight given to any individual value should decrease with its variance. That is, the more uncertainty there is in the data point, the less weight it should be given. Now, X_t is a time-invariance [*sic*] random process. Its variance is constant. Thus all its values should receive equal weight.

In advocating maximum entropy spectral analysis (see Chapter 9) in preference to nonparametric spectral analysis, Fougere (1985) stated

> Also there is never any tapering – which really ought to be called tampering – of the data with resulting loss of information!

Evidently these researchers regard tapering as an operation that effectively modifies a time series. From our viewpoint, tapering is an operation that replaces Fejér's kernel in Equation (198c) with one having better sidelobe properties. Its actions should thus be judged in the frequency domain rather than the time domain. There are some caveats and tradeoffs involved in using data tapers, but, if these are observed, tapering can be a useful way to control the bias in direct spectral estimators for an sdf with a large dynamic range. Moreover, the multitaper approach discussed in Chapter 7 largely overcomes Yuen's objection. Also, we note in Chapter 9 that, in one of its formulations, maximum entropy spectral analysis can actually be improved when used in combination with data tapering.

It should be clear from our AR(2) example that there are certain processes for which tapering with a single data taper accomplishes nothing useful (the same statement does not apply to the multitaper approach, which can be used to obtain estimates of the variability of spectral estimators even in situations where bias is not an issue – see Chapter 7). The most extreme case is a white noise process. Sloane (1969) shows that tapering such a process does little more than reduce the effective sample size; i.e., it amounts to throwing a certain portion of the data away.

[2]   We have used the normalization $\sum_{t=1}^{N} h_t^2 = 1$ (Equation (208a)) to ensure that $\hat{S}^{(d)}(\cdot)$ mimics the true sdf in that the expected value of the integral of $\hat{S}^{(d)}(\cdot)$ equals the variance of the process. An alternative normalization is to require that the sum of the squares of $h_t X_t$ be equal to that of $X_t/\sqrt{N}$, i.e., to require that h_t be such that

$$\sum_{t=1}^{N} (h_t X_t)^2 = \frac{1}{N} \sum_{t=1}^{N} X_t^2 = \hat{s}_0^{(p)}. \tag{216}$$

This makes the normalization of h_t data dependent. By applying the expectation operator to both sides of the above, we find that

$$\sum_{t=1}^{N} E\{h_t^2 X_t^2\} = s_0,$$

i.e., that the expected sum of squares of $h_t X_t$ is an unbiased estimator of the variance s_0 of the process. If $\hat{s}_0^{(p)}$ is a consistent estimator of s_0, this variance normalization (sometimes called *restoration of power* in the literature) is equivalent to (208a) for large N. Note that any (reasonable) normalization of h_t affects only the level of $\hat{S}^{(d)}(\cdot)$ and not its shape.

[3]   If $E\{X_t\} = \mu \neq 0$, we need to modify $\hat{S}^{(d)}(\cdot)$ somehow. The obvious changes are

$$\Delta t \left| \sum_{t=1}^{N} h_t (X_t - \mu) e^{-i2\pi f t \, \Delta t} \right|^2 \quad \text{or} \quad \Delta t \left| \sum_{t=1}^{N} h_t (X_t - \bar{X}) e^{-i2\pi f t \, \Delta t} \right|^2$$

accordingly as μ is either known or unknown. If both centering and tapering of a time series are needed, centering must be done before tapering (see Exercise [6.11]). There is, however, an alternative estimator to $\bar{X}$ for μ that is sometimes used in conjunction with tapering. Exercise [6.5b] shows that the periodogram $\hat{S}^{(p)}(\cdot)$ of a centered time series is zero at zero frequency. This is arguably appropriate – since zero frequency corresponds to a constant term, the power of a centered series at zero frequency should be zero. If we now taper a centered series with a nonrectangular data taper, the resulting direct estimator $\hat{S}^{(d)}(\cdot)$ need not be zero at zero frequency. Suppose, however, that we consider the following estimator of μ:

$$\tilde{\mu} \equiv \sum_{t=1}^{N} h_t X_t \Big/ \sum_{t=1}^{N} h_t. \tag{217a}$$

Note that $\tilde{\mu}$ is an unbiased estimator of μ and that it reduces to $\bar{X}$ when $\{h_t\}$ is the rectangular data taper. If we center our time series with $\tilde{\mu}$ instead of $\bar{X}$ and then taper it, our direct spectral estimator becomes

$$\Delta t \left| \sum_{t=1}^{N} h_t (X_t - \tilde{\mu}) e^{-i2\pi f t \, \Delta t} \right|^2. \tag{217b}$$

When $f = 0$ the quantity between the absolute value signs becomes

$$\sum_{t=1}^{N} h_t (X_t - \tilde{\mu}) = \sum_{t=1}^{N} h_t X_t - \sum_{t=1}^{N} h_t \left(\frac{\sum_{t=1}^{N} h_t X_t}{\sum_{t=1}^{N} h_t} \right) = 0;$$

i.e., the direct spectral estimator is again identically zero at zero frequency. We shall have occasion to use this alternative centering scheme in Section 10.11.

[4]   Since the formulae we have developed in this section show that the effect of tapering depends upon both the data taper and the true sdf, it can seem that, without knowledge of the sdf, there is no way of determining whether a data taper is required in practical applications. For fine details this is correct, but the need for tapering for a particular time series can usually be established by comparing its periodogram with a direct spectral estimate that uses a data taper with good sidelobe characteristics. If we look ahead a moment to Figure 227, we see the periodogram and a direct spectral estimate using a dpss data taper with $NW = 4$ for a simulated time series with $N = 1024$ data values. The main indicator that tapering is required here is the substantial difference in the level of the two spectral estimates in the high frequency region; moreover, as we would expect in the presence of substantial bias, the level of the periodogram is higher than that of the other estimator. A secondary indication is a marked decrease in the level of local variability of the periodogram in the high frequency region – compared with the low frequency region – when plotted on a logarithmic scale. As we shall see in Section 6.10, nonparametric spectral estimators should have the same local variability for all frequencies when plotted on a logarithmic scale. The decrease in local variability in the periodogram of Figure 227 in the high frequency region is an indication that something is not right here – evidently these estimates are being dominated by leakage of power from

other portions of the sdf. However, it is important to note several things about this simple example. First, a decrease in local variability can also be due to other causes, a common one being the presence of outliers (discordant values) in a time series – see Martin and Thomson (1982) and Chave *et al.* (1987). Second, a *lack* of decrease of local variability in regions of low power does *not* automatically imply that tapering is unnecessary. Third, the need for tapering can manifest itself as a ringing (oscillations) in a low power portion of an sdf. Finally, leakage is not limited to the high frequency region of an sdf – it can potentially occur in any portion with relatively low power. The best advice we can offer is to try different degrees of tapering and to carefully check the corresponding sdf estimates for evidence of leakage – or lack thereof.

6.5 Bias Reduction – Prewhitening

A second common method of controlling the bias in direct spectral estimators is called *prewhitening* (Press and Tukey, 1956). This technique is based upon the linear filtering theorem. Suppose that a portion X_1, $\ldots$, X_N of a stationary process $\{X_t\}$ with sdf $S_X(\cdot)$ is filtered to create

$$Y_t \equiv \sum_{u=-L}^{K} g_u X_{t+K-u}, \quad 1 \leq t \leq M \equiv N - (K + L),$$

where, for convenience, we assume $K > 0$ and $L \geq 0$. This yields a portion of length M of a stationary process $\{Y_t\}$ with sdf $S_Y(\cdot)$. The two sdf's are related by

$$S_Y(f) = \left| \sum_{u=-L}^{K} g_u e^{-i2\pi f u \Delta t} \right|^2 S_X(f).$$

Now, if $S_X(\cdot)$ has a wide dynamic range, it might be possible to choose g_u such that the dynamic range of $S_Y(\cdot)$ is much less. This means we can construct a direct spectral estimator of $S_X(\cdot)$ with low bias by using the following procedure. First, we use the filtered data to produce

$$\hat{S}_Y^{(d)}(f) \equiv \Delta t \left| \sum_{t=1}^{M} h_t Y_t e^{-i2\pi f t \Delta t} \right|^2,$$

where $\{h_t\}$ is a data taper with a fairly narrow central lobe (such as the rectangular taper or the zeroth-order dpss taper with $NW = 1$ or 2). If $S_Y(\cdot)$ has a small dynamic range, we should have $E\{\hat{S}_Y^{(d)}(f)\} \approx S_Y(f)$ to within a few decibels. This implies that

$$E \left\{ \frac{\hat{S}_Y^{(d)}(f)}{\left| \sum_{u=-L}^{K} g_u e^{-i2\pi f u \Delta t} \right|^2} \right\} \approx \frac{S_Y(f)}{\left| \sum_{u=-L}^{K} g_u e^{-i2\pi f u \Delta t} \right|^2} = S_X(f)$$

also to within a few decibels. Since we have effectively reduced the dynamic range of the sdf to be estimated and hence can now use a data taper with a narrow central lobe, this procedure could prevent the inadvertent distortion of the sdf caused by data tapers with wide central lobes (see Figures 213 and 214).

In the ideal case, the filter $\{g_u\}$ would reduce $\{X_t\}$ to white noise. It is thus called a *prewhitening filter*.

There are, however, some potential tradeoffs and problems with prewhitening. There is an inevitable decrease in the sample size of the time series used in the direct spectral estimator (from N to $M = N - (K + L)$). This decrease can adversely affect the first moment and other statistical properties of the spectral estimator if $K + L$ is large relative to N. This can happen in practice – an sdf with a large amount of structure can require a prewhitening filter of a large length to effectively reduce its dynamic range. Just such a situation occurs in exploration seismology where the power spectrum of seismic traces decays very rapidly with decreasing frequency because of the effect of analog recording filters.

There is also a 'cart and horse' problem with prewhitening. Design of a prewhitening filter requires at least some knowledge of the shape of the underlying sdf, the very thing we are trying to estimate! Sometimes it is possible to make a reasonable guess about the general shape of the sdf from previous experiments or physical theory. This is obviously not a viable general solution, particularly since spectral analysis is often used as an exploratory data analysis tool.

An alternative way of obtaining a prewhitening filter is to *estimate* it from the time series itself. One popular method for doing so is to fit an autoregressive model to the data and then use the fitted model as a prewhitening filter. The details of this approach are given in Section 9.10.

Comments and Extensions to Section 6.5

[1]   In some practical applications it is only of interest to estimate the sdf over a selected range of frequencies. If $f_{(L)}$ and $f_{(H)}$ are the lowest and highest frequencies of interest, we can subject our time series $\{X_t\}$ to a band-pass filter that attenuates all spectral components with frequencies $|f| \notin [f_{(L)}, f_{(H)}]$ – this is sometimes known as *rejection filtration* (Blackman and Tukey, 1958, pp. 42–3). If, as before, we let $\{Y_t\}$ represent our filtered data and if the gain of the band-pass filter is approximately 1 over the pass-band, then we should have

$$S_Y(f) \approx S_X(f) \text{ for } f_{(L)} \leq |f| \leq f_{(H)}$$

to a good approximation. Because of the filtering operation, estimation of $S_Y(f)$ over $|f| \in [f_{(L)}, f_{(H)}]$ should be free of leakage from frequencies outside that range. This can be particularly valuable if there is a large contribution to the variance of $\{X_t\}$ due to frequencies $|f| \notin [f_{(L)}, f_{(H)}]$. As with prewhitening, filtering of a time series to suppress certain frequencies decreases the number of data points available to estimate $S_Y(\cdot)$ from N to $N - M$, where M is the length of the band-pass filter. Since we might have to make M large to get a decent approximation to a band-pass filter, this procedure might not be worthwhile if leakage from frequencies $|f| \notin [f_{(L)}, f_{(H)}]$ in $S_X(\cdot)$ is small or can be controlled effectively with a data taper having a fairly narrow central lobe.

6.6 Statistical Properties of Direct Spectral Estimators

We investigated the first moment (bias) properties of direct spectral estimators in Section 6.4. We now want to investigate some of their other statistical properties. We begin by considering the special important case of a Gaussian white noise process $\{G_t\}$ with zero mean and variance σ^2 (the letter G is used here to emphasize the *G*aussian assumption).

Let

$$A(f) \equiv (\Delta t)^{1/2} \sum_{t=1}^{N} h_t G_t \cos\left(2\pi f t \, \Delta t\right)$$

and

$$B(f) \equiv (\Delta t)^{1/2} \sum_{t=1}^{N} h_t G_t \sin\left(2\pi f t \, \Delta t\right)$$

be the real and imaginary parts of the complex conjugate of

$$J(f) \equiv (\Delta t)^{1/2} \sum_{t=1}^{N} h_t G_t e^{-i2\pi f t \, \Delta t},$$

where $\{h_t\}$ is a data taper. Since $\{G_t\}$ has zero mean, it follows that

$$E\{A(f)\} = E\{B(f)\} = 0 \text{ for all } f.$$

By the properties of a white noise process, we have

$$\text{var}\{A(f)\} = \sigma^2 \, \Delta t \sum_{t=1}^{N} h_t^2 \cos^2(2\pi f t \, \Delta t);$$

$$\text{var}\{B(f)\} = \sigma^2 \, \Delta t \sum_{t=1}^{N} h_t^2 \sin^2(2\pi f t \, \Delta t);$$

$$\text{cov}\{A(f), A(f')\} = \sigma^2 \, \Delta t \sum_{t=1}^{N} h_t^2 \cos\left(2\pi f t \, \Delta t\right) \cos\left(2\pi f' t \, \Delta t\right);$$

$$\text{cov}\{B(f), B(f')\} = \sigma^2 \, \Delta t \sum_{t=1}^{N} h_t^2 \sin(2\pi f t \, \Delta t) \sin(2\pi f' t \, \Delta t);$$

$$\text{cov}\{A(f), B(f')\} = \sigma^2 \, \Delta t \sum_{t=1}^{N} h_t^2 \cos(2\pi f t \, \Delta t) \sin(2\pi f' t \, \Delta t).$$

The above equations simplify drastically if $h_t = 1/\sqrt{N}$ for all t (i.e., the direct spectral estimator is in fact just the periodogram) and f is one of the Fourier (or standard) frequencies (i.e., $f_k = k/(N \, \Delta t)$, where k is an integer such that $0 \le f_k \le f_{(N)}$). In this case we have

$$\text{var}\{A(f_k)\} = \begin{cases} \sigma^2 \, \Delta t/2, & \text{for } f_k \neq 0 \text{ or } f_{(N)}, \\ \sigma^2 \, \Delta t, & \text{for } f_k = 0 \text{ or } f_{(N)}; \end{cases}$$

$$\text{var}\{B(f_k)\} = \begin{cases} \sigma^2 \, \Delta t/2, & \text{for } f_k \neq 0 \text{ or } f_{(N)}, \\ 0, & \text{for } f_k = 0 \text{ or } f_{(N)}; \end{cases}$$

$$\text{cov}\{A(f_j), A(f_k)\} = 0 \text{ for all } f_j \neq f_k;$$

$$\text{cov}\{B(f_j), B(f_k)\} = 0 \text{ for all } f_j \neq f_k;$$

$$\text{cov}\{A(f_j), B(f_k)\} = 0 \text{ for all } f_j \text{ and } f_k,$$

all of which follow from the orthogonality relationships of Exercise [1.4].

Since $A(f_k)$ and $B(f_k)$ are linear combinations of Gaussian rv's, they in turn are Gaussian rv's. Because uncorrelatedness implies independence for Gaussian rv's, the $A(f_k)$ terms and the $B(f_k)$ terms are zero mean independent Gaussian rv's with variances as stated above. Recall that, if $Y_1, Y_2, \ldots, Y_\nu$ are independent zero mean, unit variance Gaussian rv's, then the rv

$$\chi_\nu^2 \equiv Y_1^2 + Y_2^2 + \cdots + Y_\nu^2$$

has a chi-square distribution with ν degrees of freedom. Since

$$A^2(f_k) + B^2(f_k) = \hat{S}_G^{(p)}(f_k),$$

i.e., the periodogram, and since for $f_k \neq 0$ or $f_{(N)}$

$$\left(\frac{2}{\sigma^2 \, \Delta t}\right)^{1/2} A(f_k) \text{ and } \left(\frac{2}{\sigma^2 \, \Delta t}\right)^{1/2} B(f_k)$$

are independent Gaussian rv's with zero mean and unit variance, it follows that

$$\frac{2}{\sigma^2 \, \Delta t} \hat{S}_G^{(p)}(f_k) \stackrel{\text{d}}{=} \chi_2^2; \text{ i.e., } \hat{S}_G^{(p)}(f_k) \stackrel{\text{d}}{=} \frac{\sigma^2 \, \Delta t}{2} \chi_2^2 \text{ for } f_k \neq 0 \text{ or } f_{(N)}.$$

Here $\stackrel{\mathrm{d}}{=}$ means 'equal in distribution,' so the statement '$Y \stackrel{\mathrm{d}}{=} c\chi_\nu^2$' means that the rv Y has the same distribution as a chi-square rv (with ν degrees of freedom) that has been multiplied by a constant c. The corresponding result for $f_k = 0$ or $f_{(N)}$ is

$$\hat{S}_G^{(p)}(f_k) \stackrel{\mathrm{d}}{=} \sigma^2 \, \Delta t \, \chi_1^2.$$

Since $S_G(f) \equiv \sigma^2 \, \Delta t$ is the sdf for $\{G_t\}$, we can rewrite the above as

$$\hat{S}_G^{(p)}(f_k) \stackrel{\mathrm{d}}{=} \begin{cases} S_G(f_k)\chi_2^2/2, & \text{for } f_k \neq 0 \text{ or } f_{(N)}; \\ S_G(f_k)\chi_1^2, & \text{for } f_k = 0 \text{ or } f_{(N)}. \end{cases}$$

Because $E\{\chi_\nu^2\} = \nu$ and $\operatorname{var}\{\chi_\nu^2\} = 2\nu$, it follows from the above that

$$E\{\hat{S}_G^{(p)}(f_k)\} = \sigma^2 \, \Delta t = S_G(f_k) \text{ for all } f_k$$

and that

$$\operatorname{var}\{\hat{S}_G^{(p)}(f_k)\} = \begin{cases} \sigma^4(\Delta t)^2 = S_G^2(f_k), & \text{for } f_k \neq 0 \text{ or } f_{(N)}; \\ 2\sigma^4(\Delta t)^2 = 2S_G^2(f_k), & \text{for } f_k = 0 \text{ or } f_{(N)}. \end{cases}$$

Moreover, all the rv's in the set $\{\hat{S}_G^{(p)}(f_k), 0 \leq k \leq \lfloor N/2 \rfloor\}$ are independent. Thus, the sampling properties of $\hat{S}_G^{(p)}(f_k)$ on the grid of Fourier frequencies are completely known and have a simple form for a zero mean Gaussian white noise process $\{G_t\}$.

For a stationary process $\{X_t\}$ with an sdf $S(\cdot)$ that is continuous over the interval $[-f_{(N)}, f_{(N)}]$ but that is not necessarily either white noise or Gaussian distributed, it can be shown – subject to the finiteness of certain high-order moments – that

$$\hat{S}^{(p)}(f) \stackrel{\mathrm{d}}{=} \begin{cases} S(f)\chi_2^2/2, & \text{for } 0 < f < f_{(N)}; \\ S(f)\chi_1^2, & \text{for } f = 0 \text{ or } f_{(N)}, \end{cases} \tag{222a}$$

asymptotically as $N \to \infty$ (since $\hat{S}^{(p)}(-f) = \hat{S}^{(p)}(f)$, we need not concern ourselves with $f < 0$). Furthermore, for $0 \leq f' < f \leq f_{(N)}$, $\hat{S}^{(p)}(f)$ and $\hat{S}^{(p)}(f')$ are asymptotically independent. We thus have

$$\operatorname{var}\{\hat{S}^{(p)}(f)\} = \begin{cases} S^2(f), & 0 < f < f_{(N)}; \\ 2S^2(f), & f = 0 \text{ or } f_{(N)}; \end{cases} \tag{222b}$$

$$\operatorname{cov}\{\hat{S}^{(p)}(f'), \hat{S}^{(p)}(f)\} = 0, \quad 0 \leq f' < f \leq f_{(N)}, \tag{222c}$$

asymptotically as $N \to \infty$ (for finite sample sizes N, these asymptotic results are useful approximations if certain restrictions are observed – see item [3] in the Comments and Extensions for this section). If we restrict ourselves just to the Fourier frequencies $f_k = k/(N \, \Delta t)$, we also find that the $\lfloor N/2 \rfloor + 1$ rv's

$$\hat{S}^{(p)}(f_0), \hat{S}^{(p)}(f_1), \ldots, \hat{S}^{(p)}(f_{\lfloor N/2 \rfloor})$$

are all approximately pairwise uncorrelated for N large enough; i.e.,

$$\text{cov}\{\hat{S}^{(p)}(f_j), \hat{S}^{(p)}(f_k)\} \approx 0, \ \ j \neq k \text{ and } 0 \leq j, k \leq \lfloor N/2 \rfloor. \quad (223a)$$

As we shall see, this last result plays an important role in the derivation of the sampling properties of spectral estimators based upon the periodogram (even though it is a large sample result, it is thought to be a reasonable approximation for finite N).

A few modifications to the above large sample theory are necessary to handle direct spectral estimators $\hat{S}^{(d)}(\cdot)$ other than the periodogram (for details, see the Comments and Extensions section). First, the variance of $\hat{S}^{(d)}(f)$ can actually be larger or smaller than that of $\hat{S}^{(p)}(f)$, depending upon the particular data taper used and the characteristics of the true underlying sdf. For processes whose acvs satisfies a summability condition and for $\{h_t\}$ that is reasonable in form, Brillinger (1981a, p. 127) shows that the large sample univariate distributional properties of $\hat{S}^{(d)}(f)$ are the same as those of $\hat{S}^{(p)}(f)$ – this implies that

$$\hat{S}^{(d)}(f) \overset{\text{d}}{=} \begin{cases} S(f)\chi_2^2/2, & \text{for } 0 < f < f_{(N)}; \\ S(f)\chi_1^2, & \text{for } f = 0 \text{ or } f_{(N)}, \end{cases} \quad (223b)$$

and

$$\text{var}\{\hat{S}^{(d)}(f)\} = \begin{cases} S^2(f), & \text{for } 0 < f < f_{(N)}; \\ 2S^2(f), & \text{for } f = 0 \text{ or } f_{(N)}. \end{cases} \quad (223c)$$

asymptotically as $N \to \infty$ (see item [3] in the Comments and Extensions for caveats on applying these asymptotic results).

Second, the grid of frequencies for which (223a) holds is no longer the grid defined by the Fourier frequencies. To see this in the case of a Gaussian process with sdf $S_G(\cdot)$, we note that it follows from Equation (230b) in the Comments and Extensions section that

$$\text{cov}\{\hat{S}_G^{(d)}(f), \hat{S}_G^{(d)}(f')\}$$

$$\approx \frac{1}{(\Delta t)^2} \left| \int_{-f_{(N)}}^{f_{(N)}} H^*(f-u)H(f'-u)S_G(u)\,du \right|^2$$

$$\leq \frac{S_{\max}^2}{(\Delta t)^2} \left| \int_{-f_{(N)}}^{f_{(N)}} |H(f-u)| \cdot |H(f'-u)|\,du \right|^2$$

$$= \frac{S_{\max}^2}{(\Delta t)^2} \left| \int_{-f_{(N)}}^{f_{(N)}} |H(f-f'+v)| \cdot |H(v)|\,dv \right|^2,$$

where $v = f' - u$ and $S_{\max} \equiv \max_f S_G(f)$ (this makes sense under our assumption that $S_G(\cdot)$ is continuous on the interval $[-f_{(N)}, f_{(N)}]$). Suppose, for example, that $H(\cdot)$ corresponds to a zeroth-order dpss. The above integrand will then be small whenever f and f' are far enough apart so that the central lobes of $|H(f - f' + \cdot)|$ and $|H(\cdot)|$ do not overlap significantly. This occurs approximately when $|f - f'| \geq W$ (or, more conservatively, when $|f - f'| \geq 2W$ – see Figure 232). If $NW = c/\Delta t$, where we assume for convenience that c is an integer, the grid of frequencies over which the direct spectral estimator $\hat{S}_G^{(d)}(\cdot)$ is approximately uncorrelated is defined by $kc/(N \Delta t)$, where the integer k ranges from 0 up to that integer K such that $Kc/(N \Delta t) \leq f_{(N)}$ and $(K + 1)c/(N \Delta t) > f_{(N)}$. For $c = 2$ (or 4) – typical values used in practice – this yields a grid size twice (or four times) as large as the Fourier frequencies; however, as Brillinger (1981b) points out, although the grid size is now larger, the pairwise covariances of the $K + 1$ rv's

$$\hat{S}_G^{(d)}(0), \hat{S}_G^{(d)}(c/(N \Delta t)), \dots, \hat{S}_G^{(d)}(Kc/(N \Delta t)) \qquad (224)$$

are 0 to a better approximation for finite N than the corresponding large sample result stated in (223a) for the periodogram. This better approximation is useful for constructing certain statistical tests and is another benefit of tapering.

Since $S(f) > 0$ typically, Equations (222b) and (223c) show that the variances of $\hat{S}^{(p)}(f)$ and $\hat{S}^{(d)}(f)$ do *not* decrease to 0 as $N \to \infty$. Thus the probability that $\hat{S}^{(p)}(f)$ or $\hat{S}^{(d)}(f)$ becomes arbitrarily close to its asymptotic expected value of $S(f)$ is 0 – in statistical terminology, neither $\hat{S}^{(p)}(f)$ nor $\hat{S}^{(d)}(f)$ is a consistent estimator of $S(f)$.

Figure 225 illustrates the inconsistency of the periodogram of a Gaussian white noise process with unit variance and unit sampling interval Δt: as the sample size increases from 64 to 1024, the periodogram shows no sign of converging to the sdf for the process, namely, $S(f) = 1$ for all $|f| \leq 1/2$ (indicated by the horizontal dashed lines). One can also see the increase in roughness of the periodogram as N increases due to the decrease in covariance between fixed nearby points.

Figure 226 shows two direct spectral estimators for one realization of size $N = 1024$ for the AR(2) process $\{X_{t,2}\}$ defined in Equation (45) (the realization itself is shown in the top plot of Figure 45). The thin curve in the top plot is the periodogram, while in the lower plot it is the direct spectral estimator that uses a zeroth-order dpss with $NW = 4$. The true sdf for $\{X_{t,2}\}$ is shown by the thick curves in both plots. We have already noted that, because the dynamic range of the sdf for this process is small, the bias in the periodogram is already negligible for a sample size as small as 64 (see Figure 201). There is thus no need for a data taper here. The smoother appearance of the direct spectral estimate with the dpss data taper is due to the increased distance

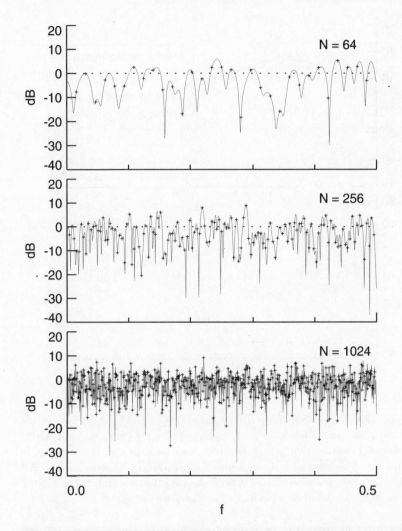

Figure 225. Inconsistency of the periodogram. The plots show the periodogram (on a decibel scale) versus frequency for samples of sizes $N = 64$, 256 and 1024 – here $\Delta t = 1$, so $f_{(N)} = 1/2$. The solid curve in each plot is the periodogram evaluated on a grid of 1024 frequencies, while the small pluses indicate its values at the Fourier frequencies. The horizontal dashed lines indicate the true sdf.

between uncorrelated spectral estimates. An alternative interpretation is that tapering in this example has effectively reduced the size of our time series without any gain in bias reduction (since there really is not any bias to reduce!). Based upon the grid size of uncorrelated spectral estimates we can argue that the statistical properties of the dpss-based

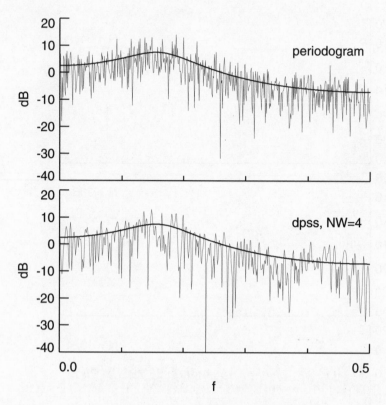

Figure 226. Variability of direct spectral estimators. The thin curve in the top plot is the periodogram for a realization of size $N = 1024$ from the AR(2) process of Equation (45) (the realization itself is shown in the top plot of Figure 45). The lower plot shows a direct spectral estimate with an $NW = 4$ dpss data taper. The thick curves in both plots are the true sdf (cf. Figure 201). All spectra are plotted on a decibel scale versus frequency.

direct spectral estimator are approximately equivalent to those of a periodogram with $1024/4 = 256$ data points. The smoother appearance can be attributed to this effective decrease in sample size. This effect is also evident in Figure 225 – as the sample size decreases, the periodogram has a smoother appearance. Finally we note that the local variability of both direct spectral estimators in Figure 226 is about the same. This supports the claim that all direct spectral estimators of $S(f)$ have approximately the same variance (see Equations (222b) and (223c)).

Figure 227 shows an example for which tapering is beneficial. The thin curve in the top plot is the periodogram for one realization of size $N = 1024$ for the AR(4) process $\{X_{t,4}\}$ defined in Equation (46a) (the

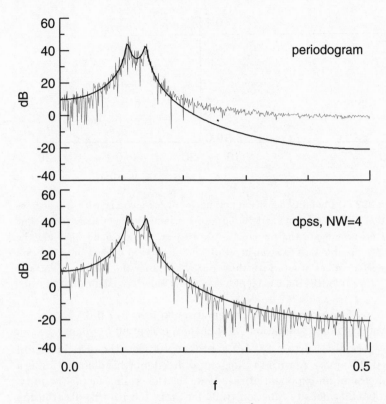

Figure 227. Bias of direct spectral estimators. The thin curve in the top plot is the periodogram for a realization of size $N = 1024$ from the AR(4) process of Equation (46a) (the realization itself is shown in the bottom plot of Figure 45). The lower plot shows a direct spectral estimate with an $NW = 4$ dpss data taper. The thick curves in both plots are the true sdf (cf. Figure 203).

realization itself is shown in the bottom plot of Figure 45). The thin curve in the lower plot is the direct spectral estimator that uses a zeroth-order dpss taper with $NW = 4$. The true sdf for $\{X_{t,4}\}$ is shown by the thick curves in both plots. The dynamic range of the sdf for this process is comparatively large, and there is substantial bias in the periodogram even for a sample size of 1024 (see the lower plot of Figure 203). The inherent variability and bias in the periodogram is quite evident here and is the source of the following comment by Tukey (1980):

> More lives have been lost looking at the raw periodogram than by any other action involving time series!

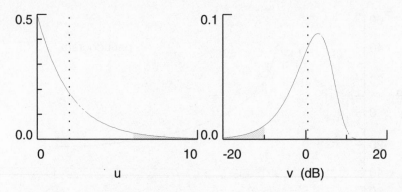

Figure 228. Probability density functions for a chi-square rv with two degrees of freedom (left-hand plot) and $10 \log_{10}$ of the same rv (right-hand plot). The shaded areas indicate the upper 5% tail area on the left-hand plot and the lower 5% tail area on the right-hand plot. The vertical dashed lines indicate the expected values of rv's with these pdf's, i.e., 2 for the left-hand plot and $10 \log_{10}(2/e^{\gamma})$ for the right-hand plot – here γ is Euler's constant $(0.5772\ldots)$.

Comments and Extensions to Section 6.6

[1]   We have noted (Equation (223b)) that, except at $f = 0$ and $f_{(N)}$, a direct spectral estimator $\hat{S}^{(d)}(f)$ approximately has the distribution of a scaled chi-square rv with two degrees of freedom (this distribution is a special case of an exponential rv). If we let the scaling factor be unity, the probability density function (pdf) for an rv U with this distribution is given by

$$f_U(u) \equiv \begin{cases} e^{-u/2}/2, & u \geq 0; \\ 0, & u < 0. \end{cases}$$

This function is shown in the left-hand plot of Figure 228. Since this pdf has an upper tail that decays at a slower rate than that of a Gaussian pdf, a random sample of variates with this distribution will typically contain 'upshoots,' i.e., a few values that appear to be unusually large compared with the bulk of the variates (see the bottom plot of Figure 41); however, when direct spectral estimates are plotted on a decibel (or logarithmic) scale, the visual appearance is quite different – we typically see prominent 'downshoots' instead (see Figures 225 to 227). The explanation for this 'paradox' is that we are actually examining variates with the pdf of $V \equiv 10 \log_{10}(U)$. This pdf is given by

$$f_V(v) = \frac{\log(10)}{20} 10^{v/10} e^{-(10^{v/10})/2}, \qquad -\infty < v < \infty,$$

and is shown in the right-hand plot of Figure 228. Note that, in contrast to $f_U(\cdot)$, this pdf has a prominent *lower* tail.

[2]   For use in Section 6.9, we develop here a more accurate expression for the covariance of direct spectral estimators than given by the qualitative 'grid size' argument presented above (Jones, 1971; Koopmans, 1974, Chapter 8; and Walden, 1990a). Suppose that $\{G_t\}$ is a stationary Gaussian process – not necessarily white noise – with zero mean and sdf $S_G(\cdot)$ and that, as before,

$$J(f) \equiv (\Delta t)^{1/2} \sum_{t=1}^{N} h_t G_t e^{-i2\pi f t \,\Delta t} \quad \text{with } \{h_t\} \longleftrightarrow H(\cdot).$$

Now,

$$\text{cov}\left\{|J(f)|^2, |J(f')|^2\right\} = \text{cov}\left\{J(f)J^*(f), J(f')J^*(f')\right\}.$$

Since $\{G_t\}$ is a Gaussian process, the distribution of $J(f)$ is complex-valued Gaussian for all f (for non-Gaussian processes, the distribution of $J(f)$ can still be reasonably approximated as such, but this depends somewhat on the effect of the taper). We can now use the Isserlis theorem (Equation (40)) to show that

$$\begin{aligned}
\text{cov}\left\{|J(f)|^2, |J(f')|^2\right\} &= \text{cov}\left\{J(f), J(f')\right\} \text{cov}\left\{J^*(f), J^*(f')\right\} \\
&\quad + \text{cov}\left\{J(f), J^*(f')\right\} \text{cov}\left\{J^*(f), J(f')\right\} \\
&= E\{J^*(f)J(f')\}E\{J(f)J^*(f')\} \\
&\quad + E\{J^*(f)J^*(f')\}E\{J(f)J(f')\} \\
&= \left|E\{J^*(f)J(f')\}\right|^2 + \left|E\{J(f)J(f')\}\right|^2 .
\end{aligned}$$
(229)

From the fact that

$$J(f) = \frac{1}{(\Delta t)^{1/2}} \int_{-f_{(N)}}^{f_{(N)}} H(f - u)\,dZ(u)$$

(see Equation (206a)), we have

$$E\{J^*(f)J(f')\} = \frac{1}{\Delta t} \int_{-f_{(N)}}^{f_{(N)}} H^*(f - u)H(f' - u)S_G(u)\,du.$$

Since $dZ(-u) = dZ^*(u)$ (see Equation (129a)), it is also true that

$$J(f) = -\frac{1}{(\Delta t)^{1/2}} \int_{-f_{(N)}}^{f_{(N)}} H(f + u)\,dZ^*(u),$$

and hence

$$E\{J(f)J(f')\} = -\frac{1}{\Delta t} \int_{-f_{(N)}}^{f_{(N)}} H(f + u)H(f' - u)S_G(u)\,du.$$

With these expressions and by substituting $f + \eta$ for f', we can now write

$$\text{cov}\,\{\hat{S}_G^{(d)}(f), \hat{S}_G^{(d)}(f + \eta)\} = \text{cov}\,\{|J(f)|^2, |J(f + \eta)|^2\}$$

$$= \frac{1}{(\Delta t)^2} \left| \int_{-f_{(N)}}^{f_{(N)}} H^*(f - u)H(f + \eta - u)S_G(u)\,du \right|^2 \quad (230a)$$

$$+ \frac{1}{(\Delta t)^2} \left| \int_{-f_{(N)}}^{f_{(N)}} H(f + u)H(f + \eta - u)S_G(u)\,du \right|^2.$$

Although this result is exact for Gaussian processes (and, as is shown in Exercise [6.13], can be computed using the inverse Fourier transforms of $H(\cdot)$ and $S_G(\cdot)$), it is unwieldy in practice, so we seek an approximation to it. Thomson (1977) points out that the second integral above is large only for f near 0 and $f_{(N)}$ – we exclude these cases here – so that the first integral is usually dominant. Suppose the central lobe of $H(\cdot)$ goes from $-W$ to W approximately. If the frequency separation η is less than $2W$ so that the central lobes of the shifted versions of $H(\cdot)$ in the first term overlap, then the above covariance depends primarily on the sdf in the domain $[f + \eta - W, f + W]$. If we take the sdf to be locally constant about f, we obtain

$$\text{cov}\,\{\hat{S}_G^{(d)}(f), \hat{S}_G^{(d)}(f + \eta)\}$$

$$\approx \frac{1}{(\Delta t)^2} \left| \int_{-f_{(N)}}^{f_{(N)}} H^*(f - u)H(f + \eta - u)S_G(u)\,du \right|^2 \quad (230b)$$

$$\approx \frac{S_G^2(f)}{(\Delta t)^2} \left| \int_{-f_{(N)}}^{f_{(N)}} H(u)H(\eta - u)\,du \right|^2 = \frac{S_G^2(f)}{(\Delta t)^4} \,|H * H(\eta)|^2,$$

where $*$ denotes convolution as defined by Equation (117a) – here we have made use of the facts that $H(\cdot)$ is periodic with period $2f_{(N)} = 1/\Delta t$ and that $H^*(-u) = H(u)$. This covariance can be written in terms of the data taper as

$$\text{cov}\,\{\hat{S}_G^{(d)}(f), \hat{S}_G^{(d)}(f + \eta)\} \approx S_G^2(f) \left| \sum_{t=1}^{N} h_t^2 e^{-i2\pi\eta t \,\Delta t} \right|^2 \equiv R(\eta, f).$$

If we let $\eta = 0$, we obtain $R(0, f) = S_G^2(f)$ because of the normalization $\sum_{t=1}^{N} h_t^2 = 1$. This implies $\text{var}\,\{\hat{S}_G^{(d)}(f)\} \approx R(0, f) = S_G^2(f)$, in agreement with our earlier quoted result for $|f| \neq 0$ or $\pm f_{(N)}$ (see Equation (223c)). Under our working assumption that $S_G(f) \approx S_G(f + \eta)$, we have var $\{\hat{S}_G^{(d)}(f)\} \approx$ var $\{\hat{S}_G^{(d)}(f+\eta)\} \approx R(0,f)$. It follows that the correlation between $\hat{S}_G^{(d)}(f)$ and $\hat{S}_G^{(d)}(f+\eta)$ is given approximately by

$$R(\eta) \equiv \frac{R(\eta, f)}{R(0, f)} = \left| \sum_{t=1}^{N} h_t^2 e^{-i2\pi\eta t \, \Delta t} \right|^2, \qquad (231a)$$

which depends only on the data taper and not upon either f or $S_G(\cdot)$ (Thomson, 1982).

The value $R(\eta)$ thus gives us an approximation to the correlation between $\hat{S}_G^{(d)}(f)$ and $\hat{S}_G^{(d)}(f+\eta)$ under the assumption that $S_G(\cdot)$ is constant in the interval $[f+\eta-W, f+W]$, where W is the half-width of the central lobe of the spectral window $\mathcal{H}(\cdot)$ associated with $\hat{S}_G^{(d)}(\cdot)$. Figure 232 shows $R(\cdot)$ for the four dpss data tapers shown in Figures 211 and 212 (here $N = 64$ and $\Delta t = 1$). The separation (or grid size) of the Fourier frequencies for this example is $1/N = 1/64$, which is indicated by a solid vertical line. For a white noise process – a case where there is no bias in the periodogram so tapering is not needed – values of the periodogram at Fourier frequencies $f_k = k/N$ and $f_{k+1} = (k+1)/N$ are uncorrelated, whereas the corresponding values for a direct spectral estimator using a dpss data taper with $NW = K$ are positively correlated with the correlation increasing as K increases (this assumes f_k and f_{k+1} are not too close to either 0 or $f_{(N)} = 1/2$).

Since $R(\cdot)$ is a $2f_{(N)}$ periodic function (see Equation (231a)), its Fourier transform pair is a sequence, say, $\{r_q\}$. To obtain an expression for r_q in terms of h_t, recall that $\{\hat{s}_\tau^{(d)}\} \longleftrightarrow \hat{S}^{(d)}(\cdot)$ (Equations (207c) and (207d)) and note that, if we replace X_t by $h_t/(\Delta t)^{1/2}$ in Equation (206c) for $\hat{S}^{(d)}(\cdot)$, the latter reduces to $R(\cdot)$. With this substitution, Equation (207d) tells us that

$$r_q \equiv \begin{cases} (1/\Delta t) \sum_{t=1}^{N-|q|} h_t^2 h_{t+|q|}^2, & |q| \le N-1; \\ 0, & |q| \ge N. \end{cases} \qquad (231b)$$

We note two facts: (1) the analog of Equation (207c) yields

$$R(\eta) = \Delta t \sum_{q=-(N-1)}^{N-1} r_q e^{-i2\pi\eta q \, \Delta t}; \qquad (231c)$$

and (2) since $\{r_q\}$ is a 'rescaled acvs' for $\{h_t^2\}$, it can be calculated efficiently using the technique outlined in Exercise [6.4]. Again, if we assume that $S_G(\cdot)$ is locally constant, then $\hat{S}_G^{(d)}(f)$ and $\hat{S}_G^{(d)}(f)/S_G(f)$ differ only by a constant in the interval $[f+\eta-W, f+W]$. Now

$$\{r_q\} \longleftrightarrow R(\cdot), \qquad (231d)$$

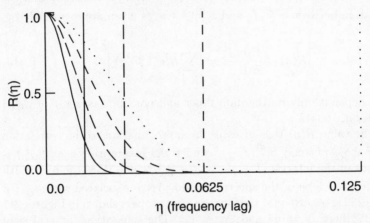

Figure 232. Correlation of direct spectral estimators. The frequency domain correlation $R(\eta)$ is plotted versus the frequency lag η for the four dpss data tapers shown in Figures 211 and 212 (recall that $N = 64$ and $\Delta t = 1$). The cases $NW = 1, 2, 4$ and 8 are shown, respectively, by the solid, long dashed, short dashed and dotted curves; the corresponding vertical lines indicate the location of W for these four cases ($1/64, 2/64, 4/64$ and $8/64$ cycles per unit time).

and
$$R(\eta) \approx \mathrm{cov}\,\{\hat{S}_G^{(d)}(f)/S_G(f), \hat{S}_G^{(d)}(f+\eta)/S_G(f)\}$$

(i.e., the direct spectral estimator is locally stationary). It follows that the 'spectrum' of the spectral estimator $\hat{S}_G^{(d)}(\cdot)$ is given by $\{S_G^2(f)r_q\}$, where q is a pseudo-lag variable. Analogous to the usual relationship between an acvs and an sdf at zero frequency (see Equation (134b)), we have

$$S_G^2(f)R(0) = S_G^2(f)\,\Delta t \sum_{q=-(N-1)}^{N-1} r_q; \text{ i.e, } R(0) = \Delta t \sum_{q=-(N-1)}^{N-1} r_q.$$

Since Equation (231a) tells us that $R(0) = 1$, we obtain

$$\Delta t \sum_{q=-(N-1)}^{N-1} r_q = 1,$$

which also follows from Equation (231c) by setting $\eta = 0$.

[3] For the practitioner, the effect of the restriction in the preceding discussion to frequencies 'not too close to 0 or Nyquist' is that – for *finite* sample sizes N from a process with zero mean – the χ_2^2 distribution for

the periodogram in Equation (222a) is only appropriate for $1/(N \Delta t) \leq f \leq f_{(N)} - 1/(N \Delta t)$. Likewise, the χ_2^2 result in Equation (223b) for general direct spectral estimators is only appropriate for $c/(N \Delta t) \leq f \leq f_{(N)} - c/(N \Delta t)$, where c is the constant defining the appropriate grid size.

Another caveat to observe in the practical application of Equations (222a) and (223b) is that the χ_1^2 results for $f = 0$ are no longer valid if we center the time series by subtracting the sample mean from each observation prior to forming the periodogram or other direct spectral estimators (as we would do if we did not know the process mean to be zero). In particular, as we noted in item [1] of the Comments and Extensions to Section 6.3, centering a time series forces the value of the periodogram at $f = 0$ to be identically equal to zero, indicating that the result $\hat{S}^{(p)}(0) \stackrel{\mathrm{d}}{=} S(0)\chi_1^2$ cannot hold (except if $S(0) = 0$).

[4] It is sometimes of interest to test the hypothesis that a particular time series can be regarded as a segment $X_1, \ldots, X_N$ of a realization of a white noise process (see Section 10.13). One of the most popular tests for white noise is known as the *cumulative periodogram test*. This test is based upon the periodogram over the set of Fourier frequencies $f_k \equiv k/(N \Delta t)$ such that $0 < f_k < f_{(N)}$, i.e., $f_1, \ldots, f_M$ with $M \equiv \lfloor (N-1)/2 \rfloor$. As noted above in this section, under the null hypothesis that $\{X_t\}$ is a Gaussian white noise process, $\hat{S}^{(p)}(f_1), \ldots, \hat{S}^{(p)}(f_M)$ constitute a set of independent and identically distributed rv's having a rescaled chi-square distribution with two degrees of freedom, i.e., an exponential distribution. If we form the normalized cumulative periodogram

$$\mathcal{P}_k \equiv \frac{\sum_{j=1}^k \hat{S}^{(p)}(f_j)}{\sum_{j=1}^M \hat{S}^{(p)}(f_j)}, \qquad k = 1, \ldots, M-1, \tag{233}$$

it can be shown (based upon the stated distribution for the $\hat{S}^{(p)}(f_j)$ terms) that the $\mathcal{P}_k$ terms have the same distribution as an ordered random sample of $M - 1$ rv's from the uniform distribution over the interval $[0, 1]$ (Bartlett, 1955). We can thus base our test for white noise on the well-known *Kolmogorov goodness of fit test* for a completely specified distribution (Conover, 1980). If we let

$$D^+ \equiv \max_{1 \leq k \leq M-1} \left(\frac{k}{M-1} - \mathcal{P}_k \right) \text{ and } D^- \equiv \max_{1 \leq k \leq M-1} \left(\mathcal{P}_k - \frac{k-1}{M-1} \right),$$

we can reject the null hypothesis of white noise at the α level of significance if the Kolmogorov test statistic $D \equiv \max(D^+, D^-)$ exceeds $D(\alpha)$, the upper $\alpha \times 100\%$ percentage point for the distribution of D under the null hypothesis. A simple approximation to $D(\alpha)$ is given by

$$\tilde{D}(\alpha) \equiv \frac{C(\alpha)}{(M-1)^{1/2} + 0.12 + 0.11/(M-1)^{1/2}},$$

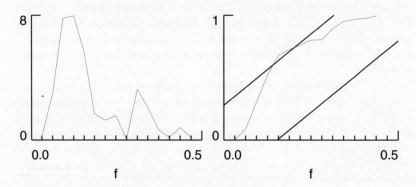

Figure 234. Periodogram (left-hand plot) and associated normalized cumulative periodogram (right-hand plot) for the first 32 values of the AR(2) series in the upper plot of Figure 45. Note that the periodogram is plotted here on a linear/linear scale.

where $C(0.10) = 1.224$, $C(0.05) = 1.358$ and $C(0.01) = 1.628$ (Stephens, 1974). For all $M \geq 7$, this approximation is good to within 1% for the three stated values of α (i.e., $0.99 < \tilde{D}(\alpha)/D(\alpha) < 1.01$ for $\alpha = 0.10$, 0.05 and 0.01). Tabulated values for $D(\alpha)$ are in Conover (1980).

We can obtain a useful graphical equivalent of this test by plotting $\mathcal{P}_k$ versus f_k for $k = 1, \ldots, M - 1$ along with the lines defined by

$$L_u(f) \equiv \frac{fN \, \Delta t - 1}{M - 1} + \tilde{D}(\alpha) \text{ and } L_l(f) \equiv \frac{fN \, \Delta t}{M - 1} - \tilde{D}(\alpha). \quad (234)$$

If any of the points given by $(f_k, \mathcal{P}_k)$ falls outside the region between these two lines, then we reject the null hypothesis at a level of significance of α; conversely, if all of the points fall inside this region, then we fail to reject the null hypothesis at the stated level.

As a concrete example, let us consider the short time series consisting of the first $N = 32$ values of the AR(2) series of Figure 45 (top plot). The left-hand plot of Figure 234 shows $\hat{S}^{(p)}(f_j)$ for this series versus $f_j = j/32$ for $j = 1, \ldots, M = 15$ (the points are connected by lines; recall also that $\Delta t = 1$). The jagged thin curve in the right-hand plot depicts the corresponding normalized cumulative periodogram $\mathcal{P}_k$ versus f_k for $k = 1, \ldots, M - 1 = 14$. The two thick lines are $L_u(\cdot)$ and $L_l(\cdot)$ corresponding to a significance level of $\alpha = 0.05$; specifically, since here $\tilde{D}(0.05) = 0.349$ (for the record, $D(0.05) = 0.349$ also), we have

$$L_u(f) = 0.349 - \frac{1}{14} + \frac{32}{14}f \text{ and } L_l(f) = -0.349 + \frac{32}{14}f.$$

Since the normalized cumulative periodogram is not entirely contained in the region between these two lines, we reject the null hypothesis of white noise at the 0.05 level of significance. Equivalently, the values of D^+ and D^- are, respectively, 0.066 and 0.392, so the value of D is 0.392, which – since it exceeds $\tilde{D}(0.05) = 0.349$ – tells us to reject the null hypothesis. We note that D^- is the value associated with $f_5 = 5/32$; i.e., $D^- = \mathcal{P}_5 - 4/14$ with $\mathcal{P}_5 = 0.6773$. (For a second example of the use of this test, see Section 10.13.)

Finally, we note that the cumulative periodogram test requires computation of the periodogram over the Fourier frequencies. When, for some sample sizes N, we cannot use an FFT algorithm directly to compute the required DFT of $X_1, \ldots, X_N$, the chirp transform algorithm can be used to obtain the required DFT efficiently (see the discussion in item [1] of the Comments and Extensions to Section 3.10).

6.7 Smoothing Direct Spectral Estimators

From the previous sections, we are forced to conclude that the 'natural' spectral estimator $\hat{S}^{(p)}(\cdot)$ is actually of limited use in estimating $S(\cdot)$ due to its poor bias and variance properties. For spectra with large dynamic range, the bias can be reduced considerably by using a direct spectral estimator $\hat{S}^{(d)}(\cdot)$ with an appropriate data taper. However, var $\{\hat{S}^{(d)}(f)\}$ is still a problem from two viewpoints. First, we need to be able to visually detect structure in our spectral estimates when plotted versus frequency – the large variability in $\hat{S}^{(d)}(\cdot)$ can mask potentially important features (in Figure 227 compare the direct spectral estimates shown by the thin curves with the true sdf shown by the thick curves). Second, certain statistical tests can be considerably strengthened by using spectral estimators with better variance properties than $\hat{S}^{(d)}(\cdot)$.

The traditional approach to this problem is to smooth $\hat{S}^{(d)}(\cdot)$ across frequencies. The justification for this procedure is the following. Suppose N is large enough so that the periodogram $\hat{S}^{(p)}(\cdot)$ is essentially an unbiased estimator of $S(\cdot)$ and is pairwise uncorrelated at the Fourier frequencies f_k (i.e., the approximations (223a) are valid). If $S(\cdot)$ is slowly varying in the neighborhood of, say, f_k, then

$$S(f_{k-M}) \approx \cdots \approx S(f_k) \approx \cdots \approx S(f_{k+M})$$

for some integer $M > 0$. Thus

$$\hat{S}^{(p)}(f_{k-M}), \ldots, \hat{S}^{(p)}(f_k), \ldots, \hat{S}^{(p)}(f_{k+M})$$

are a set of $2M + 1$ unbiased and uncorrelated estimators of the same quantity, namely, $S(f_k)$. We can thus average them to produce the estimator

$$\bar{S}(f_k) \equiv \frac{1}{2M+1} \sum_{j=-M}^{M} \hat{S}^{(p)}(f_{k-j}).$$

Under our assumptions we have

$$E\{\bar{S}(f_k)\} \approx S(f_k)$$

and

$$\text{var}\{\bar{S}(f_k)\} \approx \frac{S^2(f_k)}{2M+1} \approx \frac{\text{var}\{\hat{S}^{(p)}(f_k)\}}{2M+1}$$

using (222b) and assuming $f_{k-M} > 0$ and $f_{k+M} < f_{(N)}$ for simplicity. If we now consider increasing both the sample size N and the index k in such a way that $k/(N\,\Delta t) = f_k$ is a constant, we can then let M get large also and claim that var $\{\bar{S}(f_k)\}$ can be made arbitrarily small so that $\bar{S}(f_k)$ is a consistent estimator of $S(f_k)$.

The estimator $\{\bar{S}(f_k)\}$ is a special case of a more general spectral estimator of the form

$$\hat{S}^{(ds)}(f'_k) \equiv \sum_{j=-M}^{M} g_j \hat{S}^{(d)}(f'_{k-j}) \text{ with } f'_k \equiv \frac{k}{N'\,\Delta t}, \tag{236a}$$

where $\{g_j\}$ is a sequence of $2M+1$ smoothing coefficients with $g_{-j} = g_j$ and $g_M \neq 0$; $\hat{S}^{(d)}(\cdot)$ is a direct spectral estimator (Equation (206c)); and N' is a positive integer that controls the spacing of the frequencies over which the smoothing occurs. Typically we choose $N' \geq N$ (the sample size) so that the frequencies f'_k are at least as closely spaced as the Fourier frequencies $f_k = k/(N\,\Delta t)$. We call $\hat{S}^{(ds)}(f'_k)$ a *discretely smoothed direct spectral estimator* of $S(f'_k)$. Note that we can regard $\{g_j\}$ as the coefficients of an LTI digital filter.

Now the estimator $\{\hat{S}^{(ds)}(f'_k)\}$ is formed by smoothing the direct spectral estimator $\hat{S}^{(d)}(\cdot)$ with a *discrete* convolution over a *discrete* set of frequencies. Because $\hat{S}^{(d)}(\cdot)$ is defined for all $f \in [-f_{(N)}, f_{(N)}]$, we can also smooth it using a *continuous* convolution over a *continuous* set of frequencies. We thus consider an estimator of the form

$$\hat{S}^{(lw)}(f) = \int_{-f_{(N)}}^{f_{(N)}} V_m(f - \phi)\hat{S}^{(d)}(\phi)\,d\phi, \tag{236b}$$

where $V_m(\cdot)$ is a symmetric real-valued $2f_{(N)}$ periodic function which is square integrable over $[-f_{(N)}, f_{(N)}]$ and whose smoothing properties can be controlled by a parameter m (the rationale for the superscript (lw) is explained below). Since $\{\hat{s}_\tau^{(d)}\} \longleftrightarrow \hat{S}^{(d)}(\cdot)$ (see Equations (207c) and (207d)), we can rewrite the above as

$$\hat{S}^{(lw)}(f) = \int_{-f_{(N)}}^{f_{(N)}} V_m(f - \phi)\left(\Delta t \sum_{\tau=-(N-1)}^{N-1} \hat{s}_\tau^{(d)} e^{-i2\pi\phi\tau\,\Delta t}\right) d\phi$$

$$= \Delta t \sum_{\tau=-(N-1)}^{N-1} \left(\int_{-f_{(N)}}^{f_{(N)}} V_m(f-\phi)e^{i2\pi(f-\phi)\tau \, \Delta t} \, d\phi \right) \hat{s}_\tau^{(d)} e^{-i2\pi f \tau \, \Delta t}$$

$$= \Delta t \sum_{\tau=-(N-1)}^{N-1} v_{\tau,m} \hat{s}_\tau^{(d)} e^{-i2\pi f \tau \, \Delta t}, \tag{237a}$$

where

$$v_{\tau,m} \equiv \int_{-f_{(N)}}^{f_{(N)}} V_m(f-\phi)e^{i2\pi(f-\phi)\tau \, \Delta t} \, d\phi = \int_{-f_{(N)}}^{f_{(N)}} V_m(\phi)e^{i2\pi \phi \tau \, \Delta t} \, d\phi \tag{237b}$$

(the second integral follows from the first since $V_m(\cdot)$ is symmetric and $2f_{(N)}$ periodic). We note that $\{v_{\tau,M}\}$ and $V_m(\cdot)$ are a Fourier transform pair:

$$\{v_{\tau,m}\} \longleftrightarrow V_m(\cdot).$$

Equation (237a) tells us that we can compute $\hat{S}^{(lw)}(f)$ using a finite number of numerical operations even though it is defined in Equation (236b) via a continuous convolution over a continuous set of frequencies. We note that Equation (237a) does not involve $v_{\tau,m}$ for $|\tau| \geq N$ but that $v_{\tau,m}$ need *not* be zero for these values of τ. For theoretical purposes, we will find it convenient to define

$$w_{\tau,m} = \begin{cases} v_{\tau,m}, & |\tau| < N; \\ 0, & |\tau| \geq N, \end{cases}$$

and

$$W_m(f) \equiv \Delta t \sum_{\tau=-(N-1)}^{N-1} w_{\tau,m} e^{-i2\pi f \tau \, \Delta t} \tag{237c}$$

so that

$$\{w_{\tau,m}\} \longleftrightarrow W_m(\cdot).$$

Note that $V_m(\cdot)$ and $W_m(\cdot)$ are identical if $v_{\tau,m} = 0$ for $|\tau| \geq N$; however, even if this is not true, $V_m(\cdot)$ and $W_m(\cdot)$ are equivalent as far as Equation (236b) is concerned because we have

$$\int_{-f_{(N)}}^{f_{(N)}} W_m(f-\phi)\hat{S}^{(d)}(\phi) \, d\phi = \int_{-f_{(N)}}^{f_{(N)}} V_m(f-\phi)\hat{S}^{(d)}(\phi) \, d\phi = \hat{S}^{(lw)}(f) \tag{237d}$$

for all f; i.e., the estimator $\hat{S}^{(lw)}(\cdot)$ will be *identically* the same whether we use $V_m(\cdot)$ or $W_m(\cdot)$ (see Exercise [6.14]). The essential distinction between $V_m(\cdot)$ and $W_m(\cdot)$ is that $w_{\tau,m}$ is guaranteed to be 0 for $|\tau| \geq N$ whereas $v_{\tau,m}$ need not be so.

We thus can assume that the estimator $\hat{S}^{(lw)}(\cdot)$ can be written as

$$\hat{S}^{(lw)}(f) \equiv \int_{-f_{(N)}}^{f_{(N)}} W_m(f - \phi)\hat{S}^{(d)}(\phi)\, d\phi$$

$$= \Delta t \sum_{\tau=-(N-1)}^{N-1} w_{\tau,m}\hat{s}_\tau^{(d)} e^{-i2\pi f\tau\,\Delta t}, \tag{238a}$$

where $\{w_{\tau,m}\} \longleftrightarrow W_m(\cdot)$ with $w_{\tau,m} = 0$ for $|\tau| \geq N$. If we define

$$\hat{s}_\tau^{(lw)} \equiv \begin{cases} w_{\tau,m}\hat{s}_\tau^{(d)}, & |\tau| \leq N - 1; \\ 0, & |\tau| \geq N, \end{cases} \tag{238b}$$

we have $\{\hat{s}_\tau^{(lw)}\} \longleftrightarrow \hat{S}^{(lw)}(\cdot)$; i.e., the sequence $\{\hat{s}_\tau^{(lw)}\}$ is the estimator of the acvs corresponding to $\hat{S}^{(lw)}(\cdot)$.

We call the function $W_m(\cdot)$ defined in (237c) a *smoothing window* (although most other authors would call it a spectral window, a term we reserve for another concept); its inverse Fourier transform $\{w_{\tau,m}\}$ is called a *lag window* (other names for it are *quadratic window* and *quadratic taper*). We call $\hat{S}^{(lw)}(f)$ a *lag window spectral estimator* of $S(f)$ (see the Comments and Extensions below). In practice, we specify a lag window spectral estimator by either $V_m(\cdot)$ or $\{w_{\tau,m}\}$ – the smoothing window $W_m(\cdot)$ follows once one of these is given.

How are the spectral estimators $\{\hat{S}^{(ds)}(f_k')\}$ and $\hat{S}^{(lw)}(\cdot)$ related? Since $\{\hat{s}_\tau^{(d)}\} \longleftrightarrow \hat{S}^{(d)}(\cdot)$ (see Equation (207c)) and since $f_{k-j}' = f_k' - f_j'$, we can use Equation (236a) to see that

$$\hat{S}^{(ds)}(f_k') = \sum_{j=-M}^{M} g_j \left(\Delta t \sum_{\tau=-(N-1)}^{N-1} \hat{s}_\tau^{(d)} e^{-i2\pi f_{k-j}'\tau\,\Delta t} \right)$$

$$= \Delta t \sum_{\tau=-(N-1)}^{N-1} \left(\sum_{j=-M}^{M} g_j e^{i2\pi f_j'\tau\,\Delta t} \right) \hat{s}_\tau^{(d)} e^{-i2\pi f_k'\tau\,\Delta t}$$

$$= \Delta t \sum_{\tau=-(N-1)}^{N-1} v_{\tau,g}\hat{s}_\tau^{(d)} e^{-i2\pi f_k'\tau\,\Delta t}, \tag{238c}$$

where

$$v_{\tau,g} \equiv \sum_{j=-M}^{M} g_j e^{i2\pi f_j'\tau\,\Delta t} = \sum_{j=-M}^{M} g_j e^{i2\pi j\tau/N'}. \tag{238d}$$

Hence any discretely smoothed direct spectral estimator can be expressed as a lag window spectral estimator with a lag window given by

$$w_{\tau,g} \equiv \begin{cases} v_{\tau,g}, & |\tau| \leq N - 1; \\ 0, & |\tau| \geq N, \end{cases} \tag{238e}$$

and a corresponding smoothing window given by Equation (237c) with $w_{\tau,m}$ replaced by $w_{\tau,g}$.

Now consider any lag window spectral estimator $\hat{S}^{(lw)}(\cdot)$. Although this estimator is defined over a continuous range of frequencies from $-f_{(N)}$ to $f_{(N)}$, it is completely determined if we know $\{\hat{s}_\tau^{(lw)}\}$ because $\{\hat{s}_\tau^{(lw)}\} \longleftrightarrow \hat{S}^{(lw)}(\cdot)$. Note that we also have

$$\{\,\hat{s}_\tau^{(lw)} : \tau = -(N-1), \ldots, N\,\} \longleftrightarrow \{\,\hat{S}^{(lw)}(\tilde{f}_k) : k = -(N-1), \ldots, N\,\},$$

where, as before, $\tilde{f}_k \equiv k/(2N\,\Delta t)$ defines a grid of frequencies twice as fine as the Fourier frequencies $f_k \equiv k/(N\,\Delta t)$. Hence $\hat{S}^{(lw)}(\cdot)$ is completely determined once we know it over the finite set of frequencies $\tilde{f}_k$. Because

$$\{\,\hat{s}_\tau^{(d)} : \tau = -(N-1), \ldots, N\,\} \longleftrightarrow \{\,\hat{S}^{(d)}(\tilde{f}_k) : k = -(N-1), \ldots, N\,\}$$

implies that

$$\hat{s}_\tau^{(d)} = \frac{1}{2N\,\Delta t} \sum_{j=-(N-1)}^{N} \hat{S}^{(d)}(\tilde{f}_j) e^{i2\pi \tilde{f}_j \tau\,\Delta t},$$

we can write

$$\hat{S}^{(lw)}(\tilde{f}_k)$$

$$= \Delta t \sum_{\tau=-(N-1)}^{N-1} w_{\tau,m} \hat{s}_\tau^{(d)} e^{-i2\pi \tilde{f}_k \tau\,\Delta t}$$

$$= \sum_{\tau=-(N-1)}^{N-1} w_{\tau,m} \left(\frac{1}{2N} \sum_{j=-(N-1)}^{N} \hat{S}^{(d)}(\tilde{f}_j) e^{i2\pi \tilde{f}_j \tau\,\Delta t} \right) e^{-i2\pi \tilde{f}_k \tau\,\Delta t}$$

$$= \sum_{\tau=-(N-1)}^{N-1} w_{\tau,m} \left(\frac{1}{2N} \sum_{j=-(N-1)}^{N} \hat{S}^{(d)}(\tilde{f}_{k-j}) e^{i2\pi \tilde{f}_{k-j} \tau\,\Delta t} \right) e^{-i2\pi \tilde{f}_k \tau\,\Delta t}$$

$$= \sum_{j=-(N-1)}^{N} \left(\frac{1}{2N} \sum_{\tau=-(N-1)}^{N-1} w_{\tau,m} e^{-i2\pi \tilde{f}_j \tau\,\Delta t} \right) \hat{S}^{(d)}(\tilde{f}_{k-j})$$

$$= \sum_{j=-(N-1)}^{N} \tilde{g}_j \hat{S}^{(d)}(\tilde{f}_{k-j}), \tag{239}$$

where

$$\tilde{g}_j \equiv \frac{1}{2N} \sum_{\tau=-(N-1)}^{N-1} w_{\tau,m} e^{-i2\pi \tilde{f}_j \tau\,\Delta t} = \frac{1}{2N\,\Delta t} W_m(\tilde{f}_j)$$

(using Equation (237c)). Hence any lag window spectral estimator $\hat{S}^{(lw)}(\cdot)$ can be expressed at $f = \tilde{f}_k$ as a discretely smoothed direct spectral estimator over the set of frequencies $\{\,\tilde{f}_j : j = -(N-1), \dots, N\,\}$ with weights $\tilde{g}_j$ that are proportional to the smoothing window at frequency $\tilde{f}_j$.

From these arguments we can conclude that the class of lag window spectral estimators and the class of discretely smoothed direct spectral estimators are equivalent from a practical point of view (although they are certainly different from a computational point of view, as discussed in Section 6.16). In the following sections we will concentrate for convenience mainly on lag window spectral estimators. These can be specified via either a lag window $\{w_{\tau,m}\}$ or its Fourier transform $W_m(\cdot)$. The rationale for considering lag window estimators is that the variance of a direct spectral estimator $\hat{S}^{(d)}(\cdot)$ does not decrease as the sample size N gets large. If we let $w_{\tau,m} = 1$ for all τ in (238a), $\hat{S}^{(lw)}(\cdot)$ reduces to $\hat{S}^{(d)}(\cdot)$ (compare Equations (207c) and (238a)). This tells us that, not surprisingly, we must impose some conditions on the lag window $\{w_{\tau,m}\}$ or, equivalently, the smoothing window $W_m(\cdot)$ to ensure that $\hat{S}^{(lw)}(\cdot)$ does have smaller variance than $\hat{S}^{(d)}(\cdot)$ and is a reasonable estimator of $S(\cdot)$. The following assumptions will prove desirable.

[1] We require that $W_m(\cdot)$ be an even $2f_{(N)}$ periodic function for all choices of m. This implies that $w_{-\tau,m} = w_{\tau,m}$ for all τ. Note also that $W_m(\cdot)$ is a trigonometric polynomial, from which we have immediately that $\{w_{\tau,m}\}$ and $W_m(\cdot)$ constitute a Fourier transform pair.

[2] We require the normalization

$$\int_{-f_{(N)}}^{f_{(N)}} W_m(f)\,df = 1, \quad \text{or, equivalently, } w_{0,m} = 1, \text{ for all } m$$

(240a)

(the equivalence follows from $\{w_{\tau,m}\} \longleftrightarrow W_m(\cdot)$). This stipulation is similar to one we made in Section 5.7 to ensure that a low-pass digital filter passes unaltered a locally linear portion of a sequence. We note that, in terms of the $\tilde{g}_j$ weights, (240a) is equivalent to

$$\sum_{j=-(N-1)}^{N} \tilde{g}_j = 1.$$

[3] For any $\epsilon > 0$ and for all $|f| > \epsilon$, we require

$$W_m(f) \to 0 \quad \text{as} \quad m \to \infty \tag{240b}$$

uniformly in f. This implies that $W_m(\cdot)$ becomes more concentrated about 0 as m gets large and hence that small values of m imply more smoothing.

[4] If

$$W_m(f) \geq 0 \text{ for all } m \text{ and } f, \qquad (241a)$$

then it follows from Equation (238a) that, since $\hat{S}^{(d)}(f) \geq 0$ necessarily (consider Equation (206c)), we must have $\hat{S}^{(lw)}(f) \geq 0$. Interestingly, condition (241a) is sufficient, but not necessary, to ensure the nonnegativity of $\hat{S}^{(d)}(f)$ – see the discussion on the Daniell smoothing window in Section 6.11. As we shall see (see the discussion concerning Figure 270), there are valid reasons for considering smoothing windows for which $W_m(f) < 0$ for some values of f; so, whereas we always require items [1] to [3], condition (241a) is desirable but not required.

Note that, if condition (241a) holds, requirement (240a) implies that $W_m(\cdot)$ can be regarded as a pdf for an rv distributed over the interval $[-f_{(N)}, f_{(N)}]$.

We also need to define a bandwidth for the smoothing window $W_m(\cdot)$ in order to have some idea about the range of frequencies that influences the value of $\hat{S}^{(lw)}(f)$. When condition (241a) holds so that $W_m(\cdot)$ can be regarded as a pdf, a convenient measure of width is given by Equation (73b):

$$\beta_W \equiv \left(12 \int_{-f_{(N)}}^{f_{(N)}} f^2 W_m(f) \, df \right)^{1/2} \qquad (241b)$$

(the truncation of the integral to $|f| \leq f_{(N)}$ is appropriate since $W_m(\cdot)$ is a $2f_{(N)}$ periodic function). This definition of window bandwidth is essentially due to Grenander (1951, p. 525) although he omits the factor of 12. As noted in Section 3.4, inclusion of this factor makes the bandwidth of a rectangular smoothing window equal to its natural width (this is related to the Daniell smoothing window – see Section 6.11). A convenient computational formula for the integral above is given by

$$12 \int_{-f_{(N)}}^{f_{(N)}} f^2 W_m(f) \, df = \frac{1}{(\Delta t)^2} \left(1 + \frac{12}{\pi^2} \sum_{\tau=1}^{N-1} \frac{(-1)^\tau}{\tau^2} w_{\tau,m} \right) \qquad (241c)$$

(see Exercise [6.15]).

There are, however, two practical reasons for *not* using β_W to define the smoothing window bandwidth. The first is computational. If we use Equation (241c) to evaluate the integral in Equation (241b) and if the $w_{\tau,m}$ terms are slowly changing, we must subtract numbers that can be quite close to each other in magnitude. If sufficient care is not taken, there is a potential problem in properly computing the summation in Equation (241c) with finite precision arithmetic. The second reason is that, if the nonnegativity condition (241a) does not hold, then β_W can be imaginary because the integral in Equation (241b) can be negative.

Because of these potential problems, we prefer the following definition for the smoothing window bandwidth due to Jenkins (1961):

$$B_W \equiv \frac{1}{\int_{-f_{(N)}}^{f_{(N)}} W_m^2(f)\, df}. \tag{242a}$$

This bandwidth measure is just the equivalent width of the autocorrelation of $W_m(\cdot)$ – this is the width measure defined by Equation (85c) after the appropriate adjustment for a periodic function (in particular, the integral in the numerator of that equation becomes $\int_{-f_{(N)}}^{f_{(N)}} W(f)\, df$, which is unity by the requirement in Equation (240a)). Since $\{w_{\tau,m}\}$ and $W_m(\cdot)$ are a Fourier transform pair, Parseval's relationship states (see Equation (89c)) that

$$\int_{-f_{(N)}}^{f_{(N)}} W_m^2(\phi)\, d\phi = \Delta t \sum_{\tau=-(N-1)}^{N-1} w_{\tau,m}^2, \tag{242b}$$

so we also have

$$B_W = \frac{1}{\Delta t \sum_{\tau=-(N-1)}^{N-1} w_{\tau,m}^2}. \tag{242c}$$

In contrast to the summation in Equation (241c), the summation above involves the addition of strictly nonnegative terms; moreover, B_W is always real-valued. (See item [3] of the Comments and Extensions below for an additional way of viewing B_W.)

Comments and Extensions to Section 6.7

[1]   The lag window spectral estimator $\hat{S}^{(lw)}(\cdot)$ is known by many other names in the literature: quadratic window estimator, weighted covariance estimator, indirect nonparametric estimator, windowed periodogram estimator, smoothed periodogram estimator, Blackman–Tukey estimator, spectrograph estimator, spectral estimator of the Grenander–Rosenblatt type, and correlogram method power spectral density estimator – unfortunately this list is by no means exhaustive! To compound the confusion, some of these names are also used in the literature to refer to estimators that are slightly different from our definition of $\hat{S}^{(lw)}(\cdot)$. For example, our definition (Equation (238a)) is essentially a weighting applied to the acvs estimator $\{\hat{s}^{(d)}\}$ corresponding to *any* direct spectral estimator; other authors define it – at least implicitly – for just the special case of $\{\hat{s}^{(p)}\}$, the acvs estimator corresponding to the periodogram.

[2]   Several other measures of window bandwidth are used widely in the literature (Priestley, 1981, p. 520). For example, Parzen (1957) proposed using the equivalent width of $W_m(\cdot)$. This was defined in Equation (70b) for $L^2(-\infty, \infty)$ functions. The obvious definition for a $2f_{(N)}$ periodic function $G_p(\cdot)$ is

$$\text{width}_e\left\{G_p(\cdot)\right\} \equiv \int_{-f_{(N)}}^{f_{(N)}} G_p(f)\, df \Big/ G_p(0).$$

Since $\int_{-f_{(N)}}^{f_{(N)}} W_m(f)\, df = 1$ by assumption, Parzen's bandwidth measure reduces to

$$B_{W,P} \equiv \text{width}_e\left\{W_m(\cdot)\right\} = \frac{1}{W_m(0)} = \frac{1}{\Delta t \sum_{\tau=-(N-1)}^{N-1} w_{\tau,m}}.$$

[3]   It can be argued from the material in Sections 6.9 and 6.11 that Jenkins's smoothing window bandwidth B_W is the width of the Daniell smoothing window that yields an estimator with the same large sample variance as one employing $W_m(\cdot)$. Both Priestley (1981) and Bloomfield (1976) argue *against* use of B_W because of this link with variance – they suggest that window bandwidth should be tied instead to the bias introduced by the smoothing window. This property is enjoyed by β_W (see Equation (245)), but, as noted above, we prefer B_W mainly because of its superior practical utility. In most applications, however, as long as the nonnegativity condition (241a) holds and as long as care is taken in computing β_W, the measures β_W, B_W and $B_{W,P}$ are interchangeable.

6.8 First Moment Properties of Lag Window Estimators

We now consider the first moment properties of $\hat{S}^{(lw)}(\cdot)$. It follows from Equations (238a) and (207a) that

$$E\{\hat{S}^{(lw)}(f)\} = \int_{-f_{(N)}}^{f_{(N)}} W_m(f - f') E\{\hat{S}^{(d)}(f')\}\, df' \qquad (243a)$$

$$= \int_{-f_{(N)}}^{f_{(N)}} W_m(f - f') \left(\int_{-f_{(N)}}^{f_{(N)}} \mathcal{H}(f' - \phi) S(\phi)\, d\phi\right) df'$$

$$= \int_{-f_{(N)}}^{f_{(N)}} \left(\int_{-f_{(N)}}^{f_{(N)}} W_m(f - f') \mathcal{H}(f' - \phi)\, df'\right) S(\phi)\, d\phi$$

$$= \int_{-f_{(N)}}^{f_{(N)}} \mathcal{U}_m(f - \phi) S(\phi)\, d\phi, \qquad (243b)$$

where

$$\mathcal{U}_m(f) \equiv \int_{-f_{(N)}}^{f_{(N)}} W_m(f - f'') \mathcal{H}(f'')\, df'' \qquad (243c)$$

(the above uses the facts that $W_m(\cdot)$ and $\mathcal{H}(\cdot)$ are both periodic functions with period $2f_{(N)}$ and that $W_m(\cdot)$ is an even function). We call $\mathcal{U}_m(\cdot)$ the *spectral window* for the lag window spectral estimator $\hat{S}^{(lw)}(\cdot)$ (just as we called $\mathcal{H}(\cdot)$ the spectral window of $\hat{S}^{(d)}(\cdot)$ – compare Equations (207a) and (243b)). It can be shown (see Exercise [6.16]) that

$$\mathcal{U}_m(f) = \Delta t \sum_{\tau=-(N-1)}^{N-1} w_{\tau,m} \left(\sum_{t=1}^{N-|\tau|} h_t h_{t+|\tau|} \right) e^{-i2\pi f \tau \, \Delta t}, \qquad (244a)$$

which is useful for calculating $\mathcal{U}_m(\cdot)$ in practice.

The estimator $\hat{S}^{(lw)}(\cdot)$ has several properties that parallel those for $\hat{S}^{(d)}(\cdot)$ described in Section 6.4 – see Exercise [6.17].

Let us now assume that $\hat{S}^{(d)}(f)$ is an approximately unbiased estimator of $S(f)$ – this assumption is reasonable if we have made effective use of tapering (or prewhitening). This allows us to examine the effect of the smoothing window separately on the first moment properties of $\hat{S}^{(lw)}(\cdot)$. From Equation (243a) we then have

$$E\{\hat{S}^{(lw)}(f)\} \approx \int_{-f_{(N)}}^{f_{(N)}} W_m(f - f')S(f') \, df'. \qquad (244b)$$

By conditions (240a) and (240b), $W_m(\cdot)$ acts like a Dirac delta function as the smoothing parameter m gets large. As we shall see in Section 6.11, we cannot make m arbitrarily large if the sample size N is fixed; however, if we let $N \to \infty$, we can also let $m \to \infty$, so we can claim that $\hat{S}^{(lw)}(f)$ is an asymptotically unbiased estimator of $S(f)$:

$$E\{\hat{S}^{(lw)}(f)\} \to S(f) \text{ as } m, N \to \infty. \qquad (244c)$$

For finite sample sizes N and finite values of the smoothing parameter m, however, even if $\hat{S}^{(d)}(\cdot)$ is an approximately unbiased estimator, the smoothing window $W_m(\cdot)$ can introduce significant bias in $\hat{S}^{(lw)}(\cdot)$ by inadvertently smoothing together adjacent features in $\hat{S}^{(d)}(\cdot)$ – this happens when the true sdf is not slowly varying. This bias obviously depends upon both the true sdf and the shape of the smoothing window. We can derive an expression to quantify this as follows (Priestley, 1981, p. 458). We first define the bias due to the smoothing window alone as

$$b_W(f) \equiv \int_{-f_{(N)}}^{f_{(N)}} W_m(f - f')S(f') \, df' - S(f),$$

a reasonable definition in view of Equation (244b). Hence

$$b_W(f) = \int_{-f_{(N)}}^{f_{(N)}} W_m(f - f') \left[S(f') - S(f)\right] df'$$

by assumption (240a). Because $W_m(\cdot)$ and $S(\cdot)$ are $2f_{(N)}$ periodic and even, we find, letting $\phi = f' - f$,

$$b_W(f) = \int_{-f_{(N)}-f}^{f_{(N)}-f} W_m(-\phi)\left[S(f+\phi) - S(f)\right] d\phi$$

$$= \int_{-f_{(N)}}^{f_{(N)}} W_m(\phi)\left[S(f+\phi) - S(f)\right] d\phi.$$

Next we assume that $S(\cdot)$ can be expanded in a Taylor series about f:

$$S(f+\phi) = S(f) + \phi S'(f) + \frac{\phi^2}{2}S''(f) + o(\phi^2).$$

Since $W_m(\cdot)$ is always assumed to be an even function, we have

$$b_W(f) \approx \int_{-f_{(N)}}^{f_{(N)}} W_m(\phi)\left[\phi S'(f) + \frac{\phi^2}{2}S''(f)\right] d\phi$$

$$= \frac{S''(f)}{2}\int_{-f_{(N)}}^{f_{(N)}} \phi^2 W_m(\phi)\, d\phi = \frac{S''(f)}{24}\beta_W^2, \qquad (245)$$

where β_W is the alternative measure of smoothing window bandwidth we defined in Equation (241b). The bias due to the smoothing window in $\hat{S}^{(lw)}(f)$ is thus related to two quantities: first, the curvature (second derivative) of $S(\cdot)$ at f – a large curvature due to a peak or a valley in $S(\cdot)$ implies a large bias; and second, a measure of the bandwidth of the smoothing window – the larger this measure, the greater the bias.

Comments and Extensions to Section 6.8

[1]   We have introduced data tapers as a means of compensating for the bias (first moment) of $\hat{S}^{(p)}(\cdot)$ and lag windows as a means of decreasing the variance (second moment) of direct spectral estimators $\hat{S}^{(d)}(\cdot)$ (including the periodogram). It is sometimes claimed in the literature that a lag window can be used to control both the bias and the variance in $\hat{S}^{(p)}(\cdot)$ (see, for example, Grenander and Rosenblatt, 1984, Section 4.2); i.e., we can avoid the use of a nonrectangular data taper. The basis for this claim can be seen by considering the spectral window $\mathcal{U}_m(\cdot)$ in Equation (244a). Equation (243b) says that this spectral window describes the bias properties of a lag window spectral estimator consisting of a lag window $\{w_{\tau,m}\}$ used in combination with a data taper $\{h_t\}$. The following argument shows we can produce the same spectral window by using a different lag window in combination with just the rectangular data taper. First, we note that, in the case of a rectangular data taper, i.e., $h_t = 1/\sqrt{N}$, the spectral window $\mathcal{U}_m(\cdot)$ reduces to

$$\Delta t \sum_{\tau=-(N-1)}^{N-1} w_{\tau,m}\left(1 - |\tau|/N\right) e^{-i2\pi f \tau\,\Delta t}.$$

For any data taper $\{h_t\}$, we can write, using Equation (244a),

$$\mathcal{U}_m(f) = \Delta t \sum_{\tau=-(N-1)}^{N-1} w_{\tau,m} \frac{\sum_{t=1}^{N-|\tau|} h_t h_{t+|\tau|}}{1-|\tau|/N} \left(1-|\tau|/N\right) e^{-i2\pi f \tau \, \Delta t}$$

$$= \Delta t \sum_{\tau=-(N-1)}^{N-1} w'_{\tau,m} \left(1-|\tau|/N\right) e^{-i2\pi f \tau \, \Delta t},$$

where

$$w'_{\tau,m} \equiv w_{\tau,m} \frac{\sum_{t=1}^{N-|\tau|} h_t h_{t+|\tau|}}{1-|\tau|/N}. \tag{246}$$

Thus use of the lag window $\{w'_{\tau,m}\}$ with the rectangular data taper produces the same spectral window $\mathcal{U}_m(\cdot)$ as use of the lag window $\{w_{\tau,m}\}$ with the nonrectangular data taper $\{h_t\}$. Nuttall and Carter (1982) refer to $\{w'_{\tau,m}\}$ as a *reshaped lag window*. As we show by example in the Comments and Extensions to Section 6.11, a potential problem with this scheme is that the smoothing window associated with $\{w'_{\tau,m}\}$ (i.e., its Fourier transform) can have some undesirable properties, including prominent *negative* sidelobes. This means that $\hat{S}^{(lw)}(\cdot)$ can be negative at some frequencies for some time series. Nuttall and Carter point out that a potential benefit of lag window reshaping is the elimination of the N multiplications needed to implement tapering – this can be an important saving in certain real-time applications.

6.9 Second Moment Properties of Lag Window Estimators

We sketch here a derivation of one approximation to the large sample variance of $\hat{S}^{(lw)}(f)$ – a second, more accurate, approximation is considered in the Comments and Extensions below. We assume that the direct spectral estimator $\hat{S}^{(d)}(\cdot)$ is approximately uncorrelated on a grid of frequencies defined by $f'_k \equiv k/(N' \, \Delta t)$, where here $N' \leq N$ and k is an integer such that $0 < f'_k < f_{(N)}$ – see the discussion concerning Equation (224). From Equation (238a) and an obvious change of variable, we have

$$\hat{S}^{(lw)}(f) = \int_{-f_{(N)}-f}^{f_{(N)}-f} W_m(-\phi) \hat{S}^{(d)}(\phi+f) \, d\phi$$

$$= \int_{-f_{(N)}}^{f_{(N)}} W_m(\phi) \hat{S}^{(d)}(\phi+f) \, d\phi;$$

here we use the facts that $W_m(\cdot)$ is an even function and that both it and $\hat{S}^{(d)}(\cdot)$ are $2f_{(N)}$ periodic. From (240b) we can assume that, for large N, $W_m(\phi) \approx 0$ for all $|\phi| > J/(N' \, \Delta t) = f'_J$ for some positive integer J. Hence

$$\hat{S}^{(lw)}(f_k') \approx \int_{-f_J'}^{f_J'} W_m(\phi)\hat{S}^{(d)}(\phi + f_k')\,d\phi$$

$$\approx \sum_{j=-J}^{J} W_m(f_j')\hat{S}^{(d)}(f_j' + f_k')\frac{1}{N'\,\Delta t},$$

where we have approximated the Riemann integral by a Riemann sum. If we note that $f_j' + f_k' = f_{j+k}'$, we now have

$$\text{var}\,\{\hat{S}^{(lw)}(f_k')\} \approx \frac{1}{(N'\,\Delta t)^2}\,\text{var}\left\{\sum_{j=-J}^{J} W_m(f_j')\hat{S}^{(d)}(f_{j+k}')\right\}$$

$$\stackrel{(1)}{\approx} \frac{1}{(N'\,\Delta t)^2} \sum_{j=-J}^{J} W_m^2(f_j')\,\text{var}\left\{\hat{S}^{(d)}(f_{j+k}')\right\}$$

$$\stackrel{(2)}{\approx} \frac{1}{(N'\,\Delta t)^2} \sum_{j=-J}^{J} W_m^2(f_j')S^2(f_{j+k}')$$

$$\stackrel{(3)}{\approx} \frac{1}{(N'\,\Delta t)^2} S^2(f_k') \sum_{j=-J}^{J} W_m^2(f_j')$$

$$\stackrel{(4)}{\approx} \frac{1}{(N'\,\Delta t)^2} S^2(f_k') \sum_{j=-\lfloor N'/2\rfloor}^{\lfloor N'/2\rfloor} W_m^2(f_j')$$

$$\stackrel{(5)}{\approx} \frac{S^2(f_k')}{N'\,\Delta t} \int_{-f_{(N)}}^{f_{(N)}} W_m^2(\phi)\,d\phi,$$

where we have made use of five approximations:

(1) pairwise uncorrelatedness of the components of $\hat{S}^{(d)}(\cdot)$ defined by the grid of frequencies f_k';
(2) Equation (223c), the large sample variance of $\hat{S}^{(d)}(\cdot)$ (assuming that $f_{j+k}' \neq 0$ or $\pm\pi$);
(3) a smoothness assumption on $S(\cdot)$ that asserts that $S(f_{j+k}') \approx S(f_k')$ locally;
(4) the previous assumption that $W_m(\phi) \approx 0$ for $|\phi| > f_J'$; and
(5) an approximation of the summation by a Riemann integral.

Finally, we assume that the above holds for all frequencies (and not just the ones defined by f_k') and that N' can be related to N by $N' \approx N/C_h$, where C_h is greater than or equal to unity and depends only upon the data taper for the direct spectral estimator. (Table 248 gives the value

Data taper	C_h	See Figure
rectangular	1.00	209, left-hand plot
20% cosine	1.12	209, right-hand plot
50% cosine	1.35	210, left-hand plot
100% cosine	1.94	210, right-hand plot
$NW = 1$ dpss	1.34	211, left-hand plot
$NW = 2$ dpss	1.96	211, right-hand plot
$NW = 4$ dpss	2.80	212, left-hand plot
$NW = 8$ dpss	3.94	212, right-hand plot

Table 248. Variance inflation factor C_h for various data tapers. Note that C_h increases as the width of the central lobe of the associated spectral window increases. The values listed here are approximations – valid for large N – to the right-hand side of Equation (251b).

of C_h for some common data tapers; see item [1] of the Comments and Extensions below for a computational formula for C_h and more discussion concerning it.) With these modifications we now have

$$\text{var}\{\hat{S}^{(lw)}(f)\} \approx \frac{C_h S^2(f)}{N\,\Delta t} \int_{-f_{(N)}}^{f_{(N)}} W_m^2(\phi)\,d\phi. \tag{248a}$$

Using the definition for the smoothing window bandwidth B_W in Equation (242a), we rewrite the above as

$$\text{var}\{\hat{S}^{(lw)}(f)\} \approx \frac{C_h S^2(f)}{B_W N\,\Delta t}.$$

Thus, as one would expect, increasing (decreasing) the smoothing window bandwidth causes the variance to decrease (increase). Because $C_h \geq 1$, we can interpret it as a variance inflation factor due to tapering.

It is also useful to derive a corresponding approximation in terms of the lag window. Because of the Parseval relationship in Equation (242b), approximation (248a) can be rewritten as

$$\text{var}\{\hat{S}^{(lw)}(f)\} \approx \frac{C_h S^2(f)}{N} \sum_{\tau=-(N-1)}^{N-1} w_{\tau,m}^2. \tag{248b}$$

Since under our assumptions $\hat{S}^{(lw)}(f)$ is asymptotically unbiased, we can show that it is also a consistent estimator of $S(f)$ if its variance decreases to 0 as $N \rightarrow \infty$. From Equations (248a) and (248b), the large sample variance of $\hat{S}^{(lw)}(f)$ approaches zero if we make one of the following additional – but equivalent – assumptions:

$$\lim_{N\to\infty} \frac{1}{N\,\Delta t} \int_{-f_{(N)}}^{f_{(N)}} W_m^2(\phi)\,d\phi = 0 \ \text{ or } \ \lim_{N\to\infty} \frac{1}{N} \sum_{\tau=-(N-1)}^{N-1} w_{\tau,m}^2 = 0.$$

We can use Equation (242a) to reexpress the above in terms of B_W as

$$\lim_{N\to\infty} \frac{B_W}{\Delta f} = \infty, \ \text{ where } \ \Delta f \equiv \frac{1}{N\,\Delta t}.$$

In other words, the smoothing window bandwidth B_W must grow to cover many different Fourier frequency bins. Thus one of these three equivalent assumptions is enough to ensure that $\hat{S}^{(lw)}(f)$ is a consistent estimator of $S(f)$ (see also the discussion in Section 6.10).

We now consider the covariance between $\hat{S}^{(lw)}(f_k')$ and $\hat{S}^{(lw)}(f_{k'}')$. By the asymptotic unbiasedness of $\hat{S}^{(lw)}(f)$, it follows that

$$\text{cov}\,\{\hat{S}^{(lw)}(f_k'), \hat{S}^{(lw)}(f_{k'}')\} \approx E\{\hat{S}^{(lw)}(f_k')\hat{S}^{(lw)}(f_{k'}')\} - S(f_k')S(f_{k'}').$$

We can proceed as before to argue that

$$\hat{S}^{(lw)}(f_k') \approx \frac{1}{N'\,\Delta t} \sum_{j=-J}^{J} W_m(f_j')\hat{S}^{(d)}(f_{j+k}')$$

and

$$\hat{S}^{(lw)}(f_{k'}') \approx \frac{1}{N'\,\Delta t} \sum_{j'=-J}^{J} W_m(f_{j'}')\hat{S}^{(d)}(f_{j'+k'}'),$$

from which it follows that

$$E\{\hat{S}^{(lw)}(f_k')\hat{S}^{(lw)}(f_{k'}')\}$$

$$\approx \frac{1}{(N'\,\Delta t)^2} \sum_{j,j'=-J}^{J} W_m(f_j')W_m(f_{j'}')E\{\hat{S}^{(d)}(f_{j+k}')\hat{S}^{(d)}(f_{j'+k'}')\}.$$

If f_k' and $f_{k'}'$ are far enough apart so that $|k - k'| > 2J$ (recall that J defines the point at which $W_m(f) \approx 0$ so that $2J$ is proportional to the 'width' of $W_m(\cdot)$), we can use the asymptotic uncorrelatedness and unbiasedness of $\hat{S}^{(d)}(\cdot)$ on the grid frequencies defined by f_k' to argue that

$$E\{\hat{S}^{(lw)}(f_k')\hat{S}^{(lw)}(f_{k'}')\}$$

$$\approx \frac{1}{(N'\,\Delta t)^2} \sum_{j,j'=-J}^{J} W_m(f_j')W_m(f_{j'}')S(f_{j+k}')S(f_{j'+k'}').$$

By the same assumption on the smoothness of $S(\cdot)$ as before,

$$E\{\hat{S}^{(lw)}(f_k')\hat{S}^{(lw)}(f_{k'}')\} \approx \frac{1}{(N'\,\Delta t)^2}S(f_k')S(f_{k'}')\left(\sum_{j=-J}^{J}W_m(f_j')\right)^2$$

$$\approx S(f_k')S(f_{k'}')\left(\int_{-f_{(N)}}^{f_{(N)}}W_m(\phi)\,d\phi\right)^2$$

$$= S(f_k')S(f_{k'}').$$

Here we have used the Riemann integral to approximate the summation and assumption (240a) to show that the integral is identically 1. It now follows that

$$\text{cov}\left\{\hat{S}^{(lw)}(f_k'),\hat{S}^{(lw)}(f_{k'}')\right\} \approx 0, \quad f_k' \neq f_{k'}', \tag{250a}$$

for N large enough. By an extension of the above arguments, one can show that the same result holds for any two fixed frequencies $f' \neq f''$ for N large enough. Note carefully that uncorrelatedness need not be approximately true when f' and f'' are separated by a distance less than the 'width' of $W_m(\cdot)$, which we can conveniently assume to be measured by the smoothing window bandwidth B_W.

The important results from this and the previous section are: the asymptotic unbiasedness of $\hat{S}^{(lw)}(f)$ in (244c); an approximation in (245) to the bias of $\hat{S}^{(lw)}(f)$ due to the smoothing window; large sample approximations to the variance of $\hat{S}^{(lw)}(f)$ in (248a) and (248b); and the large sample uncorrelatedness of $\hat{S}^{(lw)}(f_k')$ and $\hat{S}^{(lw)}(f_{k'}')$ in (250a).

Comments and Extensions to Section 6.9

[1] We here derive Equation (248b) by a route that makes more explicit the form of the factor C_h. As before, let $\tilde{f}_k \equiv k/(2N\,\Delta t)$ be a set of frequencies on a grid twice as finely spaced as the Fourier frequencies. From Equation (239) we have the exact result

$$\hat{S}^{(lw)}(\tilde{f}_k) = \frac{1}{2N\,\Delta t}\sum_{j=-(N-1)}^{N}W_m(\tilde{f}_j)\hat{S}^{(d)}(\tilde{f}_{k-j}). \tag{250b}$$

This is simply the convolution of $\{W_m(\tilde{f}_j)\}$ with the 'frequency' series $\{\hat{S}^{(d)}(\tilde{f}_j)\}$. If, for large N, we assume that $W_m(\tilde{f}_j) \approx 0$ for all $\tilde{f}_j > J'/(2N\,\Delta t) \equiv f_{J'}$ for some integer J' and that in the interval $[\tilde{f}_{k-J'}, \tilde{f}_{k+J'}]$ the true sdf $S(\cdot)$ is locally constant, then the 'frequency' series $\{\hat{S}^{(d)}(\tilde{f}_j)\}$ is locally stationary. The spectrum of the resultant series $\{\hat{S}^{(lw)}(\tilde{f}_j)\}$ – locally to $\tilde{f}_k$ – is thus the product of the square of the inverse Fourier transform of $\{W_m(\tilde{f}_j)\}$ – i.e., $\{w_{q,m}^2 : q = -(N-1), \ldots, N-1\}$ – with the spectrum of $\hat{S}^{(d)}(\tilde{f}_j)$, which, under the Gaussian assumption, is $\{S^2(\tilde{f}_k)r_q\}$ (see the discussion following Equation (231d)). Hence the spectrum of $\hat{S}^{(lw)}(\tilde{f}_j)$ – locally to $\tilde{f}_k$ – is given by

$$S^2(\tilde{f}_k)w_{q,m}^2 r_q, \qquad q = -(N-1), \ldots, N-1.$$

Since a spectrum is a decomposition of a variance, the variance of $\hat{S}^{(lw)}(\tilde{f}_k)$ is, in the usual way, given by

$$\mathrm{var}\,\{\hat{S}^{(lw)}(\tilde{f}_k)\} = \Delta t\, S^2(\tilde{f}_k) \sum_{q=-(N-1)}^{N-1} w_{q,m}^2 r_q \qquad (251a)$$

(see Walden, 1990a, for details). For the rectangular taper, $h_t = 1/\sqrt{N}$ for $1 \le t \le N$, and hence from Equation (231b) we have

$$r_q = \frac{1}{\Delta t} \sum_{t=1}^{N-|q|} \frac{1}{N^2} = \frac{1}{N\,\Delta t}\left(1 - \frac{|q|}{N}\right),$$

so that

$$\mathrm{var}\,\{\hat{S}^{(lw)}(\tilde{f}_k)\} = \frac{S^2(\tilde{f}_k)}{N} \sum_{q=-(N-1)}^{N-1} w_{q,m}^2 \left(1 - \frac{|q|}{N}\right),$$

a result of some utility (see, e.g., Walden and White, 1990).

If we now simplify Equation (251a) by setting $r_q = r_0$ for all q, we obtain

$$\mathrm{var}\,\{\hat{S}^{(lw)}(\tilde{f}_k)\} \approx S^2(\tilde{f}_k) \sum_{t=1}^{N} h_t^4 \sum_{q=-(N-1)}^{N-1} w_{q,m}^2 \quad \text{since } r_0 = \frac{1}{\Delta t} \sum_{t=1}^{N} h_t^4$$

from Equation (231b). As is discussed in Walden (1990a), this approximation is really only good for relatively 'modest' data tapers, i.e., those for which the Fourier transform of $\{r_q\}$ – namely, $R(\cdot)$ of Equation (231a) – damps down to zero rapidly (see Figure 232). Comparison of the above with Equation (248b) yields

$$C_h = N \sum_{t=1}^{N} h_t^4. \qquad (251b)$$

We note that $C_h \ge 1$ by the Cauchy inequality (since we always assume the normalization $\sum_{t=1}^{N} h_t^2 = 1$) and that, as should be the case, $C_h = 1$

for the rectangular data taper $h_t = 1/\sqrt{N}$. For large N, we can compute C_h approximately for various data tapers by approximating the above summation with an integral. The results for several common data tapers are shown in Table 248.

A second justification for Equation (251b) is to note that, in the 'grid of frequencies' argument we used in this section, the spacing in frequency between uncorrelated components of the direct spectral estimator $\hat{S}^{(d)}(\cdot)$ is given approximately by $C_h/(N\,\Delta t)$ (recall that the spacing of the frequencies f_k' is $1/(N'\,\Delta t)$ and that we assumed $N' \approx N/C_h$). Since, in the Gaussian case, the correlation between $\hat{S}^{(d)}(\tilde{f}_k)$ and $\hat{S}^{(d)}(\tilde{f}_k + \eta)$ is given by $R(\eta)$ of Equation (231a), we need to determine the point at which $R(\eta)$ can be regarded as being close to zero (see Figure 232). A convenient measure of this point is the equivalent width of $R(\cdot)$, namely,

$$\text{width}_e\left\{R(\cdot)\right\} = \int_{-f_{(N)}}^{f_{(N)}} R(\eta)\,d\eta \Big/ R(0)$$

(see Equation (70b), adjusted for the case of a periodic function). However, $R(0) = 1$, and since $\{r_q\} \longleftrightarrow R(\cdot)$ implies that

$$r_q = \int_{-f_{(N)}}^{f_{(N)}} R(\eta)e^{i2\pi\eta q}\,d\eta \ \text{ and hence } \ r_0 = \int_{-f_{(N)}}^{f_{(N)}} R(\eta)\,d\eta,$$

we see that $\text{width}_e\left\{R(\cdot)\right\}$ reduces to just r_0. From Equation (231b), we have

$$r_0 = \frac{1}{\Delta t}\sum_{t=1}^{N} h_t^4.$$

Upon equating the above with $C_h/(N\,\Delta t)$, we obtain Equation (251b) again.

[2]   We give here more details concerning – and an alternative expression for – the approximation to the variance of a lag window spectral estimator given in Equation (251a). This approximation is based upon Equation (250b), from which it directly follows that

$$\text{var}\left\{\hat{S}^{(lw)}(\tilde{f}_k)\right\}$$

$$= \frac{1}{4N^2(\Delta t)^2}\sum_{j,j'=-(N-1)}^{N} W_m(\tilde{f}_j)W_m(\tilde{f}_{j'})\,\text{cov}\left\{\hat{S}^{(d)}(\tilde{f}_{k-j}),\hat{S}^{(d)}(\tilde{f}_{k-j'})\right\}.$$

From Equation (240b) we can assume that $W_m(\tilde{f}_j) \approx 0$ for all $|j| > J$. In practice we can determine J by setting $\tilde{f}_J \approx cB_W$, where B_W is the smoothing window bandwidth of Equation (242c), and c is, say, 1 or 2 (if $c = 1$, J is the integer closest to $2NB_W \Delta t$). We thus have

$$\text{var}\,\{\hat{S}^{(lw)}(\tilde{f}_k)\}$$

$$\approx \frac{1}{4N^2(\Delta t)^2} \sum_{j,j'=-J}^{J} W_m(\tilde{f}_j)W_m(\tilde{f}_{j'})\,\text{cov}\,\{\hat{S}^{(d)}(\tilde{f}_{k-j}), \hat{S}^{(d)}(\tilde{f}_{k-j'})\}.$$

Under the assumptions that we are dealing with a Gaussian stationary process, that $S(\cdot)$ is locally flat from $\tilde{f}_{k-J}$ to $\tilde{f}_{k+J}$, that $0 < \tilde{f}_{k-J}$ and $\tilde{f}_{k+J} < f_{(N)}$, that $\tilde{f}_{k-J}$ is not too near to 0 and $\tilde{f}_{k+J}$ is not too near to $f_{(N)}$, we can use the results outlined in the Comments and Extensions to Section 6.6 to simplify the above to

$$\text{var}\,\{\hat{S}^{(lw)}(\tilde{f}_k)\}$$

$$\approx \frac{S^2(\tilde{f}_k)}{4N^2(\Delta t)^2} \sum_{j=-J}^{J} W_m(\tilde{f}_j) \sum_{j'=-J}^{J} W_m(\tilde{f}_{j'})R(\tilde{f}_{j-j'}),$$

where $R(\cdot)$ is defined in Equation (231a) and depends only on the data taper $\{h_t\}$. By a change of variables similar to that used in Equation (196c), we can rewrite the above as

$$\text{var}\,\{\hat{S}^{(lw)}(\tilde{f}_k)\}$$

$$\approx \frac{S^2(\tilde{f}_k)}{4N^2(\Delta t)^2} \sum_{l=-2J}^{2J} R(\tilde{f}_l) \sum_{l'=0}^{2J-|l|} W_m(\tilde{f}_{J-l'-|l|})W_m(\tilde{f}_{J-l'}).$$

The inner summation is essentially an autocorrelation, so it can be efficiently calculated for all lags l using discrete Fourier transforms (see Walden, 1990a, for details); however, as is demonstrated in Figure 232, $R(\cdot)$ typically decreases effectively to 0 rather quickly so that we can shorten the outer summation to, say, its $2L + 1$ innermost terms:

$$\text{var}\,\{\hat{S}^{(lw)}(\tilde{f}_k)\}$$

$$\approx \frac{S^2(\tilde{f}_k)}{4N^2(\Delta t)^2} \sum_{l=-L}^{L} R(\tilde{f}_l) \sum_{l'=0}^{2J-|l|} W_m(\tilde{f}_{J-l'-|l|})W_m(\tilde{f}_{J-l'}).$$

For example, if we use a dpss data taper with $NW \leq 8$, we can conveniently let $L = \min(10, 2J)$, where $l = 10$ corresponds to $\eta = 5/64$ in Figure 232; i.e., $l/(2N\,\Delta t) = 5/64$.

6.10 Asymptotic Distribution of Lag Window Estimators

We consider here the asymptotic (large sample) distribution of $\hat{S}^{(lw)}(f)$. From Equations (238a) and (239), we can write

$$\hat{S}^{(lw)}(f) = \int_{-f_{(N)}}^{f_{(N)}} W_m(f - \phi)\hat{S}^{(d)}(\phi)\, d\phi$$

$$\approx \frac{1}{2N\,\Delta t} \sum_{j=-(N-1)}^{N} W_m(\tilde{f}_j)\hat{S}^{(d)}(f - \tilde{f}_{-j}),$$

with the approximation becoming an equality when $f = \tilde{f}_k \equiv k/(2N\,\Delta t)$ for some integer k. As noted in Section 6.6, under mild conditions the $\hat{S}^{(d)}(f - \tilde{f}_{-j})$ terms can be regarded as a set of χ^2 rv's times appropriate multiplicative constants for large N. It follows that $\hat{S}^{(lw)}(f)$ is asymptotically a linear combination of χ^2 rv's with weights that depend on the smoothing window $W_m(\cdot)$. The exact form of such a distribution is hard to determine, but there is a well-known approximation we can use here. Let us assume

$$\hat{S}^{(lw)}(f) \stackrel{\mathrm{d}}{=} a\chi_\nu^2;$$

i.e., $\hat{S}^{(lw)}(f)$ has the distribution of a constant a times a chi-square rv with ν degrees of freedom, where a and ν are both unknown. By the properties of the χ_ν^2 distribution, we have

$$E\{\hat{S}^{(lw)}(f)\} = E\{a\chi_\nu^2\} = a\nu \quad \text{and} \quad \text{var}\{\hat{S}^{(lw)}(f)\} = \text{var}\{a\chi_\nu^2\} = 2a^2\nu.$$

If we use these two expressions to derive equations for ν and a in terms of the expected value and variance of $\hat{S}^{(lw)}(f)$, we obtain

$$\nu = \frac{2\left(E\{\hat{S}^{(lw)}(f)\}\right)^2}{\text{var}\{\hat{S}^{(lw)}(f)\}} \quad \text{and} \quad a = \frac{E\{\hat{S}^{(lw)}(f)\}}{\nu}. \tag{254a}$$

Under the assumptions of the previous two sections, the large sample expected value and variance for $\hat{S}^{(lw)}(f)$ are, respectively, $S(f)$ and

$$\frac{C_h S^2(f)}{N\,\Delta t}\int_{-f_{(N)}}^{f_{(N)}} W_m^2(\phi)\, d\phi = \frac{C_h S^2(f)}{N}\sum_{\tau=-(N-1)}^{N-1} w_{\tau,m}^2$$

(see (244c), (248a) and (248b)). If we substitute these values into (254a), we find that, for large samples,

$$\nu = \frac{2N\,\Delta t}{C_h \int_{-f_{(N)}}^{f_{(N)}} W_m^2(\phi)\, d\phi} = \frac{2N}{C_h \sum_{\tau=-(N-1)}^{N-1} w_{\tau,m}^2} \quad \text{and} \quad a = \frac{S(f)}{\nu}.$$

$$\tag{254b}$$

The quantity ν is often called the *equivalent degrees of freedom* for the estimator $\hat{S}^{(lw)}(f)$. From Equation (242c) for the smoothing window bandwidth B_W, we obtain

$$\nu = \frac{2NB_W\,\Delta t}{C_h}, \tag{255a}$$

so that an increase in B_W (i.e., a greater degree of smoothing) yields an increase in ν. Using $\operatorname{var}\{a\chi_\nu^2\} = 2a^2\nu$, we have (for large samples)

$$\operatorname{var}\{\hat{S}^{(lw)}(f)\} \approx \frac{2S^2(f)}{\nu}.$$

As ν increases, this variance decreases, but, for a fixed sample size N, a decrease in the variance of $\hat{S}^{(lw)}(f)$ comes at the potential expense of an increase in its bias. Here we must be careful. The above expression for $\operatorname{var}\{\hat{S}^{(lw)}(f)\}$ is based on the assumption that $\hat{S}^{(lw)}(f)$ is approximately distributed as $a\chi_\nu^2$ with a and ν given above. If, by increasing the bandwidth B_W with the idea of making ν large and hence the variance small, we inadvertently introduce significant bias in $\hat{S}^{(lw)}(f)$, then $\hat{S}^{(lw)}(f)$ will deviate substantially from its assumed distribution, and the above expression for $\operatorname{var}\{\hat{S}^{(lw)}(f)\}$ can be misleading.

The $a\chi_\nu^2$ approximation to the distribution of $\hat{S}^{(lw)}(f)$ allows us to construct an approximate confidence interval for $S(f)$ at a fixed f. Let $Q_\nu(p)$ represent the $p \times 100\%$ percentage point of the χ_ν^2 distribution; i.e., $\mathbf{P}\left[\chi_\nu^2 \le Q_\nu(p)\right] = p$. We thus have, for $0 \le p \le 1/2$,

$$\mathbf{P}\left[Q_\nu(p) \le \chi_\nu^2 \le Q_\nu(1-p)\right] = 1 - 2p.$$

Since $\hat{S}^{(lw)}(f)$ is approximately distributed as $a\chi_\nu^2$ with $a = S(f)/\nu$, it follows that $\nu\hat{S}^{(lw)}(f)/S(f)$ is approximately distributed as χ_ν^2. Hence

$$\mathbf{P}\left[Q_\nu(p) \le \frac{\nu\hat{S}^{(lw)}(f)}{S(f)} \le Q_\nu(1-p)\right]$$

$$= \mathbf{P}\left[\frac{\nu\hat{S}^{(lw)}(f)}{Q_\nu(1-p)} \le S(f) \le \frac{\nu\hat{S}^{(lw)}(f)}{Q_\nu(p)}\right] = 1 - 2p,$$

from which it follows that

$$\left[\frac{\nu\hat{S}^{(lw)}(f)}{Q_\nu(1-p)}, \frac{\nu\hat{S}^{(lw)}(f)}{Q_\nu(p)}\right] \tag{255b}$$

ν	p					
	0.005	0.025	0.05	0.95	0.975	0.995
2	0.0100	0.0506	0.1026	5.9915	7.3778	10.5966
3	0.0717	0.2158	0.3518	7.8147	9.3484	12.8382
4	0.2070	0.4844	0.7107	9.4877	11.1433	14.8603
5	0.4117	0.8312	1.1455	11.0705	12.8325	16.7496
6	0.6757	1.2373	1.6354	12.5916	14.4494	18.5476
7	0.9893	1.6899	2.1673	14.0671	16.0128	20.2777
$\Phi^{-1}(p)$	-2.5758	-1.9600	-1.6449	1.6449	1.9600	2.5758

Table 256. Percentage points for $Q_\nu(p)$ for χ^2_ν distribution for $\nu = 2$ to 7. The bottom row gives percentage points $\Phi^{-1}(p)$ for the standard Gaussian distribution.

is a $100(1 - 2p)\%$ confidence interval for $S(f)$. Note that this confidence interval applies only to $S(f)$ at one particular value of f. We cannot get a confidence band for the entire function $S(\cdot)$ by this method.

The percentage points $Q_\nu(p)$ required to compute 90%, 95% and 99% confidence intervals are listed in Table 256 for $\nu = 2$ to 7 degrees of freedom. For larger values of ν, we can compute $Q_\nu(p)$ using the following approximation (Chambers *et al.*, 1983):

$$Q_\nu(p) \approx \tilde{Q}_\nu(p) \equiv \nu \left(1 - \frac{2}{9\nu} + \Phi^{-1}(p) \left(\frac{2}{9\nu} \right)^{1/2} \right)^3, \qquad (256)$$

where $\Phi^{-1}(p)$ is the $p \times 100\%$ percentage point of the standard Gaussian distribution and is given in the bottom line of Table 256. For all $\nu \geq 8$, the relative absolute error $|\tilde{Q}_\nu(p) - Q_\nu(p)|/Q_\nu(p)$ is less than 0.05 for all six values of p used in the table. An algorithm for computing $Q_\nu(p)$ to high accuracy for arbitrary ν and p is given by Best and Roberts (1975) and is the source of the values in Table 256.

Since $\hat{S}^{(lw)}(f)$ is approximately distributed as an $a\chi^2_\nu$ rv, we can employ – when ν is large (say greater than 30) – the usual scheme of approximating a χ^2_ν distribution by a Gaussian distribution with the same mean and variance; i.e., the rv

$$\frac{\hat{S}^{(lw)}(f) - S(f)}{S(f)(2/\nu)^{1/2}} \overset{\text{d}}{=} \frac{a\chi^2_\nu - a\nu}{(2a^2\nu)^{1/2}}$$

is approximately Gaussian distributed with zero mean and unit variance. Under these circumstances, an approximate $100(1 - 2p)\%$ confidence

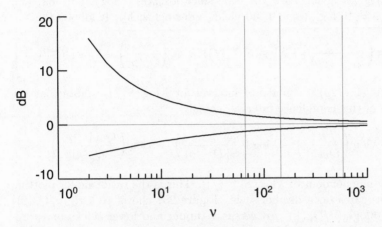

Figure 257. Upper and lower additive factors needed to create a 95% confidence interval for $10 \log_{10}(S(f))$ based upon a spectral estimator with ν equivalent degrees of freedom. Here ν ranges from 2 up to 1000. The thin vertical lines indicate $\nu = 66$, 146 and 581. These are the smallest equivalent degrees of freedom needed to achieve a 95% confidence interval for $10 \log_{10}(S(f))$ whose width is less than, respectively, 3 dB, 2 dB and 1 dB.

interval for $S(f)$ has the form

$$\left[\frac{\hat{S}^{(lw)}(f)}{1 + \Phi^{-1}(1-p)(2/\nu)^{1/2}}, \frac{\hat{S}^{(lw)}(f)}{1 + \Phi^{-1}(p)(2/\nu)^{1/2}} \right].$$

The confidence interval for $S(f)$ given in (255b) has a width of

$$\hat{S}^{(lw)}(f) \left[\frac{\nu}{Q_\nu(p)} - \frac{\nu}{Q_\nu(1-p)} \right],$$

which unfortunately depends on $\hat{S}^{(lw)}(f)$. This fact makes plots of $\hat{S}^{(lw)}(f)$ versus f difficult to interpret: the width of a confidence interval for $S(f)$ is proportional to $\hat{S}^{(lw)}(f)$ and thus varies from frequency to frequency. If we assume that $S(f) > 0$ and $\hat{S}^{(lw)}(f) > 0$, we can write

$$\mathbf{P}\left[\frac{\nu \hat{S}^{(lw)}(f)}{Q_\nu(1-p)} \leq S(f) \leq \frac{\nu \hat{S}^{(lw)}(f)}{Q_\nu(p)} \right]$$

$$= \mathbf{P}\left[\lambda \log\left(\frac{\nu \hat{S}^{(lw)}(f)}{Q_\nu(1-p)} \right) \leq \lambda \log\left(S(f) \right) \leq \lambda \log\left(\frac{\nu \hat{S}^{(lw)}(f)}{Q_\nu(p)} \right) \right],$$

where $\lambda > 0$ is a constant that allows us to use different logarithmic scales (for example, $\lambda = 1$ gives a 'log base e' scale; $\lambda = \log_{10}(e) =$

0.4343 gives a 'log base 10' scale since $\log_{10}(e) \times \log(x) = \log_{10}(x)$; while $\lambda = 10 \log_{10}(e) = 4.343$ yields a decibel scale). It follows that

$$\left[\lambda \log \left(\frac{\nu}{Q_\nu(1-p)} \right) + \lambda \log \left(\hat{S}^{(lw)}(f) \right), \lambda \log \left(\frac{\nu}{Q_\nu(p)} \right) + \lambda \log \left(\hat{S}^{(lw)}(f) \right) \right]$$
(258)

is a $100(1 - 2p)\%$ confidence interval for $\lambda \log(S(f))$. Note that the width of this confidence interval, namely

$$\lambda \log \left(\frac{\nu}{Q_\nu(p)} \right) - \lambda \log \left(\frac{\nu}{Q_\nu(1-p)} \right) = \lambda \log \left(\frac{Q_\nu(1-p)}{Q_\nu(p)} \right),$$

is now independent of $\lambda \log(\hat{S}^{(lw)}(f))$. This is the rationale for plotting sdf estimates on a decibel scale. Figure 257 shows $10 \log_{10}(\nu/Q_\nu(p))$ and $10 \log_{10}(\nu/Q_\nu(1-p))$ versus ν (upper and lower thick curves, respectively) for $p = 0.025$ and ν between 2 and 1000.

We now summarize the results of this section. Under the assumptions made here, the lag window sdf estimator $\hat{S}^{(lw)}(f)$ of $S(f)$ is approximately distributed as a χ_ν^2 rv times the constant $S(f)/\nu$, where ν – the equivalent degrees of freedom of the estimator $\hat{S}^{(lw)}(f)$ – is given by Equation (254b) and depends only upon the properties of the lag or smoothing window and the data taper. As ν gets large, $\hat{S}^{(lw)}(f)$ becomes asymptotically Gaussian distributed with mean $S(f)$ and variance $2S^2(f)/\nu$. These distributional results can be used to construct asymptotically correct confidence intervals for $S(f)$.

6.11　Examples of Lag Windows

We consider here five examples of lag and smoothing windows that can be used in practice to form a specific lag window spectral estimator $\hat{S}^{(lw)}(\cdot)$. A bewildering number of such windows has been proposed and discussed in the literature (Harris, 1978, compares 23 different classes of windows, and Priestley, 1981, gives details for 11 windows; see also Geçkinli and Yavuz, 1978). Of our five examples, we chose the first four (Bartlett, Daniell, Parzen and Papoulis) because they have been used extensively in practical applications, while the last one (the modified Daniell) is included to illustrate various points (two other lag windows – mainly of historical interest – are the subjects of Exercises [6.19] and [6.20]).

For each of the specific lag windows, we show a figure with four plots (Figures 260, 263, 265, 267 and 268). The plot with the label (a) in each figure shows an example of the lag window $w_{\tau,m}$ versus lag τ for a sample size $N = 64$ and a particular setting of the window parameter m. Plot (b) shows the corresponding smoothing window $W_m(\cdot)$ on a decibel scale versus frequency (the sampling time Δt is taken to be 1, so that the Nyquist frequency $f_{(N)}$ is $1/2$). Because $w_{-\tau,m} = w_{\tau,m}$ and $W_m(-f) =$

$W_m(f)$ by assumption, we need only plot the former for nonnegative τ and the latter for nonnegative f. Plots (c) and (d) show two different spectral windows $\mathcal{U}_m(\cdot)$ for the same m and N. From Equation (243c) we see that $\mathcal{U}_m(\cdot)$ depends upon both the smoothing window $W_m(\cdot)$ and the data taper $\{h_t\}$ through its associated spectral window $\mathcal{H}(\cdot)$. Plots (c) and (d) show $\mathcal{U}_m(\cdot)$ corresponding to, respectively, the rectangular data taper (see the left-hand plots of Figure 209) and a dpss data taper with $NW = 4$ (left-hand plots of Figure 212). We have included plots (c) and (d) to emphasize the point that the expected value of a lag window spectral estimator $\hat{S}^{(lw)}(\cdot)$ depends upon both the smoothing window $W_m(\cdot)$ and the data taper used in the corresponding direct spectral estimator $\hat{S}^{(d)}(\cdot)$.

In all five examples, the smoothing window $W_m(\cdot)$ consists of a central lobe (i.e., the one centered about zero frequency) and several sidelobes. As we note in the next section, these sidelobes ideally should be as small as possible (to minimize what is defined there as smoothing window leakage), so the magnitude of the peak in the first sidelobe relative to the magnitude of the peak in the central lobe is of interest, as is the rate of decay of the envelope formed by the peaks of the sidelobes. This envelope typically decays approximately as f^α for some $\alpha < 0$. For example, if $\alpha = -1$, a doubling of frequency corresponds to a doubling in the decay of the magnitude of the peaks in the sidelobes of $W_m(\cdot)$. On a decibel (i.e., $10 \log_{10}$) scale such a doubling corresponds to a $10 \log_{10}(2) \approx 3$ dB drop in a frequency octave (doubling of frequency). The magnitude of the envelope thus decays at 3 dB per octave. We quote this decay rate for each of the lag windows we consider because it is a good guide to the speed of the decay of sidelobes with increasing frequency (see, however, the discussion concerning the Daniell window below).

Table 269 at the end of this section lists approximations for the asymptotic variance of $\hat{S}^{(lw)}(f)$, the equivalent degrees of freedom ν, the smoothing window bandwidth B_W and the alternative bandwidth measure β_W for the first four lag windows. This table shows a number of interesting relationships. For example, for fixed m, we have $\mathrm{var}\,\{\hat{S}^{(lw)}(f)\} \to 0$ as $N \to \infty$ for lag window estimators based on these four windows. Also we can deduce that $\mathrm{var}\,\{\hat{S}^{(lw)}(f)\}$ is inversely proportional either to β_W (for the Daniell, Parzen and Papoulis windows) or to β_W^2 (for the Bartlett window). Since the bias due to the smoothing window alone is proportional to β_W^2 (see Equation (245)), there is a tradeoff between the bias and variance in lag window spectral estimators, or, to state it more picturesquely, 'bias and variance are antagonistic' (Priestley, 1981, p. 528).

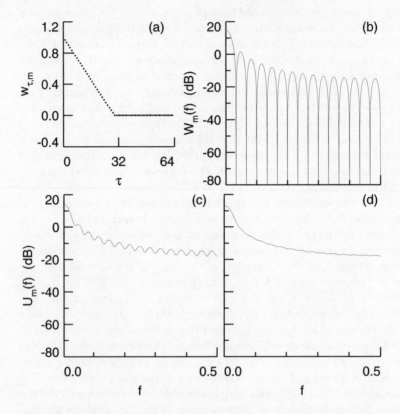

Figure 260. Bartlett window for $m = 30$ and $N = 64$ (the smoothing window bandwidth is $B_W = 0.05$). For a description of the four plots above, see the first part of this section.

- *Bartlett window* (Figure 260)

The lag window for Bartlett's spectral estimator (Bartlett, 1950) is given by

$$w_{\tau,m} = \begin{cases} 1 - |\tau|/m, & |\tau| < m; \\ 0, & |\tau| \geq m, \end{cases} \tag{260}$$

where $m \leq N$ is the window parameter. Bartlett's window applies linearly decreasing weights to the sample acvs up to lag m and zero weights thereafter. The parameter m can be interpreted as a truncation point beyond which the acvs is considered to be zero. The corresponding smoothing window is

$$W_m(f) = \Delta t \sum_{\tau=-m}^{m} \left(1 - \frac{|\tau|}{m}\right) e^{-i2\pi f \tau \, \Delta t} = \frac{\Delta t \sin^2(m\pi f \, \Delta t)}{m \sin^2(\pi f \, \Delta t)} = \mathcal{F}(f),$$

where the latter quantity is Fejér's kernel (see Equation (198b)). Note that the smoothing window is of the same form as the two-sided Cesàro partial sum we met in Section 3.7. From the plots of this function in Figure 200, we see that the width of the central lobe decreases as m increases, so the amount of smoothing decreases as m increases (a convention we maintain for the other five lag windows we discuss). Since Fejér's kernel is nonnegative everywhere, it follows that Bartlett's estimator of an sdf is also nonnegative everywhere. The magnitude of the peak of the first sidelobe of Fejér's kernel is down about 13 dB from its value at $f = 0$. The envelope of Bartlett's smoothing window decays as approximately f^{-2}, which corresponds to a decay of 6 dB per frequency octave.

Note from plots (c) and (d) in Figure 260 that the two spectral windows $\mathcal{U}_m(\cdot)$ corresponding to the default data taper and the dpss data taper both have a decay rate similar to that of the smoothing window shown in plot (b). This is due to the fact that $\mathcal{U}_m(\cdot)$ is the convolution of the smoothing window $W_m(\cdot)$ and the spectral window $\mathcal{H}(\cdot)$ associated with the data taper. In the case of the default data taper (plot (c)), both $W_m(\cdot)$ and $\mathcal{H}(\cdot)$ are in fact Fejér's kernel, so the decay rate of their convolution must be the same. In the case of the dpss data taper, a glance at its spectral window $\mathcal{H}(\cdot)$ in the lower left-hand plot of Figure 212 shows that $\mathcal{U}_m(\cdot)$ is essentially a smoothed version of Fejér's kernel and hence has the same decay rate as Fejér's kernel. Now let us make a practical interpretation. The user who selected a dpss data taper with $NW = 4$ in order to achieve low bias in a direct spectral estimate $\hat{S}^{(d)}(\cdot)$ of an sdf with a suspected large dynamic range might be shocked to find that smoothing $\hat{S}^{(d)}(\cdot)$ with Bartlett's window yields an *overall* spectral window $\mathcal{U}_m(\cdot)$ that decays at the same rate as the spectral window for the periodogram!

The rationale that Bartlett (1950) used to come up with the lag window (260) has an interesting connection to Welch's overlapped segment averaging spectral estimator (see Section 6.17). Bartlett did not apply any tapering (other than the default rectangular taper) so that the direct spectral estimator (207c) to be smoothed was just the periodogram. Bartlett argued that one could reduce the sampling fluctuations of the periodogram by

[1] splitting the original sample of N observations into N/m contiguous blocks, each block containing m observations (we assume that N/m is an integer);

[2] forming the periodogram for each block; and

[3] averaging the N/m periodograms together.

Let $\hat{S}^{(p)}_{k,m}(\cdot)$ be the periodogram for the kth block. Bartlett reasoned that $\hat{S}^{(p)}_{j,m}(f)$ and $\hat{S}^{(p)}_{k,m}(f)$ for $j \neq k$ should be approximately uncorrelated (with the approximation improving as m gets larger) so that the reduction in variance should be inversely proportional to the number of blocks N/m. Bartlett worked with the unbiased acvs estimator

$$\hat{s}_\tau^{(u)} = \frac{N}{N - |\tau|} \hat{s}_\tau^{(p)},$$

in terms of which we can write

$$\hat{S}_{k,m}^{(p)}(f) = \Delta t \sum_{\tau=-m}^{m} \left(1 - \frac{|\tau|}{m}\right) \hat{s}_{\tau,k}^{(u)} e^{-i2\pi f\tau\,\Delta t},$$

where $\{\hat{s}_{\tau,k}^{(u)}\}$ denotes the unbiased acvs estimator based upon just the data in the kth block. If we average the N/m periodogram estimates together, we get

$$\frac{1}{N/m} \sum_{k=1}^{N/m} \hat{S}_{k,m}^{(p)}(f) = \Delta t \sum_{\tau=-m}^{m} \left(1 - \frac{|\tau|}{m}\right) \left[\frac{1}{N/m} \sum_{k=1}^{N/m} \hat{s}_{\tau,k}^{(u)}\right] e^{-i2\pi f\tau\,\Delta t}$$

$$= \Delta t \sum_{\tau=-m}^{m} \left(1 - \frac{|\tau|}{m}\right) \bar{s}_\tau^{(u)} e^{-i2\pi f\tau\,\Delta t},$$

where $\bar{s}_\tau^{(u)}$ is the average of the $\hat{s}_{\tau,k}^{(u)}$ terms. Bartlett then argued that $\bar{s}_\tau^{(u)}$ ignores information concerning s_τ that is contained in pairs of data values in adjacent blocks. He suggested replacing $\bar{s}_\tau^{(u)}$ with $\hat{s}_\tau^{(u)}$, the unbiased estimator of s_τ obtained by using all N observations. This substitution yields the estimator of $S(f)$ given by

$$\Delta t \sum_{\tau=-m}^{m} \left(1 - \frac{|\tau|}{m}\right) \hat{s}_\tau^{(u)} e^{-i2\pi f\tau\,\Delta t} = \Delta t \sum_{\tau=-m}^{m} \frac{1 - |\tau|/m}{1 - |\tau|/N} \hat{s}_\tau^{(p)} e^{-i2\pi f\tau\,\Delta t}.$$

The common form for Bartlett's lag window spectral estimator (in the case of a default rectangular taper), namely,

$$\hat{S}^{(lw)}(f) = \Delta t \sum_{\tau=-m}^{m} \left(1 - \frac{|\tau|}{m}\right) \hat{s}_\tau^{(p)} e^{-i2\pi f\tau\,\Delta t},$$

can be regarded as an approximation to the above under the assumption that N is much larger than m since then

$$\frac{1 - |\tau|/m}{1 - |\tau|/N} = \left(1 - \frac{|\tau|}{m}\right) \left(1 + \frac{|\tau|}{N} + \left[\frac{|\tau|}{N}\right]^2 + \cdots\right) \approx 1 - \frac{|\tau|}{m}.$$

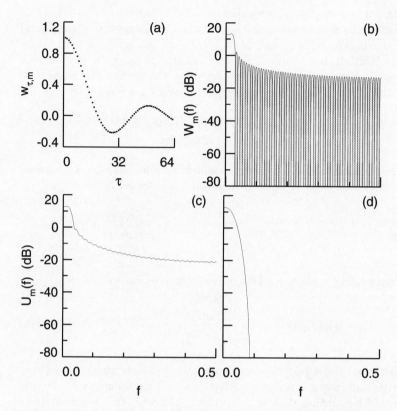

Figure 263. Daniell window for $m = 20$ and $N = 64$ ($B_W = 0.05$).

Bartlett noted that the choice of m is a compromise between reducing the variance of $\hat{S}^{(lw)}(f)$ and reducing its resolution (i.e., smearing out local features).

- *Daniell (or rectangular) window* (Figure 263)

As we have previously noted in the Comments and Extensions to Section 3.6 (see Equation (86)), the quantity

$$\frac{1}{b-a} \int_a^b f(x)\,dx$$

is often called the average value of the function $f(\cdot)$ in the interval $[a, b]$. Daniell's lag window spectral estimator (Daniell, 1946) is simply the average value of the direct spectral estimator $\hat{S}^{(d)}(\cdot)$ in an interval of length $1/(m\,\Delta t)$ around each value of f:

$$\hat{S}^{(lw)}(f) = m\,\Delta t \int_{f-1/(2m\,\Delta t)}^{f+1/(2m\,\Delta t)} \hat{S}^{(d)}(f)\,df,$$

where, for f near $\pm f_{(N)}$, we make use of the fact that $\hat{S}^{(d)}(\cdot)$ is $2f_{(N)}$ periodic. The parameter m in the Daniell window does not correspond to a truncation point – and need not be an integer – as in the case of the Bartlett lag window. It does control the degree of averaging that is applied to $\hat{S}^{(d)}(\cdot)$: the smaller m is, the greater the amount of smoothing. By comparison with Equation (236b), we see that

$$V_m(f) = \begin{cases} m\,\Delta t, & |f| \leq 1/(2m\,\Delta t); \\ 0, & \text{otherwise,} \end{cases} \tag{264a}$$

for Daniell's estimator. The corresponding lag window is given by Equation (237b) for $|\tau| < N$ and, by definition, is 0 for $|\tau| \geq N$:

$$w_{\tau,m} = \begin{cases} m\,\Delta t \int_{-1/(2m\,\Delta t)}^{1/(2m\,\Delta t)} e^{i2\pi f\tau\,\Delta t}\,df = \dfrac{\sin(\pi\tau/m)}{\pi\tau/m}, & |\tau| < N; \\ 0, & |\tau| \geq N. \end{cases}$$

The smoothing window for Daniell's spectral estimator is thus

$$W_m(f) = \Delta t \sum_{\tau=-(N-1)}^{N-1} \frac{\sin(\pi\tau/m)}{\pi\tau/m} e^{-i2\pi f\tau\,\Delta t}. \tag{264b}$$

Because the sidelobes of this smoothing window alternate between positive and negative values, we have plotted the quantity $|W_m(\cdot)|$ in Figure 263(b) rather than $W_m(\cdot)$. The negative sidelobes are shaded – unfortunately, there are so many sidelobes that this shading is barely visible. The magnitude of the peak of the first sidelobe in $|W_m(\cdot)|$ is down about 11 dB from its value at $f = 0$. The envelope of the sidelobes decays as approximately f^{-1}, which corresponds to a decay of 3 dB per frequency octave. However, since Equation (237d) implies that we can just as well regard $V_m(\cdot)$ in Equation (264a) as Daniell's smoothing window, we can also argue that the latter has *no* sidelobes! This example shows that the smoothing window $W_m(\cdot)$ has to be interpreted with some care. Because $\hat{S}^{(lw)}(\cdot)$ is the cyclic convolution of two periodic functions, one of them (namely, $\hat{S}^{(d)}(\cdot)$) with Fourier coefficients $\{\hat{s}_\tau^{(d)}\}$ that are identically 0 for lags $|\tau| \geq N$, there is an inherent ambiguity in $W_m(\cdot)$ which we have sidestepped by arbitrarily setting the Fourier coefficients of $W_m(\cdot)$ (namely, $\{w_{\tau,m}\}$) to 0 for lags $|\tau| \geq N$. Thus, the fact that a smoothing window has negative sidelobes does *not* automatically imply that the corresponding lag window spectral estimator can sometimes take on negative values!

Note that, in contrast to what happened with Bartlett's window (Figure 260), the two spectral windows $\mathcal{U}_m(\cdot)$ in Figure 263 now have a markedly different appearance. The first of these is again dominated

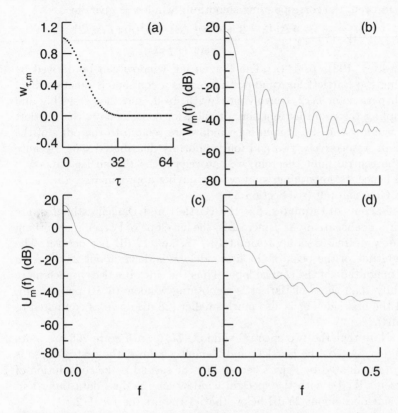

Figure 265. Parzen window for $m = 37$ and $N = 64$ ($B_W = 0.05$).

by Fejér's kernel (the spectral window for the rectangular data taper), but the second reflects the convolution of $V_m(\cdot)$ in Equation (264a) and the spectral window for the dpss data taper (lower left-hand plot of Figure 212). We can make another practical interpretation. The user who chose, say, the Daniell smoothing window because of its lack of sidelobes and used the default rectangular taper might be surprised to find that because of the effect of Fejér's kernel – basically reflecting the finiteness of the data – the *overall* spectral window $\mathcal{U}_m(\cdot)$ again decays as just f^{-2}, i.e., the same rate as the sidelobes of the periodogram!

- *Parzen window* (Figure 265)

Parzen (1961) has suggested the following lag window:

$$w_{\tau,m} = \begin{cases} 1 - 6\,(\tau/m)^2 + 6\,(|\tau|/m)^3, & |\tau| \leq m/2; \\ 2\,(1 - |\tau|/m)^3, & m/2 < |\tau| \leq m; \\ 0, & |\tau| > m. \end{cases}$$

For m even, the corresponding smoothing window is given by

$$W_m(f) = \frac{4 \Delta t \left[3 - 2 \sin^2(\pi f \Delta t)\right] \sin^4(m\pi f \Delta t/2)}{m^3 \sin^4(\pi f \Delta t)}$$

(Priestley, 1981, p. 444). The Parzen lag window can be derived by taking the Bartlett lag window (treated as a continuous function of τ) with parameter $m/2$, convolving it with itself, and then rescaling and sampling the resulting function. In fact, for continuous τ, the Bartlett lag window and the Parzen lag window are related to the pdf's of the sum of, respectively, two and four uniformly distributed rv's. Because of the central limit theorem, we can regard the Parzen lag window – and hence its smoothing window – as having approximately the shape of a Gaussian pdf (see Section 3.6).

Parzen sdf estimates, like the Bartlett and Daniell estimates, are always nonnegative. As f increases, the envelope of Parzen's smoothing window decreases as approximately f^{-4}, i.e., 12 dB per octave. The magnitude of the peak of the first sidelobe is down about 28 dB from the magnitude of the central lobe. Thus the sidelobes decay much more rapidly than those of Bartlett's smoothing window (6 dB per octave), and the first sidelobe is also much smaller (28 dB down as compared to 13 dB).

Note that the two spectral windows $\mathcal{U}_m(\cdot)$ in Figure 265 are quite different. As for the Bartlett and Daniell windows, the first of these is again dominated by Fejér's kernel, but the second is the convolution of Parzen's $W_m(\cdot)$ with the spectral window for the dpss data taper and has sidelobes about 25 dB below that of the first at $f = 1/2$.

• *Papoulis window* (Figure 267)

Papoulis (1973) found that the continuous τ analog of the following lag window produces a window with a certain minimum bias property:

$$w_{\tau,m} = \begin{cases} \dfrac{1}{\pi} |\sin(\pi\tau/m)| + (1 - |\tau|/m)\cos(\pi\tau/m), & |\tau| < m; \\ 0, & |\tau| \geq m \end{cases}$$

(Bohman, 1961, derived this window earlier in the context of characteristic functions). The rationale behind this lag window is as follows. Equation (245) tells us that the bias in a lag window spectral estimator due to the smoothing window alone is proportional to β_W^2, where β_W is defined in Equation (241b). The Papoulis window is the solution to the continuous τ analog of the following problem: for fixed $m > 0$, amongst all lag windows $\{w_{\tau,m}\}$ with $w_{\tau,m} = 0$ for $|\tau| \geq m$ and with a corresponding smoothing window $W_m(\cdot)$ such that $W_m(f) \geq 0$ for all f, find the window such that $\int_{-f_{(N)}}^{f_{(N)}} f^2 W_m(f)\,df \propto \beta_W^2$ is minimized.

The derivation of the Papoulis smoothing window is left as an exercise. Comparison of Figures 265 and 267 shows that the Parzen and the Papoulis window have quite similar characteristics.

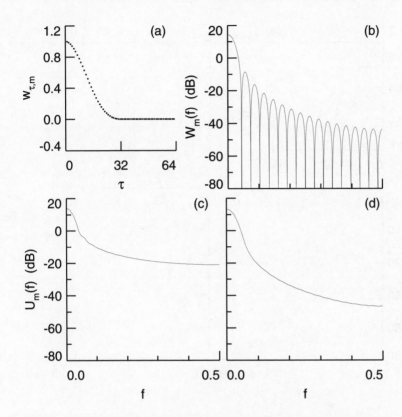

Figure 267. Papoulis window for $m = 34$ and $N = 64$ ($B_W = 0.05$).

• *Modified Daniell window* (Figure 268)
Bloomfield (1976) considers a discretely smoothed direct spectral estimator $\{\hat{S}^{(ds)}(f_k')\}$ (see Equation (236a)) with weights of the form

$$g_j = \begin{cases} 1/(2M), & |j| < M; \\ 1/(4M), & |j| = M; \\ 0, & \text{otherwise.} \end{cases}$$

We can consider these weights as being generated by sampling from a Daniell smoothing window $V_m(\cdot)$ (see Equation (264a)) with an 'end point' adjustment. The corresponding spectral estimator is termed a *modified Daniell* spectral estimator. Note that $\{g_j\}$ can be regarded as a low-pass LTI filter. Since $\sum_{j=-M}^{M} g_j = 1$, this filter has the normalization that we argued in Section 5.7 is appropriate for a smoother.

Let us assume that the frequencies involved in $\{\hat{S}^{(ds)}(f_k')\}$ are given by $f_k' = \tilde{f}_k \equiv k/(2N\,\Delta t)$. We can use Equations (238d) and (238e) to reexpress $\hat{S}^{(ds)}(\cdot)$ as a lag window spectral estimator $\hat{S}^{(lw)}(\cdot)$. The

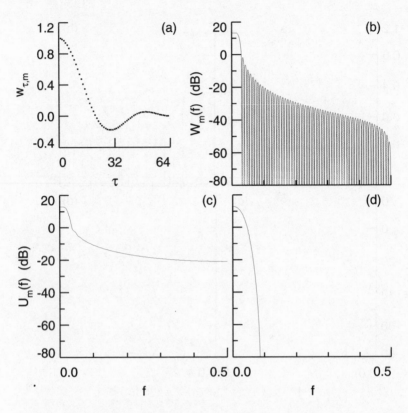

Figure 268. Modified Daniell window for $M = 3$ and $N = 64$ ($B_W = 0.05$).

corresponding lag and smoothing windows are shown, respectively, in plots (a) and (b) of Figure 268 for the case $M = 3$ and $N = 64$. As for the usual Daniell window in Figure 263, the smoothing window has negative sidelobes (shaded in plot (b)), but, because $\{\hat{S}^{(ds)}(\tilde{f}_k)\}$ is necessarily a nonnegative sequence of numbers, these sidelobes are again an artifact caused by truncation. The spectral windows for the modified Daniell window (plots (c) and (d)) closely resemble those for the usual Daniell window ((c) and (d) of Figure 263).

The device of sampling from a smoothing window to produce the weights for a discretely smoothed direct spectral estimator can obviously be applied to other smoothing windows besides the Daniell window. A variation on this idea is to generate the weights from, say, the Parzen lag window (with a renormalization to ensure that the weights sum to unity). This form of spectral estimator has been discussed in Cleveland and Parzen (1975) and Walden (1990a).

Estimator	Asymptotic variance	ν	B_W	β_W
Bartlett	$\dfrac{0.67mC_hS^2(f)}{N}$	$\dfrac{3N}{mC_h}$	$\dfrac{1.5}{m\,\Delta t}$	$\dfrac{0.92}{\sqrt{m}\,\Delta t}$
Daniell	$\dfrac{mC_hS^2(f)}{N}$	$\dfrac{2N}{mC_h}$	$\dfrac{1}{m\,\Delta t}$	$\dfrac{1}{m\,\Delta t}$
Parzen	$\dfrac{0.54mC_hS^2(f)}{N}$	$\dfrac{3.71N}{mC_h}$	$\dfrac{1.85}{m\,\Delta t}$	$\dfrac{1.91}{m\,\Delta t}$
Papoulis	$\dfrac{0.59mC_hS^2(f)}{N}$	$\dfrac{3.41N}{mC_h}$	$\dfrac{1.70}{m\,\Delta t}$	$\dfrac{1.73}{m\,\Delta t}$

Table 269. Asymptotic variance, equivalent degrees of freedom ν, smoothing window bandwidth B_W and alternative bandwidth measure β_W (used in Equation (245) for the bias due to the smoothing window alone) for four lag window spectral density estimators. The tabulated quantities are approximations to the formulae given in Equations (248a), (254b), (242a) and (241b). The quantity C_h – a variance inflation factor – depends upon the data taper used in the corresponding direct spectral estimator $\hat{S}^{(d)}(\cdot)$ (see Table 248 and Equation (251b)).

Comments and Extensions to Section 6.11

[1]　In the Comments and Extensions to Section 6.8, we introduced the idea of a reshaped lag window (see Equation (246)). Figure 270 gives an example for such a window formed from the Parzen lag window with $m = 37$ and the dpss data taper with $NW = 4$ shown in Figure 265. When used in conjunction with the rectangular data taper, the lag window in Figure 270(a) thus yields exactly the same spectral window as shown in Figure 265(d). The corresponding smoothing window is shown in plot (b), where now we plot $|W_m(\cdot)|$ rather than $W_m(\cdot)$. It has a prominent single *negative* sidelobe (indicated by the shaded area). That this lag window can lead to negative spectral estimates is demonstrated in plot (c). This plot shows two lag window spectral estimates for the first 64 values of the realization of the AR(4) process $\{X_{t,4}\}$ depicted in the bottom plot of Figure 45 (the process itself is defined in Equation (46a)). The thin curve is the lag window estimate formed using the Parzen lag window in combination with the dpss data taper; the thick curve uses the reshaped lag window with the rectangular data taper. The shaded areas under the latter curve indicate the frequencies where $\hat{S}^{(lw)}(f)$ is negative so that $|\hat{S}^{(lw)}(f)|$ is plotted rather than $\hat{S}^{(lw)}(f)$. The true sdf for the AR(4) process is shown as the thick curve in plot (d); the thin curve there is $E\{\hat{S}^{(lw)}(\cdot)\}$, which

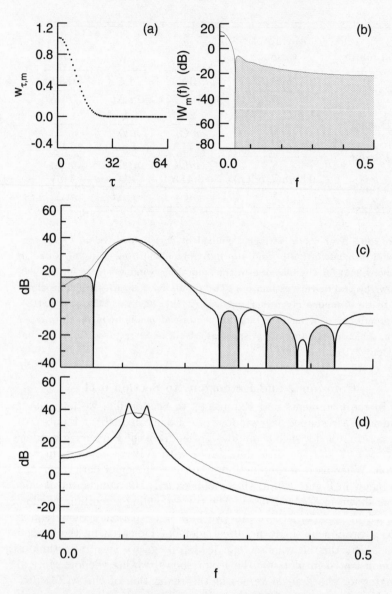

Figure 270. Reshaped lag window formed from a Parzen lag window and a dpss data taper (cf. Figure 265). Here $B_W = 0.057$.

is the result of convolving the spectral window in Figure 265(d) with the true sdf and hence is the same for both spectral estimates depicted in 270(c). (In view of Equation (243a), there are two other ways of constructing $E\{\hat{S}^{(lw)}(\cdot)\}$. First, it is the convolution of $E\{\hat{S}^{(d)}(\cdot)\}$ given by the thin curve in the upper plot of Figure 214 with the smoothing window of 265(b). Second, it is the convolution of $E\{\hat{S}^{(p)}(\cdot)\}$ shown by the thin curve in the bottom plot of Figure 202 with the smoothing window whose absolute value is shown in Figure 270(b). Note in the latter case that, even though the smoothing window has a prominent negative sidelobe, the resulting $E\{\hat{S}^{(lw)}(\cdot)\}$ is still totally positive.)

[2] As we pointed out in our discussion of the Daniell window above, smoothing a direct spectral estimator using the rectangular window is entirely equivalent to smoothing it using the corresponding smoothing window, even though the latter has sidelobes that rapidly oscillate between positive and negative values (see plot (b) of Figure 263). Because of this equivalence we can regard the Daniell smoothing window as having no sidelobes. With this substitution in mind, there is an interesting pattern in plots (c) and (d) in the figures for the Bartlett, Daniell, Parzen and Papoulis lag windows (Figures 260, 263, 265 and 267). These plots show the spectral window $\mathcal{U}_m(\cdot)$, which, from Equation (243c), is the convolution of the smoothing window $W_m(\cdot)$ and the spectral window $\mathcal{H}(\cdot)$ determined by the data taper of the underlying direct spectral estimator. The rolloff in each plot (c) reflects the slower of the rolloffs of $W_m(\cdot)$ and $\mathcal{H}(\cdot)$ for the rectangular data taper, i.e., Fejér's kernel $\mathcal{F}(\cdot)$. In all four cases, Fejér's kernel dominates the rolloff, so all of the (c) plots look similar. The rolloff in each plot (d) reflects the slower of the rolloffs of $W_m(\cdot)$ and $\mathcal{H}(\cdot)$ for a dpss data taper. Since $\mathcal{H}(\cdot)$ now damps down so fast, the rolloff is dominated by the smoothing window $W_m(\cdot)$ in all cases *except* for the Daniell window. Here the rolloff is dominated by $\mathcal{H}(\cdot)$ since $W_m(\cdot)$ can be regarded as a rectangular window with an 'infinitely fast' rolloff.

The important point to keep in mind is that the decay rate of the sidelobes of the spectral window $\mathcal{U}_m(\cdot)$ depends on both the spectral window corresponding to the data taper and the smoothing window. Thus, for example, use of the Parzen smoothing window with a sidelobe decay rate of 12 dB per octave does not imply that the sidelobes of $\mathcal{U}_m(\cdot)$ also have this decay rate *unless* the data taper for the direct spectral estimator is suitably chosen.

6.12 Choice of Lag Window

In the previous section we considered four different commonly used lag window spectral estimators, each of which can be defined via either its lag window or its smoothing window. There is also a window parameter m associated with each of these estimators. For a particular sdf estimation problem, which estimator should we use, and how should we set m? We address the first of these questions in this section, and the second in the next section.

Several different criteria have been proposed in the literature for evaluating different lag window spectral estimators. One of the more useful is based on the concept of *smoothing window leakage*. Since

$$\hat{S}^{(lw)}(f) = \int_{-f_{(N)}}^{f_{(N)}} W_m(f - \phi)\hat{S}^{(d)}(\phi)\, d\phi$$

(Equation (238a)), a lag window estimator of $S(f)$ is the result of smoothing the direct spectral estimator $\hat{S}^{(d)}(\cdot)$ with the smoothing window $W_m(\cdot)$ after the latter has been shifted so that its central lobe is centered at frequency f. Under our operational assumption that $S(\cdot)$ is slowly varying, we want $\hat{S}^{(lw)}(f)$ to be influenced mainly by values in $\hat{S}^{(d)}(\cdot)$ with frequencies 'close' to f. We define 'close' here to mean those frequencies lying within the central lobe of the shifted smoothing window $W_m(f - \cdot)$. If this smoothing window has significant sidelobes and if the dynamic range of $\hat{S}^{(d)}(\cdot)$ is large, $\hat{S}^{(lw)}(f)$ can be unduly influenced by values in $\hat{S}^{(d)}(\cdot)$ lying under one or more of the sidelobes of the smoothing window. If this in fact happens, we say that the estimate $\hat{S}^{(lw)}(f)$ suffers from smoothing window leakage.

One criterion for window selection is thus to insist that the smoothing window leakage be small. If we have two different lag window estimators whose smoothing windows have the same bandwidth B_W (a measure of the width of the central lobe of $W_m(\cdot)$), this criterion would dictate picking the window whose sidelobes are in some sense smaller. For the four commonly used smoothing windows discussed in the previous section, the sidelobes of the Bartlett window decay at a rate of 6 dB per octave (with the first sidelobe down 13 dB from the central lobe); the sidelobes of the Parzen and Papoulis windows at a rate of 12 dB per octave (with the first sidelobes down 28 and 23 dB, respectively); and arguably the Daniell smoothing window has no sidelobes. By this criterion our choices would be, from first to last, the Daniell, Parzen, Papoulis and Bartlett smoothing window.

There are valid reasons for *not* making smoothing window leakage our only criterion. First, the degree of distortion due to this leakage can be controlled somewhat by the choice of the smoothing parameter m (see Figure 275 below). Second, smoothing window leakage is a relevant consideration largely because it affects the bias in $\hat{S}^{(lw)}(f)$ due to the sidelobes of the smoothing window. Note that this source of bias is *not* the same as the smoothing window bias of Equation (245), which reflects the bias introduced by the central lobe of the smoothing window ('local' bias or loss of resolution). This is a particular problem for the very rectangular Daniell window (see, for example, Walden and White, 1984). Smoothing window leakage does not take into account either the variance of $\hat{S}^{(lw)}(f)$ or its resolution properties. Third, this leakage is a significant problem only when the direct spectral estimator has a large dynamic range. If this is not the case, the sidelobes of $W_m(\cdot)$ have little influence on $\hat{S}^{(lw)}(\cdot)$, and hence smoothing window leakage is not a relevant consideration. (See the discussion concerning the middle plot of Figure 301 for an example of smoothing window leakage.)

A second consideration is that $\hat{S}^{(lw)}(\cdot)$ should be a smoothed version of $\hat{S}^{(d)}(\cdot)$. In Section 3.6 we noted that smoothing a signal with a Gaussian kernel is preferable to smoothing with a rectangular kernel from the point of view of smoothness. This preference arises from a monotone attenuation property possessed by the Gaussian kernel but not by the rectangular kernel. This desirable property translates in the present context into a requirement that the transfer function for a smoothing window decrease monotonically in magnitude. Now the 'transfer function' here is in fact the inverse Fourier transform of the smoothing window, which is just the sequence $\{w_{\tau,m}\}$ (see Exercise [5.10]). By this smoothness requirement, we would prefer smoothing windows whose lag windows decay monotonically to 0. This is true for the Bartlett, Parzen and Papoulis lag windows (see plot (a) of Figures 260, 265 and 267), but *not* for the Daniell lag window (Figure 263) – this is not surprising because in fact the Daniell smoothing window is rectangular. A practical interpretation is that use of the Daniell smoothing window can lead to undesirable ripples in $\hat{S}^{(lw)}(\cdot)$ that are not present in $\hat{S}^{(d)}(\cdot)$.

Priestley (1981) discusses several other criteria that attempt to take into account both the bias and the variance in different lag window spectral estimators. For example, we might select that estimator such that the integrated mean square error

$$\int_{-f_{(N)}}^{f_{(N)}} E\left\{\left[\hat{S}^{(lw)}(f) - S(f)\right]^2\right\}\, df$$

is as small as possible. Unfortunately, to evaluate the above integral we need to know what $S(\cdot)$ is. This dependence upon the unknown true sdf plagues many other proposed criteria. As a result, we can only say that a particular estimator is best by a certain criterion when the true sdf is assumed to be of a particular form – this is not very helpful.

If smoothing window leakage is not an issue for the direct spectral estimate to be smoothed, our preference is for either the Parzen or Papoulis smoothing window; if leakage proves significant for these windows, our choice is then the Daniell smoothing window. However, except in cases where the dynamic range of $\hat{S}^{(d)}(\cdot)$ is large enough to make smoothing window leakage an issue, we agree with the following quote from Jenkins and Watts (1968, p. 273): '... the important question in empirical spectral analysis is the choice of [smoothing window] bandwidth and *not* the choice of [smoothing] window.'

6.13 Choice of Lag Window Parameter

In the previous section, we argued that any one of several lag window spectral estimators is a reasonable choice. Unfortunately, the choice of the window parameter m is *not* as easy to make. The main problem is that sensible choice of m depends upon the shape of the unknown sdf we want to estimate.

To illustrate the importance in selecting m properly, Figure 275 shows

$$\int_{-f_{(N)}}^{f_{(N)}} W_m(f - f')S(f')\,df'$$

for the case where $S(\cdot)$ is the sdf of the AR(4) process $\{X_{t,4}\}$ of Equation (46a); $W_m(\cdot)$ is Parzen's smoothing window; and $m = 50$, 100 and 200. The true sdf for this AR(4) process (indicated by the thick curves in Figure 275) has two sharp peaks close to each other. For $m = 50$, the amount of smoothing done by Parzen's lag window spectral estimator is such that the expected value of the estimator (given by the thin curves) does not accurately represent the true sdf in the region of the two peaks. The amount of smoothing is related to the width of the central lobe of Parzen's smoothing window (shown – centered about $f = 1/8$ – by the thin curves in the lower portion of each plot). As we increase m to 100 or 200, the width of this central lobe decreases enough so that the expected value of the estimator more accurately represents the true sdf. (Note that this example considers the bias due *only* to the smoothing window. As we have seen in Figures 213 and 214, the bias in the direct spectral estimator itself can be a significant factor for small N even if we are careful to use a good data taper.)

To make a rational choice of m, we must have some idea of the 'size' of the important features in $S(\cdot)$ – the very sdf we are trying to estimate! In many situations this is simply not known *a priori*. For example, spectral analysis is often used as an exploratory data analysis tool, in which case little is known about $S(\cdot)$ in advance.

There are cases, however, where it is possible to make an intelligent guess about the typical shape of $S(\cdot)$ either from physical arguments or from analyses of previous data of a similar nature. Let us assume that we can draw a picture of the typical sdf to be expected similar to the thick curves in Figure 275. We can define the *spectral bandwidth* B_S as roughly the width of the *smallest* feature of interest in the projected sdf $S(\cdot)$. For example, in Figure 275, B_S would be the width of either of the two prominent peaks, where we can conveniently measure the width – following standard engineering practice – as the distance between half-power points, i.e., those points $f_1 < f_0$ and $f_2 > f_0$ such that

$$S(f_1) = S(f_2) = \frac{S(f_0)}{2} \text{ and hence } B_S = f_2 - f_1.$$

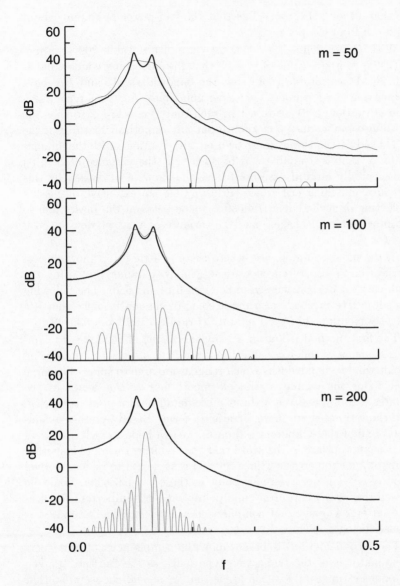

Figure 275. Effect of window parameter m on the expected value of an unbiased direct spectral estimator that has been smoothed using Parzen's lag window. The thick curve in each plot is the sdf for the AR(4) process of Equation (46a); the thin curves that more or less follow the thick curves show the component of $E\{\hat{S}^{(lw)}(\cdot)\}$ solely due to Parzen's smoothing window for $m = 50$, 100 and 200 (top to bottom); and the lower thin curves show the shape of the smoothing window itself.

Note that $10 \log_{10}(1/2) \approx -3$ so that the half-power points are about 3 dB down from the peak.

With B_S so defined, we can then use either Table 269 or Equation (242c) to find a value of m such that the smoothing window bandwidth B_W is acceptable (note that the bandwidths B_S and B_W have the same units as f, namely, cycles per unit time). A typical recommendation (Priestley, 1981) is to set m such that $B_W = B_S/2$ so that the smoothing window does not smooth out any important features in the sdf. This advice, however, must be used with caution, since the amount of data N that is available can be such that the variance of $\hat{S}^{(lw)}(\cdot)$ is unacceptably large if we were to use the value of m chosen by this bandwidth matching criterion. (Priestley has an extensive discussion on selecting m such that a tradeoff is made between the bias – due to smoothing window leakage – and the variance of a lag window spectral estimator.)

In the more common case where there is little prior knowledge of the true sdf, we can often choose the window parameter m using the two objective criteria discussed in Sections 6.14 and 6.15. The following three subjective methods also merit some discussion because they are frequently mentioned in the literature (Priestley, 1981, pp. 539–42).

The first method is known as *window closing*. The idea is to compute a sequence of different sdf estimates for the same set of data using different window bandwidths which range from large to small. For large values of B_W, the estimates will look smooth, but, as B_W decreases, the estimates will progressively exhibit more detail until a point is reached where the estimates are more 'erratic' in form. Based upon an examination of all of these different estimates, we can hope to pick a value of m that is appropriate in the sense that the resulting estimate is neither too smooth nor too erratic. Since $\hat{S}^{(lw)}(\cdot)$ is supposed to be a smoothed version of $\hat{S}^{(d)}(\cdot)$, our reference point in these comparisons should be how well the former captures – and points out – the important features of the latter. These visual comparisons can reject some estimates as obviously being 'too smooth' or 'too erratic.'

The second method is based upon the sample acvs. If our spectral estimator is of the truncation type (such as the Bartlett, Parzen or Papoulis estimators), we can argue that we should choose m so that $s_\tau \approx 0$ for all $|\tau| > m$. Thus we could plot $\hat{s}_\tau^{(p)}$ versus τ and pick m as that point where the sample acvs shows only small fluctuations around 0. Although crude, this method has evidently been quite effective in a number of problems and is arguably preferable to choosing a completely arbitrary value for m. There are several dangers inherent in this approach. First, since the sample acvs is itself highly correlated at nearby lag values, $\hat{s}_\tau^{(p)}$ often decays more slowly than the theoretical acvs. Second, sampling fluctuations can cause the sample acvs to deviate substantially from the true acvs. Third, if $S(\cdot)$ contains a large peak with a wide bandwidth and a small peak with a narrow bandwidth, the effect of the small peak will not be apparent at all in the sample acvs (Figures 148 and 149 support this assertion). The value of m selected by this method will reflect only the bandwidth of the large peak. We can thus miss some important features in the sdf by using this method.

The final subjective method is simply to let m be some fixed proportion of N, say 20% or 30%, or to let $m = \sqrt{N}$. These recommendations are rather widespread in the literature, but they are fundamentally unsound since they do not take into account the underlying process which generated the time series. Although we might have no prior knowledge of the spectral bandwidth, it is desirable to choose m based upon some properties of the process. This is precisely the aim of the previous two subjective techniques.

6.14 Estimation of Spectral Bandwidth

We have looked at the bandwidth B_W of a smoothing window and how to compute it using Equation (242c). To compute the spectral bandwidth B_S, we must have good knowledge of the sdf we are trying to estimate. Can we estimate the spectral bandwidth directly from a time series? If this were possible, it would clearly be beneficial since we could then follow the usual recommendation and set B_W to be approximately half of the estimated spectral bandwidth. The following argument shows that, if we assume the time series is drawn from a Gaussian stationary process with a dominantly unimodal sdf, it is possible to produce a satisfactory estimator for the spectral bandwidth (for details, see Walden and White, 1990).

The standard estimator of the variance s_0 of a time series drawn from a stationary process with unknown mean is

$$\hat{s}_0^{(p)} = \frac{1}{N} \sum_{t=1}^{N} \left(X_t - \bar{X} \right)^2 .$$

Let us assume for convenience that the process mean is known to be 0 so that we can take

$$\hat{s}_0^{(p)} = \frac{1}{N} \sum_{t=1}^{N} X_t^2$$

(this allows us to simplify the calculations below considerably – the resulting estimator of the spectral bandwidth is still useful even if the process mean must be estimated). The expected value of $\hat{s}_0^{(p)}$ is

$$E\{\hat{s}_0^{(p)}\} = \frac{1}{N} \sum_{t=1}^{N} E\{X_t^2\} = s_0,$$

and the variance of $\hat{s}_0^{(p)}$ is given by

$$
E\left\{\left(\hat{s}_0^{(p)} - s_0\right)^2\right\} = E\left\{\left(\frac{1}{N}\sum_{t=1}^{N}X_t^2\right)^2 - \frac{2s_0}{N}\sum_{t=1}^{N}X_t^2 + s_0^2\right\}
$$

$$
= \frac{1}{N^2}\sum_{t=1}^{N}\sum_{u=1}^{N}\left(E\{X_t^2 X_u^2\} - s_0^2\right). \tag{278a}
$$

Now

$$
\text{cov}\{X_t^2, X_u^2\} = E\{X_t^2 X_u^2\} - s_0^2,
$$

but, under the Gaussian assumption, the last expectation can be expressed in terms of $\text{cov}\{X_t, X_u\}$:

$$
\text{cov}\{X_t^2, X_u^2\} = 2\left(\text{cov}\{X_t, X_u\}\right)^2 = 2s_{t-u}^2
$$

(this follows from Equation (40) by letting $Z_1 = Z_2 = X_t$ and $Z_3 = Z_4 = X_u$). We can thus write Equation (278a) as

$$
\text{var}\{\hat{s}_0^{(p)}\} = \frac{1}{N^2}\sum_{t=1}^{N}\sum_{u=1}^{N}2s_{t-u}^2 = \frac{2}{N}\sum_{\tau=-(N-1)}^{N-1}\left(1 - \frac{|\tau|}{N}\right)s_\tau^2 \tag{278b}
$$

(this follows from Equation (189a) by replacing s_{t-u} there with $2s_{t-u}^2$).

In Section 6.10 the distribution of the spectral estimator $\hat{S}^{(lw)}(f)$ was approximated by that of a χ_ν^2 rv times a constant a. If we make the same approximation here, namely,

$$
\hat{s}_0^{(p)} \stackrel{\text{d}}{=} b\chi_\eta^2
$$

(Rice, 1945), we obtain

$$
\eta = \frac{2\left(E\{\hat{s}_0^{(p)}\}\right)^2}{\text{var}\{\hat{s}_0^{(p)}\}} = \frac{Ns_0^2}{\sum_{\tau=-(N-1)}^{N-1}\left(1 - |\tau|/N\right)s_\tau^2}
$$

as an expression for the *degrees of freedom in a time series* of length N (analogous to the degrees of freedom in a spectral estimator – see Equations (254a)). Note that both concepts involve quadratic forms (see Chapter 7), and some researchers make this point explicit by calling η the number of degrees of freedom of order 2 (Kikkawa and Ishida, 1988).

We now need to relate η somehow to the width of the dominant hump of a unimodal sdf. Suppose for the moment that $\{X_t\}$ is an ideal band-pass process with unit variance and sdf given by

$$
S(f) = \begin{cases} 1/(4W), & f' - W \leq |f| \leq f' + W; \\ 0, & \text{otherwise,} \end{cases}
$$

where $f' > 0$ is the center frequency of the passband and $2W > 0$ is the corresponding bandwidth. We assume that $f' - W > 0$ and $f' + W < f_{(N)}$. For this simple example we can relate the bandwidth $2W$ to the degrees of freedom in a sample of length N from $\{X_t\}$ by the following argument. From Parseval's theorem (see Exercise [6.6b]) and from the fact that $\hat{S}^{(p)}(\cdot)$ is $2f_{(N)}$ periodic, we can write

$$\hat{s}_0^{(p)} = \frac{1}{N\,\Delta t} \sum_{k=0}^{N-1} \hat{S}^{(p)}(f_k) \text{ with } f_k \equiv k/(N\,\Delta t)$$

$$= \frac{1}{N\,\Delta t} \sum_{k=-\lfloor (N-1)/2 \rfloor}^{\lfloor N/2 \rfloor} \hat{S}^{(p)}(f_k)$$

$$\approx \frac{1}{N\,\Delta t} \sum_{f'-W \le |f_k| \le f'+W} \hat{S}^{(p)}(f_k)$$

since $\hat{S}^{(p)}(f_k)$ should be small for those f_k such that $S(f_k) = 0$. Because $\hat{S}^{(p)}(\cdot)$ is symmetric about zero, we also have

$$\hat{s}_0^{(p)} \approx \frac{2}{N\,\Delta t} \sum_{f'-W \le f_k \le f'+W} \hat{S}^{(p)}(f_k).$$

From Section 6.6 we know that, for those f_k in the passband,

$$\hat{S}^{(p)}(f_k) \stackrel{\mathrm{d}}{=} \frac{S(f_k)}{2}\chi_2^2 = \frac{1}{8W}\chi_2^2$$

to a good approximation and that the $\hat{S}^{(p)}(f_k)$ rv's at distinct Fourier frequencies are approximately independent of each other. Since the sum of K independent χ_2^2 rv's has a χ_{2K}^2 distribution, we can conclude that

$$\hat{s}_0^{(p)} \stackrel{\mathrm{d}}{=} \frac{1}{4W N\,\Delta t}\chi_{2K}^2$$

to a good approximation, where $K \equiv 2W N\,\Delta t$ is approximately the number of f_k such that $f' - W \le f_k \le f' + W$. Hence, a sample of length N from a band-pass process with a passband of width $2W$ has degrees of freedom $\eta = 2K = 4W N\,\Delta t$, and we have the following relationship between the width and η:

$$\text{width of passband} = 2W = \frac{\eta}{2N\,\Delta t}.$$

Moreover, if the passband is far enough from both zero and Nyquist frequencies so that the smallest feature of interest in $S(\cdot)$ is in fact the passband, then we have $B_S = \eta/(2N\,\Delta t)$; if this is not true (e.g., the smallest feature might be the null region centered about zero frequency), this relationship between $\eta/(2N\,\Delta t)$ and B_S does not hold, but it is still possible to deduce B_S using $\eta/(2N\,\Delta t)$ and the center frequency of the passband.

For a time series of length N drawn from a process with a unimodal sdf, we can use the above arguments to *define* the bandwidth of the series as

$$B_T \equiv \frac{\eta}{2N\,\Delta t} = \frac{s_0^2}{2\,\Delta t\sum_{\tau=-(N-1)}^{N-1}(1-|\tau|/N)\,s_\tau^2}.$$

We can obtain an obvious estimator of this quantity from

$$\hat{B}_T \equiv \frac{(\hat{s}_0^{(p)})^2}{2\,\Delta t\sum_{\tau=-(N-1)}^{N-1}(1-|\tau|/N)(\hat{s}_\tau^{(p)})^2}.$$

Walden and White (1990) find that $\hat{B}_T$ is a biased estimator of B_T, but that the following estimator is approximately unbiased:

$$\tilde{B}_T \equiv \frac{5}{3}\hat{B}_T - \frac{1}{N\,\Delta t}. \tag{280}$$

We hope it is clear that, just as the smoothing window should be unimodal for its bandwidth measure B_W to be useful, so should the sdf for B_T to be useful. Kikkawa and Ishida (1988) say similarly that second-order parameters (such as η) '... are good measures for both low-pass and band-pass processes.' Distinct bimodality in a sdf will lead to B_S and B_T being very different since B_S will be a measure of the width of the smaller mode, while B_T will be an average over the entire sdf. If B_T is meaningful and if we replace B_S by $\tilde{B}_T$ and adopt the recommendation $B_W = \tilde{B}_T/2$ as before, the practical results are generally quite satisfactory. There are examples of the use of $\tilde{B}_T$ in Section 6.18 (see also White, 1980, and Walden and White, 1984).

6.15 Automatic Smoothing of Log Spectral Estimators

In the previous sections we have considered smoothing a direct spectral estimator by convolving it with a smoothing window. We present here an alternative approach in which we smooth $\log(\hat{S}^{(d)}(\cdot))$ instead of $\hat{S}^{(d)}(\cdot)$. This approach leads to an objective procedure for determining the amount of smoothing that is optimal in a certain mean square sense. For further details, see Wahba (1980), who originally proposed this technique for smoothing the log of the periodogram, and Sjoholm (1989), who extended it to the case of all direct spectral estimators.

For a given sdf $S(\cdot)$, let us define the *log spectral density function* as

$$C(f) \equiv \lambda\log(S(f)), \qquad |f| \le f_{(N)},$$

where λ is a constant that allows us to use different logarithmic scales (for example, if we set $\lambda = 10 \log_{10}(e)$, then $C(\cdot)$ is expressed in decibels). We assume that $C(\cdot)$ has a Fourier series representation

$$C(f) = \Delta t \sum_{\tau=-\infty}^{\infty} c_\tau e^{-i2\pi f \tau \, \Delta t}, \quad \text{where } c_\tau \equiv \int_{-f_{(N)}}^{f_{(N)}} C(f) e^{i2\pi f \tau \, \Delta t} \, df,$$

so that $\{c_\tau\} \longleftrightarrow C(\cdot)$. If $\lambda = 1$, the sequence $\{c_\tau\}$ is sometimes referred to as the *cepstrum* (Bogert *et al.*, 1963) and is in many ways analogous to the acvs.

Given observed values of $X_1, \ldots, X_N$ from a stationary process with sdf $S(\cdot)$, a natural estimator of $C(f)$ is

$$\hat{C}^{(d)}(f) \equiv \lambda \log\left(\hat{S}^{(d)}(f)\right),$$

where $\hat{S}^{(d)}(\cdot)$ is a direct spectral estimator of $S(\cdot)$. If we exclude the zero and Nyquist frequencies, we have, to a good approximation, that $\hat{S}^{(d)}(f) \stackrel{\mathrm{d}}{=} S(f)\chi_2^2/2$ (see Equation (223b)); moreover, there exists a grid of frequencies defined by $f_k' \equiv k/(N'\Delta t)$ with $N' \leq N$ such that the $\hat{S}^{(d)}(f_k')$ terms are approximately pairwise uncorrelated for $k = 0, \ldots, \lfloor N'/2 \rfloor$. We thus have

$$E\{\hat{C}^{(d)}(f)\} \approx E\{\lambda \log\left(S(f)\chi_2^2/2\right)\} = C(f) - \lambda \log(2) + \lambda E\{\log(\chi_2^2)\}$$

and

$$\mathrm{var}\{\hat{C}^{(d)}(f)\} \approx \lambda^2 \, \mathrm{var}\{\log\left(S(f)/2\right) + \log(\chi_2^2)\} = \lambda^2 \, \mathrm{var}\{\log(\chi_2^2)\}.$$

Bartlett and Kendall (1946) show that

$$E\{\log(\chi_2^2)\} = \log(2) - \gamma \quad \text{and} \quad \mathrm{var}\{\log(\chi_2^2)\} = \pi^2/6,$$

where γ is Euler's constant $(0.5772\ldots)$. These results yield

$$E\{\hat{C}^{(d)}(f)\} \approx C(f) - \lambda\gamma \quad \text{and} \quad \mathrm{var}\{\hat{C}^{(d)}(f)\} \approx \lambda^2\pi^2/6.$$

Note that the latter quantity is a *known* constant independent of $C(f)$. With the additional assumption that $\{X_t\}$ is a Gaussian process, we can argue that the $\hat{C}^{(d)}(f_k')$ rv's are also approximately pairwise uncorrelated. (Although generally $\hat{S}^{(d)}(f) > 0$, we can have $\hat{S}^{(d)}(0) = 0$ due to centering the data; see Exercise [6.5b] and the discussion concerning Equation (217a). This causes $\hat{C}^{(d)}(0)$ to be ill-defined, so we must obtain it by interpolation. One simple solution is to set $\hat{C}^{(d)}(0) = \hat{C}^{(d)}(f_1')$.)

Let us now define the following estimator of c_τ:

$$\hat{c}_\tau^{\prime(d)} \equiv \frac{1}{N'\,\Delta t} \sum_{k=-N_L}^{N_U} \hat{C}^{(d)}(f_k')e^{i2\pi f_k'\tau\,\Delta t}, \qquad \tau = -N_L,\dots,N_U, \quad (282)$$

where $N_L \equiv \lfloor (N'-1)/2 \rfloor$ and $N_U \equiv \lfloor N'/2 \rfloor$ (recall that $\lfloor x \rfloor$ is the greatest integer less than or equal to x, and note that $N_U + N_L + 1 = N'$). We can thus write

$$\hat{C}^{(d)}(f_k') = \Delta t \sum_{\tau=-N_L}^{N_U} \hat{c}_\tau^{\prime(d)} e^{-i2\pi f_k'\tau\,\Delta t}, \qquad k = -N_L,\dots,N_U,$$

and we have $\{\hat{c}_\tau^{\prime(d)}\} \longleftrightarrow \{\hat{C}^{(d)}(f_k')\}$. For a given lag window sequence $\{w_{\tau,m}\}$, let us now define the following analog to the lag window spectral estimator $\hat{S}^{(lw)}(f)$:

$$\hat{C}^{(lw)}(f) \equiv \Delta t \sum_{\tau=-N_L}^{N_U} w_{\tau,m}\hat{c}_\tau^{\prime(d)} e^{-i2\pi f\tau\,\Delta t}.$$

Just as $\hat{S}^{(lw)}(\cdot)$ is a smoothed version of $\hat{S}^{(d)}(\cdot)$, so is $\hat{C}^{(lw)}(\cdot)$ a smoothed version of $\hat{C}^{(d)}(\cdot)$.

As a measure of how well $\hat{C}^{(lw)}(\cdot)$ estimates $C(\cdot)$, let us consider the mean integrated square error:

$$E\left\{ \int_{-f_{(N)}}^{f_{(N)}} \left(\hat{C}^{(lw)}(f) - C(f) \right)^2 df \right\} = I_m + \Delta t \sum_{\substack{\tau > N_U \\ \tau < -N_L}} c_\tau^2,$$

where

$$I_m \equiv E\left\{ \Delta t \sum_{\tau=-N_L}^{N_U} \left(w_{\tau,m}\hat{c}_\tau^{\prime(d)} - c_\tau \right)^2 \right\}$$

(this is an application of Parseval's theorem; see the discussion following Equation (61) in Section 3.1). For a given lag window (such as the Parzen window), the value of I_m depends on the window parameter m. The idea is to estimate I_m by, say, $\hat{I}_m$ and then to pick m such that $\hat{I}_m$ is minimized. Note that, if c_τ is small for $\tau > N_L$, then I_m is a good approximation to the mean integrated square error. In any case, I_m is the only part of this error that depends on the lag window and is hence under our control.

Now

$$I_m = \Delta t \sum_{\tau=-N_L}^{N_U} \left[w_{\tau,m}^2 E\{(\hat{c}_\tau^{\prime(d)})^2\} - 2w_{\tau,m}c_\tau E\{\hat{c}_\tau^{\prime(d)}\} + c_\tau^2 \right].$$

From Equation (282) and the properties of $\hat{C}^{(d)}(f'_k)$, we have

$$E\{\hat{c}'^{(d)}_\tau\} = \frac{1}{N'\Delta t} \sum_{k=-N_L}^{N_U} E\{\hat{C}^{(d)}(f'_k)\}e^{i2\pi f'_k \tau \Delta t}$$

$$\approx \int_{-f_{(N)}}^{f_{(N)}} (C(f) - \lambda\gamma)\, e^{i2\pi f\tau \Delta t}\, df$$

$$= c_\tau - \frac{\lambda\gamma\delta_\tau}{\Delta t}, \quad \text{where } \delta_\tau = \begin{cases} 1, & \tau = 0; \\ 0, & \text{otherwise,} \end{cases}$$

and

$$E\{(\hat{c}'^{(d)}_\tau)^2\} = E\{\hat{c}'^{(d)}_\tau(\hat{c}'^{(d)}_\tau)^*\}$$

$$= \left(\frac{1}{N'\Delta t}\right)^2 \sum_{j=-N_L}^{N_U} \sum_{k=-N_L}^{N_U} E\{\hat{C}^{(d)}(f'_j)\hat{C}^{(d)}(f'_k)\}e^{i2\pi(f'_j - f'_k)\tau \Delta t}$$

$$\approx \left(E\{\hat{c}'^{(d)}_\tau\}\right)^2 + \frac{\lambda^2\pi^2}{6N'(\Delta t)^2} \approx \left(c_\tau - \frac{\lambda\gamma\delta_\tau}{\Delta t}\right)^2 + \frac{\lambda^2\pi^2}{6N'(\Delta t)^2}, \quad (283a)$$

so, because $w_{0,m} = 1$ always, we can write

$$I_m \approx \frac{\lambda^2\gamma^2}{\Delta t} + \Delta t \sum_{\tau=-N_L}^{N_U} \left[w^2_{\tau,m}\left(c^2_\tau + \frac{\lambda^2\pi^2}{6N'(\Delta t)^2}\right) - 2w_{\tau,m}c^2_\tau + c^2_\tau\right]$$

$$= \frac{\lambda^2\gamma^2}{\Delta t} + \Delta t \sum_{\tau=-N_L}^{N_U} c^2_\tau (1 - w_{\tau,m})^2 + \frac{\lambda^2\pi^2}{6N'\Delta t} \sum_{\tau=-N_L}^{N_U} w^2_{\tau,m}.$$

Since $1 - w_{0,m} = 0$, the value of c^2_0 does not influence the above; in view of Equation (283a), an approximately unbiased estimator of c^2_τ for $\tau \neq 0$ is given by $(\hat{c}'^{(d)}_\tau)^2 - (\lambda^2\pi^2)/[6N'(\Delta t)^2]$. This suggests the following estimator for I_m:

$$\hat{I}_m \equiv \frac{\lambda^2\gamma^2}{\Delta t} + \Delta t \sum_{\tau=-N_L}^{N_U} \left((\hat{c}'^{(d)}_\tau)^2 - \frac{\lambda^2\pi^2}{6N'(\Delta t)^2}\right)(1 - w_{\tau,m})^2$$

$$+ \frac{\lambda^2\pi^2}{6N'\Delta t} \sum_{\tau=-N_L}^{N_U} w^2_{\tau,m}. \qquad (283b)$$

We can then objectively select m by picking the value for which $\hat{I}_m$ is smallest.

Sjoholm (1989) points out that an estimate of $S(\cdot)$ that is based upon smoothing $\log(\hat{S}^{(d)}(\cdot))$ will have poor quantitative properties in that power will not be preserved: since smoothing is done in log space, a high power region can be disproportionately modified if it is adjacent to a low power region. If $\hat{C}^{(lw)}(\cdot)$ is to be used for other than qualitative purposes, the above procedure can be regarded merely as a way of establishing the appropriate smoothing window bandwidth via determining the parameter m, so that, once m is known, we can use it in computing $\hat{S}^{(lw)}(\cdot)$. Examples of the use of $\hat{I}_m$ are given in Section 6.18.

6.16 Computational Details

Here we give some details concerning the actual computation of lag window spectral estimators. We shall strive for clarity rather than to get into issues of computational efficiency. All computations are done in terms of discrete Fourier transforms (DFTs) and inverse DFTs, so we first review the following pertinent results.

Recall that the DFT for a finite sequence $g_0, \ldots, g_{M-1}$ with a sampling time of unity is the sequence $G_0, \ldots, G_{M-1}$, where

$$G_k \equiv \sum_{t=0}^{M-1} g_t e^{-i2\pi kt/M} \quad \text{and} \quad g_t = \frac{1}{M} \sum_{k=0}^{M-1} G_k e^{i2\pi kt/M}$$

(see Equations (110a) and (111a)). This relationship is summarized by the notation $\{g_t\} \longleftrightarrow \{G_k\}$. Both of these sequences are defined outside the range 0 to $M-1$ by cyclic (periodic) extension; i.e., if $s < 0$ or $s \geq M$, then $g_s \equiv g_{\text{mod}(s,M)}$ so that, for example, $g_{-1} = g_{M-1}$ and $g_M = g_0$. We define the cyclic autocorrelation of $\{g_t\}$ by

$$g^* \star g_\tau \equiv \sum_{t=0}^{M-1} g_t^* g_{t+\tau}, \qquad \tau = 0, \ldots, M-1 \tag{284a}$$

(see the discussion before Equation (85b)). If $\{h_t\}$ is another sequence of length M with DFT $\{H_k\}$, i.e., $\{h_t\} \longleftrightarrow \{H_k\}$, we define the cyclic convolution of $\{g_t\}$ and $\{h_t\}$ by

$$g * h_\tau \equiv \sum_{t=0}^{M-1} g_t h_{\tau-t}, \qquad \tau = 0, \ldots, M-1 \tag{284b}$$

(see Equation (119b)). We note that

$$\{g^* \star g_\tau\} \longleftrightarrow \{|G_k|^2\} \quad \text{and} \quad \{g * h_\tau\} \longleftrightarrow \{G_k H_k\}$$

(see Equations (85b) and (119c)). Exercise [3.14a] outlines an efficient method for computing cyclic autocorrelations and convolutions.

We start with a time series of length N that can be regarded as a realization of $X_1, X_2, \ldots, X_N$, a segment of length N of the stationary

$$\{\tilde{X}_t\} \quad \xleftrightarrow{\;(1)\;} \quad \{\tilde{X}(f'_k)\}$$

$$\Big\downarrow \text{mult} \qquad\qquad\qquad \Big\downarrow \text{conv}$$

$$\{\tilde{h}_t\tilde{X}_t\} \quad \xleftrightarrow{\;(2)\;} \quad \{\tilde{H} * \tilde{X}(f'_k)/M\}$$

$$\Big\downarrow \text{auto} \qquad\qquad\qquad \Big\downarrow \text{mod sq}$$

$$\{\tilde{s}^{(d)}_\tau\} \;=\; \{\tilde{h}\tilde{X} \star \tilde{h}_\tau\tilde{X}_\tau\} \quad \xleftrightarrow{\;(3)\;} \quad \{|\tilde{H} * \tilde{X}(f'_k)|^2/M^2\} \;=\; \{\tilde{S}^{(d)}(f'_k)\}$$

$$\Big\downarrow \text{mult} \qquad\qquad\qquad\qquad\qquad\quad \Big\downarrow \text{conv}$$

$$\{\tilde{s}^{(lw)}_\tau\} \;=\; \{\tilde{w}_{\tau,m}\tilde{s}^{(d)}_\tau\} \quad \xleftrightarrow{\;(4)\;} \quad \{\tilde{W} * \tilde{S}^{(d)}(f'_k)/M\} \;=\; \{\tilde{S}^{(lw)}(f'_k)\}$$

Figure 285. Pathways for computing $\tilde{S}^{(lw)}(\cdot)$ (adapted from Figure 1 of Van Schooneveld and Frijling, 1981).

process $\{X_t\}$ with unknown mean μ, sdf $S(\cdot)$ and sampling time Δt. Let M be any integer satisfying $M \geq 2N$. We let

$$\tilde{X}_t \equiv \begin{cases} X_{t+1} - \bar{X}, & 0 \leq t < N; \\ 0, & N \leq t < M; \end{cases} \text{ and } \tilde{h}_t \equiv \begin{cases} h_{t+1}, & 0 \leq t < N; \\ 0, & N \leq t < M, \end{cases}$$

where $\bar{X} \equiv \sum_{t=1}^{N} X_t/N$ is the sample mean, and $\{h_t\}$ is a data taper of length N with normalization $\sum_{t=1}^{N} h_t^2 = 1$. The sequences $\{\tilde{X}_t\}$ and $\{\tilde{h}_t\}$ are each of length M and are defined outside the range $t = 0$ to $M-1$ by cyclic extension. For convenience, we perform our computations assuming that the sampling time for these sequences – and all others referred to in Figure 285 – is unity so that the DFT formulae above can be used. We create these two zero padded sequences so that we can compute various noncyclic convolutions using cyclic convolution (see Exercise [3.14b]).

With $f'_k \equiv k/(M\,\Delta t)$, we denote the DFTs of $\{\tilde{X}_t\}$ and $\{\tilde{h}_t\}$ by, respectively, the sequences $\{\tilde{X}(f'_k)\}$ and $\{\tilde{H}(f'_k)\}$, both of which are of length M and indexed by k with $k = 0, \ldots, M-1$. As usual, these Fourier relationships are summarized by

$$\{\tilde{X}_t\} \longleftrightarrow \{\tilde{X}(f'_k)\} \text{ and } \{\tilde{h}_t\} \longleftrightarrow \{\tilde{H}(f'_k)\}.$$

A pathway to compute $\hat{S}^{(lw)}(f'_k)$ is depicted in Figure 285 and involves four Fourier transform pairs, the first of which we have already noted. Here are comments regarding the remaining three.

(2) The left-hand side is obtained from a term by term multiplication of the sequences $\{\tilde{h}_t\}$ and $\{\tilde{X}_t\}$, while its DFT (the right-hand side) can be obtained either directly or by convolving the sequences $\{\tilde{H}(f'_k)\}$ and $\{\tilde{X}(f'_k)\}$ using Equation (284b).

(3) The left-hand side is the autocorrelation of $\{\tilde{h}_t\tilde{X}_t\}$ and can be obtained using Equation (284a). Its DFT can be obtained either directly or by computing the squared modulus of $\{\tilde{H}*\tilde{X}(f_k')\}$ term by term. If we let $\tilde{s}_\tau^{(d)}$ be the τth element of the autocorrelation of $\{\tilde{h}_t\tilde{X}_t\}$, then we have

$$\hat{s}_\tau^{(d)} = \begin{cases} \tilde{s}_{|\tau|}^{(d)}, & |\tau| \leq N-1; \\ 0, & \text{otherwise.} \end{cases}$$

On the right-hand side, if we let $\tilde{S}^{(d)}(f_k') \equiv |\tilde{H}*\tilde{X}(f_k')|^2/M^2$, then we have

$$\hat{S}^{(d)}(f_k') = \Delta t\, \tilde{S}^{(d)}(f_k'), \qquad k = 0, \ldots, \lfloor M/2 \rfloor.$$

(4) We define the sequence $\{\tilde{w}_{\tau,m}\}$ of length M by

$$\tilde{w}_{\tau,m} = \begin{cases} w_{\tau,m}, & 0 \leq \tau < N; \\ 0, & N \leq \tau \leq M-N; \\ w_{\tau-M,M}, & M-N < \tau < M. \end{cases}$$

Its DFT is denoted by $\{\tilde{W}(f_k')\}$. The left-hand side of (4) is obtained by multiplying the sequences $\{\tilde{w}_{\tau,m}\}$ and $\{\tilde{s}_\tau^{(d)}\}$ term by term. Its DFT can be obtained directly or via the convolution of $\{\tilde{W}(f_k')\}$ and $\{\tilde{S}^{(lw)}(f_k')\}$ using Equation (284b). If we let $\tilde{s}_\tau^{(lw)}$ be the τth element of $\{\tilde{w}_{\tau,m}\tilde{s}_\tau^{(d)}\}$, then we have

$$\hat{s}_\tau^{(lw)} = \begin{cases} \tilde{s}_{|\tau|}^{(lw)}, & |\tau| \leq N-1; \\ 0, & \text{otherwise.} \end{cases}$$

On the right-hand side, if we let $\tilde{S}^{(lw)}(f_k') \equiv \tilde{W}*\tilde{S}^{(d)}(f_k')$, then we have

$$\hat{S}^{(lw)}(f_k') = \Delta t\, \tilde{S}^{(lw)}(f_k'), \qquad k = 0, \ldots, \lfloor M/2 \rfloor.$$

As shown in Figure 285, there are a number of different computational pathways that we can use to obtain $\{\hat{S}^{(lw)}(f_k')\}$ from $\{X_t\}$, each of which might be preferred under certain conditions. For example, the pathway

$$\{\tilde{X}_t\}$$
$$\downarrow \text{mult}$$
$$\{\tilde{h}_t\tilde{X}_t\} \xrightarrow{\;(2)\;} \{\tilde{H}*\tilde{X}(f_k')/M\}$$
$$\downarrow \text{mod sq}$$
$$\{\tilde{h}\tilde{X}\star\tilde{h}_\tau\tilde{X}_\tau\} \xleftarrow{\;(3)\;} \{|\tilde{H}*\tilde{X}(f_k')|^2/M^2\}$$
$$\downarrow \text{mult}$$
$$\{\tilde{w}_{\tau,m}\tilde{s}_\tau^{(d)}\} \xrightarrow{\;(4)\;} \{\tilde{W}*\tilde{S}^{(d)}(f_k')/M\}$$

requires two DFTs and one inverse DFT and allows us to easily obtain $\{\hat{S}^{(d)}(f'_k)\}$, $\{\hat{s}^{(d)}_\tau\}$, $\{\hat{s}^{(lw)}_\tau\}$ and $\{\hat{S}^{(lw)}(f'_k)\}$ (in that order). If we are not interested in examining the acvs estimates $\{\hat{s}^{(d)}_\tau\}$ and $\{\hat{s}^{(lw)}_\tau\}$, then we could use the pathway

$$\{\tilde{X}_t\}$$

$$\downarrow \text{mult}$$

$$\{\tilde{h}_t\tilde{X}_t\} \xrightarrow{(2)} \{\tilde{H}*\tilde{X}(f'_k)/M\}$$

$$\downarrow \text{mod sq}$$

$$\{|\tilde{H}*\tilde{X}(f'_k)|^2/M^2\}$$

$$\downarrow \text{conv}$$

$$\{\tilde{W}*\tilde{S}^{(d)}(f'_k)/M\}\}$$

which involves one DFT and one convolution. This second pathway might require fewer numerical operations than the first if the sequence $\{\tilde{W}(f'_k)\}$ is sufficiently short so that the convolution can be computed efficiently with Equation (284b) (in fact, we now only need $M \geq N$ instead of $M \geq 2N$ – padding with N or more zeros is only required to correctly compute $\{\hat{s}^{(d)}_\tau\}$ and $\{\hat{s}^{(lw)}_\tau\}$.) Note also that this pathway is exactly how in practice one would compute the discretely smoothed direct spectral estimator $\hat{S}^{(ds)}(\cdot)$ of Equation (236a).

As an example of these manipulations, let us consider the first $N = 20$ values of a data set related to the rotation of the earth. The entire data set (with 100 values) is shown by the small pluses in the bottom plots of Figures 172 and 173. For the record, the 20 numbers are 71, 63, 70, 88, 99, 90, 110, 135, 128, 154, 156, 141, 131, 132, 141, 104, 136, 146, 124 and 129. The sampling time for this data set is $\Delta t = 1/4$ year. Figure 288 shows eight different Fourier transform pairs of interest, one pair per row of plots: the sequence in each right-hand plot is the DFT of the sequence in the left-hand plot. The four Fourier transform pairs that are labeled in Figure 285 are shown in rows 1, 3, 4 and 6 of Figure 288. In order to use a conventional 'power of 2' FFT, we let $M = 64$ – this is the smallest power of 2 greater than $2N = 40$. The left-hand plot of the first row thus shows $\{\tilde{X}_t\}$, a sequence of length 64, the first 20 values of which are our 20 numbers centered about their sample mean (117.4), and the last 44 values of which are all zeros. The corresponding right-hand plot is its DFT – a sequence of 64 complex-valued numbers – the real parts of which are shown by the thick jagged curve, and the imaginary parts by the thin curve. The second row of plots shows $\{\tilde{h}_t\}$ (here a dpss data taper of length 20 with $NW = 4$ padded with 44 zeros) and its DFT (again this is a complex-valued sequence). The

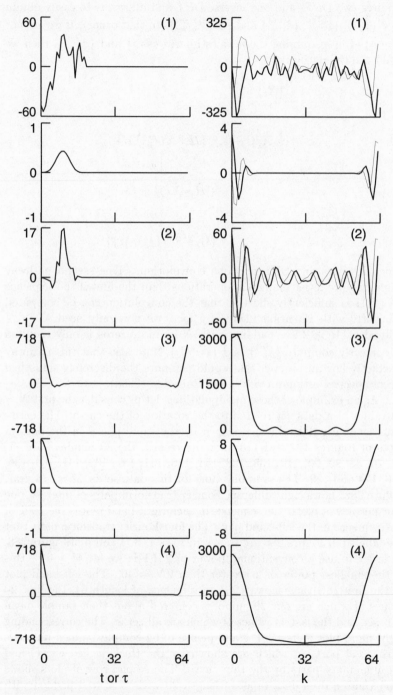

Figure 288. Illustration of computational pathways (cf. Figure 285).

left-hand plot on the third row is $\{\tilde{h}_t \tilde{X}_t\}$. The left-hand plot on the
fourth row is $\{\tilde{s}_\tau^{(d)}\}$, from which we can pick off the values for $\{\hat{s}_\tau^{(d)}\}$;
the right-hand plot is $\{\tilde{S}^{(d)}(f_k')\}$, which is real-valued. For $k = 0$ to
32, the kth value of this sequence times $\Delta t = 1/4$ yields $\hat{S}^{(d)}(f_k')$, a
direct spectral estimator of $S(\cdot)$ at frequency $f_k' = k/16$ cycles per year
(note that $k = 32$ corresponds to the Nyquist frequency of 2 cycles per
year). The left-hand plot on the fifth row is $\{\tilde{w}_{\tau,m}\}$, formed here using
a Parzen lag window with $m = 10$; the right-hand plot is the real-valued
sequence $\{\tilde{W}(f_k')\}$, the DFT of $\{\tilde{w}_{\tau,m}\}$. Finally, the left-hand plot on
the last row is $\{\tilde{w}_{\tau,m}\tilde{s}_\tau^{(d)}\}$, from which we can form $\{\hat{s}_\tau^{(lw)}\}$; the right-
hand plot is $\{\tilde{S}^{(lw)}(f_k')\}$, from which we can compute $\{\hat{S}^{(lw)}(f_k')\}$, a
smoothed version of $\{\hat{S}^{(d)}(f_k')\}$. (A final comment: the data taper and
lag window used here are *not* particularly appropriate for this short time
series – they were chosen merely to create a pleasing picture!)

6.17   Welch's Overlapped Segment Averaging

In discussing the rationale behind the Bartlett lag window of Equation (260), we introduced the idea of breaking up a time series into a
number of contiguous nonoverlapping blocks, computing a periodogram
based on the data in each block alone, and then averaging the individual periodograms together to form an overall spectral estimate. The
resulting estimator does not have the form of a lag window spectral estimator (although, as we have seen, it does approximately correspond
to a Bartlett lag window estimator), but, because it is the average of
several periodograms based upon different blocks of data, its variance
is smaller than that of the individual periodograms. Hence block averaging is a viable alternative to lag windows as a way of controlling the
inherent variability in periodograms.

Welch (1967) further developed the idea of block averaging by introducing two important modifications. First, he advocated the use of a
data taper on each block to reduce potential bias due to leakage in the
periodogram; i.e., the spectral estimator for each block is now a direct
spectral estimator as defined by Equation (206c). Second, he showed
that allowing the blocks to overlap (see Figure 290) can produce a spectral estimator with better variance properties than one using contiguous
nonoverlapping blocks. This reduction in variance occurs both because
overlapping recovers some of the information concerning the acvs contained in pairs of data values spanning adjacent nonoverlapping blocks
(a concern addressed by Bartlett, 1950), and because overlapping compensates somewhat for the effect of tapering the data in the individual
blocks; i.e., data values that are downweighted in one block can have
a higher weight in another block. In a series of papers since Welch's
work, Nuttall and Carter (1982, and references therein) have investigated in detail the statistical properties of this technique, now known in

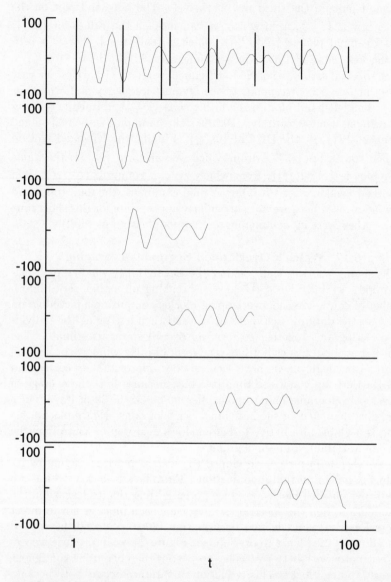

Figure 290. Breaking up a time series into overlapping blocks. The thin curve in the top plot shows a time series with 100 points (connected by lines). The ten thick vertical lines in the plot mark five blocks of $N_S = 32$ points each – lines of the same length bracket one block. Each block is plotted separately in the bottom five plots. The blocks consist of points 1 to 32, 18 to 49, 35 to 66, 52 to 83 and 69 to 100. Overlapping blocks have 15 common points, yielding a $15/32 = 47\%$ overlap percentage.

the literature as WOSA (this stands for either 'weighted overlapped segment averaging' – see Nuttall and Carter, 1982 – or 'Welch's overlapped segment averaging' – see Carter, 1987).

In what follows, we develop expressions for the mean and variance of WOSA spectral estimators, for which we need the following definitions. Let N_S represent a block size ($N_S = 32$ for the example shown in Figure 290), and let $h_1, \ldots, h_{N_S}$ be a data taper. Given a time series that is a realization of a portion $X_1, \ldots, X_N$ of a stationary process with sdf $S(\cdot)$, we define the direct spectral estimator for the block of N_S contiguous data values starting at index l as

$$\hat{S}_l^{(d)}(f) \equiv \Delta t \left| \sum_{t=1}^{N_S} h_t X_{t+l-1} e^{-i2\pi f t \, \Delta t} \right|^2, \qquad 1 \le l \le N + 1 - N_S.$$

$$(291a)$$

The WOSA spectral estimator is defined by

$$\hat{S}^{(\text{WOSA})}(f) \equiv \frac{1}{N_B} \sum_{j=0}^{N_B - 1} \hat{S}_{jn+1}^{(d)}(f), \qquad (291b)$$

where N_B is the total number of blocks to be averaged together and n is an integer-valued shift factor satisfying

$$0 < n \le N_S \text{ and } n(N_B - 1) = N - N_S$$

($n = 17$ and $N_B = 5$ in Figure 290). Note that, with these restrictions on n, the $j = 0$ block utilizes the data values $X_1, \ldots, X_{N_S}$, while the $j = N_B - 1$ block uses $X_{N-N_S+1}, \ldots, X_N$.

Let us now consider the first moment properties of $\hat{S}^{(\text{WOSA})}(\cdot)$. Since from Equation (207a) we have, for all j,

$$E\{\hat{S}_{jn+1}^{(d)}(f)\} = \int_{-f_{(N)}}^{f_{(N)}} \mathcal{H}(f - f')S(f') \, df',$$

where $\mathcal{H}(\cdot)$ is the spectral window corresponding to $h_1, \ldots, h_{N_S}$, it follows immediately that

$$E\{\hat{S}^{(\text{WOSA})}(f)\} = \int_{-f_{(N)}}^{f_{(N)}} \mathcal{H}(f - f')S(f') \, df'$$

also. Note that the first moment properties of $\hat{S}^{(\text{WOSA})}(\cdot)$ depend just on the block size N_S, the data taper $\{h_t\}$ and the true sdf $S(\cdot)$ – they do *not* depend on the total length of the time series N, the total number of blocks N_B or the shift factor n. This points out a potential pitfall in using WOSA, namely, that we must make sure that N_S is large enough so that $E\{\hat{S}^{(\text{WOSA})}(f)\} \approx S(f)$ for all f of interest.

Next, we consider the variance of $\hat{S}^{(\text{WOSA})}(f)$ under the assumption that this estimator is approximately unbiased. Since this estimator is the average of direct spectral estimators, we can obtain its variance based upon an application of Exercise [2.8]:

$$\text{var}\{\hat{S}^{(\text{WOSA})}(f)\} = \frac{1}{N_B^2} \sum_{j=0}^{N_B-1} \text{var}\{\hat{S}_{jn+1}^{(d)}(f)\}$$

$$+ \frac{2}{N_B^2} \sum_{j<k} \text{cov}\{\hat{S}_{jn+1}^{(d)}(f), \hat{S}_{kn+1}^{(d)}(f)\}.$$

For $0 < f < f_{(N)}$, Equation (223c) tells us that, for all j,

$$\text{var}\{\hat{S}_{jn+1}^{(d)}(f)\} \approx S^2(f).$$

Under the assumption that $S(\cdot)$ is locally constant about f and that f is not too close to 0 or the Nyquist frequency, we can use an argument similar to that given in the Comments and Extensions to Section 6.6 to obtain

$$\text{cov}\{\hat{S}_{jn+1}^{(d)}(f), \hat{S}_{kn+1}^{(d)}(f)\} \approx S^2(f) \left| \sum_{t=1}^{N_S} h_t h_{t+|k-j|n} \right|^2, \qquad (292a)$$

where $h_t \equiv 0$ for $t > N_S$ (see Exercise [6.22], Welch, 1967, or Thomson, 1977, Section 3.3). Note that, if the jth and kth blocks have no data values in common, the summation above is identically zero, and hence so is the covariance. We thus have

$$\text{var}\{\hat{S}^{(\text{WOSA})}(f)\} \approx \frac{S^2(f)}{N_B} \left(1 + \frac{2}{N_B} \sum_{j<k} \left| \sum_{t=1}^{N_S} h_t h_{t+|k-j|n} \right|^2\right)$$

$$= \frac{S^2(f)}{N_B} \left(1 + 2 \sum_{m=1}^{N_B-1} \left(1 - \frac{m}{N_B}\right) \left| \sum_{t=1}^{N_S} h_t h_{t+mn} \right|^2\right).$$

From an argument similar to that leading to Equation (254a), the equivalent degrees of freedom ν for $\hat{S}^{(\text{WOSA})}(f)$ are

$$\nu = \frac{2\left(E\{\hat{S}^{(\text{WOSA})}(f)\}\right)^2}{\text{var}\{\hat{S}^{(\text{WOSA})}(f)\}} \approx \frac{2N_B}{1 + 2 \sum_{m=1}^{N_B-1} \left(1 - \frac{m}{N_B}\right) \left| \sum_{t=1}^{N_S} h_t h_{t+mn} \right|^2}.$$

$$\qquad (292b)$$

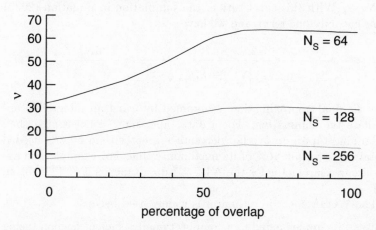

Figure 293. Degrees of freedom ν versus percentage of block overlap for sample size $N = 1024$, a Hanning data taper and block sizes $N_S = 64$, 128 and 256 (top, middle and bottom curves, respectively).

As an example, suppose that we have a time series with sample size $N = 1024$ that we want to break up into blocks of size N_S equal to either 64, 128 or 256 and that we plan to use the Hanning data taper on each block, i.e.,

$$h_t = \left(\frac{2}{3(N_S + 1)} \right)^{1/2} \left[1 - \cos\left(\frac{2\pi t}{N_S + 1} \right) \right]$$

(see Exercise [6.10] with N_S substituted for N). We can use Equation (292b) to compute approximately the equivalent degrees of freedom ν for $\hat{S}^{(\text{WOSA})}(f)$ for all acceptable shift factors n ranging from N_S (i.e., nonoverlapping blocks) down to 1 (i.e., maximally overlapped blocks, some of which have $N_S - 1$ data values in common). Figure 293 shows this approximation to ν versus the percentage of overlap $(1 - n/N_S) \times 100\%$ for block sizes $N_S = 64$ (top curve), 128 (middle) and 256 (bottom). We see that, for a fixed percentage of overlap, ν is inversely proportional to N_S; however, in practical situations, we cannot just decrease the block size arbitrarily to increase ν because typically the bias in $\hat{S}^{(\text{WOSA})}(f)$ increases as N_S decreases. We also see that each curve increases as the percentage of overlap increases from 0% on up to about 70%, after which it decreases slightly (this decrease is counterintuitive and is possibly an artifact arising because Equation (292b) is based upon an approximation to var$\{\hat{S}^{(\text{WOSA})}(f)\}$). For all three curves, the value of ν at 50% overlap is within 10% of its maximum value, so a common, operationally convenient recommendation in the engineering literature is to use a 50% overlap with the Hanning data taper, i.e., to set

$n = N_S/2$. With this substitution, the summation in Equation (292b) over m has only one term, and we have

$$\nu \approx \frac{2N_B}{1 + 2\left(1 - 1/N_B\right)\left|\sum_{t=1}^{N_S/2} h_t h_{t+N_S/2}\right|^2} \approx \frac{36N_B^2}{19N_B - 1}, \qquad (294)$$

where the final approximation is obtained by using an integral to approximate the summation. For a data taper more concentrated than the Hanning taper, an overlap percentage greater than 50% is needed to obtain ν to within 10% of its maximum value. For example, the appropriate overlap factor for the $NW = 4$ dpss taper is 65% (Thomson, 1977).

The WOSA spectral estimator is widely used because

[1] it can be implemented in a computationally efficient fashion (using FFTs of a fixed size);

[2] it can efficiently handle very long time series;

[3] there exist commercially available special purpose instruments – spectrum analyzers – that display spectral estimates based essentially upon WOSA; and

[4] a robust sdf estimator can be devised similar in spirit to WOSA except that the $\hat{S}_{jn+1}^{(d)}(f)$ terms in Equation (291b) are not just averaged together but rather are combined in such a fashion as to downweight individual estimators corresponding to blocks contaminated by outliers (for details, see Chave *et al.*, 1987).

As mentioned above, the only real potential problem with WOSA is bias caused by an insufficient block size N_S, a condition that can be guarded against if it is possible to vary the block size. The use of WOSA is illustrated in the next section in the examples concerning the AR(2) and AR(4) time series.

Comments and Extensions to Section 6.17

[1]   We have developed the theory for this section under the usual implicit assumption that the mean value of the stationary process $\{X_t\}$ is known to be zero. When the mean value is unknown (the usual case in practice), one appropriate modification is to substitute $X_l - \bar{X}$, $\ldots$, $X_{N_S+l-1} - \bar{X}$ for X_l, $\ldots$, X_{N_S+l-1} in Equation (291a), where (as usual) $\bar{X}$ is the sample mean of X_1, $\ldots$, X_N. With this modification, it is an easy exercise to show that $\hat{S}^{(\text{WOSA})}(0)$ is related to the variance of the sample means of the individual blocks of data, an interpretation of $S(0)$ that is in keeping with Equation (188).

A second way of handling an unknown process mean is to use the sample means of each block of data instead of the sample mean $\bar{X}$ of all the data. Thus, to compute $\hat{S}_l^{(d)}(f)$, we would substitute $X_l - \bar{X}_l$,

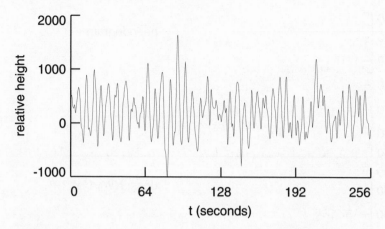

Figure 295. Ocean wave data. There are $N = 1024$ samples taken $\Delta t = 1/4$ second apart.

$\ldots, X_{N_S+l-1} - \bar{X}_l$ for $X_l, \ldots, X_{N_S+l-1}$ in Equation (291a), where $\bar{X}_l \equiv \sum_{t=1}^{N_S} X_{t+l-1}/N_S$. This procedure is particularly useful in situations where $\hat{S}_l^{(d)}(f)$ is computed in near real-time (as is done in commercial spectrum analyzers). The relative merit of these two ways of handling the process mean is evidently an open question.

[2] The acvs estimator $\{\hat{s}_\tau^{(\text{WOSA})}\}$ corresponding to $\hat{S}^{(\text{WOSA})}(\cdot)$ can be obtained readily since the sequences $\{\hat{s}_\tau^{(\text{WOSA})} : \tau = -(N_S-1), \ldots, N_S\}$ and $\{\hat{S}^{(\text{WOSA})}(\tilde{f}_k) : k = -(N_S-1), \ldots, N_S\}$ are a Fourier transform pair – here $\tilde{f}_k \equiv k/(2N_S\,\Delta t)$. If so desired, we can then smooth $\hat{S}^{(\text{WOSA})}(\cdot)$ by applying an appropriate lag window (for details, see Nuttall and Carter, 1982).

6.18 Examples of Nonparametric Spectral Analysis

• *Ocean wave data*

Figure 295 shows a plot of a time series recorded in the Pacific Ocean by a wave-follower (data courtesy of A. Jessup, Applied Physics Laboratory, University of Washington). As the wave-follower moves up and down with the waves in the ocean, it measures the surface displacement (i.e., sea level) as a function of time. The frequency response of the wave-follower is such that – mainly due to its inertia – frequencies higher than 1 Hz cannot be reliably measured. The data were originally recorded using an analog device. The signal was then low-pass filtered in analog form using an antialiasing filter with a cutoff of approximately 1 Hz and then sampled every 1/4 second, yielding a Nyquist frequency of $f_{(N)} = 2$ Hz. The plotted series consists of a 256 second portion of this data, so there are $N = 1024$ data points in all.

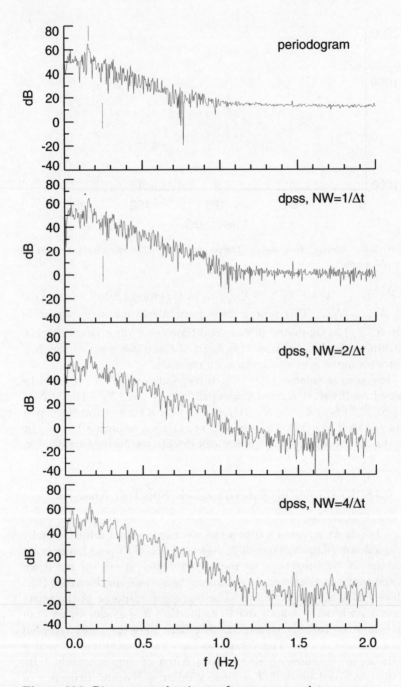

Figure 296. Direct spectral estimates for ocean wave data.

Figure 296 shows four direct spectral estimates for this ocean wave data, each evaluated on the grid of 512 Fourier frequencies. The top plot is the periodogram $\hat{S}^{(p)}(\cdot)$, while the bottom three plots are direct spectral estimates $\hat{S}^{(d)}(\cdot)$ using a dpss data taper with NW parameters of $1/\Delta t$, $2/\Delta t$ and $4/\Delta t$. For the periodogram the time series was centered by subtracting the sample mean $\bar{X} = 209.1$; for the other three estimates the series was centered using $\tilde{\mu}$ of Equation (217a). The data tapers were normalized using

$$\sum_{t=1}^{N} h_t^2 \left(X_t - \tilde{\mu} \right)^2 = \frac{1}{N} \sum_{t=1}^{N} \left(X_t - \bar{X} \right)^2$$

(see Equation (216)) rather than $\sum_{t=1}^{N} h_t^2 = 1$ to ensure that the integral of each spectral estimate is equal to the sample variance.

All four spectral estimates show a broad, low frequency peak at 0.160 Hz, corresponding to a period of 6.2 seconds (the location of this frequency is marked by a thin vertical line in the upper left-hand corner of the top plot). While the dominant features of the time series can be attributed to this broad peak and other features in the frequency range 0 to 0.2 Hz, the data were actually collected to investigate whether the rate at which the sdf decreases over the range 0.2 to 1.0 Hz is in fact consistent with calculations based upon a physical model. The range from 1 to 2 Hz is of little physical interest because it is dominated by instrumentation and preprocessing (i.e., inertia in the wave-follower and the antialiasing filter); nonetheless, it is of operational interest to examine this portion of the sdf to check that the spectral levels and shape are in accord with the frequency response claimed by the manufacturer of the wave-follower, the transfer function of the antialiasing filter, and a rough guess at the sdf for ocean waves from 1 to 2 Hz.

An examination of the high frequency portions of the sdf's in Figure 296 shows evidence of bias due to leakage in both the periodogram and the direct spectral estimate with the $NW = 1/\Delta t$ data taper. The evidence is twofold. First, the levels of these two spectral estimates at high frequencies are considerably higher than those for the $NW = 2/\Delta t$ and $NW = 4/\Delta t$ direct spectral estimates (by about 25 dB for the periodogram and 10 dB for the $NW = 1/\Delta t$ spectral estimate). Second, the local variability in the periodogram is markedly less in the high frequencies as compared to the low frequencies, an indication of leakage we have seen previously (see Figure 227). The same is true to a lesser extent for the $NW = 1/\Delta t$ spectral estimate.

As an interesting aside, in Figure 298 we have replotted the periodogram but now over a grid of frequencies twice as fine as the Fourier frequencies – recall that the plots of Figure 296 involve just the Fourier frequencies. This new plot shows an *increase* in the local variability of

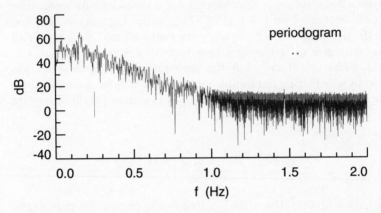

Figure 298. Periodogram for ocean wave data. Here the periodogram is plotted over a grid of frequencies twice as fine as the Fourier frequencies, whereas in the top plot of Figure 296 it is plotted over just the Fourier frequencies. The strange appearance of this plot is due to the fact that, because of leakage, the periodogram has a nearly constant value of approximately 12 dB at all high Fourier frequencies, whereas its values are considerably below 12 dB over most of the grid of high frequencies halfway between the Fourier frequencies.

the periodogram in the high frequencies as compared to the low frequencies. If we were to expand the frequency scale on the plot, we would see that the increased variability is in the form of a ringing; i.e., the sdf alternates between high and low values. Had we seen this plot first, this ringing would have told us that leakage might be a problem. The key point here is that, when plotted on a logarithmic scale (such as decibels), the periodogram and other direct spectral estimators should exhibit approximately the same local variability across all frequencies – if this is not so, it is important to understand why not. Leakage causes the ringing here, but other potential causes are outliers (see, e.g., Figure 21 of Martin and Thomson, 1982) and echoes (Bogert *et al.*, 1963).

A useful guide in assessing the variability in a direct spectral estimate is to plot crisscrosses such as shown in the lower left-hand portions of all the plots in Figures 296 and 298 (since the periodogram is a direct spectral estimate, the following results for $\hat{S}^{(d)}(\cdot)$ of course also hold for $\hat{S}^{(p)}(\cdot)$). The vertical height of each crisscross is the length (in decibels) of a 95% confidence interval for $10 \log_{10}(S(f))$ based upon $10 \log_{10}(\hat{S}^{(d)}(f))$, whereas the horizontal width is a rough measure of the distance in frequency between adjacent uncorrelated spectral estimates. These quantities are computed in the following fashion. For $0 < f < f_{(N)}$, $\hat{S}^{(d)}(f)$ is approximately distributed as a random variable with a chi-square distribution with $\nu = 2$ degrees of freedom times $S(f)/2$ (see Equation (223b)). Note that this result holds no matter what the associated data taper is. By an argument identical to that leading to Equation (255b), a 95% confidence interval for $S(f)$ is given by

$$\left[\frac{2\hat{S}^{(d)}(f)}{Q_2(0.975)}, \frac{2\hat{S}^{(d)}(f)}{Q_2(0.025)} \right].$$

By definition, $Q_2(0.025)$ is the lower 2.5% percentage point of the χ_2^2 distribution, while $Q_2(0.975)$ is the upper 2.5% percentage point. Table 256 gives $Q_2(0.025) = 0.0506$ and $Q_2(0.975) = 7.3778$, so a 95% confidence interval for $S(f)$ based upon $\hat{S}^{(d)}(f)$ is given by

$$\left[\frac{2\hat{S}^{(d)}(f)}{7.3778}, \frac{2\hat{S}^{(d)}(f)}{0.0506} \right] = \left[0.271 \times \hat{S}^{(d)}(f), 39.4 \times \hat{S}^{(d)}(f) \right].$$

Since $10 \log_{10}(2/7.3778) = -5.7$ dB and $10 \log_{10}(2/0.0506) = 16.0$ dB, a corresponding 95% confidence interval for $10 \log_{10}(S(f))$ is given by

$$\left[-5.7 + 10 \log_{10}(\hat{S}^{(d)}(f)), 16.0 + 10 \log_{10}(\hat{S}^{(d)}(f)) \right].$$

Note that the length of this confidence interval (21.7 dB) is independent of the value of $10 \log_{10}(\hat{S}^{(d)}(f))$. The vertical portions of the crisscrosses in Figures 296 and 298 would delineate a 95% confidence interval for $10 \log_{10}(S(1/4))$ if $10 \log_{10}(\hat{S}^{(d)}(1/4))$ were equal to 0. We can obtain a confidence interval for any particular $10 \log_{10}(S(f))$ by mentally moving the crisscross to the point on the plot corresponding to $10 \log_{10}(\hat{S}^{(d)}(f))$.

For the horizontal width of the crisscross, we want to display a number $\delta > 0$ such that $\hat{S}^{(d)}(f)$ and $\hat{S}^{(d)}(f + \delta)$ are approximately uncorrelated (under the restrictions both that f is positive and not too close to 0 and that $f + \delta$ is not too close to $f_{(N)}$, where here 'not too close' means $f \geq \delta$ and $f + \delta \leq f_{(N)} - \delta$). The statistical theory of Section 6.6 suggests that an appropriate measure for the periodogram is $\delta = 1/(N \Delta t)$, whereas a convenient – but somewhat conservative – measure for a direct spectral estimate based upon an $NW = k/\Delta t$ dpss data taper is $\delta = 2W = 2k/(N \Delta t)$ (cf. Figure 232). Since $\Delta t = 1/4$ second and $N = 1024$, the widths of the horizontal portions of the crisscrosses in the four plots of Figure 296 are thus, from top to bottom, 0.004 Hz (just barely visible!), 0.008 Hz, 0.016 Hz and 0.031 Hz.

Under the assumption that $S(\cdot)$ is slowly varying, these crisscrosses can give us a rough idea of the local variability we can expect to see in $\hat{S}^{(d)}(\cdot)$ across frequencies if all of the assumptions behind spectral analysis hold. If f and f' are closer together than the width of the crisscross, there should not be much variation between $\hat{S}^{(d)}(f)$ and $\hat{S}^{(d)}(f')$; on the other hand, if f and f' are farther apart than this width, $\hat{S}^{(d)}(f)$ and $\hat{S}^{(d)}(f')$ should exhibit a variation consistent with the height of the crisscross. For example, the high frequency portion of the periodogram in the upper plot of Figure 296 shows much less variability from frequency to frequency than we would expect from the height of the crisscross, while, on the finer scale of Figure 298, the local variability is both too large and on too fine a scale (i.e., rapid variations occur on a scale less than $1/(N \Delta t)$, the grid size of the Fourier frequencies). In contrast, the low frequency portion of the periodogram does not exhibit any gross departure from reasonable variability.

If we now concentrate on the bottom two plots of Figure 296, we see that the $NW = 2/\Delta t$ and $NW = 4/\Delta t$ spectral estimates look quite similar overall. Increasing the degree of tapering beyond that given by the $NW = 2/\Delta t$ data taper thus does not appear to gain us anything. We can conclude that a dpss data taper with $NW = 2/\Delta t$ is sufficient to control leakage over all frequencies for the ocean wave data (if, however, we only want to study the sdf for frequencies less than 1 Hz, the $NW = 1/\Delta t$ spectral estimate also has acceptable bias properties).

A careful examination of the $NW = 2/\Delta t$ direct spectral estimate shows one interesting feature that is barely evident in the periodogram, namely, a small peak at $f = 1.469$ Hz about 15 dB above the local background. This peak evidently corresponds to a 0.7 second resonance in the wave-follower. That this peak is not attributable to just the statistical fluctuations of direct spectral estimators can be determined by applying the F-test for periodicity discussed in Section 10.11; however, a rough indication that it is significant can be seen by mentally moving the crisscross on the plot to this peak and noting that the lower end of a 95% confidence interval for $S(1.469)$ is about 10 dB above a reasonable guess at the local level of the sdf. Other than this peak (and a smaller, less significant one at 1.715 Hz), the spectral estimate is relatively flat between 1.0 and 2.0 Hz, a conclusion that does not disagree with factors known to influence this portion of the sdf.

We now turn our attention to the frequency range 0.2 and 1.0 Hz. To illustrate a point concerning smoothing window leakage below, we will continue to concentrate on the $NW = 2/\Delta t$ direct spectral estimate even though the $NW = 1/\Delta t$ estimate is adequate for this range of frequencies. We want to smooth this direct spectral estimate in order to obtain a better visual representation of the rate at which the power in the sdf decreases over this frequency range. The solid curves in the plots of Figure 301 show three different lag window spectral estimates $\hat{S}^{(lw)}(\cdot)$, all of which are smoothed versions of the $NW = 2/\Delta t$ direct spectral estimate (the dots). The Parzen lag window with parameters $m = 150$ and $m = 55$ is used in the top two plots, whereas the Daniell lag window with parameter $m = 30$ is used in the bottom plot.

To help us assess these lag window estimates, we display a crisscross in each plot with an interpretation similar to the crisscrosses for

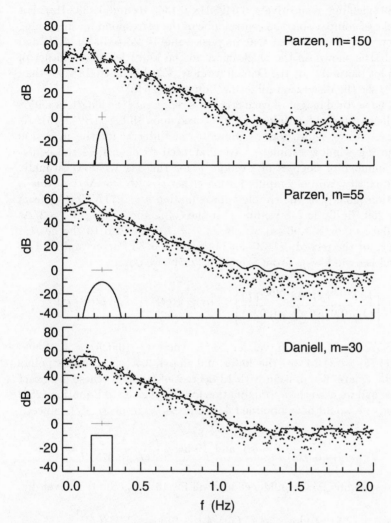

Figure 301. Lag window spectral estimates $\hat{S}^{(lw)}(\cdot)$ for ocean wave data (solid curves). The dots show the direct spectral estimate that is smoothed – it is the same in each case and is also shown in the third plot of Figure 296.

the direct spectral estimates in Figure 296. The horizontal widths of the crisscrosses in Figure 301 now depict the smoothing window bandwidth B_W, which, however, can again be interpreted as a rough measure of the distance in frequency between adjacent uncorrelated spectral estimates. This bandwidth can be computed either via its definition in Equation (242c) (yielding, from top to bottom, 0.0494 Hz, 0.1349 Hz and 0.1337 Hz) or via the simple approximating formulae given in Table 269 (yielding, respectively, 0.0493 Hz, 0.1345 Hz and 0.1333 Hz). For the sake of comparison, the central lobe of the corresponding smoothing window $W_m(\cdot)$ (shifted so that its peak value is plotted at $1/4$ Hz and -10 dB) is shown as the thick curve in the lower left-hand corner of each plot (actually, for the Daniell window, $V_m(\cdot)$ is plotted rather than $W_m(\cdot)$; see the discussion following Equation (264b)).

The vertical height of each crisscross represents the length of a 95% confidence interval for $10 \log_{10}(S(f))$ based upon $10 \log_{10}(\hat{S}^{(lw)}(f))$. As an example, we show how to compute this interval for the $m = 150$ Parzen lag window estimate. We first need to determine the equivalent number of degrees of freedom ν for this lag window estimate. This quantity can be computed using either $\nu = 2NB_W \Delta t/C_h$ (this is Equation (255a)) or the simple approximation $\nu = 3.71N/(mC_h)$ from Table 269. In Table 248 we find that the C_h factor for an $NW = 2/\Delta t$ dpss data taper is 1.96, so we obtain $\nu = 13$ (rounded to the nearest integer; for the record, $\nu = 35$ for both the $m = 55$ Parzen and $m = 30$ Daniell lag windows). From Equation (255b) we obtain

$$\left[\frac{13\,\hat{S}^{(lw)}(f)}{Q_{13}(0.975)}, \frac{13\,\hat{S}^{(lw)}(f)}{Q_{13}(0.025)}\right] = \left[0.53\,\hat{S}^{(lw)}(f), 2.60\,\hat{S}^{(lw)}(f)\right]$$

as a 95% confidence interval for $S(f)$, where $Q_{13}(0.025) = 5.01$ and $Q_{13}(0.975) = 24.74$ are the lower and upper 2.5% percentage points of a chi-square distribution with 13 degrees of freedom (these are exact values; had we used the approximation for $Q_\nu(p)$ given in Equation (256) instead, we would have obtained 4.99 and 24.74, respectively). Since

$$10 \log_{10}\left(\frac{13}{Q_{13}(0.025)}\right) = 4.1 \text{ and } 10 \log_{10}\left(\frac{13}{Q_{13}(0.975)}\right) = -2.8,$$

a corresponding 95% confidence interval for $10 \log_{10}(S(f))$ is given by

$$\left[-2.8 + 10 \log_{10}(\hat{S}^{(lw)}(f)), 4.1 + 10 \log_{10}(\hat{S}^{(lw)}(f))\right].$$

For example, at $f = 1.0$ Hz, we have $10 \log_{10}(\hat{S}^{(lw)}(1.0)) = 6.3$ dB, so a 95% confidence interval for $10 \log_{10}(S(1.0))$ is $[3.5 \text{ dB}, 10.4 \text{ dB}]$. Note that the length of the confidence interval (6.9 dB) is independent of the value of $10 \log_{10}(\hat{S}^{(lw)}(f))$. This length is plotted as the height of the crisscross in the top plot of Figure 301. A comparison between the crisscrosses in the top two plots shows the inherent tradeoff between variability (as measured by the length of a 95% confidence interval) and the smoothing window bandwidth.

Let us now consider in detail the $m = 150$ and $m = 55$ Parzen lag window estimates (solid curves in top two plots of Figure 301). The value of $m = 150$ (top plot) was chosen to achieve a smoothing window bandwidth comparable to a visual estimate of the width of the broad peak in $\hat{S}^{(d)}(\cdot)$ at $f = 0.160$ Hz. Note that this choice produces a smoothed version of $\hat{S}^{(d)}(\cdot)$ that smears out this peak somewhat (and the one at $f = 1.469$ Hz) but that is arguably still not smooth enough between 0.2 and 1.0 Hz, the range of frequencies of main interest. A value of $m = 55$ (middle plot) produces a much smoother looking estimate over this range, one that has reasonable bias and variance properties there and hence is useful for comparison with physical theory.

The $m = 55$ Parzen lag window estimate, however, does not do as well for frequencies outside 0.2 to 1.0 Hz. The peaks at 0.160 and 1.469 Hz have been smeared out almost to the point of being unrecognizable. There is also evidence of smoothing window leakage from frequencies 1.0 Hz to 2.0 Hz: the solid curve in the middle plot of Figure 301 should be a smoothed version of the $NW = 2$ direct spectral estimate (the dots), but note that this curve seems consistently *higher* than the dots (by contrast, this is not the case in the high power portion of the spectral estimate, which in this example corresponds to low frequency values). We can confirm our suspicions by computing a lag window estimator using the Daniell smoothing window with a smoothing window bandwidth B_W as close as possible to that of the $m = 55$ Parzen smoothing window – recall that the Daniell smoothing window is free of smoothing window leakage. The appropriate choice for the Daniell smoothing window is $m = 30$, and the corresponding estimate is shown in the bottom plot of Figure 301. The estimate has much better bias properties in the 1.0 to 2.0 Hz region, but, as expected, it is much less smooth than the Parzen estimate (this defect in the Daniell estimate could be corrected by smoothing it using a Parzen lag window with a narrow bandwidth). The important point here is that we should always compare $\hat{S}^{(lw)}(\cdot)$ with its corresponding $\hat{S}^{(d)}(\cdot)$ to make sure that the former is a reasonably smoothed version of the latter. (We return to an analysis of these data later on in Section 7.5, where we illustrate the multitaper approach, and in Section 9.10, where we utilize parametric models through prewhitening.)

● *Ice profile data*

Let us now consider two segments of ice profile data from the arctic regions (this analysis is adapted from an unpublished 1991 report by W. Fox, Applied Physics Laboratory, University of Washington). Ice morphology was measured by moving a profiling device along at a constant speed in a straight line roughly parallel to the surface of the ice. The distance between the profiler and the ice surface was recorded at a constant rate in time to obtain measurements spaced every $\Delta t = 1.7712$ meters apart. These measurements can thus be regarded as a 'time' series related to the variations in the surface height of the

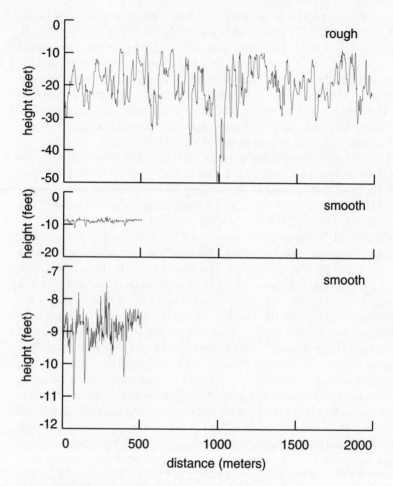

Figure 304. Two segments of ice profile data, a relatively rough one (top plot) and a shorter, relatively smooth one (middle plot – also on the bottom plot with an expanded scale). The 'time' values are really distances measured in meters, while the time series values are in feet. The distance between observations is $\Delta t = 1.7712$ meters. The rough profile has 1121 data points, while the smooth one has 288 points.

ice over distance.

The two top plots of Figure 304 show two different segments of this data on the same scale. The first segment (upper plot) has $N_r = 1121$ data points, while there are $N_s = 288$ points in the second segment (middle). These two series are, respectively, from a fairly rough profile and a fairly smooth profile. The bottom plot shows the smooth profile on an expanded scale.

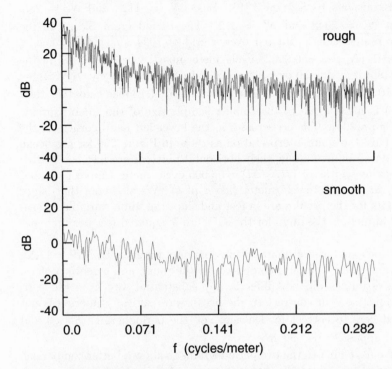

Figure 305. Periodograms for ice profile data.

One of the reasons for collecting the ice profile data was to develop a way of generating representative artificial profiles for use in simulations. It is clear from Figure 304 that it is unrealistic to regard both of these ice profiles as portions of two realizations from a single stationary process – the mean values and the variances of the two series are quite different. Superficially, however, the series in the bottom and top plots appear to have similar 'spikiness,' so one simple model worth investigating is that the spectral content of the two series is the same except for a scaling factor; i.e., if we let $S_r(\cdot)$ and $S_s(\cdot)$ represent the sdf's for the rough and smooth series, we have $S_r(f) = cS_s(f)$ over all f for some constant $c > 0$. Here we informally test this hypothesis by comparing estimated spectra for the two series (a formal statistical test can be based upon spectral ratios; for details, see Coates and Diggle, 1986, or Diggle, 1990).

Figure 305 shows the periodograms $\hat{S}_r^{(p)}(\cdot)$ and $\hat{S}_s^{(p)}(\cdot)$ for the rough and smooth profiles. In each case, these were computed using an FFT algorithm after subtracting the sample means $\bar{X}_r = -19.92$ and $\bar{X}_s = -8.93$ from their associated series and then padding the centered series with zeros so that the padded series have a number of points – say N'_r and N'_s – equal to a power of 2 (see item [1] in the Comments and Extensions to Section 3.10). Since $N_r = 1121$ and $N_s = 288$, we let $N_r' = 2048$ and $N_s' = 512$. The periodogram $\hat{S}_r^{(p)}(\cdot)$ in the upper plot is thus evaluated over a grid of 1024 nonzero frequencies spaced $1/(N_r' \Delta t) = 0.00028$ cycle/meter apart, whereas $\hat{S}_s^{(p)}(\cdot)$ in the lower plot is evaluated over 256 nonzero frequencies spaced $1/(N_s' \Delta t) = 0.00110$ cycle/meter apart. The fact that $\hat{S}_s^{(p)}(\cdot)$ looks smoother than $\hat{S}_r^{(p)}(\cdot)$ is merely due to the smaller sample size of the smooth profile (cf. Figure 225). The crisscrosses in the lower left-hand corners of the plots have the same interpretation as those in Figure 296 for the ocean wave data; unfortunately, their horizontal portions are just barely visible, having widths of $1/(N_r \Delta t) = 0.0005$ cycle/meter (upper plot) and $1/(N_s \Delta t) = 0.002$ cycle/meter (lower plot). We also note that, since the units for the profiles are in feet and since the 'time' variable is measured in meters, the units for the sdf's are in squared feet per cycle per meter.

Direct spectral estimates using dpss data tapers with $NW = 1/\Delta t$, $2/\Delta t$ and $4/\Delta t$ were also computed for the rough and smooth profiles. These estimates did not differ in any substantial way from the periodograms, so – in contrast to the ocean wave series – tapering is not needed here to correct for leakage; i.e., the periodograms $\hat{S}_r^{(p)}(\cdot)$ and $\hat{S}_s^{(p)}(\cdot)$ appear to be bias free.

We next smooth the two periodograms to allow a better comparison between them. Since the dynamic ranges of the periodograms are not very large (less than about 40 dB), smoothing window leakage should not be an issue here, so we can safely use the Parzen lag window. The two periodograms suggest that the underlying sdf's are dominantly unimodal, so we can use the estimator $\tilde{B}_T$ of Equation (280) to compute the bandwidth for the two profiles. For the rough profile we obtain $\tilde{B}_T = 0.0239$, and for the smooth profile, $\tilde{B}_T = 0.1069$. If we follow the advice of Section 6.14 and select the lag window parameter m for each profile such that the resulting smoothing window bandwidth B_W is equal to $\tilde{B}_T/2$, and if we use the approximation $B_W = 1.85/(m \Delta t)$ from Table 269 for the Parzen window, we obtain

$$m = \frac{3.7}{\tilde{B}_T \Delta t} = \begin{cases} 88, & \text{for the rough profile;} \\ 20, & \text{for the smooth profile.} \end{cases}$$

These resulting lag window estimators $\hat{S}_r^{(lw)}(\cdot)$ and $\hat{S}_s^{(lw)}(\cdot)$ are shown as the solid curves in the top two plots of Figure 307, along with the corresponding periodograms (the dots) and the usual crisscrosses depicting B_W and the lengths in decibels of 95% confidence intervals for the true sdf's at a given frequency. Using the approximation in Table 269, the equivalent degrees of freedom ν are 47 for $\hat{S}_r^{(lw)}(\cdot)$ and 53 for $\hat{S}_s^{(lw)}(\cdot)$,

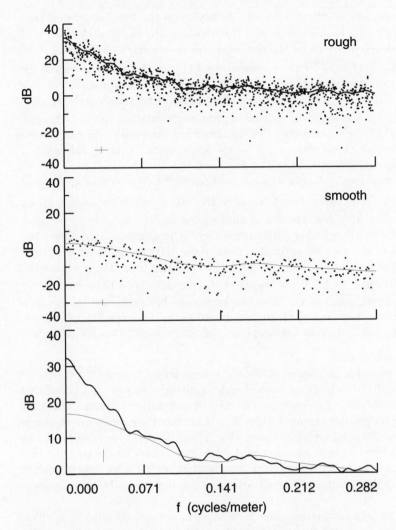

Figure 307. Lag window spectral estimates for ice profile data. For both profiles, the direct spectral estimate that is smoothed is the periodogram (the dots in the top two plots). Parzen lag window estimates (solid curves) are shown with $m = 88$ for the rough profile (top plot) and $m = 20$ for the smooth profile (middle plot). In the bottom plot, $\hat{S}_r^{(lw)}(\cdot)$ is reproduced from the top plot as the thick solid curve, while the thin solid curve shows $\hat{S}_s^{(lw)}(\cdot)$ after it has been shifted up by 13.5 dB. The thin vertical line in the lower left-hand corner of the bottom plot indicates 3.4 dB, which is close to the length of a 95% confidence interval both for $10 \log_{10}(S_r(f))$ based upon $10 \log_{10}(\hat{S}_r^{(lw)}(f))$ and for $10 \log_{10}(S_s(f))$ based upon $10 \log_{10}(\hat{S}_s^{(lw)}(f))$.

a convenient result because the variability in the two lag window estimates is thus almost the same – this explains why the lengths of the 95% confidence intervals for the two estimates are similar (3.5 dB for $\hat{S}_r^{(lw)}(\cdot)$ and 3.3 dB for $\hat{S}_s^{(lw)}(\cdot)$). A comparison between the solid curves and the dots indicates that in both cases the lag window estimate is a reasonably smoothed version of the corresponding periodogram (although arguably $\hat{S}_s^{(lw)}(\cdot)$ is too smooth: there is a prominent hump in the periodogram near 0.17 cycle per meter that has been smeared out). We can further verify that these choices for m are appropriate by using the window closing technique described in Section 6.13.

The bottom plot of Figure 307 shows $\hat{S}_r^{(lw)}(\cdot)$, the thick solid curve, and $\hat{S}_s^{(lw)}(\cdot)$, the thin curve, after the latter has been moved up by 13.5 dB – this corresponds to multiplying $\hat{S}_s^{(lw)}(\cdot)$ by a factor of 22.4. We see that, with this adjustment, there is reasonably good agreement between the two sdf estimates from about 0.05 cycle/meter up to the Nyquist frequency; however, for frequencies less than 0.05 cycle/meter, $\hat{S}_r^{(lw)}(\cdot)$ and $22.4\hat{S}_s^{(lw)}(\cdot)$ diverge, with the difference between the two sdf's being much larger than can be reasonably explained by the variability in the spectral estimates. Based upon these qualitative results, we can conclude that the simple model $S_r(f) = cS_s(f)$ is not adequate.

- *AR(2) data*

The top plot of Figure 45 shows a segment of length $N = 1024$ of one realization from the second-order autoregressive process defined by Equation (45). The periodogram $\hat{S}^{(p)}(\cdot)$ and a direct spectral estimate using a dpss data taper with $NW = 4$ for this time series are shown in Figure 226 (along with the true sdf). The differences between these two direct spectral estimates do not give any indication that the periodogram suffers from leakage, so we can assume that $\hat{S}^{(p)}(\cdot)$ is an approximately unbiased estimator of the sdf (even if we did not already know the true sdf!).

To help us interpret the periodogram, we now consider smoothing it using several different smoothing windows. We use the Parzen lag window since the dynamic range of $\hat{S}^{(p)}(\cdot)$ indicates that there should be little problem with smoothing window leakage. Figure 309 shows three Parzen lag window estimates $\hat{S}^{(lw)}(\cdot)$ with $m = 8$, 16 and 64 (thin solid curves from top to bottom plots) along with the periodogram (dots) and the true sdf (thick solid curves). Table 269 indicates that the smoothing window bandwidth for $\hat{S}^{(lw)}(\cdot)$ is approximately given by $B_W = 1.85/m$ (since $\Delta t = 1$ here) and that the equivalent degrees of freedom are approximately given by $\nu = 3799.0/m$ (since $N = 1024$ and $C_h = 1$ for the rectangular data taper – see Table 248). For $m = 8$, 16 and 64 we obtain, respectively, $B_W = 0.2312$, 0.1156 and 0.0289 and $\nu = 474.9$, 237.4 and 59.4. As usual, the crisscrosses emanating from the lower left-

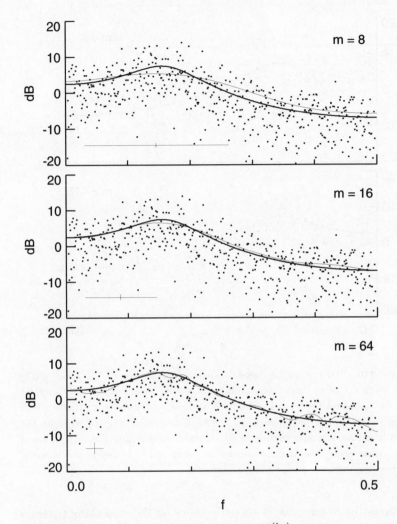

Figure 309. Lag window spectral estimates $\hat{S}^{(lw)}(\cdot)$ for AR(2) series using a Parzen smoothing window with $m = 8$, 16 and 64 (thin curves). The thick curve on each plot is the true AR(2) sdf, while the dots are the periodogram at the Fourier frequencies. The width of the crisscross emanating from each lower left-hand corner gives the smoothing window bandwidth B_W, while its height gives the length of a 95% confidence interval for $10 \log_{10}(S(f))$.

hand corner of each plot depict the smoothing window bandwidth and the length of a 95% confidence interval for $10 \log_{10}(S(f))$ based upon $10 \log_{10}(\hat{S}^{(lw)}(f))$.

A study of the three $\hat{S}^{(lw)}(\cdot)$ in Figure 309 in comparison to the true

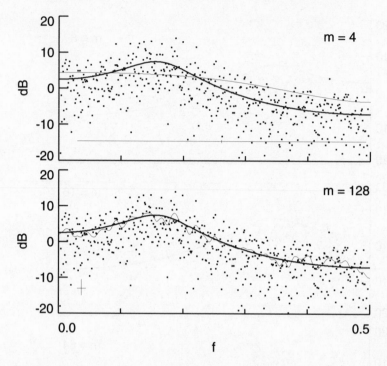

Figure 310. Two more lag window spectral estimates $\hat{S}^{(lw)}(\cdot)$ for AR(2) series using a Parzen smoothing window, now with $m = 4$ and 128 (thin solid curves). Note that the $m = 4$ estimate is a considerably oversmoothed version of the periodogram, particularly in the high frequency region, whereas the $m = 128$ estimate is an example of considerable undersmoothing. In terms of mean square error, bias is the dominant factor in the $m = 4$ estimate, whereas variance dominates the $m = 128$ estimate.

sdf shows the importance of properly choosing the smoothing parameter m: whereas the $m = 8$ estimate is oversmoothed and the $m = 64$ estimate is undersmoothed, the $m = 16$ estimate agrees with the true sdf quite well. Note in particular that a 95% confidence interval for $10 \log_{10}(S(f))$ based upon the oversmoothed $m = 8$ estimate would in fact *fail* to include the true value of the sdf over the majority of the frequencies between 0 and 1/2, whereas the opposite is true for confidence intervals based upon the $m = 16$ and $m = 64$ estimates. Note also that, had we not known the true sdf, it would have been difficult to pick one of these $\hat{S}^{(lw)}(\cdot)$ over the other two on the basis of window closing (see Section 6.13) since all three are to the eye reasonably smoothed versions of the periodogram! On the other hand, the maximum difference between any two of the three estimates is less than 3.3 dB (a factor of 2.1),

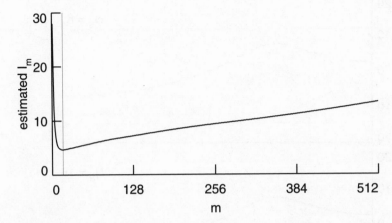

Figure 311. $\hat{I}_m$ of Equation (283b) versus m for AR(2) time series. The minimum value of the sequence $\{\hat{I}_m\}$ occurs at $m = 17$ (thin vertical line).

so the estimates all agree to within the same order of magnitude; moreover, window closing would lead us to reject values of m much below 8 or above 64 (cf. the $m = 4$ and $m = 128$ Parzen lag window estimates in Figure 310).

How well do the techniques discussed in Sections 6.14 and 6.15 for automatically choosing the smoothing parameter m do here? Since the sdf for the AR(2) series is dominantly unimodal, we can use the estimated time series bandwidth $\tilde{B}_T$ of Equation (280) to pick m. Since $\tilde{B}_T = 0.248$, setting $B_W = 1.85/m = \tilde{B}_T/2$ yields $m = 15$. This value produces a lag window estimate that is almost identical to the one for $m = 16$ and hence is quite acceptable. The second technique – Wahba's automatic smoothing criterion – picks m by finding the value that minimizes $\hat{I}_m$ of Equation (283b) versus m for $m = 0$ to 512 (see Figure 311). This procedure yields a value of $m = 17$ for the Parzen smoothing window, again producing an acceptable estimate differing very little from the $m = 16$ estimate of Figure 309. Hence, both automatic selection criteria work very well for the AR(2) series.

Finally let us illustrate the WOSA spectral estimator of Equation (291b). Figure 312 shows three $\hat{S}^{(\text{WOSA})}(\cdot)$ using a Hanning data taper and 50% overlap with block sizes of $N_S = 8$, 16 and 64 (thin solid curves from top to bottom plots) along with the true sdf (thick solid curves). Equation (294) tells us that the equivalent degrees of freedom are approximately given by, respectively, $\nu = 483.3$, 240.7 and 58.8. Note that these values are quite close to the degrees of freedom for the three lag window estimates in Figure 309, namely, $\nu = 474.9$, 237.4 and 59.4 for $m = 8$, 16 and 64, respectively. A comparison between

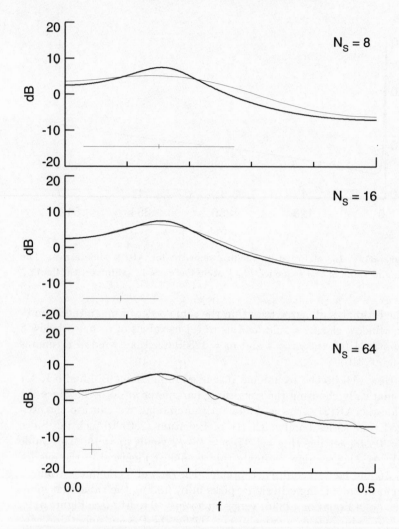

Figure 312. WOSA spectral estimates $\hat{S}^{(\text{WOSA})}(\cdot)$ for AR(2) series using a Hanning data taper (i.e., 100% cosine data taper) and 50% overlap with block sizes $N_S = 8$, 16 and 64 (thin curves from top to bottom). The thick curve on each plot is the true AR(2) sdf. The width of the crisscross emanating from the lower left-hand corner on each plot indicates the approximate distance in frequency between uncorrelated spectral estimates as measured by $C_h/(N_S \, \Delta t)$ with $C_h = 1.94$. (see item [1] of the Comments and Extensions to Section 6.9 and Table 248); its height indicates the length of a 95% confidence interval for $10 \log_{10}(S(f))$ based upon $10 \log_{10}(\hat{S}^{(\text{WOSA})}(f))$.

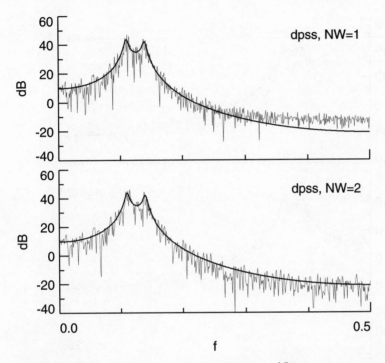

Figure 313. Two more direct spectral estimates $\hat{S}^{(d)}(\cdot)$ for AR(4) series using a dpss data taper with $NW = 1$ (thin solid curve on top plot) and $NW = 2$ (thin curve, bottom plot). The true sdf is the thick solid curve on both plots. These plots should be compared to those of Figure 227.

$\hat{S}^{(\text{WOSA})}(\cdot)$ and $\hat{S}^{(lw)}(\cdot)$ shows quite good agreement when $N_S = m$. Note that, with the WOSA estimate, there is no natural equivalent to the direct spectral estimate (although, of course, a WOSA estimate *is* the average of such estimates); i.e., lag window estimates are smoothed versions of direct spectral estimates, so we can compare the former to the latter to help us determine appropriate values for the parameter m – this is the essence of the window closing technique.

- *AR(4) data*

The AR(2) series we have just considered is a portion of a realization of a process whose sdf is rather featureless and has a low dynamic range (about 14 dB). For comparison we now do a parallel analysis using the AR(4) series, whose corresponding sdf has twin peaks and a high dynamic range (about 65 dB). This series – again of length $N = 1024$ – is shown in the bottom plot of Figure 45; the AR(4) process itself is defined in Equation (46a). The periodogram $\hat{S}^{(p)}(\cdot)$ for this time series and a direct spectral estimate $\hat{S}^{(d)}(\cdot)$ using a dpss data taper with $NW = 4$

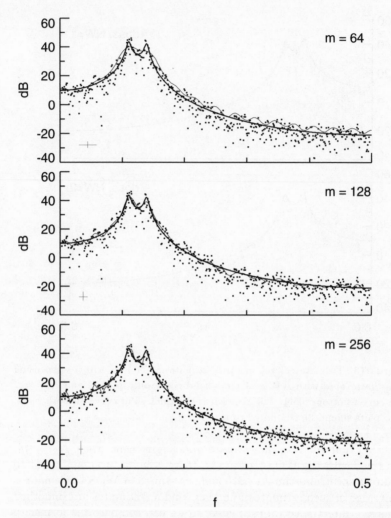

Figure 314. Lag window spectral estimates $\hat{S}^{(lw)}(\cdot)$ for AR(4) series using a Parzen smoothing window with $m = 64$, 128 and 256 (thin curves from top to bottom). The thick curve on each plot is the true AR(4) sdf, while the dots are an $NW = 2$ dpss direct spectral estimate evaluated at the Fourier frequencies. The crisscrosses in the lower left-hand corners of each plot have the same interpretation as in Figure 309.

are shown in Figure 227 (along with the true sdf). A comparison of these two estimates shows clear evidence of bias in the periodogram in the high frequency region. To determine the appropriate degree of tapering, we compute two more direct spectral estimates, one using an

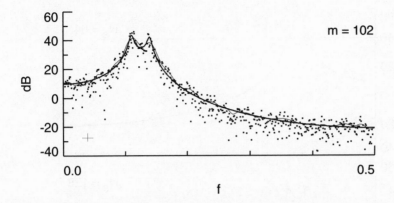

Figure 315. Lag window spectral estimate $\hat{S}^{(lw)}(\cdot)$ for AR(4) series using a Parzen smoothing window with $m = 102$ (thin solid curve).

$NW = 1$ dpss taper, and the other, an $NW = 2$ dpss taper. These estimates are shown in Figure 313 and – together with those shown in Figure 227 – indicate that an $NW = 2$ dpss taper is enough to eliminate bias due to leakage (actually, an $NW = 1.5$ taper is also adequate and hence would be preferable to the $NW = 2$ taper due to its slightly narrower central lobe, but we will stick with the latter since it is closer to the standard Hanning data taper and hence facilitates comparison with the usual WOSA spectral estimate).

To help us interpret the $NW = 2$ direct spectral estimate $\hat{S}^{(d)}(\cdot)$, we now smooth it using several different smoothing windows. We again use the Parzen lag window, but, due to the large dynamic range exhibited in $\hat{S}^{(d)}(\cdot)$, we must carefully check for smoothing window leakage. By comparing $\hat{S}^{(lw)}(\cdot)$ with a range of m to $\hat{S}^{(d)}(\cdot)$ (i.e., the window closing technique), we can determine that appropriate values for m are between approximately 100 and 300. Figure 314 shows three $\hat{S}^{(lw)}(\cdot)$ – namely, those for $m = 64$, 128 and 256 (thin solid curves in top to bottom plots) – along with the direct spectral estimate (dots) and the true sdf (thick solid curves). From Table 269, we again obtain the smoothing window bandwidth as approximately $B_W = 1.85/m$, but now the equivalent degrees of freedom are approximately given by $\nu = 1938.3/m$ since $C_h = 1.96$ for the $NW = 2$ dpss data taper – see Table 248. For $m = 64$, 128 and 256 we obtain, respectively, $B_W = 0.0289$, 0.0145 and 0.0072 and $\nu = 30.3$, 15.1 and 7.6.

The $m = 64$ lag window estimate in the top plot of Figure 314 is unsatisfactory in comparison to $\hat{S}^{(d)}(\cdot)$ – and the true sdf – both because it smears out the twin peaks too much and because there is evidence of smoothing window leakage, namely, the prominent ripples in the high frequency region of the estimate. Both the $m = 128$ and

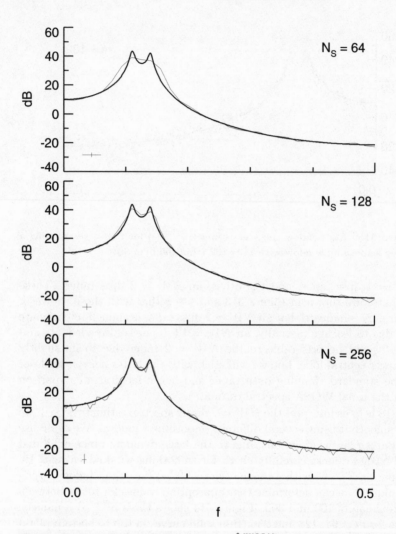

Figure 316. WOSA spectral estimates $\hat{S}^{(\mathrm{WOSA})}(\cdot)$ for AR(4) series using a Hanning data taper and 50% overlap with block sizes $N_S = 64$, 128 and 256 (thin curves from top to bottom). The thick curve on each plot is the true AR(4) sdf. The crisscrosses in the lower left-hand corners of each plot have the same interpretation as in Figure 312.

$m = 256$ estimates are better in both respects, with the former being rather better overall even though it is poorer than the latter near the twin peaks. Note that, due to these peaks, we cannot tolerate as much smoothing here as we did in the AR(2) example.

Let us now see how well the techniques of Sections 6.14 and 6.15 for automatically choosing the smoothing parameter m work. The sdf for the AR(4) series is arguably *not* dominantly unimodal, but let us look at the estimated time series bandwidth $\tilde{B}_T$ of Equation (280) anyway. Here $\tilde{B}_T = 0.036$, so equating $B_W = 1.85/m$ to $\tilde{B}_T/2$ yields $m = 102$. The corresponding spectral estimate is shown in Figure 315 and – except for the fact that the twin peaks are slightly blurred out in comparison to the $m = 128$ and $m = 256$ estimates – is quite acceptable. In contrast, Wahba's automatic smoothing criterion yields a value of $m = 47$, producing an unacceptable estimate that is slightly worse than the $m = 64$ estimate in terms of both smearing out the twin peaks and smoothing window leakage.

Finally we consider $\hat{S}^{(\text{WOSA})}(\cdot)$ of Equation (291b). Figure 316 shows three such estimates using a Hanning data taper and 50% overlap with block sizes of $N_S = 64$, 128 and 256 (thin solid curves from top to bottom plots) along with the true sdf (thick solid curves). Using Equation (294), the equivalent degrees of freedom are approximately given by, respectively, $\nu = 58.8$, 28.5 and 13.4. Note that these values are *greater* than the degrees of freedom for the three lag window estimates in Figure 314, namely, $\nu = 30.3$, 15.1 and 7.6. A comparison between $\hat{S}^{(\text{WOSA})}(\cdot)$ and $\hat{S}^{(lw)}(\cdot)$ shows that, when $N_S = m$, the bias properties of the two estimators are comparable, but that the WOSA estimate has smaller variability. As we noted in Section 6.17, this reduction in variability is due to the fact that overlapping the blocks in the WOSA estimate compensates somewhat for the effect of tapering, a theme that we return to in Chapter 7.

6.19 Comments on Complex-Valued Time Series

All of the previous sections in this chapter assumed that we are dealing with a real-valued time series. Suppose now that our time series is complex-valued and can be regarded as a realization of a portion Z_1, Z_2, ..., Z_N of a complex-valued discrete parameter stationary process $\{Z_t\}$ with zero mean and sdf $S(\cdot)$. If we can make the simplifying assumption that the real and imaginary components of $\{Z_t\}$ are uncorrelated processes with equal variances, we need only make a surprisingly few minor modifications to the material in the previous sections in order to estimate $S(\cdot)$. For example, the biased estimator of the acvs in Equation (196b) now becomes

$$
\hat{s}_\tau^{(p)} \equiv
\begin{cases}
\sum_{t=1}^{N-\tau} Z_t^* Z_{t+\tau}/N, & 0 \le \tau \le N-1; \\
\left(\hat{s}_{-\tau}^{(p)} \right)^*, & -(N-1) \le \tau < 0; \\
0, & \text{otherwise.}
\end{cases}
$$

The main result of Equation (196c) still holds, namely,

$$
\Delta t \sum_{\tau=-(N-1)}^{N-1} \hat{s}_\tau^{(p)} e^{-i2\pi f \tau \, \Delta t} = \frac{\Delta t}{N} \left| \sum_{t=1}^{N} Z_t e^{-i2\pi f t \, \Delta t} \right|^2 \equiv \hat{S}^{(p)}(f),
$$

Nonparametric Spectral Estimation

but a slightly different proof is required. All of the stated results concerning tapering and prewhitening in Sections 6.4 and 6.5 still hold, with obvious changes to Equations (207d) and (216). The main results of Section 6.6 concerning the statistical properties of direct spectral estimators also hold and, in fact, simplify somewhat in that the annoying necessity of special treatment for $f = 0$ and $f_{(N)}$ disappears; e.g., Equation (223b) now becomes

$$\hat{S}^{(d)}(f) \stackrel{\mathrm{d}}{=} \frac{S(f)}{2} \chi_2^2 \text{ for } -f_{(N)} \leq f \leq f_{(N)}$$

to a good approximation for large N, while Equation (223c) becomes

$$\mathrm{var}\,\{\hat{S}^{(d)}(f)\} \approx S^2(f) \text{ for } -f_{(N)} \leq f \leq f_{(N)}.$$

The grid of frequencies over which distinct $\hat{S}^{(d)}(f)$ variates are approximately uncorrelated now includes both positive and negative frequencies. We note that the derivation leading up to Equation (230a) is no longer valid since we made use of the fact that $dZ(-u) = dZ^*(u)$ for real-valued processes; also, we must reformulate the normalized cumulative periodogram of Equation (233) in an obvious two-sided fashion. With a few minor obvious changes, the results in Sections 6.7 to 6.12 concerning lag window spectral estimators are still valid. Finally, we note that we must plot all estimated sdf's over the set of frequencies ranging from $-f_{(N)}$ to $f_{(N)}$ instead of just from 0 to $f_{(N)}$ because the sdf is no longer an even function as in the case of real-valued processes.

6.20 Summary of Nonparametric Spectral Estimation

We assume that $X_1, X_2, \ldots, X_N$ is a sample of length N from a zero mean real-valued stationary process $\{X_t\}$ with unknown sdf $S(\cdot)$ defined over the interval $[-f_{(N)}, f_{(N)}]$, where $f_{(N)} \equiv 1/(2\,\Delta t)$ is the Nyquist frequency and Δt is the sampling interval between observations. (If $\{X_t\}$ has an unknown mean, we must replace X_t with $X_t' \equiv X_t - \bar{X}$ in all computational formulae, where $\bar{X} \equiv \sum_{t=1}^{N} X_t/N$ is the sample mean.)

[1] *Periodogram* $\hat{S}^{(p)}(\cdot)$
   a) Definition and Fourier transform pair:

$$\hat{S}^{(p)}(f) \equiv \frac{\Delta t}{N} \left| \sum_{t=1}^{N} X_t e^{-i2\pi f t\,\Delta t} \right|^2$$

$$= \Delta t \sum_{\tau=-(N-1)}^{N-1} \hat{s}_\tau^{(p)} e^{-i2\pi f\tau\,\Delta t},$$

(see (196c))

where
$$\hat{s}_\tau^{(p)} \equiv \begin{cases} \sum_{t=1}^{N-|\tau|} X_t X_{t+|\tau|}/N, & |\tau| < N; \\ 0, & |\tau| \geq N. \end{cases}$$

Note that $\{\hat{s}_\tau^{(p)}\} \longleftrightarrow \hat{S}^{(p)}(\cdot)$.

b) First moment properties:

$$E\{\hat{S}^{(p)}(f)\} = \int_{-f_{(N)}}^{f_{(N)}} \mathcal{F}(f - f')S(f')\,df',$$

$$= \Delta t \sum_{\tau=-(N-1)}^{N-1} s_\tau \left(1 - \frac{|\tau|}{N}\right) e^{-i2\pi f\tau \Delta t},$$

(see (198c) and (198a))

where $\mathcal{F}(\cdot)$ is Fejér's kernel – the spectral window for the periodogram (see Equation (198b)). For finite N, $\hat{S}^{(p)}(f)$ can be a badly biased estimator of $S(f)$; however, as $N \to \infty$, $E\{\hat{S}^{(p)}(f)\} \to S(f)$.

c) Second moment properties:

$$\text{var}\{\hat{S}^{(p)}(f)\} \approx \begin{cases} S^2(f), & \text{for } 0 < f < f_{(N)}; \\ 2S^2(f), & \text{for } f = 0 \text{ or } f_{(N)}. \end{cases} \quad \text{(see (222b))}$$

As $N \to \infty$, $\text{var}\{\hat{S}^{(p)}(f)\}$ does not decrease to 0, so $\hat{S}^{(p)}(f)$ is an inconsistent estimator of $S(f)$. Also, for $f_j \equiv j/(N\,\Delta t)$, $\text{cov}\{\hat{S}^{(p)}(f_j), \hat{S}^{(p)}(f_k)\} \approx 0$, where j and k are integers such that $j \neq k$ and $0 \leq j, k \leq \lfloor N/2 \rfloor$ (see Equation (223a)).

d) Distribution (large N):

$$\hat{S}^{(p)}(f) \stackrel{\mathrm{d}}{=} \begin{cases} S(f)\chi_2^2/2, & \text{for } 0 < f < f_{(N)}; \\ S(f)\chi_1^2, & \text{for } f = 0 \text{ or } f_{(N)}. \end{cases} \quad \text{(see (222a))}$$

[2] *Direct spectral estimator* $\hat{S}^{(d)}(\cdot)$

a) Definition and Fourier transform pair:

$$\hat{S}^{(d)}(f) \equiv \Delta t \left| \sum_{t=1}^{N} h_t X_t e^{-i2\pi f t \Delta t} \right|^2 \quad \text{(see (206c))}$$

$$= \Delta t \sum_{\tau=-(N-1)}^{N-1} \hat{s}_\tau^{(d)} e^{-i2\pi f\tau \Delta t}, \quad \text{(see (207c))}$$

where $\{h_t\}$ is a data taper with normalization $\sum_{t=1}^{N} h_t^2 = 1$ and

$$\hat{s}_\tau^{(d)} \equiv \begin{cases} \sum_{t=1}^{N-|\tau|} h_t X_t h_{t+|\tau|} X_{t+|\tau|}, & |\tau| < N; \\ 0, & |\tau| \geq N. \end{cases} \quad \text{(see (207d))}$$

Note that $\{\hat{s}_\tau^{(d)}\} \longleftrightarrow \hat{S}^{(d)}(\cdot)$.

b) First moment properties:

$$E\{\hat{S}^{(d)}(f)\} = \int_{-f_{(N)}}^{f_{(N)}} \mathcal{H}(f - f')S(f')\,df',$$

$$= \Delta t \sum_{\tau=-(N-1)}^{N-1} s_\tau \sum_{t=1}^{N-|\tau|} h_t h_{t+|\tau|} e^{-i2\pi f\tau \,\Delta t},$$

(see (207a) and (207b))

where $\mathcal{H}(\cdot)$ is the spectral window for $\hat{S}^{(d)}(\cdot)$ and is defined via $\mathcal{H}(f) = |H(f)|^2/\Delta t$ with

$$H(f) \equiv \Delta t \sum_{t=1}^{N} h_t e^{-i2\pi ft\,\Delta t}.$$ (see (206b))

Note that $\{h_t\} \longleftrightarrow H(\cdot)$. The purpose of using a data taper $\{h_t\}$ is to produce an estimator $\hat{S}^{(d)}(f)$ of $S(f)$ with better bias properties than the periodogram $\hat{S}^{(p)}(f)$. If a proper data taper can be employed, we should have $E\{\hat{S}^{(d)}(f)\} \approx S(f)$ for finite N, and, as $N \to \infty$, $E\{\hat{S}^{(d)}(f)\} \to S(f)$.

c) Second moment properties:

$$\mathrm{var}\,\{\hat{S}^{(d)}(f)\} \approx \begin{cases} S^2(f), & \text{for } 0 < f < f_{(N)}; \\ 2S^2(f), & \text{for } f = 0 \text{ or } f_{(N)}. \end{cases}$$ (see (223c))

As $N \to \infty$, $\mathrm{var}\,\{\hat{S}^{(d)}(f)\}$ does not decrease to 0, so $\hat{S}^{(d)}(f)$ is an inconsistent estimator of $S(f)$. Also, for $f_j' \equiv j/(N'\,\Delta t)$, $\mathrm{cov}\,\{\hat{S}^{(d)}(f_j'), \hat{S}^{(d)}(f_k')\} \approx 0$, where j and k are integers such that $j \neq k$ and $0 \leq j, k \leq \lfloor N'/2 \rfloor$; here $N' \approx N/C_h$, where $C_h \geq 1$ is a factor depending upon the data taper $\{h_t\}$ and given by Equation (251b) (see also Table 248).

d) Distribution (large N):

$$\hat{S}^{(d)}(f) \overset{\mathrm{d}}{=} \begin{cases} S(f)\chi_2^2/2, & \text{for } 0 < f < f_{(N)}; \\ S(f)\chi_1^2, & \text{for } f = 0 \text{ or } f_{(N)}. \end{cases}$$ (see (223b))

[3] *Lag window spectral estimator* $\hat{S}^{(lw)}(\cdot)$

a) Definition and Fourier transform pair:

$$\hat{S}^{(lw)}(f) = \Delta t \sum_{\tau=-(N-1)}^{N-1} w_{\tau,m}\hat{s}_\tau^{(d)} e^{-i2\pi f\tau\,\Delta t}$$

(see (238a))

$$= \int_{-f_{(N)}}^{f_{(N)}} W_m(f - f')\hat{S}^{(d)}(f')\,df'$$

where $\{w_{\tau,m}\}$ is a lag window with $w_{\tau,m} \equiv 0$ for $|\tau| \geq N$; $\{W_m(\cdot)\}$ is a smoothing window such that $\{w_{\tau,m}\} \longleftrightarrow W_m(\cdot)$ (see Equation (237c)); and m is a smoothing parameter that controls the degree of smoothing. Note that $\{w_{\tau,m}\hat{s}_\tau^{(d)}\} \longleftrightarrow \hat{S}^{(lw)}(\cdot)$.

b) First moment properties:

$$E\{\hat{S}^{(lw)}(f)\} = \int_{-f_{(N)}}^{f_{(N)}} \mathcal{U}_m(f - f')S(f')\,df'$$

$$= \Delta t \sum_{\tau=-(N-1)}^{N-1} w_{\tau,m}s_\tau \sum_{t=1}^{N-|\tau|} h_t h_{t+|\tau|}e^{-i2\pi f\tau\,\Delta t},$$

(see (243b) and Exercise [6.17a])

where $\mathcal{U}_m(\cdot)$ is the spectral window for $\hat{S}^{(lw)}(\cdot)$ defined by

$$\mathcal{U}_m(f) \equiv \int_{-f_{(N)}}^{f_{(N)}} W_m(f - f')\mathcal{H}(f')\,df',$$

$$= \Delta t \sum_{\tau=-(N-1)}^{N-1} w_{\tau,m} \sum_{t=1}^{N-|\tau|} h_t h_{t+|\tau|}e^{-i2\pi f\tau\,\Delta t}.$$

(see (243c) and (244a))

Note that $\{w_{\tau,m} \sum_{t=1}^{N-|\tau|} h_t h_{t+|\tau|}\} \longleftrightarrow \mathcal{U}_m(\cdot)$. The purpose of using a lag window $\{w_{\tau,m}\}$ is to produce an estimator $\hat{S}^{(lw)}(f)$ of $S(f)$ with better variance properties than the direct spectral estimator $\hat{S}^{(d)}(f)$. If a proper data taper and lag window can be employed, we should have $E\{\hat{S}^{(lw)}(f)\} \approx S(f)$ for finite N, and, as $N \to \infty$, $E\{\hat{S}^{(lw)}(f)\} \to S(f)$.

c) Second moment properties:

$$\mathrm{var}\{\hat{S}^{(lw)}(f)\} \approx \frac{C_h S^2(f)}{N\,\Delta t} \int_{-f_{(N)}}^{f_{(N)}} W_m^2(f')\,df'$$

$$\approx \frac{C_h S^2(f)}{N} \sum_{\tau=-(N-1)}^{N-1} w_{\tau,m}^2.$$

(see (248a) and (248b))

The factor C_h depends only on the data taper $\{h_t\}$ – it is defined by Equation (251b) and tabulated for a few common tapers in Table 248. Typically, as $N \to \infty$ and $m \to \infty$ in such a way that $m/N \to 0$, $\mathrm{var}\{\hat{S}^{(lw)}(f)\}$ decreases to 0, so $\hat{S}^{(lw)}(f)$ is a consistent estimator of $S(f)$. Also, for $f'_j \equiv j/(N'\,\Delta t)$, $\mathrm{cov}\{\hat{S}^{(lw)}(f'_j), \hat{S}^{(lw)}(f'_k)\} \approx 0$, where j and k are integers such that $j \neq k$ and $0 \leq j, k \leq \lfloor N'/2 \rfloor$; here $N' \equiv N/c$ where $c \geq 1$

is a factor that depends primarily upon the bandwidth B_W of the smoothing window $W_m(\cdot)$ (see Section 6.9).

d) Distribution (large N):

$$\hat{S}^{(lw)}(f) \overset{\mathrm{d}}{=} \frac{S(f)}{\nu}\chi_{\nu}^{2},$$

where ν is the equivalent degrees of freedom for $\hat{S}^{(lw)}(\cdot)$ and is defined by

$$\nu = \frac{2N\,\Delta t}{C_h \int_{-f_{(N)}}^{f_{(N)}} W_m^2(\phi)\,d\phi} = \frac{2N}{C_h \sum_{\tau=-(N-1)}^{N-1} w_{\tau,m}^2}.$$

$$\text{(see (254b))}$$

6.21 Exercises

[6.1] Suppose that $X_1, \ldots, X_N$ is a sample of size N from a white noise process with unknown mean μ and variance σ^2. With $\hat{s}_\tau^{(u)}$ and $\hat{s}_\tau^{(p)}$ defined as in Equations (190b) and (191a), show that, for $0 < |\tau| < N - 1$,

$$E\{\hat{s}_\tau^{(u)}\} = -\frac{\sigma^2}{N} \quad \text{and} \quad E\{\hat{s}_\tau^{(p)}\} = -\left(1 - \frac{|\tau|}{N}\right)\frac{\sigma^2}{N}.$$

Use this to argue a) that the magnitude of the bias of the biased estimator $\hat{s}_\tau^{(p)}$ can be less than that of the 'unbiased' estimator $\hat{s}_\tau^{(u)}$ when μ is estimated by $\bar{X}$ and b) that, for this example, the mean square error of $\hat{s}_\tau^{(p)}$ is less than that of $\hat{s}_\tau^{(u)}$ for $0 < |\tau| < N - 1$.

[6.2] Let $X_1, \ldots, X_N$ be a sample of size N from a stationary process with sdf $S(\cdot)$ defined over $[-1/2, 1/2]$ (i.e., the sampling interval Δt is taken to be 1). At lag $\tau = 0$ both the unbiased and biased estimators of the acvs reduce to

$$\hat{s}_0 \equiv \frac{1}{N}\sum_{t=1}^{N}\left(X_t - \bar{X}\right)^2.$$

a) Show that $E\{\hat{s}_0\} = s_0 - \mathrm{var}\,\{\bar{X}\} \le s_0$.

b) Show that

$$E\{\hat{s}_0\} = \int_{-1/2}^{1/2}\left(1 - \frac{1}{N}\mathcal{F}(f)\right) S(f)\,df,$$

where $\mathcal{F}(\cdot)$ is Fejér's kernel (see Equation (198b)). Hint: consider Equations (189a), (198a) and (198c).

c) Plot $(1 - \mathcal{F}(f)/N)$ versus f for, say, $N = 64$. Based upon this plot, for what kind of $S(\cdot)$ would there be a large discrepancy between $E\{\hat{s}_0\}$ and s_0?

[6.3] Construct an example to show that the sequence of numbers

$$\hat{s}_0^{(u)}, \hat{s}_1^{(u)}, \ldots, \hat{s}_{N-2}^{(u)}, \hat{s}_{N-1}^{(u)}$$

given by the 'unbiased' estimator of the acvs (see Equation (190b)) need not be a valid *theoretical* acvs for some stationary process.

[6.4] Let $X_1, \ldots, X_N$ be a portion of a stationary process with sampling interval Δt, and let $\hat{s}_\tau^{(p)}$ for $\tau = 0, \pm 1, \ldots, \pm(N-1)$ be the biased estimator of the acvs defined by either Equation (196b) – if $E\{X_t\}$ is known to be zero – or Equation (191a) – if $E\{X_t\}$ is unknown and hence estimated by the sample mean $\bar{X}$. For $|\tau| \geq N$, let $\hat{s}_\tau^{(p)} \equiv 0$. Let $f_k' \equiv k/(N'\Delta t)$, where N' is any integer greater than or equal to N.

a) Show that

$$\hat{S}^{(p)}(f_k') = \Delta t \sum_{\tau=0}^{N'-1} \left(\hat{s}_\tau^{(p)} + \hat{s}_{N'-\tau}^{(p)} \right) e^{-i2\pi k\tau/N'}$$

and hence that

$$\hat{s}_\tau^{(p)} + \hat{s}_{N'-\tau}^{(p)} = \frac{1}{N'\Delta t} \sum_{k=0}^{N'-1} \hat{S}^{(p)}(f_k') e^{i2\pi k\tau/N'}$$

(Bloomfield, 1976, Section 7.4).

b) Show how, by letting $N' \geq 2N$, we can calculate $\hat{s}_\tau^{(p)}$ for $\tau = 0$, $\ldots, N-1$ by using just discrete and/or inverse discrete Fourier transforms (Equations (110a) and (111a) with $\Delta t = 1$). Hint: study the last line of Equation (196c).

c) A discrete or inverse discrete Fourier transform of a sequence with N' terms often can be computed efficiently using a fast Fourier transform algorithm (see Section 3.10). Such an algorithm typically requires a number of arithmetic operations (additions, subtractions, multiplications and divisions) proportional to $N' \log(N')$. Compare the computational procedure for evaluating $\hat{s}_\tau^{(p)}$ for $\tau = 0, \ldots, N-1$ in part b) with direct use of Equation (191a) in terms of the number of arithmetic operations required.

[6.5] a) If C is any value independent of t, show that

$$\tilde{S}^{(p)}(f_k) \equiv \frac{\Delta t}{N} \left| \sum_{t=1}^{N} (X_t - C) e^{-i2\pi f_k t \Delta t} \right|^2$$

$$= \frac{\Delta t}{N} \left| \sum_{t=1}^{N} X_t e^{-i2\pi f_k t \Delta t} \right|^2 \equiv \hat{S}^{(p)}(f_k),$$

where k is an integer, $f_k \equiv k/(N \Delta t)$ and $0 < f_k \leq f_{(N)}$.

b) Show that, in the above,

$$\tilde{S}^{(p)}(0) = 0 \text{ when } C = \bar{X} \equiv \frac{1}{N} \sum_{t=1}^{N} X_t;$$

i.e., the periodogram of a centered time series is necessarily 0 at the origin. What does this imply about $\sum_{\tau=-(N-1)}^{N-1} \hat{s}_\tau^{(p)}$?

c) Suppose that $X_1, \ldots, X_N$ is a segment of length N of a white noise process with sdf $S(\cdot)$ and nonzero mean μ. Here we examine one of the consequences of *not* centering a time series before computing the periodogram. Show that, for

$$\hat{S}^{(p)}(f) \equiv \frac{\Delta t}{N} \left| \sum_{t=1}^{N} X_t e^{-i2\pi ft \Delta t} \right|^2,$$

we have

$$E\{\hat{S}^{(p)}(f)\} = S(f) + \mu^2 \mathcal{F}(f),$$

where $\mathcal{F}(\cdot)$ is Fejér's kernel (defined in Equation (198b)).

[6.6] Let $\hat{S}^{(p)}(\cdot)$ be the periodogram of a sample $X_1, X_2, \ldots, X_N$ of a zero mean stationary process:

$$\hat{S}^{(p)}(f) = \frac{\Delta t}{N} \left| \sum_{t=1}^{N} X_t e^{-i2\pi ft \Delta t} \right|^2.$$

a) Show that the following form of Parseval's theorem is true:

$$\int_{-f_{(N)}}^{f_{(N)}} \hat{S}^{(p)}(f) \, df = \hat{s}_0^{(p)}.$$

Thus, just as the integral of the sdf is the process variance, the integral of the periodogram is the sample variance.

b) Show that

$$\frac{1}{N \Delta t} \sum_{k=0}^{N-1} \hat{S}^{(p)}(f_k) = \frac{1}{N} \sum_{t=1}^{N} X_t^2 = \hat{s}_0^{(p)},$$

where $f_k \equiv k/(N \Delta t)$. Note that the left-hand side above can be regarded as a Riemann summation from 0 to $2f_{(N)}$ with a grid size of $1/(N \Delta t)$, which, since $\hat{S}^{(p)}(\cdot)$ is $2f_{(N)}$ periodic, can be regarded as an approximation to the integral in part a);

however, whereas a Riemann summation is usually an approximation to an integral, we see from part a) that here the finite summation is exactly equal to the corresponding integral!

c) Show that parts a) and b) still hold for a stationary process for which the process mean must be estimated using the sample mean $\bar{X}$. In this case, the periodogram becomes

$$\hat{S}^{(p)}(f) = \frac{\Delta t}{N} \left| \sum_{t=1}^{N} \left(X_t - \bar{X}\right) e^{-i2\pi f t \Delta t} \right|^2,$$

and part b) becomes

$$\frac{1}{N\,\Delta t} \sum_{k=0}^{N-1} \hat{S}^{(p)}(f_k) = \frac{1}{N} \sum_{t=1}^{N} \left(X_t - \bar{X}\right)^2 = \hat{s}_0^{(p)}.$$

d) Show that, in addition to the result stated in part b), we have

$$\frac{1}{2N\,\Delta t} \sum_{k=0}^{2N-1} \hat{S}^{(p)}(\tilde{f}_k) = \hat{s}_0^{(p)},$$

where $\tilde{f}_k \equiv k/(2N\,\Delta t)$.

[6.7] From Equation (196c) we know that, given $\hat{s}_0^{(p)}, \ldots, \hat{s}_{N-1}^{(p)}$, we can compute the periodogram $\hat{S}^{(p)}(\cdot)$ at any given frequency. Suppose that we are given the periodogram on the grid of Fourier frequencies, i.e., $\hat{S}^{(p)}(f_k)$ for $k = 0, 1, \ldots, \lfloor N/2 \rfloor$, where, as usual, $f_k \equiv k/(N\,\Delta t)$ and $\lfloor N/2 \rfloor$ is the largest integer less than or equal to $N/2$. Using these $\lfloor N/2 \rfloor + 1$ values, is it possible in general to recover $\hat{s}_0^{(p)}, \ldots, \hat{s}_{N-1}^{(p)}$?

[6.8] Verify Equation (207b). (Hint: consider what happens in Equation (196c) when X_t is replaced by $h_t X_t$.)

[6.9] Show that Equation (207c) is true.

[6.10] a) The definition for the Hanning data taper (or 100% cosine taper) in Equation (209) reduces to

$$h_t = \frac{C}{2}\left[1 - \cos\left(\frac{2\pi t}{N+1}\right)\right], \quad t = 1, 2, \ldots, N, \qquad (325)$$

where C is a normalizing constant. In this form the taper is greater than 0 for $1 \leq t \leq N$ and symmetric in the sense that $h_{N-t+1} = h_t$, so, for example, $h_N = h_1$. A second common definition is the asymmetric form

$$h_t' = \frac{C'}{2}\left[1 - \cos\left(\frac{2\pi t}{N}\right)\right], \quad t = 1, 2, \ldots, N,$$

for which $h'_N = 0$ (see, for example, Priestley, 1981, p. 562). Let

$$J(f_k) \equiv (\Delta t)^{1/2} \sum_{t=1}^{N} h'_t X_t e^{-i2\pi f_k t\,\Delta t},$$

$$I(f_k) \equiv (\Delta t)^{1/2} \sum_{t=1}^{N} X_t e^{-i2\pi f_k t\,\Delta t}, \quad \text{where } f_k \equiv \frac{k}{N\,\Delta t}.$$

Show that

$$J(f_k) = -\frac{1}{4}I(f_{k-1}) + \frac{1}{2}I(f_k) - \frac{1}{4}I(f_{k+1}).$$

Is the above valid for all k such that $0 \le f_k \le f_{(N)}$?

b) Show that, if we let $C = (8/[3(N+1)])^{1/2}$ in Equation (325), we obtain the normalization $\sum_{t=1}^{N} h_t^2 = 1$. Hint: use the identity $\cos^2(x) = (1 + \cos(2x))/2$ and Exercise [1.3c].

[6.11] Suppose that $X_1, \ldots, X_N$ is a sample of a white noise process $\{X_t\}$ with mean μ. Let $\{h_t\}$ be a data taper that satisfies Equation (208a). Compare the first moment properties of the direct spectral estimator

$$\hat{S}^{(d)}(f) = \Delta t \left| \sum_{t=1}^{N} h_t(X_t - \mu)e^{-i2\pi ft\,\Delta t} \right|^2$$

with an estimator in which the process mean is subtracted *after* the time series is tapered:

$$\tilde{S}^{(h)}(f) \equiv \Delta t \left| \sum_{t=1}^{N} (h_t X_t - \mu)e^{-i2\pi ft\,\Delta t} \right|^2.$$

[6.12] Suppose that $\{h_t\}$ is the rectangular data taper. What does $R(\eta)$ in Equation (231a) reduce to? For what values of η is $R(\eta) = 0$?

[6.13] If $\{h_t\} \longleftrightarrow H(\cdot)$ and $\{s_{\tau,G}\} \longleftrightarrow S_G(\cdot)$, show that Equation (230a) can be rewritten as

$$\operatorname{cov}\{\hat{S}_G^{(d)}(f), \hat{S}_G^{(d)}(f+\eta)\} = \operatorname{cov}\{|J(f)|^2, |J(f+\eta)|^2\}$$

$$= \left| \sum_{j=1}^{N} h_j C_j^*(f)e^{-i2\pi(f+\eta)j\,\Delta t} \right|^2 + \left| \sum_{j=1}^{N} h_j C_j(f)e^{-i2\pi(f+\eta)j\,\Delta t} \right|^2,$$

where

$$C_j(f) \equiv \sum_{k=1}^{N} h_k s_{j-k,G} e^{-i2\pi fk\,\Delta t}.$$

[6.14] Show that Equation (237d) is true.

[6.15] Verify the computational formula in Equation (241c) (hint: use Equation (237c)).

[6.16] Show that Equation (244a) is true. Does this equation still hold if we drop our assumption that $w_{\tau,m} = 0$ for $|\tau| \geq N$? If we replace $W_m(\cdot)$ with $V_m(\cdot)$ in Equation (243c), why does this not change the definition of $\mathcal{U}_m(\cdot)$? (Section 6.7 describes the distinction between $W_m(\cdot)$ and $V_m(\cdot)$.)

[6.17] a) Show that

$$E\{\hat{S}^{(lw)}(f)\} = \Delta t \sum_{\tau=-(N-1)}^{N-1} \left(w_{\tau,m} s_\tau \sum_{t=1}^{N-|\tau|} h_t h_{t+|\tau|} \right) e^{-i2\pi f \tau \, \Delta t}$$

(cf. Equation (207b)).

b) Show that

$$\int_{-f(N)}^{f(N)} E\{\hat{S}^{(lw)}(f)\} \, df = s_0 \quad \text{and} \quad \int_{-f(N)}^{f(N)} \mathcal{U}_m(f) \, df = 1$$

under the usual assumptions that

$$\sum_{t=1}^{N} h_t^2 = 1 \quad \text{and} \quad w_{0,m} = 1$$

(cf. Equations (208b) and (208c)).

[6.18] Use the approach of Section 6.10 to determine the equivalent number of degrees of freedom for the spectral estimator defined by Equation (236a).

[6.19] An early attempt to produce a spectral estimator with better variance properties than the periodogram was to truncate the summation in Equation (196c) at $m < N$ to produce

$$\hat{S}^{(tp)}(f) \equiv \Delta t \sum_{\tau=-m}^{m} \hat{s}_\tau^{(p)} e^{-i2\pi f \tau \, \Delta t}. \tag{327a}$$

The idea is to discard estimates of the acvs for which there are relatively little data. This estimator is known as the *truncated periodogram*. Show that $\hat{S}^{(tp)}(\cdot)$ can be regarded as a special case of a lag window spectral estimator, and determine its lag window and its smoothing window. Is the truncated periodogram necessarily a nonnegative estimate at all frequencies?

[6.20] Blackman and Tukey (1958, Section B.5) discuss the following class of lag windows:

$$w_{\tau,m} = \begin{cases} 1 - 2a + 2a\cos\left(\pi\tau/m\right), & |\tau| \leq m; \\ 0, & |\tau| > m, \end{cases} \tag{327b}$$

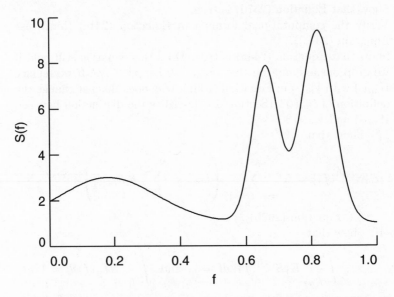

Figure 328. Hypothesized sdf. Note that $S(f)$ versus f is plotted on a linear/linear scale, not on the usual linear/decibel scale.

where $m < N$ and a is a parameter in the range $0 < a \le 1/4$. If we let $a = 0.23$, then, among all smoothing windows corresponding to lag windows of the above form, we obtain the smoothing window that minimizes the magnitude in the first sidelobe relative to the magnitude of the peak in the main lobe. This particular case is called the *Hamming lag window*. A slight variation is to let $a = 1/4$. The resulting lag window is called the *Hanning lag window*.

a) Determine the smoothing window for the lag window in (327b).

b) Show that, if we use the rectangular data taper, the corresponding lag window estimator $\hat{S}^{(lw)}(\cdot)$ can be expressed in terms of the truncated periodogram of Equation (327a) as follows:

$$\hat{S}^{(lw)}(f) = a\hat{S}^{(tp)}\left(f - 1/(2m\,\Delta t)\right) + (1 - 2a)\hat{S}^{(tp)}(f)$$
$$+ a\hat{S}^{(tp)}\left(f + 1/(2m\,\Delta t)\right).$$

[6.21] Suppose an investigator plans to collect a time series that can be modeled by a stationary process with sdf $S(\cdot)$. The sampling time Δt is set at $1/2$ second so the Nyquist frequency $f_{(N)}$ is 1. Suppose that the rough shape of $S(\cdot)$ is known from either physical theory or previous experiments and is given by Figure 328 (note that this shows $S(f)$ versus f on a *linear/linear* scale). The investigator plans to use the Parzen lag window spectral estimator.

a) What kind of data taper would you recommend if the sample size N of the collected time series is 100? Would you change your recommendation if the sample size was increased to 1000? State the reasons for your recommendations.

b) Would prewhitening be useful here? State the reasons for your answer.

c) Determine the spectral bandwidth of $S(\cdot)$ by examining Figure 328. What does this imply about the size of the window bandwidth of the Parzen smoothing window? For the data taper(s) you selected in part a) for sample sizes of $N = 100$ and 1000, determine what values of m (if any) achieve the desired window bandwidth.

[6.22] Here we derive Equation (292a) using an argument closely paralleling that leading to Equation (230b). We assume that $\{X_t\}$ is a Gaussian process. Let

$$J_l(f) \equiv (\Delta t)^{1/2} \sum_{t=1}^{N_S} h_t X_{t+l-1} e^{-i2\pi f t \, \Delta t} \quad \text{so} \quad |J_l(f)|^2 = \hat{S}_l^{(d)}(f).$$

a) Show that

$$\text{cov}\,\{\hat{S}_l^{(d)}(f), \hat{S}_m^{(d)}(f)\} = |E\{J_l^*(f)J_m(f)\}|^2 + |E\{J_l(f)J_m(f)\}|^2.$$

b) Show that

$$J_l(f) = \frac{1}{(\Delta t)^{1/2}} \int_{-f_{(N)}}^{f_{(N)}} e^{i2\pi f'(l-1)} H(f - f') \, dZ(f')$$

(cf. Equation (206a)).

c) Show that

$$E\{J_l^*(f)J_m(f)\} = \int_{-f_{(N)}}^{f_{(N)}} e^{i2\pi f'(m-l)} \mathcal{H}(f - f') S(f') \, df'$$

(recall that $\mathcal{H}(f) \equiv |H(f)|^2 / \Delta t$). If f is not near 0 or $f_{(N)}$, argue that

$$E\{J_l(f)J_m(f)\} \approx 0.$$

d) Show that, if $S(\cdot)$ is locally constant about f,

$$\text{cov}\,\{\hat{S}_l^{(d)}(f), \hat{S}_m^{(d)}(f)\} \approx S^2(f) \left| \int_{-f_{(N)}}^{f_{(N)}} e^{i2\pi f'(m-l)} \mathcal{H}(f') \, df' \right|^2$$

$$= S^2(f) \left| \sum_{t=1}^{N_S} h_t h_{t+|m-l|} \right|^2,$$

where $h_t \equiv 0$ for $t > N_S$.

[6.23] Suppose that $\{X_t\}$ is a real-valued stationary process with zero mean, sampling time Δt, acvs $\{s_\tau\}$ and sdf $S(\cdot)$. Let $a_1, \ldots, a_N$ and $b_1, \ldots, b_N$ be any two sequences of complex-valued numbers of finite length N. Define

$$A(f) \equiv \Delta t \sum_{t=1}^{N} a_t e^{-i2\pi ft\,\Delta t} \quad \text{and} \quad B(f) \equiv \Delta t \sum_{t=1}^{N} b_t e^{-i2\pi ft\,\Delta t};$$

i.e., if we define $a_t = 0$ and $b_t = 0$ for all $t < 1$ and $t > N$, then

$$\{a_t\} \longleftrightarrow A(\cdot) \quad \text{and} \quad \{b_t\} \longleftrightarrow B(\cdot).$$

a) Use the spectral representation theorem for $\{X_t\}$ to show that

$$\text{cov}\left\{\sum_{t=1}^{N} a_t X_t, \sum_{t=1}^{N} b_t X_t\right\} = \frac{1}{(\Delta t)^2} \int_{-f_{(N)}}^{f_{(N)}} A^*(f)B(f)S(f)\,df.$$

(330a)

b) Show that we also have

$$\text{cov}\left\{\sum_{t=1}^{N} a_t X_t, \sum_{t=1}^{N} b_t X_t\right\} = \sum_{t=1}^{N}\sum_{u=1}^{N} a_t^* b_u s_{t-u}. \qquad (330b)$$

c) Now suppose that $\{X_t\}$ is a Gaussian process. Show that

$$\text{cov}\left\{\left|\sum_{t=1}^{N} a_t X_t\right|^2, \left|\sum_{t=1}^{N} b_t X_t\right|^2\right\}$$

$$= \frac{1}{(\Delta t)^4}\left|\int_{-f_{(N)}}^{f_{(N)}} A^*(f)B(f)S(f)\,df\right|^2$$

$$+ \frac{1}{(\Delta t)^4}\left|\int_{-f_{(N)}}^{f_{(N)}} A(-f)B(f)S(f)\,df\right|^2$$

(330c)

(hint: generalize the argument leading to Equation (229)).

d) Show that Equation (330c) can be rewritten as

$$\text{cov}\left\{\left|\sum_{t=1}^{N} a_t X_t\right|^2, \left|\sum_{t=1}^{N} b_t X_t\right|^2\right\}$$

$$= \left|\sum_{t=1}^{N}\sum_{u=1}^{N} a_t^* b_u s_{t-u}\right|^2 + \left|\sum_{t=1}^{N}\sum_{u=1}^{N} a_t b_u s_{t-u}\right|^2.$$

(330d)

e) Use the results of part c) to verify Equation (230a) (hint: let $a_t = (\Delta t)^{1/2} h_t e^{-i2\pi ft\,\Delta t}$ and $b_t = (\Delta t)^{1/2} h_t e^{-i2\pi(f+\eta)t\,\Delta t}$).

7

Multitaper Spectral Estimation

7.0 Introduction

In Chapter 6 we introduced the important concept of tapering a time series as a way of obtaining a spectral estimator with acceptable bias properties. While tapering does reduce bias due to leakage, there is a price to pay in that the sample size is effectively reduced. When we also smooth across frequencies, this reduction translates into a loss of information in the form of an increase in variance (recall the C_h factor in Equation (248b) and Table 248). This inflated variance is acceptable in some practical applications, but in other cases it is not. The loss of information inherent in tapering can often be avoided either by prewhitening (see Sections 6.5 and 9.10) or by using Welch's overlapped segment averaging (WOSA – see Section 6.17).

In this chapter we discuss another approach to recovering information lost due to tapering. This approach was introduced in a seminal paper by Thomson (1982) and involves the use of multiple orthogonal tapers. As we shall see, multitaper spectral estimation has a number of interesting points in its favor:

[1] In contrast to either prewhitening which typically requires the careful design of a prewhitening filter or the conventional use of WOSA (i.e., a Hanning data taper with 50% overlap of blocks) which can still suffer from leakage for spectra with very high dynamic ranges, the multitaper scheme can be used in a fairly 'automatic' fashion. Hence it is useful in situations where thousands – or millions – of individual time series must be processed so that the pure volume of data precludes a careful analysis of individual series (this occurs routinely in exploration geophysics).

331

[2] The bias of multitaper spectral estimators can be broken down into two quantifiable components: the local bias (due to frequency components within a user-selectable passband $[-W, W]$) and the broad-band bias (due to components outside this passband).

[3] There is a natural nonsubjective way to define the *resolution* of multitaper estimators (it is just the bandwidth $2W$ of the passband) – this is a rather sticky concept to define for spectral estimators in general (Kay and Demeure, 1984).

[4] The tradeoff between the bias and variance for multitaper spectral estimators is also relatively easy to quantify.

[5] Although either prewhitening or WOSA can usually compensate for bias just as well as multitapering can, Thomson (1990a, p. 614) argues that there are examples of processes with highly structured sdf's for which prewhitening is inferior to multitapering; on the other hand, Bronez (1992) shows, using bounds for bias, variance and resolution, that, if a WOSA spectral estimator and a multitaper estimator are fixed so that two of the three bounds are identical, the remaining bound favors the multitaper estimator.

[6] The multitaper approach can be regarded as a way of producing direct spectral estimates with more than just 2 degrees of freedom (typically 4 to 10). Figure 257 shows that such modest increases are enough to noticeably shrink the width of a 95% confidence interval for an sdf at a fixed frequency.

[7] It is possible to obtain an 'internal' estimate of the variance of a multitaper spectral estimator by means of a technique called 'jackknifing' (Thomson and Chave, 1991).

[8] As we shall discuss in Section 10.11, there is an appealing statistical test based upon the multitaper spectral estimator for the presence of sinusoidal (line) components in a time series. The multitaper technique thus offers a unified approach to estimating both mixed spectra and sdf's (processes with mixed spectra were defined in Section 4.4 and are discussed in detail in Chapter 10).

[9] There are natural extensions to the multitaper approach for both spectral estimation of time series with missing or irregularly sampled observations and the problem of estimating multidimensional spectra (Bronez, 1985, 1986, 1988; Liu and Van Veen, 1992).

The remainder of this chapter is organized as follows. In Section 7.1 we motivate the multitaper approach by presenting a computational recipe for it and then illustrating how it works on the AR(4) time series in the lower plot of Figure 45. We investigate in Section 7.2 the statistical properties of the multitaper spectral estimator for the simple case of white noise. We formally justify the multitaper approach in Section 7.3 by showing how it arises in the context of quadratic spectral estimators. We present a second argument for the technique in Section 7.4 based upon projections and the concept of regularization. We conclude with an example in Section 7.5.

In addition to the papers by Thomson and Bronez we have already cited, the reader should consult Park *et al.* (1987b), Walden (1990b) and Mullis and Scharf (1991) for introductory discussions on multitapering. This estimation technique has been used on a number of interesting time series, including ones concerning terrestrial free oscillations (Park *et al.*, 1987a, and Lindberg and Park, 1987), the relationship between atmospheric carbon dioxide and global temperature (Kuo *et al.* 1990), oxygen isotope ratios from deep-sea cores (Thomson, 1990b) and the rotation of the earth (King, 1990). The relationship between structured covariance matrices and multitapering is discussed in Van Veen and Scharf (1990).

7.1 A Motivating Example

Here we introduce the multitaper spectral estimator $\hat{S}^{(mt)}(\cdot)$ by sketching the rationale for it and by discussing in detail the steps that must be taken to compute it. As an example, we compute $\hat{S}^{(mt)}(\cdot)$ for the AR(4) time series shown in the bottom of Figure 45. A more detailed discussion of the rationale for multitaper estimators is given in Sections 7.2 to 7.4.

Suppose we have a time series that is a realization of a portion $X_1, X_2, \ldots, X_N$ of the stationary process $\{X_t\}$ with zero mean, variance σ^2 and sdf $S(\cdot)$. We assume that the sampling interval between observations is Δt so that the Nyquist frequency is $f_{(N)} \equiv 1/(2\,\Delta t)$. For the sample size N, the *fundamental Fourier frequency* is defined to be $1/(N\,\Delta t)$ (note that this is just the smallest nonzero Fourier frequency).

As its name implies, the multitaper spectral estimator utilizes several different data tapers. In its simplest formulation, this estimator is the average of K direct spectral estimators and hence takes the form

$$\hat{S}^{(mt)}(f) \equiv \frac{1}{K}\sum_{k=0}^{K-1}\hat{S}_k^{(mt)}(f) \text{ for } \hat{S}_k^{(mt)}(f) \equiv \Delta t\left|\sum_{t=1}^{N}h_{t,k}X_t e^{-i2\pi ft\,\Delta t}\right|^2,$$

(333)

where $\{h_{t,k}\}$ is the data taper for the kth direct spectral estimator $\hat{S}_k^{(mt)}(\cdot)$ (as usual, we assume the normalization $\sum_{t=1}^{N}h_{t,k}^2 = 1$; see the discussion about Equation (208a)). In Thomson (1982) the estimator $\hat{S}_k^{(mt)}(\cdot)$ is called the kth *eigenspectrum*, but note that it is just a special case of the familiar direct spectral estimator $\hat{S}^{(d)}(\cdot)$ of Equation (206c). For each data taper, we can define an associated spectral window

$$\mathcal{H}_k(f) \equiv \Delta t\left|\sum_{t=1}^{N}h_{t,k}e^{-i2\pi ft\,\Delta t}\right|^2.$$

From Equation (207a) the first moment properties of $\hat{S}_k^{(mt)}(\cdot)$ are given by

$$E\{\hat{S}_k^{(mt)}(f)\} = \int_{-f_{(N)}}^{f_{(N)}} \mathcal{H}_k(f - f')S(f')\,df'.$$

It follows readily that

$$E\{\hat{S}^{(mt)}(f)\} = \int_{-f_{(N)}}^{f_{(N)}} \overline{\mathcal{H}}(f - f')S(f')\,df' \text{ with } \overline{\mathcal{H}}(f) \equiv \frac{1}{K}\sum_{k=0}^{K-1}\mathcal{H}_k(f).$$

(334a)

The function $\overline{\mathcal{H}}(\cdot)$ is the spectral window for the estimator $\hat{S}^{(mt)}(\cdot)$. Because the sidelobe level of $\mathcal{H}_k(\cdot)$ dictates whether or not $\hat{S}_k^{(mt)}(\cdot)$ is approximately free of leakage, the above equation tells us that the preponderance of the K spectral windows must provide good protection against leakage if $\overline{\mathcal{H}}(\cdot)$ is to produce a leakage-free $\hat{S}^{(mt)}(\cdot)$. On the other hand, the reason for averaging the K different eigenspectra is to produce an estimator of $S(f)$ with smaller variance than that of any individual $\hat{S}_k^{(mt)}(f)$. For example, if the $\hat{S}_k^{(mt)}(f)$ terms are pairwise uncorrelated with a common variance, then the variance of $\hat{S}^{(mt)}(f)$ would be smaller than that of $\hat{S}_k^{(mt)}(f)$ by a factor of $1/K$.

We thus need a set of K data tapers such that each one provides good protection against leakage and such that the resulting individual eigenspectra are as nearly uncorrelated as possible. As we shall see, if the sdf $S(\cdot)$ satisfies certain conditions, approximate uncorrelatedness follows if the data tapers are orthogonal in the sense that

$$\sum_{t=1}^{N} h_{t,j} h_{t,k} = 0 \text{ for all } j \neq k.$$

A set of K orthogonal data tapers with good leakage properties is given by portions of the *discrete prolate spheroidal sequences* (dpss) with parameter W and orders $k = 0$ to $K - 1$ if K is chosen to be less than the Shannon number $2NW\,\Delta t$ – recall that $2W$ defines the bandwidth for the concentration problems posed in Section 3.9, for which the zeroth-order dpss is the optimum solution (the reader should note that, in that section, W is expressed in standardized units so that $0 < W < 1/2$, whereas here W has physically meaningful units so that now $0 < W < 1/(2\,\Delta t) = f_{(N)}$). Hence we now define $\{h_{t,k}\}$ to be

$$h_{t,k} \equiv \begin{cases} v_{t-1,k}(N,W), & t = 1,\ldots,N; \\ 0, & \text{otherwise,} \end{cases}$$

(334b)

where $\{v_{t,k}(N,W)\}$ is the notation we used in Section 3.9 for the kth order dpss. We refer to $\{h_{t,k}\}$ as the *kth order dpss data taper*.

We can now outline a recipe for computing $\hat{S}^{(mt)}(\cdot)$.

[1] We first select the *resolution bandwidth* $2W$. Typically W is taken to be a small multiple $j > 1$ of the fundamental frequency so that $W = j/(N\,\Delta t)$ for, say, $j = 2$, 3 or 4. We note, however, that noninteger multiples are sometimes used (for example, Thomson, 1982, p. 1086, uses 3.5) and that larger values than 4 are sometimes of interest. The value of W is usually expressed indirectly via the duration $\times$ half bandwidth product $NW = j/\Delta t$ (or just $NW = j$ with Δt standardized to unity). In the example to follow, we picked $NW = 4/\Delta t$, but a reasonable choice of W must take into account the following tradeoff. If we make W large, the number of tapers with good leakage properties increases – hence we can make K large and decrease the variance of $\hat{S}^{(mt)}(f)$. On the other hand, we shall see that the resolution of $\hat{S}^{(mt)}(\cdot)$ decreases as W increases – if we make W too large, we can inadvertently smear out fine features in the sdf.

[2] With W so specified, we compute the eigenspectra $\hat{S}_k^{(mt)}(\cdot)$ for $k = 0$ to $K - 1$ with $K < 2NW\,\Delta t$. Unless we are dealing with a process that is very close to white noise, these will be all of the eigenspectra with potentially good first moment properties. Additionally, for interest, we compute here the eigenspectrum for $k = K$. For this step we obviously must first compute the dpss data tapers, a subject that is discussed in detail in Chapter 8.

[3] Finally we average together K of these eigenspectra, where K is determined by examining the individual eigenspectra for evidence of leakage.

We note here that there are more sophisticated ways of forming a multitaper spectral estimator than simply averaging the eigenspectra. These involve various weighted averages of the eigenspectra – with the weights in general being both frequency and data dependent – and can be implemented in a fairly automatic fashion. These are introduced briefly in Section 7.4 and discussed in detail in Thomson (1982).

As an example, let us construct multitaper spectral estimates for the AR(4) time series shown in the bottom of Figure 45 (see Section 6.18 for a conventional nonparametric spectral analysis of this series). Here $N = 1024$ and $\Delta t = 1$, and we have set W by letting $NW = 4$; i.e., $W = 4/N = 1/256$. Since $2NW = 8$, we need to compute the eigenspectra of orders $k = 0$ to 6 because these are the ones with potentially good bias properties. Additionally, we compute the eigenspectrum for $k = 7$. The left-hand plots of Figures 336 and 338 show the dpss data tapers $\{h_{t,k}\}$ of orders 0 to 7; the corresponding right-hand plots show the product $\{h_{t,k}X_t\}$ of these tapers with the AR(4) time series. Note carefully what happens to the values at the beginning and end of this time series as k varies from 0 to 7: for $k = 0$, these extremes are severely

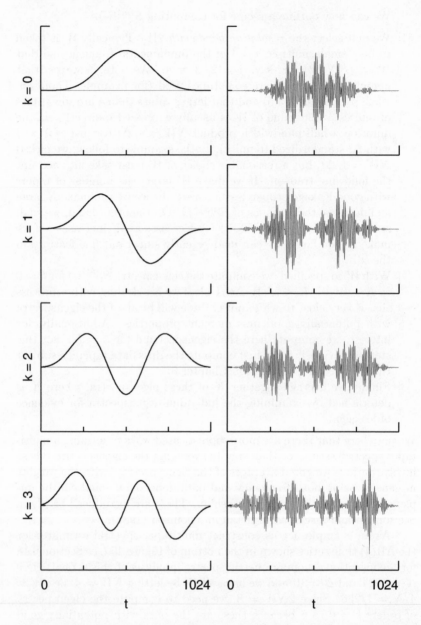

Figure 336. Discrete prolate spheroidal sequence data tapers $\{h_{t,k}\}$ (left-hand plots) and product $\{h_{t,k}X_t\}$ of these tapers and a time series $\{X_t\}$ (right-hand plots), part 1. Here $NW = 4$ with $N = 1024$. The orders k for the dpss data tapers are 0, 1, 2 and 3 (top to bottom rows). The time series is the realization of the AR(4) process shown in the bottom of Figure 45.

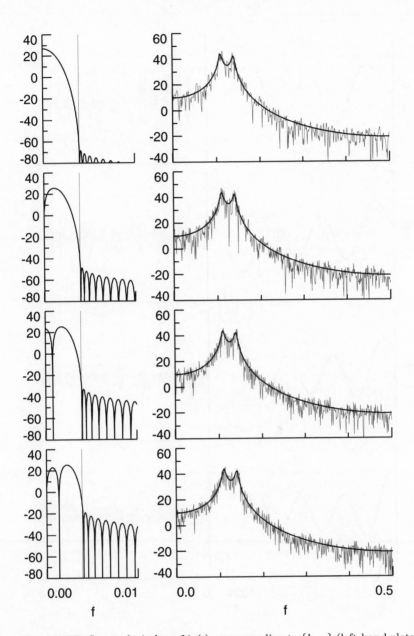

Figure 337. Spectral windows $\mathcal{H}_k(\cdot)$ corresponding to $\{h_{t,k}\}$ (left-hand plots) and eigenspectra $\hat{S}_k^{(mt)}(\cdot)$ for AR(4) time series (right-hand plots, thin curves), part 1. The orders k are the same as in Figure 336. The true sdf for the AR(4) process is the thick curve in each right-hand plot. The vertical scale for all plots is in decibels.

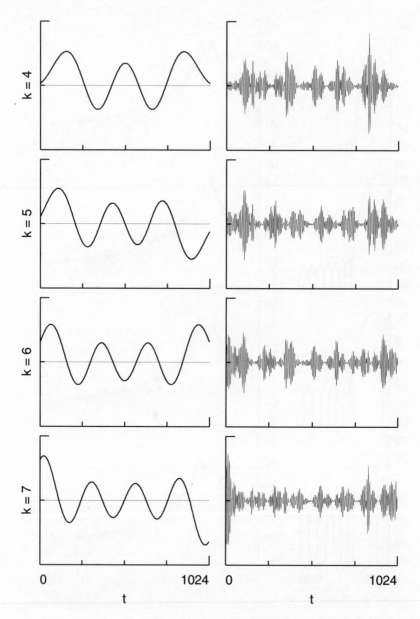

Figure 338. Discrete prolate spheroidal sequence data tapers and product of these tapers and a time series, part 2 (cf. Figure 336). Here the orders k for the tapers are 4, 5, 6 and 7 (top to bottom rows). Note that, for $k = 0$ in the top row of Figure 336, the extremes of the time series are severely attenuated, but that, as k increases, portions of the extremes are accentuated.

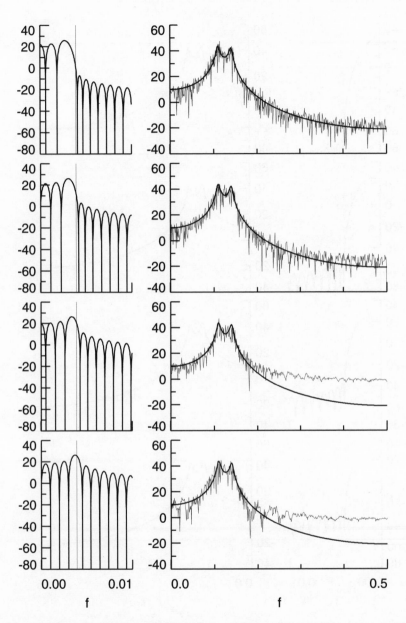

Figure 339. Spectral windows $\mathcal{H}_k(\cdot)$ corresponding to $\{h_{t,k}\}$ (left-hand plots) and eigenspectra $\hat{S}_k^{(mt)}(\cdot)$ for AR(4) time series (right-hand plots, thin curves), part 2. The orders k are the same as in Figure 338. Note the marked increase in bias in the eigenspectra as k increases.

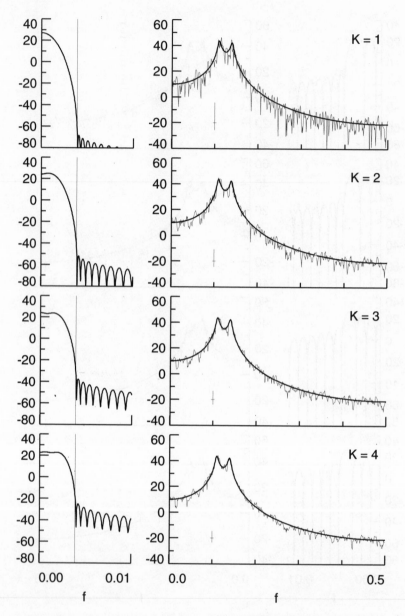

Figure 340. Multitaper spectral estimates $\hat{S}^{(mt)}(\cdot)$ formed by averaging K eigenspectra (right-hand plots) and corresponding spectral windows $\overline{\mathcal{H}}(\cdot)$ (left-hand plots) for $K = 1$, 2, 3 and 4. The thin vertical lines on the left-hand plots indicate the frequency $W = 4/N = 1/256$. The vertical scale for all plots is in decibels.

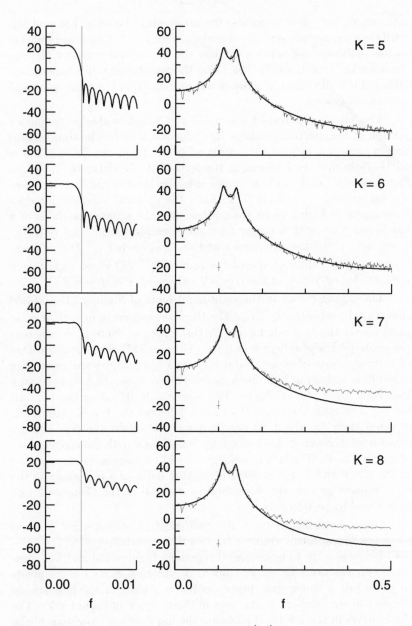

Figure 341. Multitaper spectral estimates $\hat{S}^{(mt)}(\cdot)$ formed by averaging K eigenspectra (right-hand plots) and corresponding spectral windows $\overline{\mathcal{H}}(\cdot)$ (left-hand plots) for $K = 5$, 6, 7 and 8. Note the increasing discrepancy at high frequencies between $\hat{S}^{(mt)}(\cdot)$ and the true sdf as K gets large.

attenuated, but, as k increases, the attenuation becomes less and less until the extremes are actually *accentuated* for $k = 7$. One interpretation of the multitapering scheme is that the higher order tapers pick up 'information' that is lost by just using the zeroth-order dpss taper alone (this point is discussed in more detail in item [1] of the Comments and Extensions below).

The left-hand plots of Figures 337 and 339 show the low frequency portion of the spectral windows $\mathcal{H}_k(\cdot)$ for $k = 0$ to 7. The thin vertical line in each plot indicates the location of the frequency $W = 1/256 \approx 0.004$. Note that, as k increases, the level of the sidelobes of $\mathcal{H}_k(\cdot)$ also increases until at $k = 7$ the main sidelobe level is just barely below the lowest lobe in $[-W, W]$. In fact, the spectral windows for dpss data tapers of higher order than 7 resemble the squared modulus of a band-pass filter with a center frequency outside the interval $[-W, W]$ – see item [2] of the Comments and Extensions below. Use of these tapers would yield direct spectral estimates $\hat{S}_k^{(mt)}(f)$ whose expectation is controlled by values of the true sdf *outside* $[f - W, f + W]$!

The jagged curves in the right-hand plots of Figures 337 and 339 show the eigenspectra $\hat{S}_k^{(mt)}(\cdot)$. The thick smooth curve in each of these plots shows the true sdf for the AR(4) process. Note that, whereas the eigenspectra of orders $k = 0$ to 3 (Figure 337) show no indication of leakage, those of order 4 and higher (Figure 339) show increasing evidence of leakage at high frequencies. For example, the $k = 4$ estimate (top right-hand plot of Figure 339) is about 5 dB above the true sdf for frequencies in the range 0.3 to 0.5, whereas the $k = 7$ estimate is elevated there by about 20 dB (had we not known the true sdf, we could have discovered this leakage by comparison with the eigenspectra of orders 0 to 4). The lack of homogeneity of variance across frequencies in the $k = 6$ and 7 eigenspectra is another indication of leakage in the high frequencies (see the discussion in item [4] of the Comments and Extensions to Section 6.4).

Let us now consider the multitaper spectral estimates $\hat{S}^{(mt)}(\cdot)$ and corresponding spectral windows $\overline{\mathcal{H}}(\cdot)$ for different choices of K, the number of eigenspectra to be averaged together. Figures 340 and 341 show these two functions for $K = 1$ up to 8 (the case $K = 1$ corresponds to using just a single data taper, so the two plots in the first row of Figure 340 are identical to the ones in the first row of Figure 337). The thick curves in the left-hand plots are the low frequency portions of the spectral windows, while – as before – the thin vertical lines mark the frequency $W = 1/256$. The jagged curves in the right-hand plots are the corresponding multitaper spectral estimates, and the thick smooth curve in each of these plots is the true sdf. Note that, as K increases, the level of the sidelobes in $\overline{\mathcal{H}}(\cdot)$ increases (as we shall see in Sections 7.3 and 7.4, this is related to an increase in what we refer to there as broad-band bias); note also that $\overline{\mathcal{H}}(\cdot)$ becomes noticeably closer to being constant over frequencies less than W in magnitude (this is related to a decrease in the – yet to be defined – local bias).

Let us now study how well the $\hat{S}^{(mt)}(\cdot)$ in the right-hand plots do in terms of bias and variance. For $K = 1$ up to 5, we see that there is little evidence of any bias in $\hat{S}^{(mt)}(\cdot)$ at any frequency; for $K = 6$ there is evidence of a small bias of about 3 dB for frequencies between 0.4 and 0.5; and, for $K = 7$ and 8, there is significant bias in the high frequencies. On the other hand, as K increases, we see that the variability in $\hat{S}^{(mt)}(\cdot)$ steadily decreases. We can quantify this variability in the following way. Since each eigenspectrum is simply a direct spectral estimate, we know from Equation (223b) that, if we exclude the zero and Nyquist frequencies, $\hat{S}_k^{(mt)}(f)$ is approximately distributed as a rescaled chi-square rv with 2 degrees of freedom, with the scaling factor being $S(f)/2$. If the different eigenspectra are approximately independent (as the theory in the sections following suggests they should be), then the multitaper spectral estimator $\hat{S}^{(mt)}(f)$ should approximately follow a chi-square distribution with $2K$ degrees of freedom and a scaling factor of $S(f)/(2K)$. Just as we did using the lag window spectral estimates in Section 6.10, we can use this approximation to compute a, say, 95% confidence interval for $10\log_{10}(S(f))$ based upon $10\log_{10}(\hat{S}^{(mt)}(f))$. The form of the confidence interval is the same as that shown in Equation (258) with $\hat{S}^{(mt)}(f)$ substituted for $\hat{S}^{(lw)}(f)$, ν set equal to $2K$, and λ set to $10\log_{10}(e) = 4.343$. The vertical height of the crisscross in the lower left-hand portion of the plots with $\hat{S}^{(mt)}(\cdot)$ shows the width of such a 95% confidence interval. Note that, as K increases, this height decreases and that, except in the high frequency regions of $\hat{S}^{(mt)}(\cdot)$ for $K = 7$ and 8 where bias is dominant, the amount of variability in $\hat{S}^{(mt)}(\cdot)$ is roughly consistent with these heights. The horizontal width of each crisscross is just $2W$, a natural measure of the bandwidth of $\hat{S}^{(mt)}(\cdot)$ (this measure is more appropriate for large K).

To conclude, it appears that, for this particular example, the multitaper spectral estimator with $K = 5$ data tapers gives us the best compromise between good bias and variance properties. Since the variations in the true sdf are all on a larger scale than our selected bandwidth $2W$ (compare the width of the crisscrosses with the true sdf's in Figure 340 or 341), we could obtain a $\hat{S}^{(mt)}(\cdot)$ with similar bias properties and reduced variance by increasing W by a factor of, say, 2 so that $NW = 8$. Finally, we note that this particular time series could be handled just as well using either prewhitening (all stationary AR processes can be prewhitened *perfectly*, a nicety that never happens with real data!) or a WOSA spectral estimator (see the bottom plot of Figure 316; this estimate is approximately bias free, has 7.6 degrees of freedom, and hence is roughly comparable to the $K = 4$ multitaper estimate in the lower right-hand plot of Figure 340).

Comments and Extensions to Section 7.1

[1]   We present here support for the statement that multitapering is a scheme for recovering information usually lost when but a single data taper is used (Walden, 1990b). In Section 3.9 we found that the vector

$$\mathbf{v}_k(N, W) \equiv [v_{0,k}(N,W), v_{1,k}(N,W), \ldots, v_{N-1,k}(N,W)]^T$$
$$= [h_{1,k}, h_{2,k}, \ldots, h_{N,k}]^T$$

is the kth order eigenvector for the $N \times N$ real symmetric matrix A with (t', t)th element given by $\sin[2\pi W(t' - t)]/[\pi(t' - t)]$. Now the matrix $V \equiv [\mathbf{v}_0(N,W), \mathbf{v}_1(N,W), \ldots, \mathbf{v}_{N-1}(N,W)]$ whose columns are these eigenvectors is an orthogonal matrix. Since each eigenvector is assumed to be normalized to have a unit sum of squares, we have $V^T V = I$, where I is the $N \times N$ identity matrix. Because the transpose of V is its inverse, it follows that $VV^T = I$ also. This yields the interesting result that

$$\sum_{k=0}^{N-1} h_{t,k}^2 = 1 \text{ for } t = 1, \ldots, N;$$

i.e., the sum over all orders k of the squares of the tth element of each taper is unity. The energy from all N tapered series $\{h_{t,k} X_t\}$ is thus

$$\sum_{k=0}^{N-1} \sum_{t=1}^{N} (h_{t,k} X_t)^2 = \sum_{t=1}^{N} X_t^2 \sum_{k=0}^{N-1} h_{t,k}^2 = \sum_{t=1}^{N} X_t^2;$$

i.e., the total energy using all N possible tapers equals the energy in the time series. Figure 345 shows the build-up of $\sum_{k=0}^{K-1} h_{t,k}^2$ for $t = 1, \ldots, N$ as K increases from 1 to 8 (as before, $N = 1024$, and W is set such that $NW = 4$). Loosely speaking, the plot for a particular K shows the relative influence of X_t – as t varies – on the multitaper spectral estimator $\hat{S}^{(mt)}(\cdot)$ formed using K eigenspectra. The $K = 1$ estimator (i.e., a single dpss taper) is not influenced much by values of $\{X_t\}$ near $t = 1$ or 1024. As K increases to 7, however, the amount of data in the extremes that has little influence decreases steadily, supporting the claim that multitapering is picking up information from the extremes of a time series that is lost with use of a single data taper. Increasing K further to 8 shows a breakdown in which certain values near the extremes start to have *greater* influence than the central portion of the time series (see the case $K = 9$ in the bottom plot of Figure 346).

[2]   The previous figures in this chapter refer to the dpss data tapers $\{h_{t,k}\}$ of orders $k = 0$ to 7 for $N = 1024$ and $NW = 4$. For this particular choice of W, these are all the tapers whose associated spectral

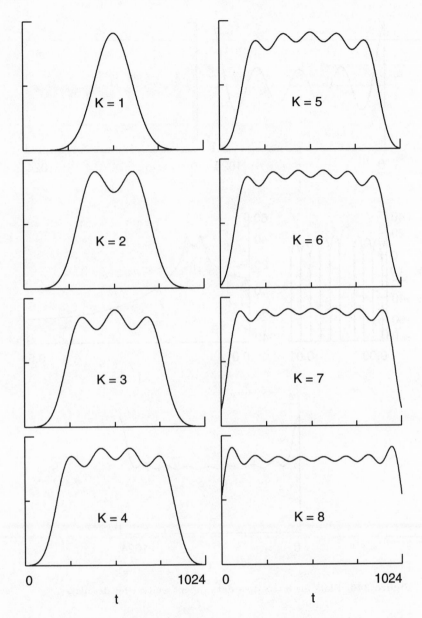

Figure 345. Decomposition of dpss taper energy across t as K – the number of tapers used in the multitapering scheme – increases (the curve on each plot is normalized such that its peak value occurs at the same relative height).

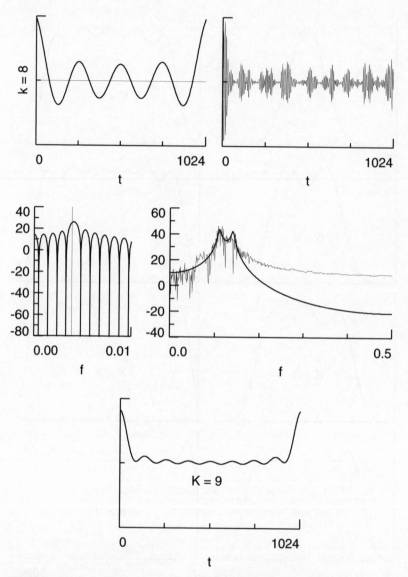

Figure 346. Plots for $k = 8$ dpss data taper (see text for details).

windows $\mathcal{H}_k(\cdot)$ have the majority of their energy concentrated in the frequency interval $[-W, W]$; i.e., we have

$$\frac{\int_{-W}^{W} \mathcal{H}_k(f)\,df}{\int_{-f_{(N)}}^{f_{(N)}} \mathcal{H}_k(f)\,df} = \lambda_k(N, W) > 1/2 \ \text{ for } \ k = 0, \ldots, 7,$$

where $\lambda_k(N, W)$ is the kth order eigenvalue defined in Section 3.9 (the normalization $\sum_{t=1}^{N} h_{t,k}^2 = 1$ implies that $\int_{-f_{(N)}}^{f_{(N)}} \mathcal{H}_k(f)\, df = 1$, so the denominator in the ratio above is just unity). The eigenvalue $\lambda_k(N, W)$ is thus a useful measure of concentration of $\mathcal{H}_k(\cdot)$ over the frequency range $[-W, W]$. For our particular example, we have the following values:

$$\lambda_0(N, W) = 0.999\,999\,999\,7 \qquad \lambda_4(N, W) = 0.999\,4$$
$$\lambda_1(N, W) = 0.999\,999\,97 \qquad \lambda_5(N, W) = 0.993$$
$$\lambda_2(N, W) = 0.999\,998\,8 \qquad \lambda_6(N, W) = 0.94$$
$$\lambda_3(N, W) = 0.999\,97 \qquad \lambda_7(N, W) = 0.70.$$

For the sake of comparison, Figure 346 shows corresponding plots for the eighth-order dpss data taper, for which $\lambda_8(N, W) = 0.30$ (for $k > 8$, the concentration measures $\lambda_k(N, W)$ are all close to zero – see Figure 110). The two plots in the top row show the taper itself and the product of the taper and the AR(4) series. Note, in particular, the accentuation of the extremes of the series. The left-hand plot in the middle row shows that the peak value of the spectral window $\mathcal{H}_8(\cdot)$ occurs just *outside* the frequency interval $[-W, W]$; the right-hand plot shows the poor leakage properties of $\hat{S}_8^{(mt)}(\cdot)$. Finally, as noted previously, the plot on the last row shows that the distribution of energy over t is skewed to accentuate the extremes of $\{X_t\}$ when $K = 9$ tapers are used to form $\hat{S}^{(mt)}(\cdot)$.

7.2 Multitapering of Gaussian White Noise

The statistical properties of the multitaper estimator $\hat{S}^{(mt)}(\cdot)$ are quite easy to derive in the case of Gaussian white noise and serve to illustrate the decrease in variability that multitapering affords. Accordingly, for this section only, let us assume that the time series we are dealing with is a realization of a portion $X_1,\, X_2,\, \ldots,\, X_N$ of a Gaussian white noise process with zero mean and unknown variance s_0. The sdf is thus just $S(f) = s_0\, \Delta t$. Standard statistical theory suggests that the best (in a number of different senses) estimator of s_0 for this model is just the sample variance

$$\frac{1}{N} \sum_{t=1}^{N} X_t^2 = \hat{s}_0^{(p)}.$$

It follows that the best estimator for the sdf is just $\hat{s}_0^{(p)}\, \Delta t$. Equation (278b) tells us that

$$\operatorname{var}\left\{\hat{s}_0^{(p)}\, \Delta t\right\} = \frac{2 s_0^2 (\Delta t)^2}{N}. \tag{347}$$

As we discussed in Section 6.3, the periodogram $\hat{S}^{(p)}(f)$ is an unbiased estimator of $S(f)$ in the case of white noise, and hence a data taper is certainly not needed to control bias. Moreover, because $S(\cdot)$ is flat, we can smooth $\hat{S}^{(p)}(\cdot)$ with a rectangular smoothing window of width $2f_{(N)}$ and height $1/(2f_{(N)}) = \Delta t$ to obtain

$$\int_{-f_{(N)}}^{f_{(N)}} \hat{S}^{(p)}(f)\,\Delta t\,df = \hat{s}_0^{(p)}\,\Delta t$$

(this follows from Exercise [6.6a]). We can thus easily recover the best spectral estimator $\hat{s}_0^{(p)}\,\Delta t$ by averaging $\hat{S}^{(p)}(\cdot)$ over $[-f_{(N)}, f_{(N)}]$. On the other hand, suppose that we use any nonrectangular taper $\{\tilde{h}_{t,0}\}$ – with the usual normalization $\sum_{t=1}^{N} \tilde{h}_{t,0}^2 = 1$ – to produce the direct spectral estimator $\hat{S}^{(d)}(\cdot)$. If we smooth this estimator, we obtain

$$\int_{-f_{(N)}}^{f_{(N)}} \hat{S}^{(d)}(f)\,\Delta t\,df = \sum_{t=1}^{N} \tilde{h}_{t,0}^2 X_t^2\,\Delta t = \hat{s}_0^{(d)}\,\Delta t.$$

Now

$$\operatorname{var}\{\hat{s}_0^{(d)}\,\Delta t\} = (\Delta t)^2 \sum_{t=1}^{N} \operatorname{var}\{\tilde{h}_{t,0}^2 X_t^2\} = 2s_0^2\,(\Delta t)^2 \sum_{t=1}^{N} \tilde{h}_{t,0}^4$$

since we have $\operatorname{var}\{X_t^2\} = 2s_0^2$ from the Isserlis theorem (Equation (40)). The Cauchy inequality, namely,

$$\left| \sum_{t=1}^{N} a_t b_t \right|^2 \le \sum_{t=1}^{N} |a_t|^2 \sum_{t=1}^{N} |b_t|^2,$$

with $a_t = \tilde{h}_{t,0}^2$ and $b_t = 1$, tells us that $\sum_{t=1}^{N} \tilde{h}_{t,0}^4 \ge 1/N$ with equality holding if and only if $\tilde{h}_{t,0} = \pm 1/\sqrt{N}$ (i.e., $\{\tilde{h}_{t,0}\}$ is – to within an innocuous change of sign – the rectangular data taper). We can conclude that $\operatorname{var}\{\hat{s}_0^{(p)}\,\Delta t\} < \operatorname{var}\{\hat{s}_0^{(d)}\,\Delta t\}$ for any nonrectangular data taper. Use of a single data taper on white noise yields a smoothed spectral estimator with increased variance and thus amounts to throwing away a certain portion of the data (Sloane, 1969).

We now show that we can compensate for this loss through multitapering. Accordingly, let $\{\tilde{h}_{t,0}\}$, $\{\tilde{h}_{t,1}\}$, ..., $\{\tilde{h}_{t,N-1}\}$ be *any* set of orthonormal data tapers, each of length N (note that these need not be dpss tapers). If we let $\tilde{V}$ be the $N \times N$ matrix whose columns are formed from these tapers, i.e.,

$$\tilde{V} \equiv \begin{bmatrix} \tilde{h}_{1,0} & \tilde{h}_{1,1} & \cdots & \tilde{h}_{1,N-1} \\ \tilde{h}_{2,0} & \tilde{h}_{2,1} & \cdots & \tilde{h}_{2,N-1} \\ \vdots & \vdots & \ddots & \vdots \\ \tilde{h}_{N,0} & \tilde{h}_{N,1} & \cdots & \tilde{h}_{N,N-1} \end{bmatrix},$$

orthonormality says that $\tilde{V}^T\tilde{V} = I$, where I is the $N \times N$ identity matrix. Since the transpose of $\tilde{V}$ is thus its inverse, we also have $\tilde{V}\tilde{V}^T = I$; i.e.,

$$\sum_{k=0}^{N-1} \tilde{h}_{t,k}\tilde{h}_{u,k} = \begin{cases} 1, & \text{if } t = u; \\ 0, & \text{otherwise} \end{cases} \tag{349a}$$

(see item [1] of the Comments and Extensions to the previous section). Let

$$\tilde{S}_k^{(mt)}(f) \equiv \Delta t \left| \sum_{t=1}^{N} \tilde{h}_{t,k}X_t e^{-i2\pi ft\,\Delta t} \right|^2 \tag{349b}$$

be the kth direct spectral estimator, and consider the multitaper estimator formed by averaging all N of the $\tilde{S}_k^{(mt)}(f)$:

$$\tilde{S}^{(mt)}(f) \equiv \frac{1}{N} \sum_{k=0}^{N-1} \tilde{S}_k^{(mt)}(f).$$

We can write

$$\tilde{S}^{(mt)}(f) = \frac{\Delta t}{N} \sum_{k=0}^{N-1} \left(\sum_{t=1}^{N} \tilde{h}_{t,k}X_t e^{-i2\pi ft\,\Delta t} \right) \left(\sum_{u=1}^{N} \tilde{h}_{u,k}X_u e^{i2\pi fu\,\Delta t} \right)$$

$$= \frac{\Delta t}{N} \sum_{t=1}^{N} \sum_{u=1}^{N} X_t X_u \left(\sum_{k=0}^{N-1} \tilde{h}_{t,k}\tilde{h}_{u,k} \right) e^{-i2\pi f(t-u)\,\Delta t}.$$

From Equation (349a), the term in the parentheses is unity if $t = u$ and is zero otherwise, so we obtain

$$\tilde{S}^{(mt)}(f) = \frac{\Delta t}{N} \sum_{t=1}^{N} X_t^2 = \hat{s}_0^{(p)}\,\Delta t. \tag{349c}$$

A multitaper spectral estimator using N orthogonal data tapers is thus identical to the best spectral estimator for a white noise process, so multitapering effectively restores the information normally lost when but a single taper is used.

We note that the above argument is valid for *any* set of orthogonal data tapers. We can get away with such a cavalier choice only in the very special case of white noise because then any data taper yields an unbiased direct spectral estimator. For nonwhite processes, this is certainly not the case – we must then insist upon a set of orthogonal tapers, at least some of which provide decent leakage protection. With

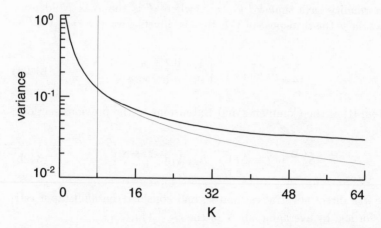

Figure 350. Decrease in var $\{\hat{S}^{(mt)}(f)\}$ for $f = 1/4$ as the number K of eigenspectra used to form the multitaper spectral estimator increases from 1 up to the sample size $N = 64$ (thick curve). Here a Gaussian white noise process is assumed with unit variance and a sampling time $\Delta t = 1$. The data tapers are dpss tapers with $NW = 4$. The thin curve indicates the decrease in variance to be expected if the eigenspectra were all uncorrelated. The thin vertical line marks the Shannon number $2NW = 8$.

this restriction, we are led (via the concentration measure) to the dpss data tapers as a natural set of choice.

Finally, let us study the rate at which the variance of the multitaper estimator decreases as the number of eigenspectra K that are averaged together increases. To do so, we now assume that we use the dpss data tapers $\{h_{t,k}\}$ to form the eigenspectra. The variance of $\hat{S}^{(mt)}(f)$ can be expressed as follows (see Exercise [2.8]):

$$\operatorname{var}\{\hat{S}^{(mt)}(f)\} = \operatorname{var}\left\{ \frac{1}{K} \sum_{k=0}^{K-1} \hat{S}_k^{(mt)}(f) \right\}$$

$$= \frac{1}{K^2} \sum_{j=0}^{K-1} \sum_{k=0}^{K-1} \operatorname{cov}\{\hat{S}_j^{(mt)}(f), \hat{S}_k^{(mt)}(f)\}.$$

Exercise [7.1b] indicates how to compute the covariance terms in the above expression for a Gaussian white noise process. The exercise also shows that var $\{\hat{S}^{(mt)}(f)\}$ depends on f. As a specific example, let us set $f = 1/4$. The thick curve in Figure 350 shows var $\{\hat{S}^{(mt)}(1/4)\}$ versus the number of eigenspectra K for $N = 64$, $NW = 4$, $\Delta t = 1$ and $s_0 = 1$. Note that, under these conditions, Equation (223c) says that var $\{\hat{S}^{(mt)}(1/4)\} \approx 1$ for $K = 1$ because $S(1/4) = s_0 \Delta t = 1$, while Equations (349c) and (347) tell us that var $\{\hat{S}^{(mt)}(1/4)\} = 2/64 \approx$

0.03 for $K = N = 64$. The extremes of the thick curve agree with these values. The thin curve shows var $\{\hat{S}^{(mt)}(f)\}$ versus K under the incorrect assumption that cov $\{\hat{S}_j^{(mt)}(f), \hat{S}_k^{(mt)}(f)\}$ is zero for all $j \neq k$. The thin vertical line marks the Shannon number $2NW = 8$. We see that, for $K \leq 8$, the variance decreases at a rate consistent with cov $\{\hat{S}_j^{(mt)}(f), \hat{S}_k^{(mt)}(f)\} = 0$ to a good approximation for all $j \neq k$ such that $0 \leq j, k < 8$. As K increases beyond the Shannon number, the rate of decrease in variance becomes slower than that implied by uncorrelatedness of the $\hat{S}_k^{(mt)}(f)$ terms. For nonwhite processes, this result suggests that the modest decrease in the variance of $\hat{S}^{(mt)}(f)$ afforded by including high-order eigenspectra will be readily offset by the poor bias properties of these eigenspectra.

7.3 Quadratic Spectral Estimators and Multitapering

We now present two complementary approaches to justifying the multitaper spectral estimator. The first is due to Bronez (1985, 1988) and is based upon a study of the first and second moments of quadratic spectral estimators (defined below); the second is closely related to the integral equation approach in Thomson's 1982 paper and is discussed in the next section.

Suppose that $X_1, \ldots, X_N$ is a segment of length N of a real-valued stationary process with zero mean and sdf $S(\cdot)$. For a fixed frequency f such that $0 \leq f \leq f_{(N)}$, let us define the complex-valued process

$$Z_t \equiv X_t e^{-i2\pi f t \, \Delta t}.$$

The results of Exercise [5.7a] tell us that $\{Z_t\}$ is a stationary process with zero mean and sdf given by

$$S_Z(f') = S(f + f') \text{ for } |f'| \leq f_{(N)}$$

(recall that $S(\cdot)$ is a periodic function with period $2f_{(N)}$). In particular, $S_Z(0) = S(f)$, so we can obtain an estimate of $S(f)$ via an estimate of $S_Z(\cdot)$ at zero frequency.

Let $\mathbf{Z}$ be an N-dimensional column vector whose tth element is Z_t, and let $\mathbf{Z}^H$ be its Hermitian transpose; i.e., $\mathbf{Z}^H \equiv (Z_1^*, \ldots, Z_N^*)$, where the asterisk denotes complex conjugation (for what follows, the reader should note that, if A is a matrix with real-valued elements, its Hermitian transpose A^H is equal to its usual transpose A^T). If X_t has units of, say, meters and if the sampling interval Δt is, say, 1 second, then the sdf $S(\cdot)$ has units of (meters)^2 per Hertz or, equivalently, seconds $\times$ (meters)^2/cycle. Since $S(\cdot)$ is real-valued, an estimator for $S(f)$ with the correct dimensionality is

$$\hat{S}^{(q)}(f) \equiv \hat{S}_Z^{(q)}(0) \equiv \Delta t \sum_{s=1}^{N} \sum_{t=1}^{N} Z_s^* Q_{s,t} Z_t = \Delta t \, \mathbf{Z}^H Q \mathbf{Z}, \qquad (351)$$

where $Q_{s,t}$ is the (s,t)th element of the $N \times N$ matrix Q of weights. For simplicity, we assume that $Q_{s,t}$ is real-valued; with this proviso, there is no loss of generality in assuming that Q is a symmetric matrix (since $\mathbf{Z}^H Q \mathbf{Z} = \mathbf{Z}^H \left(Q + Q^H \right) \mathbf{Z}/2$ with $Q + Q^H$ being necessarily symmetric). We assume that $Q_{s,t}$ does not depend on $\{Z_t\}$. An estimator of form (351) is known in the literature as a *quadratic estimator*. Note that, if the matrix Q is positive semidefinite, we have the desirable property $\hat{S}^{(q)}(f) \geq 0$.

We have already met an important example of a quadratic estimator in Chapter 6, namely, the lag window spectral estimator

$$\hat{S}^{(lw)}(f) \equiv \Delta t \sum_{\tau=-(N-1)}^{N-1} w_{\tau,m} \hat{s}_\tau^{(d)} e^{-i2\pi f \tau \, \Delta t}, \tag{352a}$$

where $\{w_{\tau,m}\}$ is a lag window;

$$\hat{s}_\tau^{(d)} \equiv \begin{cases} \sum_{t=1}^{N-|\tau|} h_t X_t h_{t+|\tau|} X_{t+|\tau|}, & \tau = 0, \pm 1, \ldots, \pm(N-1); \\ 0, & \text{otherwise;} \end{cases}$$

and $\{h_t\}$ is a data taper. Since Equation (352a) can be rewritten as

$$\hat{S}^{(lw)}(f) = \Delta t \sum_{s=1}^{N} \sum_{t=1}^{N} h_s X_s h_t X_t w_{t-s,m} e^{-i2\pi f(t-s) \, \Delta t}$$

$$= \Delta t \sum_{s=1}^{N} \sum_{t=1}^{N} Z_s^* h_s w_{t-s,m} h_t Z_t,$$

it follows that $\hat{S}^{(lw)}(f)$ can be expressed as a quadratic estimator with $Q_{s,t} = h_s w_{t-s,m} h_t$. Since a lag window spectral estimator can sometimes be negative (consider the reshaped lag window estimate of Figure 270(c)), it follows that the matrix Q for such an estimator need not be positive semidefinite. By letting $w_{\tau,m} = 1$ for all τ, a lag window spectral estimator reduces to a direct spectral estimator

$$\hat{S}^{(d)}(f) \equiv \Delta t \left| \sum_{t=1}^{N} h_t X_t e^{-i2\pi f t \, \Delta t} \right|^2, \tag{352b}$$

for which $Q_{s,t} = h_s h_t$. Since a direct spectral estimator can never be negative, the matrix Q for such an estimator must be positive semidefinite. The WOSA spectral estimator $\hat{S}^{(\text{WOSA})}(\cdot)$ of Equation (291b) is another example of a quadratic estimator whose Q matrix is necessarily positive semidefinite.

We want to find a quadratic estimator $\hat{S}^{(q)}(f)$ for $S(f)$ with good bias and variance properties. To ensure the condition $\hat{S}(f) \geq 0$, we insist from now on that Q in Equation (351) be positive semidefinite. If Q were positive definite, its rank would be N; since we are only assuming it to be positive semidefinite, its rank – call it K – can be taken to satisfy $1 \leq K \leq N$ (the case $K = 0$ is uninteresting). Based upon some standard results in linear algebra (see Exercise [7.2]), we can decompose Q as

$$Q = AA^T = AA^H,$$

where A is a real-valued $N \times K$ matrix such that $A^T A$ is a $K \times K$ diagonal matrix; i.e., if we let the N-dimensional column vector $\mathbf{a}_k$ represent the kth column of A, we have $\mathbf{a}_j^T \mathbf{a}_k = 0$ for $j \neq k$ so that the columns of A are mutually orthogonal. We can now write

$$\hat{S}^{(q)}(f) = \Delta t \, \mathbf{Z}^H A A^T \mathbf{Z} \tag{353a}$$

$$= \Delta t \, \mathbf{Z}^H \begin{bmatrix} \mathbf{a}_1 & \mathbf{a}_2 & \cdots & \mathbf{a}_K \end{bmatrix} \begin{bmatrix} \mathbf{a}_1^T \\ \mathbf{a}_2^T \\ \vdots \\ \mathbf{a}_K^T \end{bmatrix} \mathbf{Z}$$

$$= \Delta t \begin{bmatrix} \mathbf{Z}^H \mathbf{a}_1 & \mathbf{Z}^H \mathbf{a}_2 & \cdots & \mathbf{Z}^H \mathbf{a}_K \end{bmatrix} \begin{bmatrix} \mathbf{a}_1^T \mathbf{Z} \\ \mathbf{a}_2^T \mathbf{Z} \\ \vdots \\ \mathbf{a}_K^T \mathbf{Z} \end{bmatrix}$$

$$= \Delta t \sum_{k=1}^{K} \mathbf{Z}^H \mathbf{a}_k \mathbf{a}_k^T \mathbf{Z} = \Delta t \sum_{k=1}^{K} (\mathbf{a}_k^T \mathbf{Z})^* \mathbf{a}_k^T \mathbf{Z} = \Delta t \sum_{k=1}^{K} \left| \mathbf{a}_k^T \mathbf{Z} \right|^2$$

$$= \Delta t \sum_{k=1}^{K} \left| \sum_{t=1}^{N} a_{t,k} Z_t \right|^2 = \frac{\Delta t}{K} \sum_{k=0}^{K-1} \left| \sum_{t=1}^{N} \tilde{h}_{t,k} X_t e^{-i2\pi f t \, \Delta t} \right|^2,$$

where $a_{t,k}$ is the tth element of $\mathbf{a}_k$, and we define $\tilde{h}_{t,k} \equiv a_{t,k+1} \sqrt{K}$. Note that the above can be written as

$$\hat{S}^{(q)}(f) = \frac{1}{K} \sum_{k=0}^{K-1} \hat{S}_k^{(q)}(f) \text{ with } \hat{S}_k^{(q)}(f) \equiv \Delta t \left| \sum_{t=1}^{N} \tilde{h}_{t,k} X_t e^{-i2\pi f t \, \Delta t} \right|^2.$$
$$\tag{353b}$$

We have thus the important result that all quadratic estimators with a real-valued, symmetric, positive semidefinite matrix Q of weights can be written as an average of K direct spectral estimators $\hat{S}_k^{(q)}(f)$ (K being the rank of Q), the kth one of which uses $\{\tilde{h}_{t,k}\}$ as its data taper; moreover, because the $\mathbf{a}_k$ vectors are mutually orthogonal, it follows that the data tapers $\{\tilde{h}_{t,j}\}$ and $\{\tilde{h}_{t,k}\}$ are also orthogonal for $j \neq k$.

We need expressions for the expected value of $\hat{S}^{(q)}(f)$. It follows from Equations (353b) and (207a) that

$$E\{\hat{S}^{(q)}(f)\} = \frac{1}{K} \sum_{k=0}^{K-1} E\{\hat{S}_k^{(q)}(f)\} = \frac{1}{K} \sum_{k=0}^{K-1} \int_{-f_{(N)}}^{f_{(N)}} \widetilde{\mathcal{H}}_k(f - f')S(f')\, df'$$

$$= \int_{-f_{(N)}}^{f_{(N)}} \widetilde{\mathcal{H}}(f - f')S(f')\, df', \quad \text{(354a)}$$

where

$$\widetilde{\mathcal{H}}_k(f) \equiv \Delta t \left| \sum_{t=1}^{N} \tilde{h}_{t,k} e^{-i2\pi f t\, \Delta t} \right|^2 \quad \text{and} \quad \widetilde{\mathcal{H}}(f) \equiv \frac{1}{K} \sum_{k=0}^{K-1} \widetilde{\mathcal{H}}_k(f). \quad \text{(354b)}$$

The function $\widetilde{\mathcal{H}}(\cdot)$ is the spectral window of the quadratic spectral estimator $\hat{S}^{(q)}(\cdot)$, while $\widetilde{\mathcal{H}}_k(\cdot)$ is the spectral window of the kth direct spectral estimator.

It is also convenient to have a 'covariance domain' expression for $E\{\hat{S}^{(q)}(f)\}$. If we let $\{s_\tau\}$ represent the acvs for $\{X_t\}$, it follows from Exercise [5.7a] that the acvs $\{s_{\tau,Z}\}$ for $\{Z_t\}$ is given by $s_{\tau,Z} = s_\tau e^{-i2\pi f\tau}$. If we let Σ_Z be the $N \times N$ covariance matrix for $Z_1, \ldots, Z_N$ (i.e., its (j, k)th element is given by $s_{j-k,Z}$), it follows from Exercise [7.3] that

$$E\{\hat{S}^{(q)}(f)\} = \Delta t \operatorname{tr}\{Q\Sigma_Z\} = \Delta t \operatorname{tr}\{AA^T\Sigma_Z\} = \Delta t \operatorname{tr}\{A^T\Sigma_Z A\}, \quad \text{(354c)}$$

where $\operatorname{tr}\{\cdot\}$ is the matrix trace operator. If we equate the above to Equation (354a), we obtain

$$\int_{-f_{(N)}}^{f_{(N)}} \widetilde{\mathcal{H}}(f - f')S(f')\, df' = \Delta t \operatorname{tr}\{A^T\Sigma_Z A\}. \quad \text{(354d)}$$

We can make substantial progress in obtaining a quadratic estimator with quantifiably good first moment properties for a wide class of sdf's if we adopt a slightly different definition of what we mean by 'good.' If we were to use the same approach as in Chapter 6, we would take this to mean $E\{\hat{S}^{(q)}(f)\} \approx S(f)$. Here we adopt a criterion that incorporates the important notion of *resolution*: for a selected $W > 0$, we want

$$E\{\hat{S}^{(q)}(f)\} \approx \frac{1}{2W} \int_{f-W}^{f+W} S(f')\, df' \equiv \overline{S}(f), \quad \text{(354e)}$$

where $2W > 0$ defines a *resolution bandwidth*. Note that $\overline{S}(f)$ is the average value of $S(\cdot)$ over the interval from $f - W$ to $f + W$ (see Equation (86)).

The requirement that (354e) holds rather than just $E\{\hat{S}^{(q)}(f)\} \approx S(f)$ can be justified from two points of view. First, we noted in Section 5.6 that one interpretation of $S(f)$ is that $2S(f)\,df$ is approximately the variance of the process created by passing $\{X_t\}$ through a sufficiently narrow band-pass filter. If we were to actually use such a filter with bandwidth $2W$ and center frequency f, the spectral estimate we would obtain via the sample variance of the output from the filter would be more closely related to $\overline{S}(f)$ than to $S(f)$. The requirement $E\{\hat{S}^{(q)}(f)\} \approx \overline{S}(f)$ thus incorporates the limitations imposed by the filter bandwidth. Second, because we must base our estimate of the function $S(\cdot)$ on the finite sample $X_1, \ldots, X_N$, we face an inherently ill-posed problem in the sense that, with a finite number of observations, we cannot hope to determine $S(\cdot)$ over an infinite number of frequencies unless $S(\cdot)$ is sufficiently smooth. Since $\overline{S}(\cdot)$ is a smoothed version of $S(\cdot)$, it should be easier to find an approximately unbiased estimator for $\overline{S}(f)$ than for $S(f)$; moreover, if $S(\cdot)$ is itself sufficiently smooth, then $\overline{S}(\cdot)$ and $S(\cdot)$ should be approximately equal for W small enough. In effect, we are redefining the function to be estimated from one that need not be smooth, namely, $S(\cdot)$, to one that is guaranteed to be smooth to a certain degree, namely, $\overline{S}(\cdot)$.

With these comments in mind, let us see what constraints we need to impose on the matrix A in Equation (353a) so that $\hat{S}^{(q)}(f)$ is an approximately unbiased estimator of $\overline{S}(f)$, i.e., that $E\{\hat{S}^{(q)}(f)\} = \overline{S}(f)$ to some level of approximation. Can we insist upon having

$$E\{\hat{S}^{(q)}(f)\} = \overline{S}(f) = \frac{1}{2W} \int_{f-W}^{f+W} S(f')\,df'$$

for all possible $\overline{S}(\cdot)$? From Equation (354a) we would then need

$$\widetilde{\mathcal{H}}(f') = \begin{cases} 1/(2W), & |f'| \le W; \\ 0, & \text{otherwise.} \end{cases} \tag{355}$$

Since $\widetilde{\mathcal{H}}(\cdot)$ is the average of the $\widetilde{\mathcal{H}}_k(\cdot)$ terms, each one of which is proportional to the modulus squared of the Fourier transform of a finite sequence of values $\{\tilde{h}_{t,k}\}$, we cannot attain the simple form stated in (355) (see the discussion at the beginning of Section 5.8). We must therefore be satisfied with unbiasedness to some level of approximation. The approach that we take is, first, to insist that $\hat{S}^{(q)}(f)$ be an unbiased estimator of $\overline{S}(f)$ for white noise processes and, second, to develop bounds for two components of the bias for colored noise.

If $\{X_t\}$ is a white noise process with variance σ^2, then $S(f) = \sigma^2\,\Delta t$ for all f, so

$$\overline{S}(f) = \frac{1}{2W} \int_{f-W}^{f+W} \sigma^2\,\Delta t\,df' = \sigma^2\,\Delta t.$$

If $\hat{S}^{(q)}(f)$ is to be an unbiased estimator of $\overline{S}(f)$ for white noise, Equation (354a) tells us that

$$E\{\hat{S}^{(q)}(f)\} = \sigma^2 \, \Delta t \int_{-f_{(N)}}^{f_{(N)}} \widetilde{\mathcal{H}}(f - f') \, df' = \sigma^2 \, \Delta t,$$

i.e., that

$$\int_{-f_{(N)}}^{f_{(N)}} \widetilde{\mathcal{H}}(f - f') \, df' = \int_{-f_{(N)}}^{f_{(N)}} \widetilde{\mathcal{H}}(f') \, df = 1$$

(see Equation (208c)). Since Σ_Z for a white noise process is a diagonal matrix, all of whose elements are equal to σ^2, it follows from Equation (354d) that the above requirement is equivalent to

$$\mathrm{tr}\{A^T A\} = 1.$$

For a colored noise process let us define the bias, b, in the estimator $\hat{S}^{(q)}(f)$ as

$$b\{\hat{S}^{(q)}(f)\} \equiv E\{\hat{S}^{(q)}(f)\} - \overline{S}(f)$$

$$= \int_{-f_{(N)}}^{f_{(N)}} \widetilde{\mathcal{H}}(f - f')S(f') \, df' - \frac{1}{2W} \int_{f-W}^{f+W} S(f) \, df.$$

It is impossible to give any sort of reasonable bound for $b\{\hat{S}^{(q)}(f)\}$ since the second integral above does not depend upon the matrix A. It is useful, however, to split the bias into two components, one due to frequencies from $f - W$ to $f + W$ – the *local bias* – and the other, to frequencies outside this interval – the *broad-band bias*. Accordingly, we write

$$b\{\hat{S}^{(q)}(f)\} = b^{(l)}\{\hat{S}^{(q)}(f)\} + b^{(b)}\{\hat{S}^{(q)}(f)\},$$

where

$$b^{(l)}\{\hat{S}^{(q)}(f)\} \equiv \int_{f-W}^{f+W} \widetilde{\mathcal{H}}(f - f')S(f') \, df' - \frac{1}{2W} \int_{f-W}^{f+W} S(f) \, df$$

$$= \int_{f-W}^{f+W} \left(\widetilde{\mathcal{H}}(f - f') - \frac{1}{2W} \right) S(f') \, df'$$

and

$$b^{(b)}\{\hat{S}^{(q)}(f)\} \equiv \int_{-f_{(N)}}^{f-W} \widetilde{\mathcal{H}}(f - f')S(f') \, df' + \int_{f+W}^{f_{(N)}} \widetilde{\mathcal{H}}(f - f')S(f') \, df'.$$

To obtain useful upper bounds on the local and broad-band bias, let us assume that $S(\cdot)$ is bounded by $S_{\max}$, i.e., that $S(f) \leq S_{\max} < \infty$

for all f in the interval $[-f_{(N)}, f_{(N)}]$. For the local bias we then have

$$\left| b^{(l)}\{\hat{S}^{(q)}(f)\} \right| \leq \int_{f-W}^{f+W} \left| \tilde{\mathcal{H}}(f-f') - \frac{1}{2W} \right| S(f')\, df'$$

$$\leq S_{\max} \int_{-W}^{W} \left| \tilde{\mathcal{H}}(-f'') - \frac{1}{2W} \right| df'',$$

where $f'' \equiv f' - f$. We thus can take the quantity

$$\int_{-W}^{W} \left| \tilde{\mathcal{H}}(f'') - \frac{1}{2W} \right| df'' \tag{357}$$

as a useful measure of the magnitude of the local bias in $\hat{S}^{(q)}(f)$. Note that we can control the local bias by approximating the unattainable ideal $\tilde{\mathcal{H}}(\cdot)$ of Equation (355) as closely as possible over the resolution band.

Let us now obtain a bound for the broad-band bias (usually a more important source of concern than the local bias). With $f'' \equiv f' - f$ as before, we have

$$b^{(b)}\{\hat{S}^{(q)}(f)\} \leq S_{\max} \left(\int_{-f_{(N)}}^{f-W} \tilde{\mathcal{H}}(f-f')\, df' + \int_{f+W}^{f_{(N)}} \tilde{\mathcal{H}}(f-f')\, df' \right)$$

$$= S_{\max} \left(\int_{-f_{(N)}}^{f_{(N)}} \tilde{\mathcal{H}}(f-f')\, df' - \int_{f-W}^{f+W} \tilde{\mathcal{H}}(f-f')\, df' \right)$$

$$= S_{\max} \left(\int_{-f_{(N)}}^{f_{(N)}} \tilde{\mathcal{H}}(f-f')\, df' - \int_{-W}^{W} \tilde{\mathcal{H}}(-f'')\, df'' \right).$$

By considering Equation (354d) for a white noise process with unit variance, we can rewrite the first integral above as

$$\int_{-f_{(N)}}^{f_{(N)}} \tilde{\mathcal{H}}(f-f')\, df' = \mathrm{tr}\,\{A^T A\}$$

(this follows because Σ_Z then becomes the identity matrix). The second integral can be rewritten using (354d) again, but this time with an sdf $S^{(bl)}(\cdot)$ for band-limited white noise; i.e.,

$$S^{(bl)}(f) = \begin{cases} \Delta t, & |f| \leq W; \\ 0, & W < |f| \leq f_{(N)}. \end{cases}$$

The corresponding acvs $s_\tau^{(bl)}$ at lag τ is given by Equation (134a):

$$s_\tau^{(bl)} = \int_{-f_{(N)}}^{f_{(N)}} e^{i2\pi f'\tau\,\Delta t} S^{(bl)}(f')\, df'$$

$$= \int_{-W}^{W} e^{i2\pi f'\tau\,\Delta t}\Delta t\, df' = \frac{\sin(2\pi W\tau\,\Delta t)}{\pi\tau},$$

where, as usual, this ratio is defined to be $2W \Delta t$ when $\tau = 0$. If we let $\Sigma^{(bl)}$ be the $N \times N$ matrix whose (j, k)th element is $s_{j-k}^{(bl)}$, we obtain from Equation (354d) with $f = 0$

$$\int_{-W}^{W} \widetilde{\mathcal{H}}(-f'') \, df'' = \operatorname{tr} \{ A^T \Sigma^{(bl)} A \}.$$

We thus have that

$$b^{(b)} \{ \hat{S}^{(q)}(f) \} \leq S_{\max} \left(\operatorname{tr} \{ A^T A \} - \operatorname{tr} \{ A^T \Sigma^{(bl)} A \} \right),$$

so we can take the quantity

$$\operatorname{tr} \{ A^T A \} - \operatorname{tr} \{ A^T \Sigma^{(bl)} A \} \tag{358}$$

to be a useful measure of the broad-band bias in $\hat{S}^{(q)}(f)$.

Suppose for the moment that our only criterion for choosing the matrix A is that our broad-band bias measure be made as small as possible, subject to the constraint that the resulting estimator $\hat{S}^{(q)}(f)$ be unbiased for white noise processes. This means that we want to

$$\text{maximize } \operatorname{tr} \{ A^T \Sigma^{(bl)} A \} \text{ subject to } \operatorname{tr} \{ A^T A \} = 1.$$

Exercise [7.4] indicates that, if $\Sigma^{(bl)}$ were positive definite with a distinct largest eigenvalue, then a solution to this maximization problem would be to set A equal to a normalized eigenvector associated with the largest eigenvalue of the matrix $\Sigma^{(bl)}$. In fact, $\Sigma^{(bl)}$ satisfies the stated condition: because its (j, k)th element is just $\sin{[2\pi W(j - k) \Delta t]}/[\pi(j - k)]$, its eigenvalues and eigenvectors are the solutions to the problem posed previously by Equation (105b) in Section 3.9 (note that W on the left-hand side of (105b) must be replaced by $W \Delta t$; however, we use the same notation for the eigenvalues $\lambda_k(N, W)$ as before). From that discussion, we know that all of the eigenvalues are positive (and hence $\Sigma^{(bl)}$ is positive definite) and that there is only one eigenvector corresponding to the largest eigenvalue $\lambda_0(N, W)$. Moreover, the elements of the normalized eigenvector corresponding to $\lambda_0(N, W)$ can be taken to be a finite subsequence of the zeroth-order dpss, namely, $v_{0,0}(N, W)$, $v_{1,0}(N, W)$, $\ldots$, $v_{N-1,0}(N, W)$. Because of the definition for the zeroth-order dpss data taper $\{h_{t,0}\}$ in Equation (334b), the solution to the maximization problem is thus to set A to the $N \times 1$ vector whose elements are $h_{1,0}$, $h_{2,0}$, $\ldots$, $h_{N,0}$, i.e., the nonzero portion of $\{h_{t,0}\}$. If we denote this vector as $\mathbf{h}_0$, the constraint $\operatorname{tr} \{ \mathbf{h}_0^T \mathbf{h}_0 \} = 1$ implies that

$$\sum_{t=1}^{N} h_{t,0}^2 = 1,$$

consistent with the usual normalization for a data taper (see Equation (208a), and recall that the trace of a scalar is just the scalar itself). Note also that, because $\mathbf{h}_0$ is an eigenvector of $\Sigma^{(bl)}$ corresponding to the eigenvalue $\lambda_0(N, W)$, we have

$$\Sigma^{(bl)}\mathbf{h}_0 = \lambda_0(N, W)\mathbf{h}_0.$$

Using this fact, we see that the minimum value of the broad-band bias measure in (358) is simply

$$\begin{aligned}
\text{tr}\,\{\mathbf{h}_0^T\mathbf{h}_0\} - \text{tr}\,\{\mathbf{h}_0^T\Sigma^{(bl)}\mathbf{h}_0\} &= 1 - \text{tr}\,\{\mathbf{h}_0^T(\lambda_0(N, W)\mathbf{h}_0)\} \\
&= 1 - \text{tr}\,\{\lambda_0(N, W)\mathbf{h}_0^T\mathbf{h}_0\} \\
&= 1 - \lambda_0(N, W).
\end{aligned}$$

In summary, with the constraint $\text{tr}\,\{A^TA\} = 1$, the broad-band bias measure (358) is minimized by setting A equal to the N-dimensional vector $\mathbf{h}_0$ whose elements are the nonzero portion of a zeroth-order dpss data taper. The resulting quadratic estimator $\hat{S}^{(q)}(\cdot)$ has a weight matrix $Q = \mathbf{h}_0\mathbf{h}_0^T$ of rank $K = 1$, and the estimator amounts to a direct spectral estimator of the form of Equation (352b). Note that this method of specifying a quadratic estimator obviously does *not* take into consideration either the variance of the estimator or its local bias, both of which we now examine in more detail.

Suppose now that we wish to obtain a quadratic estimator (with a positive semidefinite weight matrix $Q = AA^T$) that has good variance properties, subject to the mild condition that the estimator be unbiased in the case of white noise, i.e., that $\text{tr}\,\{A^TA\} = 1$. From Equation (353b) we know that a rank K quadratic estimator $\hat{S}^{(q)}(f)$ can be written as the average of K direct spectral estimators $\hat{S}_k^{(q)}(f)$ employing a set of K orthonormal data tapers. In Section 7.2 we found that, for Gaussian white noise, the variance of such an estimator is minimized by using *any* set of $K = N$ orthonormal data tapers. This result is in direct conflict with the recommendation we came up with for minimizing our broad-band bias measure, namely, that we use just a single zeroth-order dpss data taper.

An obvious compromise estimator that attempts to balance this conflict between variance and broad-band bias is to use as many members K of the set of orthonormal dpss data tapers as possible. With this choice, our quadratic estimator $\hat{S}^{(q)}(f)$ becomes the multitaper estimator $\hat{S}^{(mt)}(f)$ defined by Equations (333) and (334b). Let us now discuss how this compromise affects the broad-band bias measure, the variance and the local bias measure for the resulting quadratic estimator.

[1] If we use K orthonormal dpss data tapers, the broad-band bias

measure of Equation (358) becomes

$$1 - \frac{1}{K} \sum_{k=0}^{K-1} \lambda_k(N, W) \qquad (360)$$

(this is Exercise [7.5]). Because the $\lambda_k(N, W)$ terms are all close
to unity as long as we set K to an integer less than the Shannon
number $2NW \, \Delta t$, the broad-band bias measure must be close to
zero if we impose the restriction $K < 2NW \, \Delta t$.

[2] For Gaussian white noise, we saw in Figure 350 that, for $K \leq$
$2NW \, \Delta t$, the variance of the average $\hat{S}^{(mt)}(f)$ of K direct spectral
estimators $\hat{S}_k^{(mt)}(f)$ is consistent with the approximation that all
K of the $\hat{S}_k^{(mt)}(f)$ rv's are pairwise uncorrelated. Since $\hat{S}_k^{(mt)}(f)$ is
approximately distributed as $S(f)\chi_2^2/2$ for $k = 0, \dots, K-1$ (as long
as f is not too close to 0 or $f_{(N)}$), this figure suggests that, with K so
chosen, the estimator $\hat{S}^{(mt)}(f)$ should be approximately distributed
as $S(f)\chi_{2K}^2/2K$. The variance of $\hat{S}^{(mt)}(f)$ is hence approximately
$S^2(f)/K$. For colored Gaussian processes, this approximation is
still good as long as $S(f)$ does not vary too rapidly over the interval
$[f - W, f + W]$ (see the next section or Thomson, 1982, Section IV).

[3] The local bias measure of Equation (357) is small if the spectral
window for $\hat{S}^{(mt)}(\cdot)$ is as close to $1/(2W)$ as possible over the resolution band. This window is denoted as $\overline{\mathcal{H}}(\cdot)$ in Equation (334a)
and is shown in the left-hand plots of Figures 340 and 341 for $K = 1$
to 8, $N = 1024$ and $W = 4/N$. Since $10 \log_{10}(1/(2W)) = 21$ dB
here, these plots indicate that, as K increases, the spectral window
$\overline{\mathcal{H}}(\cdot)$ becomes closer to $1/(2W)$ over the resolution band and hence
that the local bias decreases as K increases.

In summary, we have shown that all quadratic estimators with a symmetric real-valued positive semidefinite weight matrix Q of rank K can
be written as an average of K direct spectral estimators with orthogonal
data tapers. The requirement that the quadratic estimator be unbiased
for white noise is satisfied if the tapers are in fact orthonormal. We then
redefine (or 'regularize') our spectral estimation problem so that the
quantity to be estimated is $S(f)$ averaged over an interval of selectable
width $2W$ (i.e., $\overline{S}(f)$ of Equation (354e)) rather than just $S(f)$ itself.
This redefinition allows us to profitably split the bias of a quadratic
estimator into two parts, denoted as the local bias and the broad-band
bias. Minimization of a measure of broad-band bias dictates that we set
$K = 1$, with the single data taper being a zeroth-order dpss data taper;
on the other hand, the variance of a quadratic estimator is minimized
in the special case of Gaussian white noise by choosing K to be N, but
any set of N orthonormal data tapers yields the same minimum value.

The obvious compromise for colored noise is thus to use as many of the low-order members of the family of dpss data tapers as possible, yielding a multitaper spectral estimator $\hat{S}^{(mt)}(f)$. The broad-band bias measure suggests restricting K to be less than the Shannon number $2NW\,\Delta t$. With K thus restricted, the estimator $\hat{S}^{(mt)}(f)$ follows approximately a rescaled χ^2_{2K} distribution (provided that f is not too close to either 0 or $f_{(N)}$ and that $S(f)$ does not vary too rapidly over the interval $[f - W, f + W]$). The variance of $\hat{S}^{(mt)}(f)$ is inversely proportional to K – increasing K thus decreases the variance. A bound on the magnitude of the local bias also decreases as K increases, so an increase in K above unity improves both the variance and local bias of $\hat{S}^{(mt)}(f)$ at the cost of increasing its broad-band bias. (The reader should consult Bronez (1985, 1988) for an extension of the development in this section to the case of continuous parameter complex-valued stationary processes sampled at irregular time intervals.)

7.4 Regularization and Multitapering

In this section we present a rationale for the multitaper spectral estimator that closely follows Thomson's original 1982 approach. It makes extensive use of the spectral representation theorem and presents the spectral estimation problem as a search for a calculable approximation to a desirable quantity that unfortunately cannot be calculated from observable data alone. We begin with a review of some key concepts.

A way of representing any stationary process $\{X_t\}$ with zero mean is given by the spectral representation theorem introduced in Chapter 4:

$$X_t = \int_{-1/2}^{1/2} e^{i2\pi ft}\, dZ(f)$$

(this is Equation (130b); for convenience, we assume that $\Delta t = 1$ in this section, yielding a Nyquist frequency of $f_{(N)} = 1/2$). The increments of the orthogonal process $\{Z(f)\}$ define the sdf of $\{X_t\}$ as

$$S(f)\, df = E\{|dZ(f)|^2\}. \tag{361a}$$

Thomson (1982) suggests – purely for convenience – changing the definition of $dZ(\cdot)$ by a phase factor, which of course does not affect (361a):

$$X_t = \int_{-1/2}^{1/2} e^{i2\pi f[t-(N+1)/2]}\, dZ(f). \tag{361b}$$

Given $X_1, \ldots, X_N$, we want to relate the Fourier transform of these N values to the spectral representation for $\{X_t\}$. For convenience, we shall work with the following phase-shifted Fourier transform:

$$\Upsilon(f) \equiv \sum_{t=1}^{N} e^{-i2\pi f[t-(N+1)/2]} X_t. \tag{361c}$$

Thus, from Equations (361b) and (361c), we have

$$\Upsilon(f) = \int_{-1/2}^{1/2} \left(\sum_{t=1}^{N} e^{-i2\pi(f-f')[t-(N+1)/2]} \right) dZ(f'). \qquad (362a)$$

The summation in the parentheses is proportional to Dirichlet's kernel (see Exercise [1.3c]). Hence we arrive at a Fredholm integral equation of the first kind:

$$\Upsilon(f) = \int_{-1/2}^{1/2} \frac{\sin[N\pi(f-f')]}{\sin[\pi(f-f')]} dZ(f') \qquad (362b)$$

(Thomson, 1982). Since $dZ(\cdot)$ is by no stretch of the imagination a smooth function, it is not possible to solve for $dZ(\cdot)$ using standard inverse theory. We adopt an approach that emphasizes regularization of the spectral estimation problem and uses the same building blocks as Thomson. This leads us to consider projecting the Fourier transform $\Upsilon(\cdot)$ onto an appropriate set of basis functions, to which we now turn.

We introduced the discrete prolate spheroidal wave functions (dpswf's) in Section 3.9, where we denoted them as $U_k(\cdot; N, W)$, but here we simplify the notation to just $U_k(\cdot)$. As indicated in Exercise [3.12],

$$U_k(f) = (-1)^k \epsilon_k \sum_{t=0}^{N-1} v_{t,k}(N, W) e^{-i2\pi f[t-(N-1)/2]},$$

where

$$\epsilon_k \equiv \begin{cases} 1, & \text{if } k \text{ is even;} \\ \sqrt{-1}, & \text{if } k \text{ is odd;} \end{cases}$$

and $\{v_{t,k}(N, W)\}$ is the kth order dpss. If we define the kth order dpss data taper $\{h_{t,k}\}$ as in Equation (334b), we have

$$U_k(f) = (-1)^k \epsilon_k \sum_{t=1}^{N} h_{t,k} e^{-i2\pi f[t-(N+1)/2]}, \quad k = 0, \ldots, N-1. \quad (362c)$$

If we recall that the $U_k(\cdot)$ functions are orthogonal over $[-W, W]$, i.e.,

$$\int_{-W}^{W} U_j(f) U_k(f)\, df = \begin{cases} \lambda_k, & j = k; \\ 0, & \text{otherwise,} \end{cases} \qquad (362d)$$

then the rescaled functions $U_k(\cdot)/\sqrt{\lambda_k}$ are orthonormal:

$$\int_{-W}^{W} \frac{U_j(f)}{\sqrt{\lambda_j}} \frac{U_k(f)}{\sqrt{\lambda_k}}\, df = \begin{cases} 1, & j = k; \\ 0, & \text{otherwise} \end{cases}$$

(here λ_k is shorthand notation for $\lambda_k(N, W)$ defined in Section 3.9). Slepian (1978, Section 4.1) shows that the finite-dimensional space of functions of the form of Equation (361c) is spanned over the interval $[-W, W]$ by the rescaled dpswf's $U_k(\cdot)/\sqrt{\lambda_k}$. Hence we can write $\Upsilon(\cdot)$ as

$$\Upsilon(f) = \sum_{k=0}^{N-1} \Upsilon_k \frac{U_k(f)}{\sqrt{\lambda_k}} \text{ for } \Upsilon_k \equiv \int_{-W}^{W} \Upsilon(f) \frac{U_k(f)}{\sqrt{\lambda_k}} \, df$$

(see the discussion in item [2] of the Comments and Extensions to Section 5.1). If, for a given frequency f', we now define a new function $\Upsilon_{f'}(\cdot)$ via $\Upsilon_{f'}(f) \equiv \Upsilon(f + f')$, we can express it as

$$\Upsilon_{f'}(f) = \sum_{k=0}^{N-1} \Upsilon_k(f') \frac{U_k(f)}{\sqrt{\lambda_k}} \text{ for } \Upsilon_k(f') \equiv \int_{-W}^{W} \Upsilon_{f'}(f) \frac{U_k(f)}{\sqrt{\lambda_k}} \, df;$$

i.e., we have

$$\Upsilon(f + f') = \sum_{k=0}^{N-1} \Upsilon_k(f') \frac{U_k(f)}{\sqrt{\lambda_k}} \text{ for } \Upsilon_k(f') = \int_{-W}^{W} \Upsilon(f + f') \frac{U_k(f)}{\sqrt{\lambda_k}} \, df.$$

$$(363a)$$

If we substitute the expression for $\Upsilon(\cdot)$ given in Equation (361c) into the above (after switching the roles of the variables f and f'), we obtain

$$\Upsilon_k(f) = \int_{-W}^{W} \left(\sum_{t=1}^{N} e^{-i2\pi(f'+f)[t-(N+1)/2]} X_t \right) \frac{U_k(f')}{\sqrt{\lambda_k}} \, df'$$

$$= \sum_{t=1}^{N} e^{-i2\pi f[t-(N+1)/2]} X_t \int_{-W}^{W} e^{-i2\pi f'[t-(N+1)/2]} \frac{U_k(f')}{\sqrt{\lambda_k}} \, df'.$$

From Exercise [3.12] we have the following relationship between the kth order dpss $\{v_{t,k}(N, W)\}$ and the kth order dpswf $U_k(\cdot)$:

$$v_{t-1,k}(N, W) = \frac{(-1)^k}{\epsilon_k \sqrt{\lambda_k}} \int_{-W}^{W} e^{i2\pi f'[t-(N+1)/2]} \frac{U_k(f')}{\sqrt{\lambda_k}} \, df'.$$

Because both $v_{t-1,k}(N, W)$ and $U_k(\cdot)$ are real-valued and because $\epsilon_k^* = 1/\epsilon_k$, we also have

$$v_{t-1,k}(N, W) = \frac{(-1)^k \epsilon_k}{\sqrt{\lambda_k}} \int_{-W}^{W} e^{-i2\pi f'[t-(N+1)/2]} \frac{U_k(f')}{\sqrt{\lambda_k}} \, df'. \quad (363b)$$

This yields

$$\Upsilon_k(f) = \frac{\sqrt{\lambda_k}}{(-1)^k \epsilon_k} \sum_{t=1}^{N} e^{-i2\pi f[t-(N+1)/2]} v_{t-1,k}(N, W) X_t$$

$$= \epsilon_k \sqrt{\lambda_k} \sum_{t=1}^{N} h_{t,k} X_t e^{-i2\pi f[t-(N+1)/2]}. \quad (363c)$$

(here we use the fact that $1/((-1)^k \epsilon_k) = \epsilon_k$). Thus the projection $\Upsilon_k(\cdot)$ of the phase-shifted Fourier transform $\Upsilon(\cdot)$ onto the kth basis function $U_k(\cdot)/\sqrt{\lambda_k}$ is just the (phase-shifted and rescaled) Fourier transform of $X_1, \ldots, X_N$ multiplied by the kth order dpss data taper.

It is shown in Exercise [7.6] that another representation for $\Upsilon_k(f)$ is

$$\Upsilon_k(f) = \sqrt{\lambda_k} \int_{-1/2}^{1/2} U_k(f') \, dZ(f + f'). \qquad (364a)$$

In Equation (363a) the phase-shifted Fourier transform $\Upsilon_k(\cdot)$ is 'seen' through the dpswf with smoothing carried out only over the interval $[-W, W]$, while, in the alternative representation of Equation (364a), $dZ(\cdot)$ is seen through the dpswf with smoothing carried out over the whole Nyquist interval $[-1/2, 1/2]$.

For $\Delta t = 1$ (as we assume in this section), we have

$$\hat{S}_k^{(mt)}(f) \equiv \left| \sum_{t=1}^{N} h_{t,k} X_t e^{-i2\pi f t} \right|^2$$

from Equation (333). The right-hand side of the above can be rewritten as

$$\left| \epsilon_k \sum_{t=1}^{N} h_{t,k} X_t e^{-i2\pi f[t-(N+1)/2]} \right|^2.$$

It now follows from Equation (363c) that

$$\hat{S}_k^{(mt)}(f) = \left| \Upsilon_k(f)/\sqrt{\lambda_k} \right|^2. \qquad (364b)$$

As we noted in Section 7.1, the estimator $\hat{S}_k^{(mt)}(\cdot)$ is called the kth *eigenspectrum*. The terminology 'eigenspectrum' is motivated by the fact that the taper $\{h_{t,k}\}$ is an eigenvector for Equation (105b). The taper $\{h_{t,k}\}$ has the usual normalization $\sum_{t=1}^{N} h_{t,k}^2 = 1$ (this follows from Parseval's theorem of Equation (89c) because the functions $U_k(\cdot)$ are orthonormal over the interval $[-1/2, 1/2]$). When $\{X_t\}$ is a white noise process, we noted in Section 7.2 that $\hat{S}_k^{(mt)}(f)$ is an unbiased estimator of $S(f)$; i.e., $E\{\hat{S}_k^{(mt)}(f)\} = S(f)$ for all f and k.

We are now in a position to consider a method of 'regularizing' the spectral estimation problem. Our approach is first to introduce a set of desirable (but unobservable) projections and then to outline how these can best be approximated by the $\Upsilon_k(\cdot)$ terms – recall that these were formed in Equation (363a) by projecting the rescaled Fourier transform $\Upsilon(\cdot)$ onto the kth basis function $U_k(\cdot)/\sqrt{\lambda_k}$. In Equation (364a) the projection $\Upsilon_k(\cdot)$ was then expressed in terms of $dZ(\cdot)$, which determines the sdf via Equation (361a). Now consider the equivalent *direct* projection for $dZ(\cdot)$ onto the same kth basis function:

$$Z_k(f) \equiv \int_{-W}^{W} \frac{U_k(f')}{\sqrt{\lambda_k}}\, dZ(f + f'). \tag{365a}$$

Comparison of Equations (364a) and (365a) shows that – apart from different uses of the scaling factor $\sqrt{\lambda_k}$ – the difference between $\Upsilon_k(\cdot)$ and $Z_k(\cdot)$ lies in the integration limits. Of course, there is a large practical difference – $\Upsilon_k(\cdot)$ is calculable from $X_1, \ldots, X_N$ via Equation (363c), while $Z_k(\cdot)$ is not! Also $dZ(\cdot)$ cannot be written as a finite linear combination of the dpswf's: unlike $\Upsilon(\cdot)$, $dZ(\cdot)$ does not fall in an N-dimensional space due to its highly discontinuous nature. Nevertheless, we are still at liberty to project $dZ(\cdot)$ onto each of the N dpswf's to obtain $Z_k(\cdot)$ in Equation (365a).

Note that Equation (365a) can be rewritten in terms of a convolution:

$$Z_k(f) = (-1)^k \int_{f-W}^{f+W} \frac{U_k(f - f')}{\sqrt{\lambda_k}}\, dZ(f'),$$

so that clearly $Z_k(\cdot)$ represents smoothing $dZ(\cdot)$ by a rescaled kth order dpswf over an interval of width $2W$. For a chosen W, $Z_k(\cdot)$ represents the best – in the sense that the dpswf's maximize the concentration measure (see Section 3.9) – glimpse at dZ in each orthogonal direction. It is straightforward to show that

$$E\{|Z_k(f)|^2\} = \int_{f-W}^{f+W} \left[\frac{U_k(f - f')}{\sqrt{\lambda_k}} \right]^2 S(f')\, df'. \tag{365b}$$

For a general spectral shape this expected value can be interpreted as a smoothed version of the spectrum, with the smoothing by the scaled dpswf (squared) over the interval of width $2W$. When the true sdf is that of a white noise process, we see, using Equation (362d), that

$$E\{|Z_k(f)|^2\} = S(f);$$

i.e., $|Z_k(f)|^2$ is an unbiased (but uncalculable) estimator of $S(f)$ for white noise.

Thomson's 1982 approach puts emphasis on finding a calculable expression close to $Z_k(f)$. Why is $|Z_k(f)|^2$ so appealing when its expected value is a smoothed version of the sdf? The answer illuminates a key difficulty with spectral estimation. Estimation of $S(\cdot)$ from a time series $X_1, \ldots, X_N$ is an ill-posed problem because, given a finite number N of observations, we cannot uniquely determine $S(\cdot)$ over the infinite number of points for which it is defined. As we indicated in the previous section, the problem can be 'regularized' by instead calculating an average of the function over a small interval, i.e., the integral convolution of the unknown quantity with a good smoother. The smoother in Equation (365b) is good because it smooths only over the main lobe of width $2W$ and thus avoids sidelobe leakage.

Let us now establish a connection between the projections $Z_k(\cdot)$ and $\Upsilon_k(\cdot)$. As has already been pointed out, $Z_k(\cdot)$ cannot be computed from the data whereas $\Upsilon_k(\cdot)$ can. Following the approach of Thomson (1982), we introduce the weight $b_k(f)$ and look at the difference

$$Z_k(f) - b_k(f)\Upsilon_k(f).$$

What weight should we use to make the right-hand side – incorporating the calculable but defective quantity $\Upsilon_k(f)$ – most like the uncalculable but desirable (leakage-free) $Z_k(f)$? From Equations (365a) and (364a) we can write this difference in the following way:

$$
\begin{aligned}
Z_k(f) &- b_k(f)\Upsilon_k(f) \\
&= \int_{-W}^{W} \frac{U_k(f')}{\sqrt{\lambda_k}}\, dZ(f+f') - b_k(f)\sqrt{\lambda_k} \int_{-1/2}^{1/2} U_k(f')\, dZ(f+f') \\
&= \left(\frac{1}{\sqrt{\lambda_k}} - b_k(f)\sqrt{\lambda_k} \right) \int_{-W}^{W} U_k(f')\, dZ(f+f') \\
&\quad - b_k(f)\sqrt{\lambda_k} \int_{f \notin [-W,W]} U_k(f')\, dZ(f+f'),
\end{aligned}
$$

where the second integral is over all frequencies in the disjoint intervals $[-1/2, -W]$ and $[W, 1/2]$. Since both these integrals are with respect to $dZ(f)$, they are uncorrelated over these disjoint domains of integration (because for $f \neq f'$ we know $E\{dZ^*(f)\, dZ(f')\} = 0$). Thus the mean square error between $Z_k(f)$ and $b_k(f)\Upsilon_k(f)$, namely,

$$\mathrm{mse}_k\,(f) \equiv E\{|Z_k(f) - b_k(f)\Upsilon_k(f)|^2\},$$

is given by

$$
\left(\frac{1}{\sqrt{\lambda_k}} - b_k(f)\sqrt{\lambda_k} \right)^2 E\left\{ \left| \int_{-W}^{W} U_k(f')\, dZ(f+f') \right|^2 \right\}
$$
$$
+ b_k^2(f)\lambda_k E\left\{ \left| \int_{f \notin [-W,W]} U_k(f')\, dZ(f+f') \right|^2 \right\}. \tag{366}
$$

Let us look at the first expectation above. We have

$$E\left\{\left|\int_{-W}^{W} U_k(f')\,dZ(f+f')\right|^2\right\}$$

$$= \int_{-W}^{W}\int_{-W}^{W} U_k(f')U_k(f'')E\{dZ^*(f+f')\,dZ(f+f'')\}$$

$$= \int_{-W}^{W} U_k^2(f')S(f+f')\,df' \approx S(f)\int_{-W}^{W} U_k^2(f')\,df' = \lambda_k S(f),$$

provided the sdf $S(\cdot)$ is slowly varying in $[f-W, f+W]$ (the approximation becomes an equality for a white noise sdf).

The second expectation is the expected value of the part of the smoothed $dZ(\cdot)$ outside the primary smoothing band $[-W, W]$. As in the previous section, this can be described as the broad-band bias of the kth eigenspectrum. This bias depends on details of the sdf outside $[f-W, f+W]$, but, as a useful approximation, let us consider its average value – in the sense of Equation (86) – over the frequency interval $[-1/2, 1/2]$ of unit length:

$$\int_{-1/2}^{1/2} E\left\{\left|\int_{f\notin[-W,W]} U_k(f')\,dZ(f+f')\right|^2\right\}\,df$$

$$= \int_{f\notin[-W,W]} U_k^2(f')\int_{-1/2}^{1/2} S(f+f')\,df\,df'$$

$$= \sigma^2\int_{f\notin[-W,W]} U_k^2(f')\,df',$$

where we use the fundamental properties that $S(\cdot)$ is periodic with period 1 and that $\int_{-1/2}^{1/2} S(f)\,df = \sigma^2 = \text{var}\{X_t\}$. Because the $U_k(\cdot)$ functions are orthonormal over $[-1/2, 1/2]$ and orthogonal over $[-W, W]$, we have

$$\int_{f\notin[-W,W]} U_k^2(f')df' = \int_{-1/2}^{1/2} U_k^2(f')\,df' - \int_{-W}^{W} U_k^2(f')\,df' = 1 - \lambda_k.$$

Hence the average value of the broad-band bias for general spectra is given by $(1-\lambda_k)\sigma^2$; for white noise the second expectation in Equation (366) need not be approximated by averaging over $[-1/2, 1/2]$, but can be evaluated directly, giving $(1-\lambda_k)\sigma^2$ again, so that this expression is exact for the broad-band bias for white noise. For general spectra a useful approximation to the mean square error is thus

$$\text{mse}_k(f) \approx \left(\frac{1}{\sqrt{\lambda_k}} - b_k(f)\sqrt{\lambda_k}\right)^2 \lambda_k S(f) + b_k^2(f)\lambda_k(1-\lambda_k)\sigma^2, \quad (367)$$

but note that this approximation is an equality for a white noise sdf. To find the value of $b_k(f)$ that minimizes the approximate mean square error, we differentiate with respect to $b_k(f)$ and set the result to zero to obtain

$$b_k(f) = \frac{S(f)}{\lambda_k S(f) + (1 - \lambda_k)\sigma^2}. \tag{368a}$$

With this optimum value for $b_k(f)$, we can substitute it into the approximation to the mean square error to obtain

$$\text{mse}_k(f) \approx \frac{S(f)(1 - \lambda_k)\sigma^2}{\lambda_k S(f) + (1 - \lambda_k)\sigma^2}. \tag{368b}$$

Let us now consider the special case in which $\{X_t\}$ is white noise; i.e., $S(f) = \sigma^2$. From Equation (368a) we see that $b_k(f)$ is identically unity for all frequencies. Hence, for white noise,

$$\Upsilon_k(f) \text{ estimates } Z_k(f)$$

in a minimum mean square error sense. Since the approximation in Equation (367) is an equality in the case of white noise, we have

$$\text{mse}_k(f) = \left(\frac{1}{\sqrt{\lambda_k}} - \sqrt{\lambda_k} \right)^2 \lambda_k \sigma^2 + \lambda_k(1 - \lambda_k)\sigma^2 = (1 - \lambda_k)\sigma^2;$$

i.e., the mean square error is identical to the broad-band bias. If k is less than $2NW - 1$ and hence $\lambda_k \approx 1$, then the mean square error is negligible.

What is the best way to combine the individual eigenspectra in the special case of white noise? From the expression for $\hat{S}_k^{(mt)}(\cdot)$ in Equation (363c) we have

$$\lambda_k \hat{S}_k^{(mt)}(f) = |\Upsilon_k(f)|^2.$$

Since $\Upsilon_k(f)$ estimates $Z_k(f)$, it follows that

$$\frac{1}{K} \sum_{k=0}^{K-1} \lambda_k \hat{S}_k^{(mt)}(f) \text{ estimates } \frac{1}{K} \sum_{k=0}^{K-1} |Z_k(f)|^2.$$

What is $E\{\hat{S}_k^{(mt)}(f)\}$? By definition, this can be obtained by rescaling $E\{|\Upsilon_k(f)|^2\}$. From Equation (364a) it follows that

$$E\{|\Upsilon_k(f)|^2\} = E\{\Upsilon_k^*(f)\Upsilon_k(f)\}$$

$$= \lambda_k \int_{-1/2}^{1/2} \int_{-1/2}^{1/2} U_k(f')U_k(f'')E\{dZ^*(f + f')\, dZ(f + f'')\}$$

$$= \lambda_k \int_{-1/2}^{1/2} U_k^2(f')S(f + f')\, df'$$

$$= \lambda_k \int_{-1/2}^{1/2} U_k^2(f - f')S(f')\, df'. \tag{368c}$$

For a white noise process, because of the orthonormality of the $U_k(\cdot)$ functions over $[-1/2, 1/2]$, it follows that

$$E\{\hat{S}_k^{(mt)}(f)\} = S(f).$$

Hence a natural spectrum estimator when the sdf is white – based on the first few eigenspectra (i.e., those with least sidelobe leakage) – is given by

$$\bar{S}^{(mt)}(f) \equiv \frac{\sum_{k=0}^{K-1} \lambda_k \hat{S}_k^{(mt)}(f)}{\sum_{k=0}^{K-1} \lambda_k}. \tag{369a}$$

The denominator makes the expression unbiased. Note also that $K \approx \sum_{k=0}^{K-1} \lambda_k$ provided K is chosen to be less than or equal to $2NW - 1$ so that each λ_k for $k = 0, \ldots, K-1$ is close to unity.

The estimator in Equation (369a) is considered here to be the initial or basic spectrum estimator resulting from the theory. It is intuitively attractive since, as the order k of the dpss tapers increases, the corresponding eigenvalues will decrease, and the eigenspectra will become more contaminated with leakage; i.e., $\text{mse}_k(f) = (1 - \lambda_k)\sigma^2$ will become larger. The eigenvalue weights in Equation (369a) will help to lessen the contribution of the higher leakage eigenspectra. In practice this effect will be negligible provided K is chosen no larger than $2NW - 1$.

The estimator in Equation (369a) can be refined to take into account a colored sdf, as follows. If we combine Equations (364b), (368c) and (369a), we obtain

$$E\{\bar{S}^{(mt)}(f)\} = \int_{-1/2}^{1/2} \left(\frac{1}{\sum_{k=0}^{K-1} \lambda_k} \sum_{k=0}^{K-1} \lambda_k U_k^2(f - f') \right) S(f')\, df'. \tag{369b}$$

Note that this expectation integrates over the full band $[-1/2, 1/2]$ and not just $[-W, W]$. If $S(\cdot)$ has a large dynamic range, notable spectral leakage could occur in this smoothing expression due to sidelobes of the dpswf outside $[-W, W]$ for larger values of k, e.g., those values of k approaching $2NW - 1$. Hence Equation (369a) might not be a satisfactory estimator for nonwhite sdf's. By way of contrast, the integration range for the expectation of $|Z_k(f)|^2$ in Equation (365b) is only $[-W, W]$, so that such sidelobe problems are avoided for $|Z_k(f)|^2$. Now

$$b_k^2(f)|\Upsilon_k(f)|^2 \text{ estimates } |Z_k(f)|^2,$$

where $b_k(f)$ will take the general form of Equation (368a), involving the true sdf $S(f)$. Hence

$$\frac{1}{K} \sum_{k=0}^{K-1} b_k^2(f)\lambda_k \hat{S}_k^{(mt)}(f) \text{ estimates } \frac{1}{K} \sum_{k=0}^{K-1} |Z_k(f)|^2$$

for a nonwhite sdf. By noting that $K \approx \sum_{k=0}^{K-1} b_k^2(f)\lambda_k$ provided K does not exceed $2NW - 1$, we arrive at the *adaptive multitaper spectral estimator*

$$\hat{S}^{(amt)}(f) \equiv \frac{\sum_{k=0}^{K-1} b_k^2(f)\lambda_k \hat{S}_k^{(mt)}(f)}{\sum_{k=0}^{K-1} b_k^2(f)\lambda_k}. \tag{370a}$$

This is, of course, of the same form as Equation (369a) when the weights are set to unity. It cannot however be claimed that Equation (370a) is exactly unbiased since $E\{\hat{S}_k^{(mt)}(f)\}$ is not identically $S(f)$ for a non-white sdf (see Equation (368c)), but it will generally be close provided $S(f)$ does not vary rapidly over the interval $[f - W, f + W]$. The arguments leading to Equation (370a) justify the same result in Thomson (1982, Equation (5.3)), but the definition of the weights $b_k(f)$ used here is more appealing since these weights are unity for a white noise sdf.

Now the weight formula of Equation (368a) involves the true unknown spectrum and variance. Spectral estimation via Equation (370a) must thus be carried out in an iterative fashion such as the following. To proceed, we assume $NW \geq 2$. We start with a spectral estimate of the form of Equation (369a) with K set equal to 1 or 2. This initial estimate involves only the one or two tapers with lowest sidelobe leakage and hence will preserve rapid spectral decays. This spectral estimate is then substituted – along with the estimated variance – into Equation (368a) to obtain the weights for orders $k = 0, \ldots, K - 1$ with K typically $2NW - 1$. These weights are then substituted into Equation (370a) to obtain the new spectral estimate with K again $2NW - 1$. The spectral estimate given by Equation (370a) is next substituted back into Equation (368a) to get new weights and so forth. Usually two executions of Equation (370a) are sufficient. Thomson (1982) describes a method for estimating the broad-band bias (rather than its average over $[-1/2, 1/2]$) within the iterative stage, but this additional complication often leads to very marginal changes and suffers from additional estimation problems.

An 'equivalent degrees of freedom' argument says that $\hat{S}^{(amt)}(f)$ is approximately equal in distribution to the rv $S(f)\chi_\nu^2/\nu$, where the chi-square rv has degrees of freedom ν given by

$$\nu = \frac{2\left(\sum_{k=0}^{K-1} b_k^2(f)\lambda_k\right)^2}{\sum_{k=0}^{K-1} b_k^4(f)\lambda_k^2} \tag{370b}$$

for $f \neq 0$ or $f_{(N)}$ (this is Exercise [7.8]).

Comments and Extensions to Section 7.4

[1]   The estimators of Equations (369a) and (370a) differ significantly from each other in two important properties. First, the weights in the spectral estimator of Equation (369a) do not depend on frequency – as a result, Parseval's theorem is satisfied in expected value (this is also true for the simple average of Equation (333)). To see this, use Equation (369b) to write

$$
\int_{-1/2}^{1/2} E\{\bar{S}^{(mt)}(f)\}\, df
$$

$$
= \sum_{k=0}^{K-1} \lambda_k \int_{-1/2}^{1/2} \frac{1}{\sum_{k=0}^{K-1} \lambda_k} \left(\int_{-1/2}^{1/2} U_k^2(f - f')\, df \right) S(f')\, df'.
$$

The integral in the parentheses is unity, so we obtain

$$
\int_{-1/2}^{1/2} E\{\bar{S}^{(mt)}(f)\}\, df = \int_{-1/2}^{1/2} S(f')\, df' = \sigma^2,
$$

as required. In contrast, Exercise [7.10] shows that Parseval's theorem is in general *not* satisfied exactly in expected value for the adaptive multitaper spectral estimator of Equation (370a).

Second, the simply weighted estimator of Equation (369a) (and also Equation (333)) has approximately constant variance across all the Fourier frequencies (excluding 0 and Nyquist), whereas the adaptive weighting scheme of Equation (370a) can give rise to appreciable variations of the variance throughout the spectrum, making interpretation of the plot of the spectral estimate more difficult (see the example in the next section).

7.5 Multitaper Spectral Analysis of Ocean Wave Data

As an example of the multitaper approach to spectral analysis, let us reconsider the ocean wave data of Section 6.18 (this time series is plotted in Figure 295, and various nonparametric spectral estimates for the series are shown in Figures 296 and 301). As discussed previously, we are mainly interested in the rate at which the sdf decreases over the range 0.2 to 1.0 Hz, but the sdf at higher frequencies is also of some marginal interest. Because there are unlikely to be sharp spectral features in these frequency ranges (an assumption that is verified by our analysis in Section 6.18), we can set the resolution bandwidth $2W$ to be fairly large. We thus let $W = 4/(N \Delta t)$ initially, yielding $2W = 0.03125$ Hz (recall that $N = 1024$ and $\Delta t = 1/4$ second). Since $2NW \Delta t - 1 = 7$, we compute the eigenspectra $\hat{S}_k^{(mt)}(\cdot)$ for orders $k = 0, 1, \ldots, 6$. A comparison of the high-order eigenspectra with the order $k = 0$ and $k = 1$

eigenspectra indicates that $\hat{S}_6^{(mt)}(\cdot)$ has unacceptable bias in the high frequency range (i.e., 1 Hz $\leq f \leq$ 2 Hz) but that all the eigenspectra of lower order than 6 are acceptable. We thus form the simple multitaper estimate $\hat{S}^{(mt)}(\cdot)$ of Equation (333) by averaging $K = 6$ eigenspectra together. This estimate is plotted in Figure 373(a), along with a criss-cross in the lower left-hand corner depicting the bandwidth $2W$ and the length (in decibels) of a 95% confidence interval for $10 \log_{10}(S(f))$ based on $10 \log_{10}(\hat{S}^{(mt)}(f))$ with $2K = 12$ degrees of freedom.

If we now let $W = 6/(N \Delta t)$ and hence increase the resolution bandwidth to $2W = 0.046875$ Hz, we find that there are $K = 10$ eigenspectra with acceptable bias properties. When these are averaged together, we obtain the multitaper estimate $\hat{S}^{(mt)}(\cdot)$ shown by the thick curve in Figure 373(b). This estimate has $2K = 20$ degrees of freedom. A comparison between this estimate and the one for $W = 4/(N \Delta t)$ in the top plot shows that the $W = 6/(N \Delta t)$ estimate is smoother in appearance, as we would expect. It is also of interest to compare this estimate with the $m = 150$ Parzen lag window estimate $\hat{S}^{(lw)}(\cdot)$ shown by the thick curve in the top plot of Figure 301. This lag window estimate has a bandwidth of 0.0494 Hz, which is nearly the same as the bandwidth for the $W = 6/(N \Delta t)$ multitaper spectral estimate ($2W = 0.046875$ Hz). We have replotted $\hat{S}^{(lw)}(\cdot)$ as a thin curve in Figure 373(b), but it is just barely visible because it is in such good agreement with $\hat{S}^{(mt)}(\cdot)$ for $W = 6/(N \Delta t)$. Thus, for this time series, lag window and multitaper spectral estimates are quite comparable.

Next, we examine how well the adaptive multispectral estimator $\hat{S}^{(amt)}(\cdot)$ of Equation (370a) performs here. With $W = 4/(N \Delta t)$ as in Figure 373(a), we obtain the adaptive estimate shown as the thick curve in Figure 373(c). A comparison of this estimate with $\hat{S}^{(mt)}(\cdot)$ of Figure 373(a) indicates very good agreement for frequencies between 0 and 1 Hz. For higher frequencies, the estimate $\hat{S}^{(amt)}(\cdot)$ shows more structure than $\hat{S}^{(mt)}(\cdot)$. What accounts for this difference? The estimator $\hat{S}^{(amt)}(\cdot)$ is based upon weighted averages of the eigenspectra $\hat{S}_0^{(mt)}(\cdot)$, $\ldots, \hat{S}_6^{(mt)}(\cdot)$. As indicated by Equation (370b), the degrees of freedom ν for $\hat{S}^{(amt)}(\cdot)$ are frequency-dependent – we have plotted ν as a function of frequency in Figure 373(d). Note that ν drops down to about 6 for f between 1 and 2 Hz, which is half the degrees of freedom associated with $\hat{S}^{(mt)}(\cdot)$ in the top plot. At least part of the additional structure in $\hat{S}^{(amt)}(\cdot)$ can be attributed to this decrease in ν. To explore this point further, the upper and lower thin curves in Figure 373(c) show upper and lower 95% confidence limits for each $10 \log_{10}(S(f))$ based upon $10 \log_{10}(\hat{S}^{(amt)}(f))$ (the thick curve). As ν decreases in the high frequency region, these confidence intervals increase in width, indicating an increase in variability in $\hat{S}^{(amt)}(\cdot)$. The apparent additional spectral structure is in accordance with this increased variability.

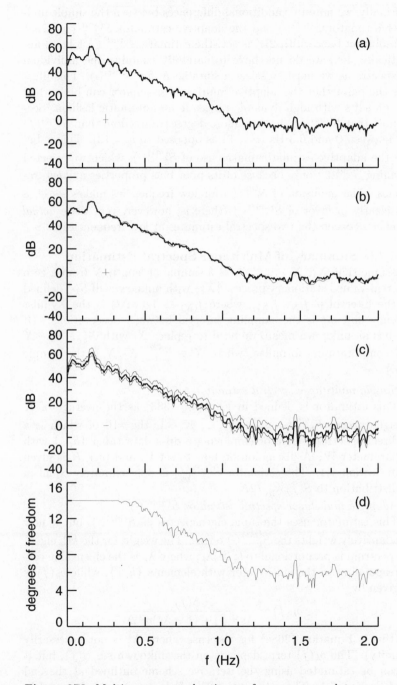

Figure 373. Multitaper spectral estimates for ocean wave data.

Finally, we note two additional differences between the simple multitaper estimator $\hat{S}^{(mt)}(\cdot)$ and the adaptive estimator $\hat{S}^{(amt)}(\cdot)$. Once the resolution bandwidth $2W$ is set, the estimate $\hat{S}^{(amt)}(\cdot)$ follows automatically; i.e., we do not have to carefully examine the individual eigenspectra as we must to select a suitable K for $\hat{S}^{(mt)}(\cdot)$. This illustrates the point that the adaptive multitaper estimator can be 'turned loose' on sdf's with high dynamic ranges in an automatic fashion. Second, note that $\hat{S}^{(amt)}(f)$ has two more degrees of freedom than $\hat{S}^{(mt)}(f)$ for f between 0 and 0.3 Hz ($\nu = 14$ as opposed to $\nu = 12$). This is because the adaptive estimator makes use of $\hat{S}_6^{(mt)}(\cdot)$, which we rejected in forming $\hat{S}^{(mt)}(\cdot)$ only because of its poor bias properties at *high* frequencies. The inclusion of $\hat{S}_6^{(mt)}(\cdot)$ for low frequencies makes sense, a point clearly in favor of $\hat{S}^{(amt)}(\cdot)$ (there is, however, very little *actual* difference between the two spectral estimates at low frequencies).

7.6 Summary of Multitaper Spectral Estimation

We assume that X_1, X_2, ..., X_N is a sample of length N from a zero mean real-valued stationary process $\{X_t\}$ with unknown sdf $S(\cdot)$ defined over the interval $[-f_{(N)}, f_{(N)}]$, where $f_{(N)} \equiv 1/(2\,\Delta t)$ is the Nyquist frequency and Δt is the sampling interval between observations. (If $\{X_t\}$ has an unknown mean, we need to replace X_t with $X_t' \equiv X_t - \bar{X}$ in all computational formulae, where $\bar{X} \equiv \sum_{t=1}^{N} X_t/N$ is the sample mean.)

[1] *Simple multitaper spectral estimator $\hat{S}^{(mt)}(\cdot)$*
This estimator is defined in Equation (333) as the average of K eigenspectra $\hat{S}_k^{(mt)}(\cdot)$, $k = 0$, ..., $K - 1$, the kth of which is a direct spectral estimator employing a dpss data taper $\{h_{t,k}\}$ with parameter W. A discussion on how to set W and pick K is given in Section 7.1. The estimator $\hat{S}^{(mt)}(f)$ is approximately equal in distribution to $S(f)\chi_{2K}^2/2K$.

[2] *Adaptive multitaper spectral estimator $\hat{S}^{(amt)}(\cdot)$*
This estimator uses the same eigenspectra as $\hat{S}^{(mt)}(\cdot)$, but it now adaptively weights the $\hat{S}_k^{(mt)}(\cdot)$ terms. The weight for the kth eigenspectrum is proportional to $b_k^2(f)\lambda_k$, where λ_k is the eigenvalue corresponding to the eigenvector with elements $\{h_{t,k}\}$, while $b_k(f)$ is given by

$$b_k(f) = \frac{S(f)}{\lambda_k S(f) + (1 - \lambda_k)\sigma^2\,\Delta t}$$

(this is Equation (368a) for the case where Δt is not necessarily unity). The $b_k(f)$ term depends on the unknown sdf $S(f)$, but it can be estimated using the iterative scheme outlined at the end of Section 7.4. The estimator $\hat{S}^{(mt)}(f)$ is approximately equal in distribution to $S(f)\chi_\nu^2/\nu$, where ν is given by Equation (370b).

7.7 Exercises

[7.1] Suppose that $X_1, \ldots, X_N$ is a sample of a zero mean white noise process $\{X_t\}$ with variance s_0 and sampling time Δt. Let $\{\tilde{h}_{t,j}\}$ and $\{\tilde{h}_{t,k}\}$ be any two orthonormal data tapers; i.e.,

$$\sum_{t=1}^{N} \tilde{h}_{t,j} \tilde{h}_{t,k} = 0 \quad \text{and} \quad \sum_{t=1}^{N} \tilde{h}_{t,j}^2 = \sum_{t=1}^{N} \tilde{h}_{t,k}^2 = 1.$$

For $l = j$ or k, define

$$\tilde{J}_l(f) \equiv (\Delta t)^{1/2} \sum_{t=1}^{N} \tilde{h}_{t,l} X_t e^{-i2\pi f t \, \Delta t} \quad \text{so that} \quad \left| \tilde{J}_l(f) \right|^2 = \tilde{S}_l^{(mt)}(f)$$

(see Equation (349b)).

a) Show that $\text{cov}\{\tilde{J}_j(f), \tilde{J}_k(f)\} = 0$.

b) Under the additional assumption that $\{X_t\}$ is a Gaussian process, use Equation (330d) to show that, for $l = j$ or k,

$$\text{var}\{\tilde{S}_l^{(mt)}(f)\} = s_0^2 (\Delta t)^2 \left(1 + \left| \sum_{t=1}^{N} \tilde{h}_{t,l}^2 e^{-i4\pi f t \, \Delta t} \right|^2 \right)$$

and that

$$\text{cov}\{\tilde{S}_j^{(mt)}(f), \tilde{S}_k^{(mt)}(f)\} = s_0^2 (\Delta t)^2 \left| \sum_{t=1}^{N} \tilde{h}_{t,j} \tilde{h}_{t,k} e^{-i4\pi f t \, \Delta t} \right|^2.$$

What do the above equations reduce to if $f = 0$, $f_{(N)}/2$ or $f_{(N)}$?

[7.2] Let Q be an $N \times N$ symmetric matrix of real-valued numbers. The following standard results can be found in any number of books on matrix theory (see, for example, Graybill, 1983):

a) There exists an $N \times N$ orthogonal matrix P such that $P^T Q P = D_N$, where D_N is an $N \times N$ diagonal matrix with diagonal elements $d_1, \ldots, d_N$. Each d_j is an eigenvalue of Q and is necessarily real-valued. We can assume that they are ordered such that $d_1 \geq d_2 \geq \cdots \geq d_{N-1} \geq d_N$.

b) Because P is an orthogonal matrix, we have $P^T P = P P^T = I$, where I is the $N \times N$ identity matrix. This implies that we can write $Q = P D_N P^T$.

c) If Q is positive semidefinite with rank K, then the eigenvalues of Q are all nonnegative; the number of positive eigenvalues is equal to K so that $d_1 \geq d_2 \geq \cdots \geq d_K > 0$ and $d_j = 0$

for $K + 1 \leq j \leq N$; and we can write $Q = P\sqrt{D_N}\sqrt{D_N}P^T$, where $\sqrt{D_N}$ refers to the diagonal matrix whose elements are the square roots of the elements of D_N; i.e., $\sqrt{D_N}\sqrt{D_N} = D_N$.

Using these facts, prove that, if Q is positive semidefinite with rank $1 \leq K \leq N$, we can write $Q = AA^T$, where A is an $N \times K$ matrix such that $A^T A = D_K$, with D_K being a $K \times K$ diagonal matrix whose diagonal elements are the nonzero eigenvalues of Q, namely, $d_1, \ldots, d_K$.

[7.3] Show that the first part of Equation (354c) holds, namely,

$$E\{\hat{S}^{(q)}(f)\} = \Delta t \,\operatorname{tr}\{Q\Sigma_Z\}.$$

[7.4] Suppose that Σ is an $N \times N$ symmetric positive definite matrix with eigenvalues $\lambda_0, \lambda_1, \ldots, \lambda_{N-1}$ (ordered from largest to smallest). Show that, if the largest eigenvalue λ_0 is distinct (i.e., $\lambda_0 > \lambda_1 \geq \cdots \geq \lambda_{N-1}$) and if A is any $N \times K$ matrix such that $\operatorname{tr}\{A^T A\} = 1$, then $\operatorname{tr}\{A^T \Sigma A\}$ is maximized when A is a normalized eigenvector associated with the eigenvalue λ_0. Hints:

    a) the eigenvalues of Σ must be positive; and

    b) if $\mathbf{V}_0, \mathbf{V}_1, \ldots, \mathbf{V}_{N-1}$ are an orthonormal set of eigenvectors corresponding to the eigenvalues $\lambda_0, \lambda_1, \ldots, \lambda_{N-1}$, then each column of A can be expressed as a unique linear combination of the $\mathbf{V}_k$ terms.

[7.5] If A is an $N \times K$ matrix whose kth column is the rescaled dpss data taper $\{h_{t,k}/\sqrt{K}\}$ of order $(k-1)$, verify that the broad-band bias measure of Equation (358) can be rewritten as

$$\operatorname{tr}\{A^T A\} - \operatorname{tr}\{A^T \Sigma^{(bl)} A\} = 1 - \frac{1}{K}\sum_{k=0}^{K-1}\lambda_k(N, W)$$

(this result was stated in Equation (360)).

[7.6] Show that Equation (364a) holds. Hints: start with the expression for $\Upsilon_k(f')$ in Equation (363a), use Equation (362a), interchange the order of the integrals and the summation such that the innermost integral equals the integral on the right-hand side of Equation (363b), and then use the complex conjugate of Equation (362c) (recall that the dpswf $U_k(\cdot)$ is a real-valued function). Finally, use the fact that both $dZ(\cdot)$ and $U_k(\cdot)$ are periodic functions with a period of unity.

[7.7] If we make use of the fact that the $U_k(\cdot)$ functions are orthonormal over $[-1/2, 1/2]$, we can replace the definition of the projections $\Upsilon_k(f')$ in Equation (363a) with

$$\tilde{\Upsilon}_k(f') \equiv \int_{-1/2}^{1/2} \Upsilon(f + f')U_k(f)\,df.$$

a) Show that $\tilde{\Upsilon}_k(f') = \Upsilon_k(f)/\sqrt{\lambda_k}$.

b) Show that the $\tilde{\Upsilon}_k(f')$ projections lead to the same multitaper spectral estimators we obtained using $\Upsilon_k(f')$, i.e., Equation (369a) for white noise and Equation (370a) for colored noise.

[7.8] Under the assumption that the $\hat{S}_k^{(mt)}(f)$ terms in Equation (370a) are uncorrelated and have the usually assumed distribution for direct spectral estimators (see Equation (223b)), use the 'equivalent degrees of freedom' argument of Section 6.10 to show that the adaptive multitaper estimator $\hat{S}^{(amt)}(f)$ is approximately distributed as $S(f)\chi_\nu^2/\nu$, where ν is given by Equation (370b).

[7.9] Suppose that the eigenspectra $\hat{S}_k^{(mt)}(f)$ used in forming the multitaper estimator $\hat{S}^{(mt)}(f)$ of Equation (333) are approximately unbiased and uncorrelated (this is a reasonable assertion with restrictions on f, the number of tapers and an assumption that the true sdf does not vary too rapidly). Describe a way of estimating the variability in the multitaper estimator that does not make use of the argument that $\hat{S}^{(mt)}(f)$ is proportional to a chi-square random variable with $2K$ degrees of freedom (Thomson and Chave, 1991).

[7.10] Show that, for the adaptive multitaper spectral estimator of Equation (370a), we have

$$\int_{-1/2}^{1/2} E\{\hat{S}^{(amt)}(f)\}\, df$$
$$= \int_{-1/2}^{1/2} S(f') \left(\sum_{k=0}^{K-1} \lambda_k \int_{-1/2}^{1/2} \frac{b_k^2(f)}{\sum_{k=0}^{K-1} \lambda_k b_k^2(f)} U_k^2(f - f')\, df \right) df'.$$

Since the term in the parentheses above is not unity in general, it follows that, in contrast to both $\hat{S}^{(mt)}(\cdot)$ and $\bar{S}^{(mt)}(\cdot)$, Parseval's theorem is in general *not* satisfied exactly in expected value for $\hat{S}^{(amt)}(\cdot)$.

8

Calculation of Discrete
Prolate Spheroidal Sequences

8.0 Introduction

We saw in Section 3.9 that the sequence of length N with the highest concentration of energy in the frequency interval $[-W, W]$ is the eigenvector $\mathbf{v}_0(N, W)$ corresponding to the largest eigenvalue $\lambda_0(N, W)$ in the equation

$$A\mathbf{g} = \lambda_k(N, W)\mathbf{g}, \tag{378}$$

where A is an N x N matrix whose (t, t')th element is $\sin[2\pi W(t' - t)]/[\pi(t' - t)]$; $\mathbf{g}$ is an N-dimensional vector; and W lies between 0 and $1/2$. The sequence of length N that is orthogonal to this one with the highest concentration of energy in $[-W, W]$ is the eigenvector $\mathbf{v}_1(N, W)$ corresponding to the second largest eigenvalue $\lambda_1(N, W)$, etc. There are N of these eigenvectors in all, $\mathbf{v}_k(N, W)$, $k = 0, \ldots, N - 1$, and they are ordered by their eigenvalues $1 > \lambda_0(N, W) > \lambda_1(N, W) > \cdots > \lambda_{N-1}(N, W) > 0$. The first $2NW - 1$ eigenvalues $\lambda_k(N, W)$ are all near unity and then fall to zero. We shall here call the vectors $\mathbf{v}_0(N, W)$, $\mathbf{v}_1(N, W)$, $\ldots$, $\mathbf{v}_{N-1}(N, W)$ discrete prolate spheroidal sequences (dpss's). Strictly speaking, this terminology refers to the doubly infinite sequences of Equation (103b); however, the term seems to be in fairly common usage when referring to the finite length equivalent.

The ability to accurately and rapidly calculate these discrete prolate spheroidal sequences is a clear necessity in order to implement the multitaper approach to spectral estimation we discussed in the previous chapter.

An obvious method for finding the dpss's is to solve Equation (378) for eigenvectors $\mathbf{v}_0(N, W)$, $\ldots$, $\mathbf{v}_{K-1}(N, W)$, where $K < N$ is the number of dpss's required (typically $K \approx 2NW - 1$). In practice this is, unfortunately, not quite as simple as might be expected. In Section 8.1

378

we look at calculating the dpss's directly from Equation (378), first with no regard to the spread of the eigenvalues, and then taking this knowledge (see, e.g., Figure 110) into account. In Section 8.2 we consider a method that uses numerical integration and enables rapid and simple calculation for the zeroth-order dpss. In Section 8.3 we look at a very efficient direct tridiagonal formulation due to Slepian (1978) that avoids Equation (378) altogether. Finally, we consider in Section 8.4 a family of simple – but effective – approximations to the dpss's known as the trig prolates.

Before proceeding, we note that, if **g** is an eigenvector of Equation (378), then so is $c\mathbf{g}$, where c is any nonzero constant. We thus need to impose a scaling and polarity convention. We choose that given by Slepian (1978, equations 19 and 20). First, energy is set to unity for each order k:

$$\sum_{t=0}^{N-1} v_{t,k}^2(N,W) = 1.$$

For symmetric tapers ($k = 0, 2, 4, \ldots$), the average of the taper elements is made positive,

$$\sum_{t=0}^{N-1} v_{t,k}(N,W) > 0,$$

while for skew-symmetric tapers ($k = 1, 3, 5, \ldots$), the taper is made to start with a positive first lobe:

$$\sum_{t=0}^{N-1} (N - 1 - 2t) v_{t,k}(N,W) > 0.$$

This convention was used for the dpss's shown in Figures 106, 108, 336, 338 and 346.

8.1 Calculating DPSSs from Defining Equation

In solving the symmetric eigenproblem of Equation (378) for $k = 0$, $\ldots, K - 1$, we can either use software that finds all the eigenvectors and eigenvalues or use specialized software for which subsets can be specified. For small N, let us take $K = N$ and use the former approach to estimate all the eigenvectors and eigenvalues. Here we can use the FORTRAN routines TDIAG and LRVT given in Griffiths and Hill (1985, p. 110). The first routine reduces the real symmetric matrix A to tridiagonal form using Householder's reduction, and then LRVT determines all the eigenvalues and eigenvectors of this real, symmetric, tridiagonal matrix using QL decomposition. Equivalent routines are given in Press *et al.* (1986) as TRED2 and TQLI. The EISPACK system (Smith *et al.*, 1976) is the origin of most FORTRAN eigenproblem software. Clear

k	$\lambda_k(31, 6/31)$	$\lambda_k(31, 7/31)$	$\lambda_k(31, 8/31)$
0	0.9999999999999997	1.000000000000007	1.000000000000002
1	0.9999999999999769	0.9999999999999933	1.000000000000001
2	0.9999999999978725	0.9999999999999921	0.9999999999999945
3	0.9999999998764069	0.9999999999998824	0.9999999999999908

Table 380. The first four computed eigenvalues using double precision when (a) $N = 31$, $NW = 6$, (b) $N = 31$, $NW = 7$, (c) $N = 31$, $NW = 8$. For $NW = 7$ and 8 the numbers are not all less than one as they should be.

discussion of the numerical methods is given, for example, in Press *et al.* (1986, pp. 350–363) and references therein.

For any fixed level of accuracy·at which computations are to be carried out, the quantity PRECIS must be specified in the routines TDIAG and LRVT mentioned above. This is defined as the smallest positive number x such that $1 + x > 1$ in the floating point arithmetic of the computer. On the particular computer we used, there are three levels of precision, denoted as single, double and quadruple precision, yielding values for PRECIS of 1.19×10^{-07}, 2.77×10^{-17} and 1.93×10^{-34}, respectively. It was pointed out earlier that the first few eigenvalues of (378) are close to unity. Clearly, if the difference between 1 and $\lambda_k(N, W)$ is of the same order of magnitude as PRECIS, then the numerical estimate of the eigenvalue can be in error. In particular, the *calculated* eigenvalue can be greater than unity; since this is *theoretically* impossible, the corresponding calculated eigenvectors can be in error. By moving to a higher level of computational accuracy (e.g., changing from double to quadruple precision), PRECIS is decreased, and the problem can be alleviated.

Let us give an example where we run into numerical problems. Suppose that we use double precision arithmetic so that PRECIS is 2.77×10^{-17}. For $N = 31$ and $NW = 6$, the first four calculated eigenvalues are listed in the second column of Table 380. The first calculated eigenvalue differs from unity by 3×10^{-16}, still an order of magnitude greater than PRECIS. The third column of the table gives the calculated eigenvalues for $NW = 7$. The first eigenvalue is now greater than unity and must therefore be incorrect. Increasing NW to 8 gives the calculated eigenvalues of the final column of the table, where the first two eigenvalues are now greater than unity and hence unacceptable.

The four computed dpss's corresponding to the computed eigenvalues in the final column of Table 380 (i.e., the $NW = 8$ case) are given by

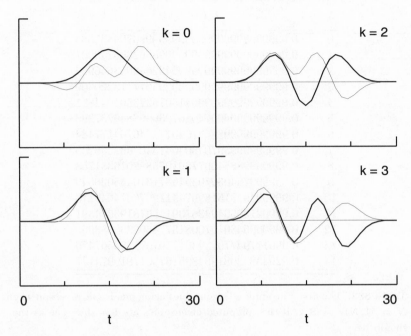

Figure 381. Effect of numerical precision on computation of dpss's. The thin and thick curves show the difference between using, respectively, double and quadruple precision in computing the dpss's of orders $k = 0$, 1, 2 and 3 for $N = 31$ and $NW = 8$.

the thin curves in the plots of Figure 381. The alternate symmetry and skew-symmetry of the dpss's for $k = 0$ to 3 fail because of the numerical errors in the computed eigenvalues. By changing to quadruple precision the problem can be alleviated since now PRECIS is 1.93×10^{-34}. The first 15 computed eigenvalues for $NW = 8$ are given in Table 382 – note that now all eigenvalues are less than unity. The first four computed dpss's for $NW = 8$ and quadruple precision are given as the thick curves in Figure 381. Note that the alternate symmetry and skew-symmetry are reestablished.

Clearly a more stable method is desirable. One approach is to use *inverse iteration*, the implementation of which for dpss's – including code – is fully discussed in Bell *et al.* (1993). We describe this method first for the zeroth-order dpss and then for other orders. Let

$$B_0 \equiv A - I = A - \lambda_{-1}I,$$

where I is the Nth order identity matrix, and we define $\lambda_{-1} = 1$. The subscript 0 indicates that we are looking first at the zeroth-order case. Eigenvectors $\mathbf{v}$ of A satisfy $A\mathbf{v} = \lambda\mathbf{v}$ and therefore also satisfy

k	$\lambda_k(31, 8/31)$
0	0.99999999999999999999990704835383
1	0.99999999999999999996662306749001
2	0.99999999999999999432644053636964
3	0.99999999999999939295977113220309
4	0.99999999999995415079273501044881
5	0.99999999999974016703096884973293
6	0.99999999998853764677484670417484
7	0.99999999957048040415404681655238
8	0.99999998854097769107588201633175 3
9	0.99999973403970394687131044502054 9
10	0.99999494315189674411967607404911 3
11	0.99992136564943673015597819367439 1
12	0.99900954801070081377860833508396 5
13	0.99017794770739381759707135356742 6
14	0.92943822081984805167361781997413 7

Table 382. The first 15 computed eigenvalues using quadruple precision when $N = 31, NW = 8$. All the computed eigenvalues are less than one as they should be.

$(A - \ I)\mathbf{v} = (\lambda - \lambda_{-1})\mathbf{v}$. For the zeroth-order dpss the eigenvector of A corresponding to the eigenvalue nearest 1 is required. For B_0 the eigenvectors are the same as for A, but now the eigenvalues are $\lambda - \lambda_{-1}$. Hence the eigenvalue λ nearest to $\lambda_{-1} = 1$ will correspond to the smallest eigenvalue of B_0. This eigenvalue and corresponding eigenvector can be found by the method of inverse iteration (Parlett, 1980, p. 62; Gourlay and Watson, 1973, p. 57). The iterative scheme is

$$B_0\mathbf{u}_0^{(j+1)} = \mathbf{v}_0^{(j)} \text{ followed by } \mathbf{v}_0^{(j+1)} = \mathbf{u}_0^{(j+1)}/\|\mathbf{u}_0^{(j+1)}\|,$$

where $\mathbf{v}_0^{(0)}$ consists of a vector of elements $1/\sqrt{N}$, and $\|\mathbf{V}\|^2 \equiv \mathbf{V}^T\mathbf{V}$ is the squared norm of the vector $\mathbf{V}$. Since B_0 is a *symmetric* Toeplitz matrix, we can rapidly solve $B_0\mathbf{u}_0^{(j+1)} = \mathbf{v}_0^{(j)}$ for $\mathbf{u}_0^{(j+1)}$ using a Levinson-type method (Bell *et al.*, 1993). (Each row of a symmetric Toeplitz matrix contains shifted versions of the same elements, and hence code for solving the system usually only requires one row of the matrix. Storage requirements are thus low, even for large N.) This scheme produces on convergence the eigenvector corresponding to the smallest eigenvalue of B_0:

$$B_0\mathbf{v}_0 = \gamma_0\mathbf{v}_0$$

so that $\lambda_0 = \lambda_{-1} + \gamma_0 = 1 + \gamma_0$.

Suppose that convergence is deemed to have taken place when

$$\mathbf{v}_0^{(L+1)} \approx \mathbf{v}_0^{(L)} \quad \text{or} \quad \mathbf{v}_0^{(L+1)} \approx -\mathbf{v}_0^{(L)}.$$

Now $\gamma_0 < 0$ since $\lambda_0 < 1$, and hence $\mathbf{v}_0^{(L+1)}$ and $\mathbf{v}_0^{(L)}$ are of opposite polarity. Hence we take $\gamma_0 = -1/\|\mathbf{u}_0^{(L)}\|$. Let $\sigma_0 = \gamma_0$ so that also $\lambda_0 = 1 + \sigma_0$.

For orders $k > 0$ a modified method is required. For $k = 1$ we need to find the eigenvector corresponding to the next to largest eigenvector, λ_1. Define the matrix

$$B_1 = A - (1 + \sigma_0)I = A - \lambda_0 I.$$

Since for eigenvalues λ and eigenvectors $\mathbf{v}$ of A,

$$B_1 \mathbf{v} = (\lambda - \lambda_0)\mathbf{v},$$

the eigenvalue of A closest to λ_0 (besides λ_0 itself) will correspond to the smallest eigenvalue of B_1 (besides zero). What we require though is λ_1, which although close to λ_0 is distinct; the corresponding eigenvectors are orthogonal. We use the modified iterative scheme:

$$B_1 \mathbf{u}_1^{(j+1)} = \mathbf{v}_1^{(j)} \quad \text{and} \quad \mathbf{v}_1^{(j+1)} = \frac{\mathbf{u}_1^{(j+1)} - \left([\mathbf{u}_1^{(j+1)}]^T \mathbf{v}_0 \right) \mathbf{v}_0}{\left\| \mathbf{u}_1^{(j+1)} - \left([\mathbf{u}_1^{(j+1)}]^T \mathbf{v}_0 \right) \mathbf{v}_0 \right\|}.$$

The subtraction of the projection of $\mathbf{u}_1^{(j+1)}$ onto $\mathbf{v}_0$ from $\mathbf{u}_1^{(j+1)}$ means that the scheme will end up finding the eigenvector whose eigenvalue is closest to λ_0 consistent with the corresponding eigenvector being orthogonal to $\mathbf{v}_0$. In other words, on convergence

$$B_1 \mathbf{v}_1 = \gamma_1 \mathbf{v}_1$$

with $\lambda_1 = \lambda_0 + \gamma_1$, and since $\lambda_1 < \lambda_0$ we know $\gamma_1 < 0$. In the same way as above, we take $\gamma_1 = -1/\|\mathbf{u}_1^{(L)}\|$. Let $\sigma_1 = \gamma_0 + \gamma_1$. Since $\gamma_0 = \lambda_0 - 1$ and $\gamma_1 = \lambda_1 - \lambda_0$ we have $\lambda_1 = 1 + (\gamma_0 + \gamma_1) = 1 + \sigma_1$.

For $k > 1$ taking

$$B_k = A - (1 + \sigma_{k-1})I = A - \lambda_{k-1} I,$$

the iterative scheme uses

$$B_k \mathbf{u}_k^{(j+1)} = \mathbf{v}_k^{(j)} \quad \text{and} \quad \mathbf{v}_k^{(j+1)} = \frac{\mathbf{u}_k^{(j+1)} - \sum_{\ell=0}^{k-1} \left([\mathbf{u}_k^{(j+1)}]^T \mathbf{v}_\ell \right) \mathbf{v}_\ell}{\left\| \mathbf{u}_k^{(j+1)} - \sum_{\ell=0}^{k-1} \left([\mathbf{u}_k^{(j+1)}]^T \mathbf{v}_\ell \right) \mathbf{v}_\ell \right\|},$$

which ensures that $\mathbf{v}_k$ will be orthogonal to $\mathbf{v}_0, \ldots, \mathbf{v}_{k-1}$. Note that $\lambda_k = 1 + \sum_{\ell=0}^{k} \gamma_\ell = 1 + \sigma_k$.

For finding the eigenvector corresponding to λ_k it follows from Parlett (1980, p. 63) that the convergence factor is $|\lambda_k - \lambda_{k-1}|/|\lambda_{k+1} - \lambda_{k-1}| = \gamma_k/(\gamma_k + \gamma_{k+1})$. The convergence rate is thus very rapid for the first few eigenvalues and then decreases. For the example of Table 380(a), the convergence rate of *inverse* iteration for finding the first eigenvalue/vector is approximately 0.013, while if instead forward iteration was used, the convergence rate would be (Parlett, 1980, p. 60) $\lambda_1/\lambda_0 \approx 1$, i.e., very much slower.

8.2 Calculating DPSSs from Numerical Integration

The seminal paper by Thomson (1982) on the use of discrete prolate spheroidal sequences for spectral estimation gave a method for calculating the sequences that makes use of numerical integration. We outline this method here.

The prolate spheroidal wave function (pswf) $\psi_k(\cdot\,; c)$ – defined for *continuous* time – satisfies the following orthonormal condition (Section 3.5, property [4]):

$$\int_{-1}^{1} \frac{\psi_k^2(x; c)}{\lambda_k(c)} \, dx = 1,$$

where $c = \pi N W$ (i.e., π times the duration–bandwidth product with the sampling interval Δt taken to be unity) and $\lambda_k(c)$ is the continuous analog of $\lambda_k(N, W)$. For large N we can write

$$\int_{-1}^{1} \frac{\psi_k^2(x; c)}{\lambda_k(c)} \, dx \approx \frac{2}{N} \sum_{t=0}^{N-1} \frac{\psi_k^2([2t + 1 - N]/N; c)}{\lambda_k(c)}.$$

If we compare this summation to the normalization condition for a dpss, namely,

$$\sum_{t=0}^{N-1} v_{t,k}^2(N, W) = 1,$$

this suggests the approximation

$$v_{t,k}(N, W) \approx \pm \left(\frac{2}{N \lambda_k(c)} \right)^{1/2} \psi_k\left([2t + 1 - N]/N; c\right); \qquad (384)$$

i.e., we approximate the dpss in terms of the continuous pswf. Thomson in fact sampled the continuous pswf using $\psi_k([2t - N]/N, c)$ rather than $\psi_k([2t + 1 - N]/N, c)$; however, the latter corresponds to a better integration approximation (midpoint rule).

The continuous pswf satisfies

$$\lambda_k(c)\psi_k(x;c) = \int_{-1}^{1} \frac{\sin\left(c[x-y]\right)}{\pi[x-y]}\psi_k(y;c)\,dy$$

(see Equation (77)). The integral can be approximated by a Gauss–Legendre quadrature formula using J points (in the examples below we set $J = 32$, which is sufficiently accurate for most cases – see Thomson, 1982, p. 1091). We obtain

$$\lambda_k(c)\psi_k(x;c) \approx \sum_{j=1}^{J} w_j \frac{\sin\left(c[x-x_j]\right)}{\pi(x-x_j)}\psi_k(x_j;c), \qquad (385a)$$

where the $\{x_j\}$ and $\{w_j\}$ are, respectively, the Gauss–Legendre abscissae and weights. These can be either computed directly using, for example, subroutine GAULEG from Press *et al.* (1986, p. 125) – thus enabling other values of J to be tried – or obtained from Table 25.4 of Abramowitz and Stegun (1964, pp. 916–9). By setting x equal to x_l in the above, we obtain

$$\lambda_k(c)\Psi_{l,k} \approx \sum_{j=1}^{J} K_{l,j}\Psi_{j,k}, \qquad (385b)$$

where

$$\Psi_{j,k} \equiv (w_j)^{1/2}\psi_k(x_j;c) \quad\text{and}\quad K_{l,j} \equiv (w_l w_j)^{1/2}\frac{\sin\left(c[x_l-x_j]\right)}{\pi[x_l-x_j]}$$

(as usual, when $j = l$, we take $K_{l,l} = w_l c/\pi$). Since $K_{l,j}$ is real-valued and since $K_{j,l} = K_{l,j}$, the vector with elements $\Psi_{1,k}, \ldots, \Psi_{J,k}$ is the kth eigenvector for the real, symmetric, $J \times J$ matrix defined by $K_{l,j}$. Once this eigenvector has been computed, we can use Equations (384) and (385a) to obtain

$$v_{t,k}(N,W) \approx C_k \sum_{j=1}^{J}(w_j)^{1/2}\frac{\sin\left(c\left[([2t+1-N]/N)-x_j\right]\right)}{\pi\left[([2t+1-N]/N)-x_j\right]}\Psi_{j,k} \qquad (385c)$$

for $t = 0, \ldots, N - 1$, where C_k is a normalizing constant chosen to satisfy Slepian's convention.

If $K - 1$ is the maximum order of dpss required, the J eigenvectors of Equation (385b) can be found using the same routines referenced in Section 8.1, and then $\mathbf{v}_k(N,W)$ can be constructed for $k = 0, \ldots, K-1$ using (385c). Having approximated the dpss $\mathbf{v}_k(N,W)$, the corresponding $\lambda_k(N,W)$ can then be approximated from Equation (378). Small inaccuracies in the approximation of Equation (385c) mean that the first

few eigenvalues approximated from Equation (378) are often somewhat
unreliable (e.g., they can turn out to be slightly greater than unity);
however, these numbers are often of little interest as we know they are
virtually unity.

Just as for the direct approach of Section 8.1, this method will become unsatisfactory for certain parameter combinations. For example
for $N = 79$, $NW = 7.9$ and for $N = 1000$, $NW = 8$, serious errors were
found in the tapers when using double precision. Even using quadruple
precision did not clearly sort out all the problems. Errors arise through
both the approximation involved and because of the unstable eigenproblem in Equation (385b). For parameter combinations leading to stable
estimation, e.g., $N = 99$, $NW = 4$, changing J from 32 to 64 changed
the computed taper values only in the 11th or 12th decimal place.

For the zeroth-order dpss only (a useful data taper by itself, as
we have seen in Chapter 6), Walden (1989) used the result in Equation (384) – which relates the dpss to the continuous prolate spheroidal
wave function – to improve on an approximation of Kaiser (1966, 1974)
and obtained

$$v_{t,0}(N, W) \approx C_0' \frac{I_0 \left(\pi W (N - 1) \left(1 - ([2t + 1 - N]/N)^2 \right)^{1/2} \right)}{I_0(\pi W (N - 1)),} \qquad (386)$$

$t = 0, \ldots, N - 1$, where C_0' is a scaling constant to satisfy Slepian's
convention, and I_0 is the modified Bessel function of the first kind and
zeroth order (although the denominator Bessel function does not depend
on t, dividing the numerator Bessel function by it – *before* determining
C_0' – eliminates numerical instabilities, particularly when N is large).
The advantage of Equation (386) is that short and simple subroutines exist for calculating I_0; see, for example, the FORTRAN function BESSJ0
in Section 6.4 of Press *et al.* (1986).

8.3 Tridiagonal Formulation

It is a remarkable result that all dpss's $\mathbf{v}_k(N, W)$, $k = 0, \ldots, N - 1$,
satisfy the difference equation

$$t(N - t)v_{t-1,k}(N, W)/2$$
$$+ \left[([N - 1 - 2t]/2)^2 \cos (2\pi W) - \theta_k(N, W) \right] v_{t,k}(N, W)$$
$$+ (t + 1)(N - 1 - t)v_{t+1,k}(N, W)/2 = 0$$

for $t = 0, 1, \ldots, N - 1$ with $v_{-1,k}(N, W) = v_{N,k}(N, W) \equiv 0$ (Slepian,
1978, p. 1379). As a result, it follows that $\mathbf{v}_k(N, W)$ is an eigenvector
of the symmetric tridiagonal matrix with diagonal elements

$$([N - 1 - 2t]/2)^2 \cos (2\pi W), \qquad t = 0, \ldots, N - 1$$

and off-diagonal elements

$$t(N-t)/2, \qquad t = 1, \ldots, N-1.$$

The $\theta_k(N, W)$ terms are real and distinct and, in fact, are the eigenvalues for the tridiagonal matrix. Note that these are *not* the same as the eigenvalues $\lambda_k(N, W)$ of Equation (378); in particular, they are not bunched near unity or zero as are the $\lambda_k(N, W)$ eigenvalues. Hence calculation of the dpss's using the symmetric tridiagonal matrix is numerically stable (also, because the defining matrix is here tridiagonal, storage is also efficient even for very large N). Solving for the dpss's in this way obviously does not provide the $\lambda_k(N, W)$ eigenvalues. The approach is to solve for the eigenvectors and (unwanted) eigenvalues of the tridiagonal matrix equation. Having thus accurately calculated $\mathbf{v}_k(N, W)$ – subject only to computational inaccuracies but no other approximation – the matching eigenvalues $\{\lambda_k(N, W)\}$ can be obtained as outlined in Equation (378).

Since in general N can be large, and only the first few eigenvectors (dpss's) are required for tapering, it is clearly wasteful to compute *all* the eigenvectors. As described on pp. 43–5 and 78–9 of the EISPACK documentation (Smith *et al.*, 1976), it is possible to compute just selected blocks of eigenvectors using special calls to two EISPACK routines. As a specific example, suppose that $N = 64$ and that we wish to compute the eight eigenvectors $\mathbf{v}_k(N, W)$, $k = 0, \ldots, 7$. First, after filling the N-dimensional vectors D, E and E2 with the diagonal elements, off-diagonal elements and squares of the off-diagonal elements of the tridiagonal matrix, we evoke TRIDIB with the variables M11 and M set to 57 and 8, respectively (Smith *et al.*, 1976, pp. 501–2). This tells EISPACK to prepare to extract the 8 eigenvectors associated with the 57th to 64th *smallest* eigenvalues, i.e., the eight *largest* eigenvalues. The actual extraction is done by evoking TINVIT immediately after TRIDIB – the required eigenvectors $\mathbf{v}_k(N, W)$ are returned in reverse order in the two-dimensional array Z (Smith *et al.*, 1976, pp. 448–50). A check can be made that the polarity convention of Section 8.0 is satisfied.

This method, and the inverse iteration method of Section 8.1, are the recommended methods.

8.4 Substitutes for the DPSSs

Greenhall (1990) sought to bypass the problems associated with computation of the dpss's by deriving more easily computable substitutes. Consider the time-limited trigonometric polynomial of degree $\leq M$:

$$\xi(t; M) = \begin{cases} \sum_{j=-M}^{M} g_j e^{i2\pi jt}, & \text{if } |t| < 1/2; \\ 0, & \text{otherwise}; \end{cases}$$

this function has Fourier transform

$$\Xi(f;M) = \sum_{j=-M}^{M} g_j \int_{-1/2}^{1/2} e^{-i2\pi t(f-j)} \, dt = \sum_{j=-M}^{M} g_j \frac{\sin{(\pi[f-j])}}{\pi[f-j]}$$

(see Equation (65b)). Greenhall sought the coefficients, say $\{g_j(q)\}$, that minimize

$$\gamma^2(q) \equiv \frac{\int_{|f|>q} |\Xi(f;M)|^2 \, df}{\int_{-\infty}^{\infty} |\Xi(f;M)|^2 \, df}.$$

Note this *minimization* of leakage outside the pass-band $[-q, q]$ is equivalent to *maximizing* concentration in the pass-band. By expressing $\Xi(f;M)$ in terms of its Fourier transform pair $\xi(t;M)$ we can write

$$\gamma^2(q) = \frac{\sum_{j=-M}^{M} \sum_{k=-M}^{M} g_j(q) \mathcal{A}_{j,k}(q) g_k^*(q)}{\sum_{j=-M}^{M} |g_j(q)|^2},$$

where $\mathcal{A}_{j,k}(\cdot)$ is the kernel defined by

$$\mathcal{A}_{j,k}(q) \equiv \int_{|f|>q} \frac{\sin{(\pi[f-j])}\sin{(\pi[f-k])}}{\pi^2(f-j)(f-k)} \, df, \qquad j,k = -M, \ldots, M.$$

Note that $\mathcal{A}_{j,k}(0) = \delta_{j,k}$ leading to the simplification of the denominator of $\gamma^2(q)$. Analogous to Equation (105b), we see that the sequence $\{g_j(q)\}$ that minimizes $\gamma^2(q)$ must satisfy

$$\sum_{j=-M}^{M} \mathcal{A}_{j,k}(q) g_j(q) = \zeta_l(M, q) g_k(q), \qquad k = -M, \ldots, M.$$

To each eigenvalue $\zeta_0(M, q) > \zeta_1(M, q) > \cdots > \zeta_{2M}(M, q)$, there is a corresponding eigenvector $\mathbf{c}_k(M, q)$ whose elements are the sequence $\{\, c_{j,k}(M, q) : j = -M, \ldots, M \,\}$. The elements of $\mathbf{c}_{2M}(M, q)$ form the sequence with minimum leakage, corresponding to the smallest eigenvalue $\zeta_{2M}(M, q)$. The sequences that are the elements of $\mathbf{c}_{2M-1}(M, q)$, $\mathbf{c}_{2M-2}(M, q)$, etc., give rise to increasing leakage. The trigonometric polynomials (eigenfunctions) corresponding to these eigenvectors – namely, $\xi_k(t; M, q), k = 0, \ldots, 2M$ – are defined (to within a factor of $\pm i$ – see below) by

$$\xi_k(t; M, q) \equiv \begin{cases} \sum_{j=-M}^{M} c_{j,k}(M, q) e^{i2\pi jt}, & \text{if } |t| < 1/2; \\ 0, & \text{otherwise,} \end{cases}$$

with Fourier transforms $\Xi_k(\cdot; M, q)$. These polynomials are orthonormal and called *trig prolates* by Greenhall (1990), because they can be regarded as finite-dimensional analogs of the prolate spheroidal wave functions (pswf's) of Section 3.5. These eigenfunctions are either even or odd, the odd ones being multiplied by $\pm i$ to make them real-valued.

Since each trig prolate $\xi_k(\cdot; M, q)$ is orthonormal over $[-1/2, 1/2]$, we have

$$\int_{-1/2}^{1/2} \xi_k^2(t; M, q)\, dt = 1.$$

If we approximate the integral using the midpoint rule of numerical integration (used also in Section 8.2), we obtain

$$\int_{-1/2}^{1/2} \xi_k^2(t; M, q)\, dt \approx \frac{1}{N} \sum_{t=0}^{N-1} \xi_k^2 \left([2t - (N-1)]/2N; M, q\right).$$

We can hence construct an orthogonal set of $2M+1$ sequences for use as data tapers. Adopting the same normalization condition as for a dpss, namely,

$$\sum_{t=0}^{N-1} v_{t,k}^2(N, W) = 1,$$

we define, for $t = 0, \ldots, N-1$ and $k = 0, \ldots, 2M$,

$$u_{t,k}(N, M, W) \equiv \frac{1}{\sqrt{N}} \xi_k \left([2t - (N-1)]/2N; M, q\right), \qquad (389a)$$

where $W = q/N$. The choice of sampling positions ensures that the basis functions $e^{i2\pi jt}$ remain orthogonal when so sampled, and the sampled trig prolates are themselves orthogonal.

Greenhall (1990) set $M = q$ and, for $q = 2$, 3, 4 and 5, used numerical integration to compute the kernel $\mathcal{A}_{j,k}(\cdot)$. He then solved for the eigenvalues and eigenvectors using EISPACK. For any q in this range he gives coefficients $\{ a_{j,k}(q) : j = 0, \ldots, q \}$ so that, for even k,

$$\xi_k(t; q, q) = a_{0,k}(q) + 2 \sum_{j=1}^{q} a_{j,k}(q) \cos(2\pi jt), \qquad (389b)$$

and, for odd k,

$$\xi_k(t; q, q) = 2 \sum_{j=1}^{q} a_{j,k}(q) \sin(2\pi jt), \qquad (389c)$$

for all k such that the leakage is less than 1%. The coefficients for $q = 2$, 3 and 4 are reproduced in Table 390 (with some sign changes to ensure that the $\{u_{t,k}(N, M, W)\}$ sequences follow Slepian's dpss conventions).

Numerical studies by Greenhall (1990) showed that, if the sampled trig prolates were used as substitutes for the dpss's, sidelobe leakage

q	k	$j=0$	$j=1$	$j=2$	$j=3$	$j=4$
2	0	0.8202108	0.4041691	0.0165649		
	1	—	−0.7007932	−0.0942808		
3	0	0.7499700	0.4596063	0.0867984	0.0007513	
	1	—	−0.6507499	−0.2765560	−0.0064282	
	2	0.4969513	−0.3050683	−0.5312499	−0.0350227	
	3	—	−0.2731233	0.6397174	0.1271430	
4	0	0.6996910	0.4830013	0.1473918	0.0141997	0.0000368
	1	—	−0.5927723	−0.3805986	−0.0613650	−0.0003329
	2	0.4783016	−0.1666510	−0.5724443	−0.1736202	−0.0022015
	3	—	−0.3540569	0.4929565	0.3626279	0.0117722
	4	0.3862293	−0.3223025	0.0856254	0.5584413	0.0484379

Table 390. Coefficients $a_{j,k}(q)$ for generating $\{u_{t,k}(N, M, W)\}$, an approximation to the dpss $\{ v_{t,k}(N, W) : t = 0, \ldots, N-1 \}$. For a selected duration–bandwidth product $NW = q$ and order k, the appropriate row of tabulated coefficients is used in either Equation (389b) (if k is even) or Equation (389c) (if k is odd) to define $\xi_k(\cdot; q, q)$, which is then sampled via Equation (389a) to produce the desired $u_{t,k}(N, M, W)$ (Greenhall, 1990).

is higher, but often marginally so; for the range of cases examined, he found a worst-case penalty of 5.4 dB greater leakage than would be achieved using the equivalent bandwidth dpss. Since $q = NW$, Table 390 enables computation of the sampled trig prolates for $NW = 2$, 3 and 4. For other values of NW, the initial overhead in numerical computation is large, indeed greater than finding the dpss's directly using inverse iteration (Section 8.2) or the tridiagonal system (Section 8.3).

8.5 Exercises

[8.1] Show that, given the eigenvector $\mathbf{v}_k(N, W)$ for Equation (378) with elements $v_{0,k}(N, W)$, $v_{1,k}(N, W)$, ..., $v_{N-1,k}(N, W)$, we can compute the corresponding eigenvalue $\lambda_k(N, W)$ efficiently using

$$\lambda_k(N, W) = 2 \left(W q_0 + 2 \sum_{\tau=1}^{N-1} \frac{\sin(2\pi W \tau)}{\pi \tau} q_\tau \right),$$

where

$$q_\tau \equiv \sum_{t=0}^{N-\tau-1} v_{t,k}(N, W) v_{t+\tau,k}(N, W).$$

Note that $\{q_\tau\}$ has the form of an acvs and hence can be computed using FFTs.

9

Parametric Spectral Estimation

9.0 Introduction

In this chapter we discuss the basic theory behind parametric spectral density function (sdf) estimation. The main idea is simple. Suppose the discrete parameter stationary process $\{X_t\}$ has an sdf $S(\cdot)$ that is completely determined by K parameters $a_1, a_2, \ldots, a_K$:

$$S(f) = S(f; a_1, a_2, \ldots, a_K).$$

Given a time series that can be regarded as a realization of this process, suppose we can estimate the parameters of $S(\cdot)$ from these data by, say, $\hat{a}_1, \hat{a}_2, \ldots, \hat{a}_K$. If these parameter estimates are reasonable, then

$$\hat{S}(f) \equiv S(f; \hat{a}_1, \hat{a}_2, \ldots, \hat{a}_K)$$

should be a reasonable estimate of $S(f)$.

9.1 Notation

In what follows, it is important to keep in mind what basic assumptions are in effect. Accordingly, in this chapter we adopt these notational conventions for the following discrete parameter stationary processes, all of which are assumed to be real-valued and have zero mean:

[1] $\{X_t\}$ represents an arbitrary such process;
[2] $\{Y_t\}$, an *autoregressive* process of finite order p;
[3] $\{G_t\}$, a *Gaussian* process; and
[4] $\{H_t\}$, a *Gaussian autoregressive* process of finite order p.

All of these processes are assumed to have an autocovariance sequence (acvs) $\{s_k\}$ that is defined on the integers, and an sdf $S(\cdot)$ that is defined on the interval $[-f_{(N)}, f_{(N)}]$ – here, as elsewhere, $f_{(N)} \equiv 1/(2\,\Delta t)$ is the Nyquist frequency, and Δt is the sampling interval between values in the process.

391

9.2 The Autoregressive Model

The most widely used form of parametric sdf estimation uses an autoregressive model of order p, denoted as $\text{AR}(p)$, as the underlying functional form for $S(\cdot)$. Recall that a stationary $\text{AR}(p)$ process $\{Y_t\}$ with zero mean satisfies the equation

$$Y_t = \phi_{1,p}Y_{t-1} + \phi_{2,p}Y_{t-2} + \cdots + \phi_{p,p}Y_{t-p} + \epsilon_t, \qquad (392a)$$

where $\phi_{1,p}$, $\phi_{2,p}$, $\ldots$, $\phi_{p,p}$ are p fixed coefficients, and $\{\epsilon_t\}$ is a white noise process with zero mean and variance σ_p^2. The process $\{\epsilon_t\}$ is often called the *innovations process* associated with the $\text{AR}(p)$ process, and σ_p^2 is called the *innovations variance*. Since we will have to refer to $\text{AR}(p)$ models of different orders, we include the order p of the process in the notation for the parameters. The above parameterization is the same as that used by Box and Jenkins (1976), but the reader should be aware that there is another common way of writing an $\text{AR}(p)$ model, namely,

$$Y_t + \gamma_{p,1}Y_{t-1} + \gamma_{p,2}Y_{t-2} + \cdots + \gamma_{p,p}Y_{t-p} = \epsilon_t,$$

where $\gamma_{p,j} = -\phi_{j,p}$. This convention is used, for example, by Priestley (1981) and is prevalent in the engineering literature. Equation (392a) emphasizes the analogy of an $\text{AR}(p)$ model to a multiple linear regression model, but it is only an analogy and not a correspondence: if we regard an $\text{AR}(p)$ model as a regression model, we find that the 'independent' – supposedly nonrandom – variables are in fact lagged copies of the 'dependent' random variable.

The sdf for a stationary $\text{AR}(p)$ process is given by

$$S(f) = \frac{\sigma_p^2 \, \Delta t}{\left| 1 - \sum_{j=1}^p \phi_{j,p} e^{-i2\pi f j \, \Delta t} \right|^2}, \qquad |f| \leq f_{(N)} \qquad (392b)$$

(cf. Equation (168b) with $\Delta t = 1$). Here we have $p + 1$ parameters, namely, the $\phi_{j,p}$ terms and σ_p^2, all of which we must estimate to produce an $\text{AR}(p)$ sdf estimate. We note in passing that the $\phi_{j,p}$ terms cannot be chosen arbitrarily if $\{Y_t\}$ is to be a stationary process. It can be shown that $\{Y_t\}$ can be considered to be a stationary process if all of the roots of the polynomial equation $1 - \sum_{j=1}^p \phi_{j,p} z^{-j} = 0$ lie *inside* the unit circle in the complex plane (an equivalent condition is that the roots of $1 - \sum_{j=1}^p \phi_{j,p} z^j = 0$ lie *outside* the unit circle; see Priestley, 1981, p. 133). For example, for $p = 1$, this requirement implies that $|\phi_{1,1}| < 1$.

The rationale for this particular class of parametric sdf's is six-fold.

[1] One can show that any continuous sdf $S(\cdot)$ can be approximated arbitrarily well by an $\text{AR}(p)$ sdf if p is chosen large enough (Anderson, 1971, p. 411). Thus the class of AR processes is rich enough

to approximate a wide range of processes. Unfortunately 'large enough' can well mean an order p that is too large compared to the amount of available data.

[2] There exist efficient algorithms for fitting AR(p) models to time series. This might seem like a strange justification, but, since the early days of spectral analysis, recommended methodology has often been governed by what can in practice be calculated with commonly available computers.

[3] For a Gaussian process $\{G_t\}$ with autocovariances known up to lag p, the maximum entropy spectrum is identical to that of an AR(p) process. We discuss the principle of maximum entropy and comment upon its applicability to physical time series further below.

[4] A side effect of fitting an AR(p) process to a time series for the purpose of sdf estimation is that we have simultaneously estimated a linear predictor and found a linear prewhitening filter for the series. The importance of prewhitening has been discussed in Section 6.5. In Section 9.10 we describe an overall approach to sdf estimation that views the role of parametric sdf estimation as providing a good method for determining appropriate prewhitening filters.

[5] For certain phenomena a physical argument can be made that an AR model is appropriate. A leading example is the acoustic properties of human speech (Rabiner and Schafer, 1978, Chapter 3).

[6] Pure sinusoidal variations can be expressed as an AR model with $\sigma_p^2 = 0$. This fact – and its implications – are discussed in more detail in Section 10.14.

We have already encountered two examples of AR(p) processes in Chapter 2: the AR(2) process of Equation (45) and the AR(4) process of Equation (46a). The sdf's for these two processes are plotted as thick curves in, respectively, Figures 201 and 202. One realization of each process is shown in Figure 45.

To form an AR(p) sdf estimate from a given set of data, we have two problems: first, to determine the order p that is most appropriate for the data and, second, to estimate the parameters $\phi_{1,p}, \ldots, \phi_{p,p}$ and σ_p^2. Typically we determine p by examining how well a range of AR models fits our data (as judged by some criterion), so it is necessary first to assume that p is known and to learn how to estimate the parameters for an AR(p) model.

9.3 The Yule–Walker Equations

The oldest method of estimating the parameters for a stationary AR(p) process $\{Y_t\}$ with zero mean and acvs given by $s_k = E\{Y_t Y_{t+k}\}$ is based upon matching lagged moments, for which we need to express the parameters in terms of the acvs. To do so, we first take Equation (392a) and multiply both sides of it by Y_{t-k} to get

$$Y_t Y_{t-k} = \sum_{j=1}^{p} \phi_{j,p} Y_{t-j} Y_{t-k} + \epsilon_t Y_{t-k}. \tag{394a}$$

If we take the expectation of both sides, we have, for $k > 0$,

$$s_k = \sum_{j=1}^{p} \phi_{j,p} s_{k-j}. \tag{394b}$$

Here we use the plausible fact that $E\{\epsilon_t Y_{t-k}\} = 0$: although Y_{t-k} obviously depends upon $\epsilon_{t-k}, \epsilon_{t-k-1}, \ldots$ (in fact, if $\sigma_p^2 > 0$, it can be written as an infinite linear combination of these rv's), Y_{t-k} is uncorrelated with values of the white noise sequence that occur after time $t - k$. Let $k = 1, 2, \ldots, p$ in Equation (394b) and recall that $s_{-j} = s_j$ to get the following p equations, known as the *Yule–Walker equations*:

$$
\begin{aligned}
s_1 &= \phi_{1,p} s_0 + \phi_{2,p} s_1 + \cdots + \phi_{p,p} s_{p-1} \\
s_2 &= \phi_{1,p} s_1 + \phi_{2,p} s_0 + \cdots + \phi_{p,p} s_{p-2} \\
&\ \vdots \qquad\quad \vdots \qquad\quad \vdots \qquad \ddots \qquad \vdots \\
s_p &= \phi_{1,p} s_{p-1} + \phi_{2,p} s_{p-2} + \cdots + \phi_{p,p} s_0
\end{aligned}
\tag{394c}
$$

or, in matrix notation,

$$\gamma_p = \Gamma_p \mathbf{\Phi}_p,$$

where $\gamma_p \equiv [s_1, s_2, \ldots, s_p]^T$; $\mathbf{\Phi}_p \equiv [\phi_{1,p}, \phi_{2,p}, \ldots, \phi_{p,p}]^T$; and

$$
\Gamma_p \equiv
\begin{bmatrix}
s_0 & s_1 & \cdots & s_{p-1} \\
s_1 & s_0 & \cdots & s_{p-2} \\
\vdots & \vdots & \ddots & \vdots \\
s_{p-1} & s_{p-2} & \cdots & s_0
\end{bmatrix}.
\tag{394d}
$$

If the covariance matrix Γ_p is positive definite (as is always the case in practical applications), we can now solve for $\phi_{1,p}, \phi_{2,p}, \ldots, \phi_{p,p}$ in terms of the lag 0 to lag p values of the acvs:

$$\mathbf{\Phi}_p = \Gamma_p^{-1} \gamma_p. \tag{394e}$$

Given a time series that is a realization of a portion $X_1, X_2, \ldots, X_N$ of *any* discrete parameter stationary process with zero mean and sdf $S(\cdot)$, one possible way to fit an AR(p) model to it is to replace s_k in Γ_p and γ_p with

$$\hat{s}_k^{(p)} \equiv \frac{1}{N} \sum_{t=1}^{N-|k|} X_t X_{t+|k|}$$

to produce estimates $\tilde{\Gamma}_p$ and $\tilde{\gamma}_p$. (How reasonable this procedure is for an arbitrary stationary process depends on how well it can be approximated by a stationary AR(p) process.) If the time series is not known to have zero mean (the usual case), we redefine our acvs estimator to be

$$\hat{s}_k^{(p)} \equiv \frac{1}{N} \sum_{t=1}^{N-|k|} (X_t - \bar{X})(X_{t+|k|} - \bar{X}),$$

where $\bar{X}$ is the sample mean. We noted in item [2] of the Comments and Extensions to Section 6.2 that, with the above form for $\hat{s}_k^{(p)}$, a realization of the sequence $\{\hat{s}_k^{(p)}\}$ is positive definite if and only if the realizations of $X_1, \ldots, X_N$ are not all exactly the same number (Newton, 1988, p. 165). This very mild condition holds in all practical applications. The positive definiteness of the sequence $\{\hat{s}_k^{(p)}\}$ in turn implies that the matrix $\tilde{\Gamma}_p$ is positive definite. Hence we can obtain estimates for AR(p) coefficients from

$$\tilde{\boldsymbol{\Phi}}_p = \tilde{\Gamma}_p^{-1}\tilde{\gamma}_p. \tag{395a}$$

We are not quite done: we still need to estimate σ_p^2 somehow. We can develop an equation to do so in the following way. If we let $k = 0$ in Equation (394a) and take expectations, we get

$$s_0 = \sum_{j=1}^p \phi_{j,p} s_j + E\{\epsilon_t Y_t\}.$$

From the fact that $E\{\epsilon_t Y_{t-j}\} = 0$ for $j > 0$, it follows that

$$E\{\epsilon_t Y_t\} = E\left\{\epsilon_t\Big(\sum_{j=1}^p \phi_{j,p} Y_{t-j} + \epsilon_t\Big)\right\} = \sigma_p^2$$

and hence

$$\sigma_p^2 = s_0 - \sum_{j=1}^p \phi_{j,p} s_j. \tag{395b}$$

This equation suggests that we estimate σ_p^2 by

$$\tilde{\sigma}_p^2 \equiv \hat{s}_0^{(p)} - \sum_{j=1}^p \tilde{\phi}_{j,p} \hat{s}_j^{(p)}. \tag{395c}$$

We call the estimators $\tilde{\boldsymbol{\Phi}}_p$ and $\tilde{\sigma}_p^2$ the *Yule–Walker estimators* of the AR(p) parameters. An important property of these estimators is that the fitted AR(p) process has a theoretical acvs that is *identical* to the sample acvs up to lag p. This forced agreement is of questionable value: as we have seen in Chapter 6, the Fourier transform of $\{\hat{s}_k^{(p)}\}$ is the periodogram, which can suffer from severe bias.

Now that we have estimators for $\phi_{j,p}$ and σ_p^2, we can estimate the sdf of $\{X_t\}$ by

$$\tilde{S}(f) \equiv \frac{\tilde{\sigma}_p^2 \, \Delta t}{\left|1 - \sum_{j=1}^p \tilde{\phi}_{j,p} e^{-i2\pi f j \, \Delta t}\right|^2}.$$

This estimator depends only on the sample acvs up to lag p. Under certain reasonable assumptions on an arbitrary stationary process $\{X_t\}$, Berk (1974) has shown that $\tilde{S}(f)$ is a consistent estimator of $S(f)$ for all f if the order p of the approximating AR process is allowed to increase as the sample size N increases. Unfortunately, this result is not particularly informative if one has a time series of fixed length N.

Equations (394c) and (395b) can be combined to produce the so-called *augmented Yule–Walker equations*:

$$\begin{bmatrix} s_0 & s_1 & \cdots & s_p \\ s_1 & s_0 & \cdots & s_{p-1} \\ \vdots & \vdots & \ddots & \vdots \\ s_p & s_{p-1} & \cdots & s_0 \end{bmatrix} \begin{bmatrix} 1 \\ -\phi_{1,p} \\ \vdots \\ -\phi_{p,p} \end{bmatrix} = \begin{bmatrix} \sigma_p^2 \\ 0 \\ \vdots \\ 0 \end{bmatrix}.$$

The first row above follows from Equation (395b), while the remaining p rows are just transposed versions of Equations (394c). This formulation is sometimes useful for expressing the estimation problem so that all the AR(p) parameters can be found simultaneously using a standard routine for solving a Toeplitz system of equations (see below).

Given the acvs of $\{X_t\}$ or its estimator up to lag p, Equations (394e) and (395a) formally require matrix inversions to obtain the values of the actual or estimated AR(p) coefficients. As we shall show in the next section, there is an interesting alternative way to relate these coefficients to the true or estimated acvs that avoids matrix inversion and clarifies the relationship between AR sdf estimation and the so-called maximum entropy method (see Section 9.6).

Comments and Extensions to Section 9.3

[1]   There is no reason to insist upon the substitution of the usual biased estimator of the acvs into Equation (394e) to produce estimated values for the AR coefficients. Any other reasonable estimates for the acvs can be used – the only requirement is that the estimated sequence be positive definite so that the matrix inversion in (394e) can be done. For example, if the process mean is assumed to be 0, the direct spectral

estimator $\hat{S}^{(d)}(\cdot)$ of Equation (206c) yields an estimate of the acvs of the form

$$\hat{s}_k^{(d)} \equiv \sum_{t=1}^{N-|k|} h_t X_t h_{t+|k|} X_{t+|k|}$$

(see Equation (207d)), where $\{h_t\}$ is the data taper used with $\hat{S}^{(d)}(\cdot)$. As is true for $\{\hat{s}_k^{(p)}\}$, realizations of the sequence $\{\hat{s}_k^{(d)}\}$ are always positive definite in practical applications. Since proper use of tapering ensures that the first moment properties of $\hat{S}^{(d)}(\cdot)$ are better overall than those of the periodogram, it makes some sense to use $\{\hat{s}_k^{(d)}\}$ – the inverse Fourier transform of $\hat{S}^{(d)}(\cdot)$ – rather than the usual biased estimator $\{\hat{s}_k^{(p)}\}$ – the inverse Fourier transform of the periodogram. For an example, see [1] of the Comments and Extensions to Section 9.4.

9.4   The Levinson–Durbin Recursions

The Levinson–Durbin recursions are an alternative way of solving for the AR coefficients in Equation (394c). Here we follow Papoulis (1985) and derive the equations that define the recursions by considering the following problem: given values of $X_{t-1}, X_{t-2}, \ldots, X_{t-k}$ of a stationary process $\{X_t\}$ with zero mean, how can we predict the value of X_t? A mathematically tractable solution is to find that *linear* function of $X_{t-1}, X_{t-2}, \ldots, X_{t-k}$, say

$$\overrightarrow{X}_t(k) \equiv \sum_{j=1}^{k} \phi_{j,k} X_{t-j},$$

such that the *mean square linear prediction error*

$$P_k \equiv E\left\{ \left(X_t - \overrightarrow{X}_t(k) \right)^2 \right\} = E\left\{ \left(X_t - \sum_{j=1}^{k} \phi_{j,k} X_{t-j} \right)^2 \right\}$$

is minimized. We call $\overrightarrow{X}_t(k)$ the *best linear predictor* of X_t, given X_{t-1}, $\ldots, X_{t-k}$. (For reasons that will become clear in the next few paragraphs, we are purposely using the same symbols for denoting the coefficients above and the coefficients of an AR(k) process, which $\{X_t\}$ need *not* be. The reader should regard the above as a new definition of $\phi_{j,k}$ for the time being – we will show below that this agrees with our old definition if in fact $\{X_t\}$ is an AR(k) process. Also note that $\overrightarrow{X}_t(k)$ is a scalar quantity: it is not a vector as the arrow over the 'X' might suggest to readers familiar with textbooks on physics. The arrow is meant to connote 'forward prediction.')

If we denote the prediction error associated with the best linear predictor by

$$\overrightarrow{\epsilon_t}(k) \equiv X_t - \overrightarrow{X_t}(k), \tag{398a}$$

we can then derive the following result. For any set of real-valued numbers $\psi_1, \ldots, \psi_k$, define

$$P_k(\psi_1, \ldots, \psi_k) \equiv E\left\{ \left(X_t - \sum_{l=1}^{k} \psi_l X_{t-l} \right)^2 \right\}.$$

Since $P_k(\cdot)$ is a quadratic function of the ψ_l terms and since the best linear predictor is defined as that set of ψ_l terms that minimizes the above, we can find the $\phi_{l,k}$ terms by differentiating the above equation and setting the derivatives to zero:

$$\frac{\delta P_k}{\delta \psi_j} = -2E\left\{ \left(X_t - \sum_{l=1}^{k} \psi_l X_{t-l} \right) X_{t-j} \right\} = 0, \qquad 1 \leq j \leq k. \tag{398b}$$

Since $\psi_j = \phi_{j,k}$ for all j at the solution point and since

$$X_t - \sum_{l=1}^{k} \phi_{l,k} X_{t-l} = \overrightarrow{\epsilon_t}(k),$$

Equation (398b) reduces to the so-called *orthogonality principle*, namely,

$$E\{\overrightarrow{\epsilon_t}(k) X_{t-j}\} = 0, \qquad 1 \leq j \leq k. \tag{398c}$$

In words, the orthogonality principle states that the prediction error is uncorrelated with all of the rv's utilized in the prediction.

If we rearrange Equations (398b), we are led to a series of equations that allows us to solve for $\phi_{l,k}$, namely,

$$E\{X_t X_{t-j}\} = \sum_{l=1}^{k} \phi_{l,k} E\{X_{t-l} X_{t-j}\}, \qquad 1 \leq j \leq k,$$

or, equivalently in terms of the acvs,

$$s_j = \sum_{l=1}^{k} \phi_{l,k} s_{j-l}, \qquad 1 \leq j \leq k. \tag{398d}$$

Comparison of the above with the Yule–Walker equations (394c) shows that they are identical! Thus, Equation (394e) here shows that, if the covariance matrix is positive definite, the $\phi_{j,k}$ terms are necessarily unique and hence $\overrightarrow{X}_t(k)$ is unique. This uniqueness leads to a useful corollary to the orthogonality principle, which is the subject of Exercise [9.1].

The reader should note that the $\phi_{l,k}$ terms of the best linear predictor are uniquely determined by the covariance properties of the process $\{X_t\}$ – we have not discussed so far the practical problem of estimating these coefficients for a process with an unknown covariance structure.

For the Yule–Walker equations for an AR(p) process, the innovations variance σ_p^2 can be related to values of the acvs and the AR(p) coefficients (Equation (395b)); by an analogous argument, the mean square linear prediction error P_k can be expressed in terms of the acvs and the $\phi_{j,k}$ terms:

$$P_k = E\{\overrightarrow{\epsilon}_t^2(k)\} = E\left\{\left(X_t - \overrightarrow{X}_t(k)\right)^2\right\} = E\left\{\overrightarrow{\epsilon}_t(k)\left(X_t - \overrightarrow{X}_t(k)\right)\right\}$$

$$= E\{\overrightarrow{\epsilon}_t(k)X_t\} = E\left\{\left(X_t - \overrightarrow{X}_t(k)\right)X_t\right\} = s_0 - \sum_{j=1}^{k} \phi_{j,k}s_j,$$

(399a)

where we have appealed to the orthogonality principle in order to go from the first line to the second. From the first expression on the second line, we note the important fact that, since both $\overrightarrow{\epsilon}_t(k)$ and X_t have zero mean,

$$P_k = E\{\overrightarrow{\epsilon}_t(k)X_t\} = \text{cov}\{\overrightarrow{\epsilon}_t(k), X_t\};$$

(399b)

i.e., the mean square linear prediction error is just the covariance between the prediction error $\overrightarrow{\epsilon}_t(k)$ and the quantity being predicted, namely, X_t.

To summarize our discussion to this point, the Yule–Walker equations arise in two related problems:

[1] For a stationary AR(p) process $\{Y_t\}$ with zero mean, acvs $\{s_k\}$ and coefficients $\phi_{1,p}$, $\phi_{2,p}$, $\ldots$, $\phi_{p,p}$, the Yule–Walker equations relate the coefficients to the acvs; an auxiliary equation (395b) gives the innovations variance in terms of $\{s_k\}$ and the AR(p) coefficients. Sampling versions of these equations allow us to use an AR(p) process to approximate a time series that can be regarded as a portion of an arbitrary stationary process $\{X_t\}$.

[2] For a stationary process $\{X_t\}$ with zero mean and acvs $\{s_k\}$, the Yule–Walker equations relate the coefficients of the best linear predictor of X_t, given the k most recent prior values, to the acvs; an auxiliary equation (399a) gives the mean square linear prediction error in terms of $\{s_k\}$ and the coefficients of the best linear predictor.

The above implies that, for the AR(p) process $\{Y_t\}$, the mean square linear prediction error P_p of its pth order linear predictor is identical to its innovations variance σ_p^2.

Before we get to the heart of our derivation of the Levinson–Durbin recursions, we note the following seemingly trivial fact: if $\{X_t\}$ is a stationary process, then so is $\{X_{-t}\}$, the process with time reversed. Since

$$E\{X_{-t}X_{-t+k}\} = E\{X_t X_{t+k}\} = s_k,$$

both $\{X_t\}$ and $\{X_{-t}\}$ have the same acvs. It follows from Equation (398d) that $\overleftarrow{X}_t(k)$, the best linear 'predictor' of X_t given the next k future values, can be written as

$$\overleftarrow{X}_t(k) \equiv \sum_{j=1}^{k} \phi_{j,k} X_{t+j},$$

where $\phi_{j,k}$ is *exactly* the same coefficient occurring in the best linear predictor of X_t, given the k most recent prior values. The orthogonality principle applied to the reversed process tells us that

$$E\{\overleftarrow{\epsilon}_t(k)X_{t+j}\} = 0, \qquad 1 \le j \le k, \text{ where } \overleftarrow{\epsilon}_t(k) \equiv X_t - \overleftarrow{X}_t(k).$$

From now on we refer to $\overrightarrow{X}_t(k)$ and $\overleftarrow{X}_t(k)$, respectively, as the *forward* and *backward* predictors of X_t of length k. We call $\overrightarrow{\epsilon}_t(k)$ and $\overleftarrow{\epsilon}_t(k)$ the corresponding *forward* and *backward prediction errors*. It follows from symmetry that

$$E\{\overleftarrow{\epsilon}_t^{\,2}(k)\} = E\{\overrightarrow{\epsilon}_t^{\,2}(k)\} = P_k$$

and that the analog of Equation (399b) is

$$P_k = E\{\overleftarrow{\epsilon}_{t-k}(k)X_{t-k}\} = \text{cov}\,\{\overleftarrow{\epsilon}_{t-k}(k), X_{t-k}\}. \tag{400a}$$

The Levinson–Durbin algorithm follows from an examination of the following equation:

$$\overrightarrow{\epsilon}_t(k) = \overrightarrow{\epsilon}_t(k-1) - \theta_k \overleftarrow{\epsilon}_{t-k}(k-1), \tag{400b}$$

where θ_k is a constant to be determined. This equation is by no means obvious, but it is clearly plausible: $\overrightarrow{\epsilon}_t(k)$ depends upon one more variable than $\overrightarrow{\epsilon}_t(k-1)$, namely, X_{t-k}, upon which $\overleftarrow{\epsilon}_{t-k}(k-1)$ obviously depends. To prove (400b), we first note that, by the orthogonality principle,

$$E\{\overrightarrow{\epsilon}_t(k-1)X_{t-j}\} = 0 \text{ and } E\{\overleftarrow{\epsilon}_{t-k}(k-1)X_{t-j}\} = 0$$

for $j = 1, \ldots, k-1$, and, hence, for *any* constant θ_k,

$$E\left\{\left[\overrightarrow{\epsilon}_t(k-1) - \theta_k \overleftarrow{\epsilon}_{t-k}(k-1)\right] X_{t-j}\right\} = 0, \qquad j = 1, \ldots, k-1. \tag{400c}$$

The equation above will also hold for $j = k$ if we let

$$\theta_k = \frac{E\{\overrightarrow{\epsilon_t}(k-1)X_{t-k}\}}{E\{\overleftarrow{\epsilon}_{t-k}(k-1)X_{t-k}\}} = \frac{E\{\overrightarrow{\epsilon_t}(k-1)X_{t-k}\}}{P_{k-1}}$$

(this follows from Equation (400a)). With this choice of θ_k and with $d_t(k) \equiv \overrightarrow{\epsilon_t}(k-1) - \theta_k \overleftarrow{\epsilon}_{t-k}(k-1)$, we have

$$E\{d_t(k)X_{t-j}\} = 0, \qquad 1 \le j \le k.$$

The corollary to the orthogonality principle stated in Exercise [9.1] now tells us that $d_t(k)$ is in fact the same as $\overrightarrow{\epsilon_t}(k)$, thus completing the proof of Equation (400b).

For later use, we note the time-reversed version of Equation (400b), namely,

$$\overleftarrow{\epsilon}_{t-k}(k) = \overleftarrow{\epsilon}_{t-k}(k-1) - \theta_k \overrightarrow{\epsilon_t}(k-1). \tag{401}$$

We are now ready to extract the Levinson–Durbin recursions. It readily follows from Equation (400b) that

$$\overrightarrow{X_t}(k) = \overrightarrow{X_t}(k-1) + \theta_k\left(X_{t-k} - \overleftarrow{X}_{t-k}(k-1)\right);$$

the definitions of $\overrightarrow{X_t}(k)$ and $\overrightarrow{X_t}(k-1)$ further yield

$$\sum_{j=1}^{k} \phi_{j,k}X_{t-j} = \sum_{j=1}^{k-1} \phi_{j,k-1}X_{t-j} + \theta_k\left(X_{t-k} - \sum_{j=1}^{k-1} \phi_{j,k-1}X_{t-k+j}\right).$$

Equating coefficients of X_{t-j} (and appealing to the fundamental theorem of algebra) yields

$$\phi_{j,k} = \phi_{j,k-1} - \theta_k\phi_{k-j,k-1}, \qquad 1 \le j \le k-1, \ \text{and} \ \phi_{k,k} = \theta_k.$$

Given $\phi_{1,k-1}, \ldots, \phi_{k-1,k-1}$ and $\phi_{k,k}$, we can therefore compute $\phi_{1,k}, \ldots, \phi_{k-1,k}$. Moreover, $\phi_{k,k}$ can be expressed using $\phi_{1,k-1}, \ldots, \phi_{k-1,k-1}$ and the acvs up to lag k: if we multiply both sides of Equation (400b) by X_{t-k} and take expectations, we have

$$E\{\overrightarrow{\epsilon_t}(k)X_{t-k}\} = E\{\overrightarrow{\epsilon_t}(k-1)X_{t-k}\} - \phi_{k,k}E\{\overleftarrow{\epsilon}_{t-k}(k-1)X_{t-k}\}.$$

Let us consider the three expectations in this equation: first,

$$E\{\overrightarrow{\epsilon_t}(k)X_{t-k}\} = 0$$

by the orthogonality principle; second,

$$E\{\overrightarrow{\epsilon_t}(k-1)X_{t-k}\} = E\left\{\left(X_t - \sum_{j=1}^{k-1}\phi_{j,k-1}X_{t-j}\right)X_{t-k}\right\}$$

$$= s_k - \sum_{j=1}^{k-1}\phi_{j,k-1}s_{k-j};$$

and third,

$$E\{\overleftarrow{\epsilon}_{t-k}(k-1)X_{t-k}\} = P_{k-1}$$

by Equation (400a). With these substitutions we have

$$\phi_{k,k} = \frac{s_k - \sum_{j=1}^{k-1}\phi_{j,k-1}s_{k-j}}{P_{k-1}}.$$

We have now completed our derivation of what is known in the literature as the *Levinson–Durbin recursions* or the *Levinson recursions*, which we can summarize as follows. Suppose $\phi_{1,k-1}, \ldots, \phi_{k-1,k-1}$ and P_{k-1} are known. We can calculate $\phi_{1,k}, \ldots, \phi_{k,k}$ and P_k using the following three equations:

$$\phi_{k,k} = \frac{s_k - \sum_{j=1}^{k-1}\phi_{j,k-1}s_{k-j}}{P_{k-1}}; \tag{402a}$$

$$\phi_{j,k} = \phi_{j,k-1} - \phi_{k,k}\phi_{k-j,k-1}, \quad 1 \le j \le k-1; \tag{402b}$$

$$P_k = s_0 - \sum_{j=1}^{k}\phi_{j,k}s_j. \tag{402c}$$

We can initiate the recursions by solving Equations (398d) and (399a) explicitly for the case $k = 1$:

$$\phi_{1,1} = s_1/s_0 \text{ and } P_1 = s_0 - \phi_{1,1}s_1. \tag{402d}$$

There is an important variation on the Levinson–Durbin recursions (evidently due to Burg, 1975, p. 14). The difference between the two recursions is only in Equation (402c) for updating P_k, but it is noteworthy for three reasons: first, it is a numerically better way of calculating P_k because it is not so sensitive to rounding errors; second, it requires fewer numerical operations and actually speeds up the recursions slightly; and third, it emphasizes the central role of $\phi_{k,k}$, the so-called *kth order partial autocorrelation coefficient* (see item [3] in the Comments and Extensions). To derive the alternative equation to (402c), we multiply both sides of Equation (400b) by X_t, recall that $\theta_k = \phi_{k,k}$, and take expectations to get

$$E\{\overrightarrow{\epsilon_t}(k)X_t\} = E\{\overrightarrow{\epsilon_t}(k-1)X_t\} - \phi_{k,k}E\{\overleftarrow{\epsilon}_{t-k}(k-1)X_t\}.$$

Using Equation (399b) we can rewrite the above as

$$P_k = P_{k-1} - \phi_{k,k}E\{\overleftarrow{\epsilon}_{t-k}(k-1)X_t\}.$$

Now

$$E\{\overleftarrow{\epsilon}_{t-k}(k-1)X_t\} = E\Big\{\Big(X_{t-k} - \sum_{j=1}^{k-1}\phi_{j,k-1}X_{t-k+j}\Big)X_t\Big\}$$

$$= s_k - \sum_{j=1}^{k-1}\phi_{j,k-1}s_{k-j} = \phi_{k,k}P_{k-1},$$

where the last equality follows from Equation (402a). The alternative to Equation (402c) is thus

$$P_k = P_{k-1}(1 - \phi_{k,k}^2),\tag{403}$$

and the Levinson–Durbin recursions are equivalent to equations (402a), (402b) and (403).

Note that Equation (403) tells us that, if $P_{k-1} > 0$, we must have $|\phi_{k,k}| \leq 1$ to ensure that P_k is nonnegative and, if in fact $|\phi_{k,k}| = 1$, the mean square linear prediction error P_k is 0. This implies that we could predict the process perfectly in the mean square sense with a kth order linear predictor. Since a nontrivial linear combination of the rv's of the process $\{X_t\}$ thus has zero variance, the covariance matrix for $\{X_t\}$ is in fact positive semidefinite instead of positive definite (however, this cannot happen if $\{X_t\}$ is a purely continuous stationary process with nonzero variance; i.e., the derivative of its integrated spectrum (the sdf) exists – see Papoulis, 1985, for details).

We can now summarize explicitly the Levinson–Durbin recursive solution to the Yule–Walker equations for estimating the parameters of an AR(p) model from a sample acvs. Although this method avoids the matrix inversion in Equation (395a), the two solutions are necessarily identical: the Levinson–Durbin recursions simply take advantage of the Toeplitz structure of $\hat{\Gamma}_p$ (see the discussion following Equation (38)) to solve the problem more efficiently than brute force matrix inversion can. On a digital computer, however, the two solutions can be disturbingly different due to the vagaries of rounding error. We begin by solving Equations (395a) and (395c) explicitly for an AR(1) model to get

$$\tilde{\phi}_{1,1} = \hat{s}_1^{(p)}/\hat{s}_0^{(p)} \text{ and } \tilde{\sigma}_1^2 = \hat{s}_0^{(p)} - \tilde{\phi}_{1,1}\hat{s}_1^{(p)} = \hat{s}_0^{(p)}(1 - \tilde{\phi}_{1,1}^2).$$

For $k = 2, \ldots, p$, we then recursively evaluate

$$\tilde{\phi}_{k,k} = \frac{\hat{s}_k^{(p)} - \sum_{j=1}^{k-1} \tilde{\phi}_{j,k-1} \hat{s}_{k-j}^{(p)}}{\tilde{\sigma}_{k-1}^2}; \tag{404a}$$

$$\tilde{\phi}_{j,k} = \tilde{\phi}_{j,k-1} - \tilde{\phi}_{k,k} \tilde{\phi}_{k-j,k-1}, \quad 1 \le j \le k-1; \tag{404b}$$

$$\tilde{\sigma}_k^2 = \tilde{\sigma}_{k-1}^2 (1 - \tilde{\phi}_{k,k}^2) \tag{404c}$$

to obtain finally $\tilde{\phi}_{1,p}, \ldots, \tilde{\phi}_{p,p}$ and $\tilde{\sigma}_p^2$. (The intermediate quantities σ_k^2, $k < p$, are *defined* to be P_k, in view of the fact that $P_p = \sigma_p^2$ for an AR(p) process.)

As an example of autoregressive sdf estimation, we reconsider the time series shown in the lower plot of Figure 45. This series is a realization of the AR(4) process defined in Equation (46a) – this process has been cited in the literature as posing a difficult case for sdf estimation. Figure 405 shows the result of using the Yule–Walker method to fit an AR(4) model to the first 64 values in this time series (top plot), to the first 256 values (middle) and, finally, to the entire series (bottom). In each case the biased estimator of the acvs was calculated from the appropriate segment of data and used as input to the Levinson–Durbin recursions. In each of the plots the thin curve is the sdf corresponding to the fitted AR(4) model, whereas the thick curve is the true AR(4) sdf for the process from which the time series was drawn. We see that the sdf estimates improve with increasing sample size N, but that there is still significant deviation from the true sdf even for $N = 1024$ – particularly in the region of the twin peaks, which collapse incorrectly to a single peak in the estimated sdf's.

Figure 406 shows the effect on the spectral estimates of increasing the order of the fitted model to 8. Although the process that generated this time series is an AR(4) process, we generally get a much better fit to the true sdf by using a higher order model for this example – particularly in the low frequency portion of the sdf and around the twin peaks.

Comments and Extensions to Section 9.4

[1]   The Yule–Walker estimator is usually defined in terms of the biased acvs estimator $\{\hat{s}_k^{(p)}\}$; however, as noted in the Comments and Extensions to Section 9.3, there is no reason why we cannot use other positive definite estimates of the acvs such as $\{\hat{s}_k^{(d)}\}$. Figures 407 and 408 illustrate the possible benefits of doing so (see also Zhang, 1992). These show sdf estimates for the same data used in Figures 405 and 406, but now the Yule–Walker estimator uses $\{\hat{s}_k^{(d)}\}$ corresponding to a direct spectral estimator with $NW = 1$ (in Figure 407) and $NW = 2$ (in Figure 408)

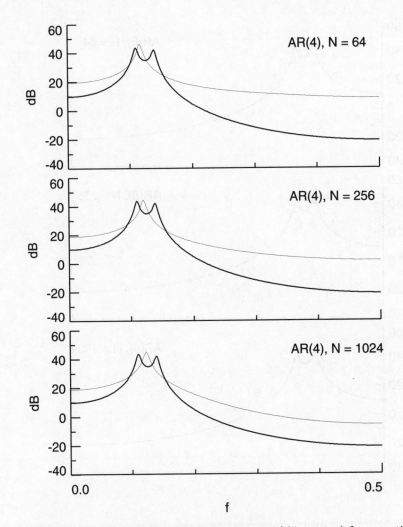

Figure 405. Yule–Walker AR(4) sdf estimates (thin curves) for a portion of length 64, 256 and 1024 of the realization of the AR(4) process shown in the lower plot of Figure 45 (the process is defined in Equation (46a)). The thick curve on each plot is the true sdf.

dpss data tapers. The improvements over the usual Yule–Walker estimates are rather dramatic. (If we increase the degree of tapering to, say, an $NW = 4$ dpss taper, the estimates deteriorate slightly, but the maximum difference over all frequencies between these estimates and the corresponding $NW = 2$ estimates in Figure 408 is less than 3 dB. An estimate with excessive tapering is thus still much better than the

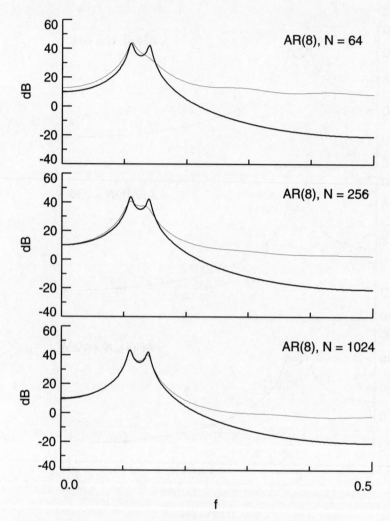

Figure 406. Yule–Walker AR(8) sdf estimates (see Figure 405).

usual Yule–Walker estimate.)

[2]   Note that we did *not* need to assume Gaussianity in order to derive the Levinson–Durbin recursions. The recursions are also *not* tied explicitly to AR(p) processes, since they can also be used to find the coefficients of the best linear predictor of X_t, given the p prior values of the process, when the only assumption on $\{X_t\}$ is that it is a discrete parameter stationary process with zero mean. The recursions are simply an efficient method of solving a system of equations that possesses a Toeplitz structure. The $p \times p$ matrix associated with such a

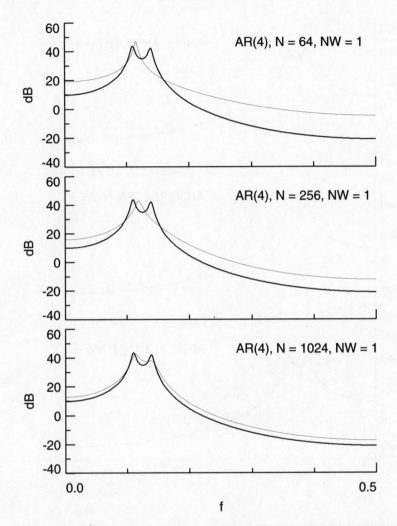

Figure 407. Yule–Walker AR(4) sdf estimates (thin curves) for a portion of length 64, 256 and 1024 of the realization of the AR(4) process shown in the bottom plot of Figure 45. The thick curve in each plot is the true sdf. Instead of the usual biased acvs estimator, the acvs estimator $\{\hat{s}_k^{(d)}\}$ corresponding to a direct spectral estimator with an $NW = 1$ dpss data taper was used.

system (for example, Γ_p in Equation (394d)) consists of at most p distinct elements ($s_0, \ldots, s_{p-1}$ in this example). A computational analysis of the recursions shows that they require $O(p^2)$ or fewer operations to solve the equations instead of the $O(p^3)$ operations required by conventional methods using direct matrix inversion. The recursions also

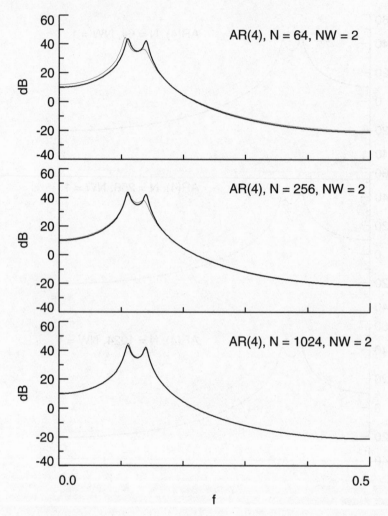

Figure 408. Yule–Walker AR(4) sdf estimates using the acvs estimate $\{\hat{s}_k^{(d)}\}$ corresponding to a direct spectral estimator with an $NW = 2$ dpss data taper.

require less storage space on a computer (being proportional to p rather than p^2). An excellent discussion of Toeplitz matrices, their numerical properties and important refinements to the Levinson–Durbin recursions is given in Bunch (1985). Of particular interest, he discusses new algorithms for solving Toeplitz systems of equations that require just $O(p \log^2(p))$ operations. In practice, however, for p approximately less than 2000 (a rather high order for an AR model!), the Levinson–Durbin recursions are still the faster method due to the fact that these new algorithms require much more computer code. A discussion of the application of the recursions to other problems in time series analysis is given by Morettin (1984).

[3] As a sequence indexed by k, $\{\phi_{k,k}\}$ is often called the *partial autocorrelation sequence* (pacs) or the *reflection coefficient sequence*. The former terminology arises from the fact that $\phi_{k,k}$ is the correlation between X_t and X_{t-k} after they have been 'adjusted' by the intervening $k-1$ values of the process. The adjustment takes the form of subtracting off predictions based upon $X_{t-k+1}, \ldots, X_{t-1}$; i.e., the adjusted values are

$$X_t - \overrightarrow{X_t}(k-1) = \overrightarrow{\epsilon_t}(k-1) \text{ and } X_{t-k} - \overleftarrow{X}_{t-k}(k-1) = \overleftarrow{\epsilon}_{t-k}(k-1)$$

(for $k = 1$ we define $\overrightarrow{X_t}(0) = 0$ and $\overleftarrow{X}_{t-1}(0) = 0$). The claim is thus that

$$\frac{\text{cov}\left\{\overrightarrow{\epsilon_t}(k-1), \overleftarrow{\epsilon}_{t-k}(k-1)\right\}}{\left(\text{var}\left\{\overrightarrow{\epsilon_t}(k-1)\right\} \text{var}\left\{\overleftarrow{\epsilon}_{t-k}(k-1)\right\}\right)^{1/2}} = \phi_{k,k} \tag{409a}$$

(Exercise [9.3] is to show that this is true).

Ramsey (1974) has investigated a number of properties of the pacs. He has shown that, if $\phi_{p,p} \neq 0$ and $\phi_{k,k} = 0$ for all $k > p$ for a Gaussian stationary process, the process is necessarily an AR(p) process. Equation (403) then tells us that

$$P_{p-1} > P_p = P_{p+1} = \cdots. \tag{409b}$$

We use this fact later on as a basis for selecting the order of an AR model for a time series.

[4] The Levinson–Durbin recursions allow us to build up the coefficients for the one-step-ahead best linear predictor of order k in terms of the order $k - 1$ coefficients (combined with the order $k - 1$ mean square linear prediction error and the acvs up to lag k). It is also possible to go in the other direction: given the order k coefficients, we can determine the corresponding order $k - 1$ quantities. To do so, note that we can write Equation (402b) both as

$$\phi_{j,k-1} = \phi_{j,k} + \phi_{k,k}\phi_{k-j,k-1} \text{ and } \phi_{k-j,k-1} = \phi_{k-j,k} + \phi_{k,k}\phi_{j,k-1}$$

for $1 \leq j \leq k-1$. If, for $\phi_{k-j,k-1}$ in the left-hand equation, we substitute its value in the right-hand equation and solve for $\phi_{j,k-1}$, we get the order $k - 1$ coefficients in terms of the order k coefficients:

$$\phi_{j,k-1} = \frac{\phi_{j,k} + \phi_{k,k}\phi_{k-j,k}}{1 - \phi_{k,k}^2}, \qquad 1 \leq j \leq k - 1.$$

We can invert Equation (403) to get the order $k - 1$ mean square linear prediction error:

$$P_{k-1} = \frac{P_k}{1 - \phi_{k,k}^2}.$$

One use for this step-down procedure is in determining whether, given particular values for the coefficients $\phi_{1,p}$, ..., $\phi_{p,p}$, the AR(p) process of Equation (392a) is in fact stationary. If we use these coefficients in the step-down procedure to recursively obtain the partial autocorrelations $\phi_{p-1,p-1}$, $\phi_{p-2,p-2}$, ..., $\phi_{1,1}$, then the AR(p) process is stationary if $|\phi_{k,k}| < 1$ for $k = 1, \ldots, p$ (Papoulis, 1985; Newton, 1988).

The step-down procedure has also been used by Kay (1981b) in stating an efficient algorithm for simulating Gaussian ARMA(p, q) processes.

[5]   A more general way to predict X_t given the k most recent prior values of $\{X_t\}$ is to find that function of $X_{t-1}, X_{t-2}, \ldots, X_{t-k}$ – call it $g(X_{t-1}, \ldots, X_{t-k})$ – such that the *mean square prediction error*

$$M_k \equiv E\{[X_t - g(X_{t-1}, \ldots, X_{t-k})]^2\}$$

is minimized. The form of $g(\cdot)$ turns out to be an appealing one (for a derivation of this result, see Priestley, 1981, p. 76):

$$g(X_{t-1}, \ldots, X_{t-k}) = E\{X_t | X_{t-1}, \ldots, X_{t-k}\};$$

i.e., under the mean square error criterion, the predictor of X_t, given $X_{t-1}, \ldots, X_{t-k}$, is simply the conditional mean of X_t, given $X_{t-1}, \ldots, X_{t-k}$. We call $g(\cdot)$ the *best predictor* of X_t, given $X_{t-1}, \ldots, X_{t-k}$. As simple and appealing as $g(\cdot)$ is, it is unfortunately mathematically intractable in many cases of interest. An important exception is the case of a stationary Gaussian process $\{G_t\}$, for which the best predictor is identical to the best linear predictor:

$$E\{G_t | G_{t-1}, \ldots, G_{t-k}\} = \overrightarrow{G_t}(k) \equiv \sum_{j=1}^{k} \phi_{j,k} G_{t-j}.$$

For a non-Gaussian process, the best predictor need not be linear. Moreover, for these processes, the symmetry between the forward and backward best *linear* predictors of X_t does not necessarily carry through to the forward and backward best predictors. Rosenblatt (1985, p. 52) gives an example of a rather extreme case of asymmetry, in which the best forward predictor corresponds to the best linear predictor and has a prediction error variance of 4, whereas the best backward predictor is nonlinear and has a prediction error variance of 0 (i.e., it predicts perfectly with probability 1)!

[6] It is also possible to derive the Levinson–Durbin recursions using a direct linear algebra argument. As before, let us write the Yule–Walker equations in matrix form

$$\Gamma_p \Phi_p = \gamma_p, \tag{411a}$$

where $\gamma_p \equiv [s_1, s_2, \ldots, s_p]^T$; $\Phi_p \equiv [\phi_{1,p}, \phi_{2,p}, \ldots, \phi_{p,p}]^T$; and

$$\Gamma_p \equiv \begin{bmatrix} s_0 & s_1 & \cdots & s_{p-1} \\ s_1 & s_0 & \cdots & s_{p-2} \\ \vdots & \vdots & \ddots & \vdots \\ s_{p-1} & s_{p-2} & \cdots & s_0 \end{bmatrix}.$$

The trick is to solve for Φ_{p+1} using the solution Φ_p. Suppose for the moment that it is possible to express $\phi_{j,p+1}$ for $j \leq p$ in the following manner:

$$\phi_{j,p+1} = \phi_{j,p} - \phi_{p+1,p+1}\theta_{j,p},$$

where $\Theta_p \equiv [\theta_{1,p}, \theta_{2,p}, \ldots, \theta_{p,p}]^T$ is a p-dimensional vector that we must determine. With this recursion, Equation (411a) becomes the following for order $p + 1$:

$$\begin{bmatrix} & & & s_p \\ & \Gamma_p & & \vdots \\ & & & s_1 \\ s_p & \cdots & s_1 & s_0 \end{bmatrix} \begin{bmatrix} \Phi_p - \phi_{p+1,p+1}\Theta_p \\ \phi_{p+1,p+1} \end{bmatrix} = \begin{bmatrix} \gamma_p \\ s_{p+1} \end{bmatrix}.$$

If we separate the first p equations from the last one, we have

$$\Gamma_p \Phi_p - \phi_{p+1,p+1}\Gamma_p \Theta_p + \phi_{p+1,p+1} \begin{bmatrix} s_p \\ \vdots \\ s_1 \end{bmatrix} = \gamma_p \tag{411b}$$

and

$$\begin{bmatrix} s_p & \cdots & s_1 \end{bmatrix}(\Phi_p - \phi_{p+1,p+1}\Theta_p) + s_0\phi_{p+1,p+1} = s_{p+1}. \tag{411c}$$

If we use the fact that Φ_p satisfies Equation (411a), we conclude that the following condition is sufficient for Equation (411b) to hold:

$$\Gamma_p \Theta_p = \begin{bmatrix} s_p \\ \vdots \\ s_1 \end{bmatrix}.$$

If we chose $\theta_{j,p} = \phi_{p+1-j,p}$, then the condition above is equivalent to

$$
\begin{bmatrix}
s_0 & s_1 & \cdots & s_{p-1} \\
s_1 & s_0 & \cdots & s_{p-2} \\
\vdots & \vdots & \ddots & \vdots \\
s_{p-1} & s_{p-2} & \cdots & s_0
\end{bmatrix}
\begin{bmatrix}
\phi_{p,p} \\
\phi_{p-1,p} \\
\vdots \\
\phi_{1,p}
\end{bmatrix}
=
\begin{bmatrix}
s_p \\
s_{p-1} \\
\vdots \\
s_1
\end{bmatrix} .
$$

If we reverse the order of the rows in the equation above, we have that it is equivalent to

$$
\begin{bmatrix}
s_{p-1} & s_{p-2} & \cdots & s_0 \\
s_{p-2} & s_{p-3} & \cdots & s_1 \\
\vdots & \vdots & \ddots & \vdots \\
s_0 & s_1 & \cdots & s_{p-1}
\end{bmatrix}
\begin{bmatrix}
\phi_{p,p} \\
\phi_{p-1,p} \\
\vdots \\
\phi_{1,p}
\end{bmatrix}
=
\begin{bmatrix}
s_1 \\
s_2 \\
\vdots \\
s_p
\end{bmatrix} .
$$

If we now reverse the order of the columns, we get Equation (411a), which is assumed to hold. Hence our choice of $\mathbf{\Theta}_p$ satisfies Equation (411b). Equation (411c) can be satisfied by solving for $\phi_{p+1,p+1}$. In summary, if $\mathbf{\Phi}_p$ satisfies Equation (411a),

$$
\phi_{p+1,p+1} = \frac{s_{p+1} - \sum_{j=1}^{p} \phi_{j,p} s_{p+1-j}}{s_0 - \sum_{j=1}^{p} \phi_{j,p} s_j} ;
$$

$$
\phi_{j,p+1} = \phi_{j,p} - \phi_{p+1,p+1}\phi_{p+1-j,p}
$$

for $j = 1, \ldots, p$, then $\Gamma_{p+1}\mathbf{\Phi}_{p+1} = \gamma_{p+1}$.

[7] The Levinson–Durbin recursions are closely related to the *modified Cholesky decomposition* of a positive definite covariance matrix Γ_N for a portion $\mathbf{X} \equiv [X_1, X_2, \ldots, X_N]^T$ of a stationary process (Therrien, 1983; Newton, 1988). Such a decomposition exists for any positive definite symmetric $N \times N$ matrix. In the case of Γ_N (defined as in Equation (394d)), the decomposition states that we can write

$$
\Gamma_N = L_N D_N L_N^T, \tag{412a}
$$

where L_N is an $N \times N$ lower triangular matrix with 1 as each of its diagonal elements, and D_N is an $N \times N$ diagonal matrix, each of whose diagonal elements are positive. Note that this decomposition implies both

$$
\Gamma_N^{-1} = L_N^{-T} D_N^{-1} L_N^{-1} \quad \text{and} \quad L_N^{-1}\Gamma_N L_N^{-T} = D_N \tag{412b}
$$

(note that $L_N^{-T} \equiv (L_N^{-1})^T = (L_N^T)^{-1}$). Because of the special structure of Γ_N, it can be shown that L_N^{-1} is the lower triangular matrix given by

$$
L_N^{-1} = \begin{bmatrix}
1 & & & & & \\
-\phi_{1,1} & 1 & & & & \\
-\phi_{2,2} & -\phi_{1,2} & 1 & & & \\
\vdots & \vdots & \ddots & \ddots & & \\
-\phi_{k,k} & -\phi_{k-1,k} & \cdots & -\phi_{1,k} & 1 & \\
\vdots & \vdots & & \ddots & \ddots & \ddots \\
-\phi_{N-1,N-1} & -\phi_{N-2,N-1} & \cdots & & \cdots & -\phi_{1,N-1} & 1
\end{bmatrix},
$$

while D_N has P_0, P_1, ..., P_{N-1} as its diagonal elements (here $\phi_{j,k}$ and P_k are obtained via the Levinson–Durbin recursions, and we define $P_0 = s_0$, the process variance). Note that, with $\overrightarrow{\epsilon_1}(0) \equiv X_1$, we have

$$
L_N^{-1}\mathbf{X} = \left[\overrightarrow{\epsilon_1}(0), \overrightarrow{\epsilon_2}(1), \overrightarrow{\epsilon_3}(2), \ldots, \overrightarrow{\epsilon_k}(k-1), \ldots, \overrightarrow{\epsilon_N}(N-1) \right]^T.
$$

The covariance matrix of $L_N^{-1}\mathbf{X}$ is simply $L_N^{-1}\Gamma_N L_N^{-T} = D_N$; i.e., the matrix L_N^{-1} transforms $\mathbf{X}$ into a vector of uncorrelated prediction errors with variances given by the diagonal elements of D_N.

For the AR(p) process $\{Y_t\}$ of Equation (392a), the matrix L_N^{-1} specializes to

$$
L_N^{-1} = \begin{bmatrix}
1 & & & & & \\
-\phi_{1,1} & 1 & & & & \\
-\phi_{2,2} & -\phi_{1,2} & 1 & & & \\
\vdots & \vdots & & \ddots & & \\
-\phi_{p,p} & \cdots & & -\phi_{1,p} & 1 & \\
0 & -\phi_{p,p} & \cdots & & -\phi_{1,p} & 1 \\
\vdots & & \ddots & \ddots & & \ddots & \ddots \\
0 & & \cdots & 0 & -\phi_{p,p} & \cdots & -\phi_{1,p} & 1
\end{bmatrix}, \qquad (413a)
$$

while the upper p diagonal elements of D_N are σ_0^2, σ_1^2, ..., σ_{p-1}^2, and the lower $N - p$ elements are all equal to σ_p^2 (note that $\sigma_0^2 = s_0$). With $\mathbf{Y} \equiv [Y_1, Y_2, \ldots, Y_N]^T$, we have

$$
L_N^{-1}\mathbf{Y} = \left[\overrightarrow{\epsilon_1}(0), \overrightarrow{\epsilon_2}(1), \overrightarrow{\epsilon_3}(2), \ldots, \overrightarrow{\epsilon_p}(p-1), \epsilon_{p+1}, \epsilon_{p+2}, \ldots, \epsilon_N \right]^T.
$$

$$
(413b)
$$

These facts will prove useful when we discuss maximum likelihood estimation in Section 9.8.

9.5 Burg's Algorithm

The Yule–Walker estimators of the AR(p) parameters are only one of many proposed and studied in the literature. Since the late 1960s, an estimation technique based upon *Burg's algorithm* has been widely used and discussed (particularly in the engineering and geophysical literature). The popularity of this method is due to a number of reasons.

[1] Burg's algorithm can be regarded as a variation on the solution of the Yule–Walker equations via the Levinson–Durbin recursions and, like them, is computationally efficient and recursive. If we recall that the AR(p) process that is fit to a time series via the Yule–Walker method has a theoretical acvs that agrees *identically* with that of the sample acvs (i.e., the biased estimator of the acvs), we can regard Burg's algorithm as an alternative method for estimating the acvs of a time series. Burg's modification is intuitively reasonable and, as we note below, relies heavily on the relationship of the recursions to the prediction error problem discussed above.

[2] Burg's algorithm arose from his work with the maximum entropy principle in connection with sdf estimation. We describe and criticize this principle in the next section, but it is important to realize that, as shown below, Burg's algorithm can be justified without appealing to entropy arguments.

[3] Like the Yule–Walker method, Burg's algorithm yields an estimated AR(p) process that must be stationary for *any* observed time series (ignoring the vagaries of computer rounding error). This property is often mentioned in descriptions of Burg's algorithm. If sdf estimation is the only reason that a researcher is fitting an AR(p) model to his or her data, the relevance of this property is not particularly strong.

[4] A number of Monte Carlo studies and experience with actual data indicate that, particularly for short time series, Burg's algorithm produces more reasonable estimates than the Yule–Walker method (see, for example, Lysne and Tjøstheim, 1987). As we note in Section 10.16, it is not, unfortunately, free of problems of its own.

The key to Burg's algorithm is the realization of the central role of $\phi_{k,k}$, the kth order partial autocorrelation coefficient, in the Levinson–Durbin recursions: if we have $\phi_{1,k-1}, \ldots, \phi_{k-1,k-1}$ and $\sigma_{k-1}^2 = P_{k-1}$, Equations (402b) and (402c) tell us that we need only determine $\phi_{k,k}$ to be able to calculate the remaining order k terms (namely, $\phi_{1,k}, \ldots, \phi_{k-1,k}$ and $\sigma_k^2 = P_k$). In the Yule–Walker scheme, Equation (404a) is used to calculate an estimate of $\phi_{k,k}$. Note that it utilizes the estimated acvs up to lag p. Burg's algorithm uses a different approach that estimates $\phi_{k,k}$ by minimizing a certain sum of squares of *observed* forward and backward prediction errors. Given Burg's estimators $\bar{\phi}_{1,k-1}, \ldots, \bar{\phi}_{k-1,k-1}$ of the coefficients for a model of order $k-1$, these are defined, respectively, by

$$\overrightarrow{e}_t(k-1) \equiv X_t - \sum_{j=1}^{k-1} \bar{\phi}_{j,k-1} X_{t-j}, \qquad k \leq t \leq N,$$

and

$$\overleftarrow{e}_{t-k}(k-1) \equiv X_{t-k} - \sum_{j=1}^{k-1} \bar{\phi}_{j,k-1} X_{t-k+j}, \qquad k+1 \leq t \leq N+1,$$

when we have observed a time series of length N that is a realization of a portion $X_1, X_2, \ldots, X_N$ of a discrete parameter stationary process with zero mean. Suppose for the moment that $\bar{\phi}_{k,k}$ is *any* estimate of $\phi_{k,k}$ and that the remaining $\bar{\phi}_{j,k}$ terms are generated in a manner analogous to Equation (404b). We then have

$$\overrightarrow{e}_t(k) = X_t - \sum_{j=1}^{k} \bar{\phi}_{j,k} X_{t-j}$$

$$= X_t - \sum_{j=1}^{k-1} \left(\bar{\phi}_{j,k-1} - \bar{\phi}_{k,k} \bar{\phi}_{k-j,k-1} \right) X_{t-j} - \bar{\phi}_{k,k} X_{t-k}$$

$$= X_t - \sum_{j=1}^{k-1} \bar{\phi}_{j,k-1} X_{t-j} - \bar{\phi}_{k,k} \left(X_{t-k} - \sum_{j=1}^{k-1} \bar{\phi}_{k-j,k-1} X_{t-j} \right)$$

$$= \overrightarrow{e}_t(k-1) - \bar{\phi}_{k,k} \left(X_{t-k} - \sum_{j=1}^{k-1} \bar{\phi}_{j,k-1} X_{t-k+j} \right)$$

$$= \overrightarrow{e}_t(k-1) - \bar{\phi}_{k,k} \overleftarrow{e}_{t-k}(k-1), \quad k+1 \leq t \leq N. \qquad (415a)$$

We recognize this as a sampling version of Equation (400b). In a similar way, we can derive a sampling version of Equation (401):

$$\overleftarrow{e}_{t-k}(k) = \overleftarrow{e}_{t-k}(k-1) - \bar{\phi}_{k,k} \overrightarrow{e}_t(k-1), \qquad k+1 \leq t \leq N. \quad (415b)$$

Equations (415a) and (415b) allow us to calculate the observed order k forward and backward prediction errors in terms of the order $k-1$ errors. These equations do not depend on *any* particular property of $\bar{\phi}_{k,k}$ and hence are valid no matter how we determine $\bar{\phi}_{k,k}$ as long as the remaining $\bar{\phi}_{j,k}$ terms are generated as dictated by Equation (404b). Burg's idea was to estimate $\phi_{k,k}$ such that the order k observed prediction errors are as small as possible by the appealing criterion that

$$SS_k(\bar{\phi}_{k,k}) \equiv \sum_{t=k+1}^{N} \overrightarrow{e}_t^{\,2}(k) + \overleftarrow{e}_{t-k}^{\,2}(k)$$

$$= \sum_{t=k+1}^{N} \left[\overrightarrow{e_t}(k-1) - \bar{\phi}_{k,k} \overleftarrow{e}_{t-k}(k-1) \right]^2$$

$$+ \left[\overleftarrow{e}_{t-k}(k-1) - \bar{\phi}_{k,k} \overrightarrow{e_t}(k-1) \right]^2$$

$$= A_k - 2\bar{\phi}_{k,k} B_k + A_k \bar{\phi}_{k,k}^2 \tag{416a}$$

be as small as possible, where

$$A_k \equiv \sum_{t=k+1}^{N} \overrightarrow{e_t}^{2}(k-1) + \overleftarrow{e}_{t-k}^{2}(k-1) \tag{416b}$$

and

$$B_k \equiv 2 \sum_{t=k+1}^{N} \overrightarrow{e_t}(k-1) \overleftarrow{e}_{t-k}(k-1). \tag{416c}$$

Since $SS_k(\cdot)$ is a quadratic function of $\bar{\phi}_{k,k}$, we can differentiate it and set the resulting expression to 0 to find the desired value of $\bar{\phi}_{k,k}$, namely,

$$\bar{\phi}_{k,k} \equiv B_k / A_k. \tag{416d}$$

(Note that $\bar{\phi}_{k,k}$ has a natural interpretation as an estimator of the left-hand side of Equation (409a).) With $\bar{\phi}_{k,k}$ so determined, we can estimate the remaining $\bar{\phi}_{j,k}$ using an equation analogous to Equation (404b) – the second of the usual Levinson–Durbin recursions – and the corresponding observed prediction errors using Equations (415a) and (415b).

From the above description, it should be obvious that, with the proper initialization, we can apply Burg's algorithm recursively to work our way up to estimates for the coefficients of an AR(p) model. In the same spirit as the recursive step, we can initialize the algorithm by finding that value of $\bar{\phi}_{1,1}$ that minimizes

$$SS_1(\bar{\phi}_{1,1}) = \sum_{t=2}^{N} \overrightarrow{e_t}^{2}(1) + \overleftarrow{e}_{t-1}^{2}(1)$$

$$= \sum_{t=2}^{N} \left(X_t - \bar{\phi}_{1,1} X_{t-1} \right)^2 + \left(X_{t-1} - \bar{\phi}_{1,1} X_t \right)^2. \tag{416e}$$

This is equivalent to Equation (416a) with $k = 1$ if we adopt the conventions $\overrightarrow{e_t}(0) \equiv X_t$ and $\overleftarrow{e}_{t-1}(0) \equiv X_{t-1}$. We can interpret these as meaning that, with no prior (future) observations, we should predict (backward predict) X_t (X_{t-1}) by 0, the assumed known process mean.

Burg's algorithm also specifies a way of estimating the innovations variance σ_p^2. It is done recursively using an equation analogous to Equation (404c), namely,

$$\bar{\sigma}_k^2 = \bar{\sigma}_{k-1}^2 \left(1 - \bar{\phi}_{k,k}^2 \right). \tag{416f}$$

This assumes that, at the kth step, $\bar{\sigma}^2_{k-1}$ is available. For $k = 1$ there is obviously a problem in that we have not defined what $\bar{\sigma}^2_0$ is. If we follow the logic of the previous paragraph and consider the zeroth-order predictor of our time series to be 0 (the known process mean), an obvious estimator of the zeroth-order mean square linear prediction error is just

$$\bar{\sigma}^2_0 \equiv \frac{1}{N} \sum_{t=1}^{N} X_t^2 = \hat{s}_0^{(p)},$$

the sample variance of the time series.

As an example of the application of Burg's algorithm to autoregressive sdf estimation, Figure 418 shows the Burg AR(4) sdf estimates for the same series used to produce the Yule–Walker AR(4) and AR(8) sdf estimates of Figures 405 and 406. A comparison of these three figures shows that Burg's method is in these cases clearly superior to the Yule–Walker method (at least in the standard formulation using the biased estimator of the acvs).

Comments and Extensions to Section 9.5

[1]   We have previously noted that the *theoretical* kth order partial autocorrelation coefficient $\phi_{k,k}$ can be interpreted as a correlation coefficient and thus must lie in the interval $[-1, 1]$. One important property of Burg's algorithm is that the corresponding *estimated* value $\bar{\phi}_{k,k}$ is also necessarily in that range. To see this, note that

$$0 \leq \left(\overrightarrow{e_t}(k-1) \pm \overleftarrow{e}_{t-k}(k-1) \right)^2$$
$$= \overrightarrow{e}_t^{\,2}(k-1) \pm 2\overrightarrow{e_t}(k-1)\overleftarrow{e}_{t-k}(k-1) + \overleftarrow{e}_{t-k}^{\,2}(k-1)$$

implies that

$$\left| 2\overrightarrow{e_t}(k-1)\overleftarrow{e}_{t-k}(k-1) \right| \leq \overrightarrow{e}_t^{\,2}(k-1) + \overleftarrow{e}_{t-k}^{\,2}(k-1)$$

and hence that

$$|B_k| \equiv \left| 2 \sum_{t=k+1}^{N} \overrightarrow{e_t}(k-1)\overleftarrow{e}_{t-k}(k-1) \right|$$
$$\leq \sum_{t=k+1}^{N} \overrightarrow{e}_t^{\,2}(k-1) + \overleftarrow{e}_{t-k}^{\,2}(k-1) \equiv A_k.$$

Since $A_k \geq 0$ (with equality rarely occurring in practical cases) and $\bar{\phi}_{k,k} = B_k/A_k$, we have $|\bar{\phi}_{k,k}| \leq 1$ as claimed.

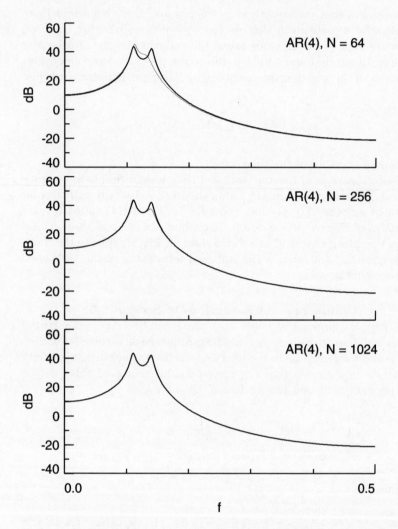

Figure 418. Burg AR(4) sdf estimates (thin curves) for a portion of length 64, 256 and 1024 of the realization of the AR(4) process shown in the bottom plot of Figure 45. The thick curve in each plot is the true sdf. This figure should be compared with Figures 405 and 406.

[2] We have described the estimator of σ_k^2 commonly associated with Burg's algorithm. There is a second obvious estimator. Since σ_k^2 is the kth order mean square linear prediction error, this estimator is

$$\bar{\nu}_k^2 \equiv \frac{\mathrm{SS}_k(\bar{\phi}_{k,k})}{2(N-k)}.$$

In the numerator the sum of squares of Equation (416a) is evaluated at its minimum value. Are $\bar{\nu}_k^2$ and $\bar{\sigma}_k^2$ in fact different? The answer is 'yes,' but usually the difference is small. It can be shown (see Exercise [9.5]) that

$$\bar{\nu}_k^2 = \left(1 - \bar{\phi}_{k,k}^2\right) \left(\bar{\nu}_{k-1}^2 + \frac{2\bar{\nu}_{k-1}^2 - \overrightarrow{e}_k^2(k-1) - \overleftarrow{e}_{N-k+1}^2(k-1)}{2(N-k)}\right).$$

(419a)

Comparison of this equation with the definition of $\bar{\sigma}_k^2$ in Equation (416f) shows that, even if $\bar{\nu}_{k-1}^2$ and $\bar{\sigma}_{k-1}^2$ are identical, $\bar{\nu}_k^2$ and $\bar{\sigma}_k^2$ need not be the same. However, since $\bar{\nu}_{k-1}^2$, $\overrightarrow{e}_k^2(k-1)$ and $\overleftarrow{e}_{N-k+1}^2(k-1)$ can all be regarded as estimators of the same parameter, σ_k^2, the last term in the parentheses should usually be small. For large N, we thus have

$$\bar{\nu}_k^2 \approx \bar{\nu}_{k-1}^2 \left(1 - \bar{\phi}_{k,k}^2\right),$$

in agreement with the third equation of the Levinson–Durbin recursions. The relative merits of these two estimators are unknown, but, as far as sdf estimation is concerned, use of one or the other will only affect the level of the resulting estimate and not its shape.

[3]   There is a second – but equivalent – way of expressing Burg's algorithm that is both more succinct and also clarifies the relationship between this algorithm and the Yule–Walker method (Newton, 1988, Section 3.4). Before presenting it, we need to establish some notation. Let us define a circular shift operator $\mathcal{L}$ and a subvector extraction operator $\mathcal{M}_{j,k}$ as follows: if $\mathbf{V}_1 \equiv [v_1, v_2, \ldots, v_{M-1}, v_M]^T$ is any M-dimensional column vector of real numbers, then

$$\mathcal{L}\mathbf{V}_1 \equiv [v_M, v_1, v_2, \ldots, v_{M-1}]^T$$

$$\mathcal{M}_{j,k}\mathbf{V}_1 \equiv [v_j, v_{j+1}, \ldots, v_{k-1}, v_k]^T$$

(we assume that $1 \leq j < k \leq M$). If $\mathbf{V}_2$ is any vector similar to $\mathbf{V}_1$, we define $\langle \mathbf{V}_1, \mathbf{V}_2 \rangle \equiv \mathbf{V}_1^T \mathbf{V}_2$ to be their inner product. We also define the squared norm of $\mathbf{V}_1$ to be $\|\mathbf{V}_1\|^2 \equiv \langle \mathbf{V}_1, \mathbf{V}_1 \rangle = \mathbf{V}_1^T \mathbf{V}_1$.

In order to fit an AR(p) model to $X_1, \ldots, X_N$ using Burg's algorithm, we first set up the following two vectors of length $M = N + p$:

$$\overrightarrow{\mathbf{e}}(0) \equiv [X_1, X_2, \ldots, X_N, \underbrace{0, \ldots, 0}_{p \text{ of these}}]^T$$

$$\overleftarrow{\mathbf{e}}(0) \equiv \mathcal{L}\overrightarrow{\mathbf{e}}(0) = [0, X_1, X_2, \ldots, X_N, \underbrace{0, \ldots, 0}_{p-1 \text{ of these}}]^T.$$

As before, we define $\bar{\sigma}_0^2 = \hat{s}_0^{(p)}$. For $k = 1, \ldots, p$, we then recursively compute

$$\bar{\phi}_{k,k} = \frac{2\langle \mathcal{M}_{k+1,N}\overrightarrow{\mathbf{e}}(k-1), \mathcal{M}_{k+1,N}\overleftarrow{\mathbf{e}}(k-1) \rangle}{\|\mathcal{M}_{k+1,N}\overrightarrow{\mathbf{e}}(k-1)\|^2 + \|\mathcal{M}_{k+1,N}\overleftarrow{\mathbf{e}}(k-1)\|^2}$$

(419b)

$$\bar{\sigma}_k^2 = \bar{\sigma}_{k-1}^2 \left(1 - \bar{\phi}_{k,k}^2\right)$$

$$\overrightarrow{\mathbf{e}}(k) = \overrightarrow{\mathbf{e}}(k-1) - \bar{\phi}_{k,k}\overleftarrow{\mathbf{e}}(k-1)$$

$$\overleftarrow{\mathbf{e}}(k) = \mathcal{L}(\overleftarrow{\mathbf{e}}(k-1) - \bar{\phi}_{k,k}\overrightarrow{\mathbf{e}}(k-1)).$$

It can be shown that this procedure yields the Burg estimators of $\phi_{k,k}$ and σ_k^2 for $k = 1, \ldots, p$ (if so desired, the remaining $\bar{\phi}_{j,k}$ can be generated using an equation similar to Equation (404b)). The $N - p$ forward prediction errors $\overrightarrow{e}_{p+1}(p), \ldots, \overrightarrow{e}_N(p)$ are given by the elements of $\mathcal{M}_{p+1,N}\overrightarrow{\mathbf{e}}(p)$; the backward prediction errors $\overleftarrow{e}_1(p), \ldots, \overleftarrow{e}_{N-p}(p)$ are given by $\mathcal{M}_{p+2,N+1}\overleftarrow{\mathbf{e}}(p)$.

The Yule–Walker estimators $\tilde{\phi}_{k,k}$ and $\tilde{\sigma}_k^2$ can be generated by a scheme that is *identical* to the above, with one key modification: Equation (419b) becomes

$$\tilde{\phi}_{k,k} = \frac{2\langle \overrightarrow{\mathbf{e}}(k-1), \overleftarrow{\mathbf{e}}(k-1)\rangle}{\|\overrightarrow{\mathbf{e}}(k-1)\|^2 + \|\overleftarrow{\mathbf{e}}(k-1)\|^2} = \frac{\langle \overrightarrow{\mathbf{e}}(k-1), \overleftarrow{\mathbf{e}}(k-1)\rangle}{\|\overrightarrow{\mathbf{e}}(k-1)\|^2} \quad (420)$$

since $\|\overleftarrow{\mathbf{e}}(k-1)\|^2 = \|\overrightarrow{\mathbf{e}}(k-1)\|^2$ (all 'overbars' in the three equations below (419b) should also be changed to 'tildes'). Note that, whereas Burg's algorithm uses an inner product tailored to involve just the actual data values, the inner product of the Yule–Walker estimator is influenced by the p zeros used to construct $\overrightarrow{\mathbf{e}}(0)$ and $\overleftarrow{\mathbf{e}}(0)$ (a similar interpretation of the Yule–Walker estimator appears in the context of least squares theory – see item [1] in the Comments and Extensions to Section 9.7). This finding supports Burg's contention (discussed in the next section) that the Yule–Walker method implicitly assumes $X_t = 0$ for $t < 1$ and $t > N$. Finally we note that the above formulation allows us to compute the Yule–Walker estimators *directly* from a time series, thus bypassing the need to first compute $\hat{s}_k^{(p)}$ as required by Equation (404a).

9.6 The Maximum Entropy Argument

Burg's algorithm was an outgrowth of his work on maximum entropy spectral analysis (MESA). Burg (1967) criticized the use of lag window spectral estimators of the form

$$\Delta t \sum_{k=-(N-1)}^{N-1} w_{k,m}\hat{s}_k^{(p)}e^{-i2\pi f k \Delta t}$$

because, first, we effectively force the sample acvs to zero by multiplying it by a lag window $\{w_{k,m}\}$ and, second, we assume that $s_k = 0$ for $|k| \geq N$. To quote from Burg (1975),

While window theory is interesting, it is actually a problem that has been artificially induced into the estimation problem by the assumption that $s_k = 0$ for $|k| \geq N$ and by the willingness to change perfectly good data by the weighting function. If one were not blinded by the mathematical elegance of the conventional approach, making unfounded assumptions as to the values of unmeasured data and changing the data values that one knows would be totally unacceptable from a common sense and, hopefully, from a scientific point of view. To overcome these problems, it is clear that a completely different philosophical approach to spectral analysis is required . . . .

While readily understood, [the] conventional window function approach produces spectral estimates which are negative and/or spectral estimates which do not agree with their known autocorrelation values. These two affronts to common sense were the main reasons for the development of maximum entropy spectral analysis . . . .

Burg's 'different philosophical approach' was to apply the principle of maximum entropy. In physics entropy is a measure of disorder; for a stationary process, it can be usefully defined in terms of the predictability of the process. To be specific, let $\{X_t\}$ be any stationary process with zero mean and sdf $S(\cdot)$. Suppose that we know the 'infinite past' of this process starting at time $t - 1$ (i.e., X_{t-1}, X_{t-2}, ...) and that we want to predict X_t. If we consider only predictors that are linear combinations of the infinite past, say,

$$\hat{X}_t \equiv \sum_{u=1}^{\infty} b_u X_{t-u},$$

we can then find that set of coefficients $\{b_u\}$ such that the prediction error variance

$$\eta^2 \equiv E\{(X_t - \hat{X}_t)^2\}$$

is minimized. Let $\tilde{X}_t$ denote the predictor formed from the linear combination that achieves this minimum; $\tilde{X}_t$ is called the best linear predictor of X_t, given the infinite past. A theorem due to Szegő and Kolmogorov (see, for example, Priestley, 1981, p. 741) states that

$$\eta^2 = \frac{1}{\Delta t} \exp\left(\Delta t \int_{-f_{(N)}}^{f_{(N)}} \log\left(S(f)\right) \, df\right),$$

so that η^2 depends only on the sdf of the process. For the purposes of our discussion, we can now take the entropy of the sdf $S(\cdot)$ to be given by

$$H\left\{S(\cdot)\right\} = \int_{-f_{(N)}}^{f_{(N)}} \log\left(S(f)\right) \, df. \tag{421}$$

Thus large entropy is equivalent to a large prediction error variance. (In fact, the concept of entropy is usually defined in such a way that the entropy measure we have just stated is only valid for the case of stationary *Gaussian* processes.)

We can now state Burg's maximum entropy argument. If we have *perfect* knowledge of the acvs for a process just out to lag p, we know only that $S(\cdot)$ lies within a certain class of sdf's. One way to select a particular member of this class is to pick that sdf, call it $\tilde{S}(\cdot)$, that maximizes the entropy subject to the constraints

$$\int_{-f_{(N)}}^{f_{(N)}} \tilde{S}(f)e^{i2\pi fk\,\Delta t}\,df = s_k, \qquad 0 \leq k \leq p. \tag{422}$$

To quote again from Burg (1975),

> Maximum entropy spectral analysis is based on choosing the spectrum which corresponds to the most random or the most unpredictable time series whose autocorrelation function agrees with the known values.

The solution to this constrained maximization problem is

$$\tilde{S}(f) = \frac{\sigma_p^2\,\Delta t}{\left|1 - \sum_{k=1}^{p} \phi_{k,p}e^{-i2\pi fk\,\Delta t}\right|^2}, \qquad |f| \leq f_{(N)},$$

where $\phi_{k,p}$ and σ_p^2 are the solutions to the augmented Yule–Walker equations of order p; i.e., the sdf is an AR(r) sdf with $r \leq p$ (r is not necessarily equal to p since $\phi_{r+1,p}, \ldots, \phi_{p,p}$ could possibly be equal to 0). To see that this is true, let us consider the AR(p) process $\{Y_t\}$ with sdf $S_Y(\cdot)$ and innovations variance σ_p^2. For such a process the best linear predictor of Y_t given the infinite past is just

$$\tilde{Y}_t \equiv \sum_{k=1}^{p} \phi_{k,p} Y_{t-k},$$

and the corresponding prediction error variance is just the innovations variance σ_p^2. Let $\{X_t\}$ be any other stationary process whose acvs agrees with that of $\{Y_t\}$ up to lag p, and let $S(\cdot)$ denote its sdf. We know that the best linear predictor of order p for this process is given by

$$\hat{X}_t \equiv \sum_{k=1}^{p} \phi_{k,p} X_{t-k},$$

since the $\phi_{k,p}$ terms depend only on the acvs of $\{X_t\}$ up to lag p; moreover, the associated prediction error variance must be σ_p^2. The prediction error variance of the best linear predictor of X_t given the infinite past must necessarily be no smaller than that of the pth order predictor. Hence we have

$$\sigma_p^2 = \frac{1}{\Delta t} \exp\left(\Delta t \int_{-f_{(N)}}^{f_{(N)}} \log\left(S_Y(f)\right) df\right)$$

$$\geq \frac{1}{\Delta t} \exp\left(\Delta t \int_{-f_{(N)}}^{f_{(N)}} \log\left(S(f)\right) df\right),$$

which implies that

$$H\{S_Y(\cdot)\} = \int_{-f_{(N)}}^{f_{(N)}} \log\left(S_Y(f)\right) df \geq \int_{-f_{(N)}}^{f_{(N)}} \log\left(S(f)\right) df = H\{S(\cdot)\};$$

i.e., the entropy for a class of stationary processes – all of whose members are known to have the same acvs out to lag p – is maximized by an AR(p) process with the specified acvs out to lag p (Brockwell and Davis, 1991, Section 10.6).

Burg's original idea for MESA (1967) was to *assume* that the biased *estimator* of the acvs up to some lag p was in fact equal to the *true* acvs. The maximum entropy principle under these assumptions leads to a procedure that is identical to the Yule–Walker estimation method. Burg (1968) rather quickly abandoned this approach. He argued that it was not justifiable because, from his viewpoint, the usual biased estimator for the acvs makes the implicit assumption that $X_t = 0$ for all $t < 1$ and $t > N$ (for example, with this assumption, the limits in Equation (196b) defining $\hat{s}_k^{(p)}$ can be modified to range from $t = -\infty$ to $t = \infty$). Burg's algorithm was his attempt to overcome this defect by estimating the acvs in a different manner. There is still a gap in this logic, because true MESA requires *exact* knowledge of the acvs up to a certain lag, whereas Burg's algorithm is just another way of *estimating* the acvs from the data. (Another claim that is often made is that Burg's algorithm implicitly extends the time series – through predictions – outside the range $t = 1$ to N so that it is the 'most random' one consistent with the observed data, i.e., closest to white noise. There are formidable problems in making this statement rigorous.)

The proponents and critics of the maximum entropy principle often evoke some interesting philosophical ideas. For example, Parzen (1983) says that

> ... the maximum entropy principle provides no insight into how to identify an optimal order ...,

i.e., the lag p below (above) which we have perfect (no) knowledge of the acvs, whereas Jaynes (1982) says that the maximum entropy principle is,

... to the best of the writer's knowledge, the only theoretical relation that *does* determine a definite AR order for us. It tells us that
given data, the optimal spectrum estimate corresponds to an AR
model whose order is the maximum lag for which we have relevant
data.

It is, to say the least, difficult to reconcile these two points of view! (A
good discussion of other claims and counterclaims concerning maximum
entropy can be found in Makhoul's 1986 article with the intriguing title
'Maximum Confusion Spectral Analysis.')

Maximum entropy spectral analysis is equivalent to AR spectral
analysis when (a) the constraints on the entropy maximization are in
terms of low-order values of the acvs and (b) Equation (421) is used
as the measure of entropy. However, as Jaynes (1982) points out, the
maximum entropy principle is more general so they need not be equivalent. If either the constraints (or 'given data') are in terms of quantities
other than just the acvs or the measure of entropy is different from the
one appropriate for stationary Gaussian processes, the resulting maximum entropy spectrum can have a different analytical form than the
AR model. For example, recall from Section 6.15 that the cepstrum for
an sdf $\tilde{S}(\cdot)$ is the sequence $\{c_k\}$ defined by

$$c_k \equiv \int_{-f_{(N)}}^{f_{(N)}} \log\left(\tilde{S}(f)\right) e^{i2\pi fk\,\Delta t}\,df \tag{424}$$

(Bogert *et al.*, 1963). If we maximize the entropy (421) over $\tilde{S}(\cdot)$ subject
to the $p+1$ constraints of Equation (422) and q additional constraints
that Equation (424) holds for known values $c_1, \ldots, c_q$, then – subject
to an existence condition – the maximum entropy principle yields an
ARMA(p,q) sdf (Lagunas-Hernández *et al.*, 1984; Makhoul, 1986, Section 8). Ihara (1984) and Franke (1985) discuss other sets of equivalent
constraints that yield an ARMA spectrum as the maximum entropy
spectrum.

Is the maximum entropy principle a good criterion for sdf estimation? Figure 425 shows an example adapted from an article entitled
'Spectrum Estimation: an Impossibility?' (Nitzberg, 1979). Here we
assume that the acvs of a certain process is known to be

$$s_k = 1 - \frac{|k|}{8} \text{ for } |k| \leq 8 = p.$$

The left-hand plots in this figure show four different possible extrapo-
lations of the acvs for $|k| > p$. The upper three left-hand plots assume
that the acvs is given by

$$s_k = \begin{cases} \alpha\left(1 - \left||k| - 16\right|/8\right), & 8 < |k| \leq 24; \\ 0, & |k| > 24, \end{cases}$$

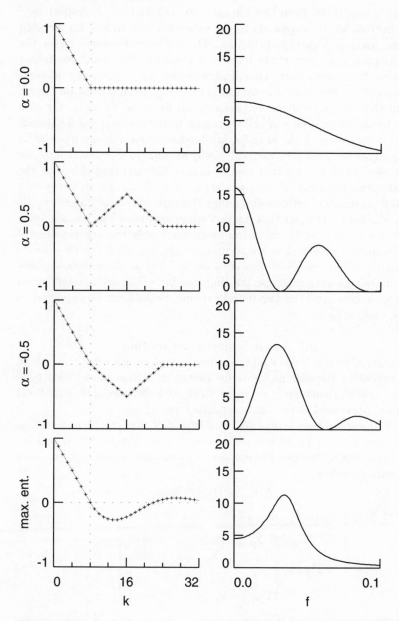

Figure 425. Example of four acvs's (left-hand column) with identical values up to lag 8 and low frequency portions of corresponding sdf's (right-hand column – these are plotted on a linear/linear scale). The maximum entropy extension is shown on the bottom row.

where, going down from the top, $\alpha = 0$, $1/2$ and $-1/2$, respectively. The bottom left-hand plot shows the extension (out to $k = 32$) dictated by the maximum entropy principle. The right-hand column shows the low frequency portions of the four corresponding sdf's, which are rather different from each other. This illustrates the point that many different sdf's can have the same acvs up to a certain lag. Even if we grant Burg's claim that we know the acvs perfectly up to lag p (we never do with real data!), there is no compelling reason to believe that the maximum entropy extension of the acvs for lags greater than p is any more valid than any other arbitrary method – why should we *a priori* regard the acvs and sdf in the bottom row of Figure 425 as being closer to the 'truth' than the other three pairs shown there? Burg's criticism that window estimators effectively change 'known' low-order values of the acvs is offset by the fact that his procedure estimates lags beyond some arbitrary order p by an extension based upon only the first p values of the estimated acvs and hence completely ignores what our time series might be telling us about the acvs beyond lag p. Nonetheless, there are many processes with spectra that can be well approximated by an AR(p) process, and the maximum entropy method can be expected to work well for these.

9.7 Least Squares Estimation

In addition to the Yule–Walker estimates and Burg's algorithm, there are two other popular methods for fitting an AR(p) model to a time series, namely, least squares (ls) methods and the maximum likelihood method (discussed in the next section).

Suppose we have a time series of length N that can be regarded as a portion $Y_1, \ldots, Y_N$ of one realization of a stationary AR(p) process with zero mean. We can formulate an appropriate ls model in terms of our data as follows:

$$\mathbf{Y}_{(f)} = \mathcal{Y}_{(f)}\mathbf{\Phi} + \epsilon_{(f)},$$

where $\mathbf{Y}_{(f)} \equiv [Y_{p+1}, Y_{p+2}, \ldots, Y_N]^T$;

$$\mathcal{Y}_{(f)} \equiv \begin{bmatrix} Y_p & Y_{p-1} & \cdots & Y_1 \\ Y_{p+1} & Y_p & \cdots & Y_2 \\ \vdots & \vdots & \ddots & \vdots \\ Y_{N-1} & Y_{N-2} & \cdots & Y_{N-p} \end{bmatrix};$$

$\mathbf{\Phi} \equiv [\phi_{1,p}, \phi_{2,p}, \ldots, \phi_{p,p}]^T$; and $\epsilon_{(f)} \equiv [\epsilon_{p+1}, \epsilon_{p+2}, \ldots, \epsilon_N]^T$ (the need for the subscript '(f)' in our notation will become apparent in a moment). We can thus estimate $\mathbf{\Phi}$ by finding the $\mathbf{\Phi}$ for which

$$\mathrm{SS}_{(f)}(\mathbf{\Phi}) \equiv \|\mathbf{Y}_{(f)} - \mathcal{Y}_{(f)}\mathbf{\Phi}\|^2 = \sum_{t=p+1}^{N}\left(Y_t - \sum_{k=1}^{p}\phi_{k,p}Y_{t-k}\right)^2 \qquad (426)$$

is minimized – as before, $\|\mathbf{V}\|^2 \equiv \mathbf{V}^T \mathbf{V}$ for a vector $\mathbf{V}$. If we denote the vector that minimizes the above as $\hat{\boldsymbol{\Phi}}_{(f)}$, standard ls theory tells us that it is given by

$$\hat{\boldsymbol{\Phi}}_{(f)} \equiv \left(\mathcal{Y}_{(f)}^T \mathcal{Y}_{(f)} \right)^{-1} \mathcal{Y}_{(f)}^T \mathbf{Y}_{(f)}.$$

We can estimate the innovations variance σ_p^2 by the usual estimator of the residual variation, namely,

$$\hat{\sigma}_{(f)}^2 \equiv \frac{\|\mathbf{Y}_{(f)} - \mathcal{Y}_{(f)} \hat{\boldsymbol{\Phi}}_{(f)}\|^2}{N - 2p},$$

where the divisor arises because there are effectively $N - p$ observations and p parameters to be estimated.

The estimator $\hat{\boldsymbol{\Phi}}_{(f)}$ is known in the literature as the *forward least squares estimator* of $\boldsymbol{\Phi}$ to contrast it with two other closely related estimators. If we recall that a stationary $\mathrm{AR}(p)$ process also has a 'time-reversed' formulation, we can reformulate the ls problem as

$$\mathbf{Y}_{(b)} = \mathcal{Y}_{(b)} \boldsymbol{\Phi} + \epsilon_{(b)},$$

where $\mathbf{Y}_{(b)} \equiv [Y_1, Y_2, \ldots, Y_{N-p}]^T$;

$$\mathcal{Y}_{(b)} \equiv \begin{bmatrix} Y_2 & Y_3 & \cdots & Y_{p+1} \\ Y_3 & Y_4 & \cdots & Y_{p+2} \\ \vdots & \vdots & \ddots & \vdots \\ Y_{N-p+1} & Y_{N-p+2} & \cdots & Y_N \end{bmatrix};$$

and $\epsilon_{(b)}$ is a vector of uncorrelated errors. The function of $\boldsymbol{\Phi}$ to be minimized is now

$$\mathrm{SS}_{(b)}(\boldsymbol{\Phi}) \equiv \|\mathbf{Y}_{(b)} - \mathcal{Y}_{(b)} \boldsymbol{\Phi}\|^2 = \sum_{t=1}^{N-p} \left(Y_t - \sum_{k=1}^{p} \phi_{k,p} Y_{t+k} \right)^2.$$

The *backward least squares estimator* of $\boldsymbol{\Phi}$ is thus given by

$$\hat{\boldsymbol{\Phi}}_{(b)} \equiv \left(\mathcal{Y}_{(b)}^T \mathcal{Y}_{(b)} \right)^{-1} \mathcal{Y}_{(b)}^T \mathbf{Y}_{(b)},$$

and the appropriate estimate for the innovations variance is

$$\hat{\sigma}_{(b)}^2 = \frac{\|\mathbf{Y}_{(b)} - \mathcal{Y}_{(b)} \hat{\boldsymbol{\Phi}}_{(b)}\|^2}{N - 2p}.$$

Finally, we can define a *forward/backward least squares estimator* $\hat{\boldsymbol{\Phi}}_{(fb)}$ of $\boldsymbol{\Phi}$ as that vector minimizing

$$\mathrm{SS}_{(fb)}(\boldsymbol{\Phi}) \equiv \mathrm{SS}_{(f)}(\boldsymbol{\Phi}) + \mathrm{SS}_{(b)}(\boldsymbol{\Phi}). \tag{427}$$

For a specified order p, this estimator can be computed as efficiently as Burg's algorithm using a specialized algorithm due to Marple (1980). In terms of spectral estimation, Monte Carlo studies indicate that the forward/backward ls estimator generally performs better than the Yule–Walker, Burg, forward ls and backward ls estimators (see Marple, 1987, and Kay, 1988, and references therein). The chief disadvantage of $\hat{\boldsymbol{\Phi}}_{(fb)}$ is that the estimated AR model *need not* be stationary (this can also be a problem with either $\hat{\boldsymbol{\Phi}}_{(f)}$ or $\hat{\boldsymbol{\Phi}}_{(b)}$, but, as we have noted previously, the Yule–Walker and Burg estimators are guaranteed to produce coefficients that correspond to a stationary AR process). Nonstationarity can be a concern if, say, the fitted model is used to make long-term predictions of the process $\{Y_t\}$. For spectral estimation, however, the nonstationary estimated coefficients $\hat{\boldsymbol{\Phi}}_{(fb)}$ can still be substituted into Equation (392b) to produce an sdf estimator that has all of the properties of a valid sdf (i.e., it is nonnegative everywhere, symmetric about the origin and integrates to a finite number).

Comments and Extensions to Section 9.7

[1] It should be noted that what we have defined as the Yule–Walker estimator in Section 9.3 is often called the least squares estimator in the statistical literature. To see the origin of this terminology, consider extending the time series $Y_1, Y_2, \ldots, Y_N$ by appending p dummy observations – each identically equal to zero – before and after it. The forward ls model for this extended time series of length $N + 2p$ is

$$\mathbf{Y}_{(\text{YW})} = \mathcal{Y}_{(\text{YW})}\boldsymbol{\Phi} + \boldsymbol{\epsilon}_{(\text{YW})},$$

where

$$\mathbf{Y}_{(\text{YW})} \equiv [Y_1, Y_2, \ldots, Y_N, \underbrace{0, \ldots, 0}_{p \text{ of these}}]^T;$$

$$\mathcal{Y}_{(\text{YW})} \equiv \begin{bmatrix} 0 & 0 & \cdots & 0 & 0 \\ Y_1 & 0 & \cdots & 0 & 0 \\ Y_2 & Y_1 & \cdots & 0 & 0 \\ \vdots & \vdots & \ddots & \vdots & \vdots \\ Y_{p-1} & Y_{p-2} & \cdots & Y_1 & 0 \\ Y_p & Y_{p-1} & \cdots & Y_2 & Y_1 \\ \vdots & \vdots & \ddots & \vdots & \vdots \\ Y_N & Y_{N-1} & \cdots & Y_{N-p+2} & Y_{N-p+1} \\ 0 & Y_N & \cdots & Y_{N-p+3} & Y_{N-p+2} \\ \vdots & \vdots & \ddots & \vdots & \vdots \\ 0 & 0 & \cdots & Y_N & Y_{N-1} \\ 0 & 0 & \cdots & 0 & Y_N \end{bmatrix};$$

and $\epsilon_{(\text{YW})} \equiv [\epsilon'_1, \ldots, \epsilon'_p, \epsilon_{p+1}, \epsilon_{p+2}, \ldots, \epsilon_N, \epsilon'_{N+1}, \ldots, \epsilon'_{N+p}]^T$ – here the primes are needed because the first and last p elements of this vector are *not* members of the innovations process $\{\epsilon_t\}$. Since

$$\frac{1}{N}\mathcal{Y}_{(\text{YW})}^T \mathcal{Y}_{(\text{YW})} = \begin{bmatrix} \hat{s}_0^{(p)} & \hat{s}_1^{(p)} & \cdots & \hat{s}_{p-1}^{(p)} \\ \hat{s}_1^{(p)} & \hat{s}_0^{(p)} & \cdots & \hat{s}_{p-2}^{(p)} \\ \vdots & \vdots & \ddots & \vdots \\ \hat{s}_{p-1}^{(p)} & \hat{s}_{p-2}^{(p)} & \cdots & \hat{s}_0^{(p)} \end{bmatrix} = \hat{\Gamma}_p$$

and $\mathcal{Y}_{(\text{YW})}^T \mathbf{Y}_{(\text{YW})}/N = [\hat{s}_1^{(p)}, \hat{s}_2^{(p)}, \ldots, \hat{s}_p^{(p)}]^T = \hat{\gamma}_p$, it follows that the forward ls estimator for this extended time series is

$$\left(\mathcal{Y}_{(\text{YW})}^T \mathcal{Y}_{(\text{YW})}\right)^{-1} \mathcal{Y}_{(\text{YW})}^T \mathbf{Y}_{(\text{YW})} = \hat{\Gamma}_p^{-1}\hat{\gamma}_p,$$

which is identical to the Yule–Walker estimator.

[2]   Some of the AR estimators that we have defined so far in this chapter are known in the engineering literature under a different name. For the record, we note the following correspondences (Kay and Marple, 1981):

$$\text{Yule–Walker} \Longleftrightarrow \text{autocorrelation method}$$
$$\text{forward least squares} \Longleftrightarrow \text{covariance method}$$
$$\text{forward/backward least squares} \Longleftrightarrow \text{modified covariance method.}$$

9.8 Maximum Likelihood Estimation

We now consider maximum likelihood estimation of the $p+1$ parameters $\mathbf{\Phi}_p \equiv [\phi_{1,p}, \phi_{2,p}, \ldots, \phi_{p,p}]^T$ and σ_p^2 in the AR(p) model. To do so, we assume that our observed time series is a realization of a portion H_1, $\ldots$, H_N of a Gaussian AR(p) process with zero mean. Given these observations, the likelihood function for the unknown parameters is

$$L(\mathbf{\Phi}_p, \sigma_p^2 \mid \mathbf{H}) \equiv \frac{1}{(2\pi)^{N/2}|\Gamma_N|^{1/2}} e^{-\mathbf{H}^T \Gamma_N^{-1}\mathbf{H}/2},$$

where $\mathbf{H} \equiv [H_1, \ldots, H_N]^T$, and Γ_N is the covariance matrix for $\mathbf{H}$ (i.e., its (j,k)th element is s_{j-k}; see Equation (394d)). For particular values of $\mathbf{\Phi}_p$ and σ_p^2, we can compute $L(\mathbf{\Phi}_p, \sigma_p^2 \mid \mathbf{H})$ by assuming that $\{H_t\}$ has these as its true parameter values. The maximum likelihood estimators (mle's) are defined to be those values of the parameters that maximize $L(\mathbf{\Phi}_p, \sigma_p^2 \mid \mathbf{H})$ or, equivalently, that minimize the (rescaled and relocated) log likelihood function

$$l(\mathbf{\Phi}_p, \sigma_p^2 \mid \mathbf{H}) \equiv -2 \log\left(L(\mathbf{\Phi}_p, \sigma_p^2 \mid \mathbf{H})\right) - N \log(2\pi)$$
$$= \log(|\Gamma_N|) + \mathbf{H}^T \Gamma_N^{-1}\mathbf{H}. \tag{429}$$

To obtain the mle's, we must be able to compute $l(\boldsymbol{\Phi}_p, \sigma_p^2 \mid \mathbf{H})$ for particular choices of $\boldsymbol{\Phi}_p$ and σ_p^2, a task which at first glance appears to be formidable due to the necessity of finding the determinant for – and inverting – the $N \times N$-dimensional matrix Γ_N. Fortunately, the Toeplitz structure of Γ_N allows us to simplify the above expression considerably, as we now show (see, for example, Newton, 1988).

We first assume that Γ_N is positive definite so that we can make use of the modified Cholesky decomposition for it stated in Equation (412a), namely,

$$\Gamma_N = L_N D_N L_N^T,$$

where L_N is a lower triangular matrix, all of whose diagonal elements are 1, while D_N is a diagonal matrix. Because $\{H_t\}$ is assumed to be an AR(p) process, the first p diagonal elements of D_N are $\sigma_0^2, \sigma_1^2, \ldots, \sigma_{p-1}^2$, while the remaining $N - p$ elements are all equal to σ_p^2. Because the determinant of a product of square matrices is equal to the product of the individual determinants, and because the determinant of a triangular matrix is equal to the product of its diagonal elements, we now have

$$|\Gamma_N| = |L_N| \cdot |D_N| \cdot |L_N^T| = \sigma_p^{2(N-p)} \prod_{j=0}^{p-1} \sigma_j^2. \tag{430}$$

The modified Cholesky decomposition immediately yields

$$\Gamma_N^{-1} = L_N^{-T} D_N^{-1} L_N^{-1},$$

where, for an AR(p) process, the matrix L_N^{-1} takes the form indicated by Equation (413a). Hence we have

$$\mathbf{H}^T \Gamma_N^{-1} \mathbf{H} = \mathbf{H}^T L_N^{-T} D_N^{-1} L_N^{-1} \mathbf{H} = (L_N^{-1} \mathbf{H})^T D_N^{-1} (L_N^{-1} \mathbf{H}).$$

Now the first p elements of $L_N^{-1} \mathbf{H}$ are simply $\overrightarrow{\epsilon_1}(0), \overrightarrow{\epsilon_2}(1), \ldots, \overrightarrow{\epsilon_p}(p-1)$, where, in accordance with earlier definitions,

$$\overrightarrow{\epsilon_1}(0) \equiv H_1 \quad \text{and} \quad \overrightarrow{\epsilon_t}(t-1) \equiv H_t - \sum_{j=1}^{t-1} \phi_{j,t-1} H_{t-j}, \qquad t = 2, \ldots, p.$$

The last $N - p$ elements follow from the assumed model for $\{H_t\}$ and are given by

$$\epsilon_t = H_t - \sum_{j=1}^{p} \phi_{j,p} H_{t-j}, \qquad t = p+1, \ldots, N$$

(see Equation (413b)). Hence we can write

$$\mathbf{H}^T \Gamma_N^{-1} \mathbf{H} = \sum_{t=1}^{p} \frac{\overrightarrow{\epsilon_t}^2(t-1)}{\sigma_{t-1}^2} + \sum_{t=p+1}^{N} \frac{\epsilon_t^2}{\sigma_p^2}.$$

Combining this and Equation (430) with Equation (429) yields

$$l(\mathbf{\Phi}_p, \sigma_p^2 \mid \mathbf{H})$$

$$= \sum_{j=0}^{p-1} \log\left(\sigma_j^2\right) + (N-p) \log\left(\sigma_p^2\right) + \sum_{t=1}^{p} \frac{\overrightarrow{\epsilon_t}^2(t-1)}{\sigma_{t-1}^2} + \sum_{t=p+1}^{N} \frac{\epsilon_t^2}{\sigma_p^2}.$$

If we note that we can write

$$\sigma_j^2 = \sigma_p^2 \lambda_j \quad \text{with} \quad \lambda_j \equiv \left(\prod_{k=j+1}^{p} (1 - \phi_{k,k}^2) \right)^{-1}$$

for $j = 0, \ldots, p-1$ (this follows from Exercise [9.2c] since $P_j = \sigma_j^2$), we now obtain a useful form for the log likelihood function:

$$l(\mathbf{\Phi}_p, \sigma_p^2 \mid \mathbf{H}) = \sum_{j=0}^{p-1} \log\left(\lambda_j\right) + N \log\left(\sigma_p^2\right) + \frac{\text{SS}_{(ml)}(\mathbf{\Phi}_p)}{\sigma_p^2}, \qquad (431a)$$

where

$$\text{SS}_{(ml)}(\mathbf{\Phi}_p) \equiv \sum_{t=1}^{p} \frac{\overrightarrow{\epsilon_t}^2(t-1)}{\lambda_{t-1}} + \sum_{t=p+1}^{N} \epsilon_t^2. \qquad (431b)$$

We can obtain an expression for the mle $\hat{\sigma}_{(ml)}^2$ of σ_p^2 by differentiating Equation (431a) with respect to σ_p^2 and setting it to zero. This shows that, no matter what the mle $\hat{\mathbf{\Phi}}_{(ml)}$ of $\mathbf{\Phi}_p$ turns out to be, the estimator $\hat{\sigma}_{(ml)}^2$ is given by

$$\hat{\sigma}_{ml}^2 \equiv \text{SS}_{(ml)}(\hat{\mathbf{\Phi}}_{(ml)})/N. \qquad (431c)$$

The parameter σ_p^2 can thus be eliminated from Equation (431a), yielding what Brockwell and Davis (1991) refer to as the *reduced likelihood*:

$$l(\mathbf{\Phi}_p \mid \mathbf{H}) \equiv \sum_{j=0}^{p-1} \log\left(\lambda_j\right) + N \log\left(\text{SS}_{(ml)}(\mathbf{\Phi}_p)/N\right) + N. \qquad (431d)$$

We can determine $\hat{\mathbf{\Phi}}_{(ml)}$ by finding the value of $\mathbf{\Phi}_p$ that minimizes the reduced likelihood.

Let us now specialize to the case $p = 1$. To determine $\phi_{1,1}$, we must minimize

$$l(\phi_{1,1} \mid \mathbf{H}) = -\log\left(1 - \phi_{1,1}^2\right) + N \log\left(\mathrm{SS}_{(ml)}(\phi_{1,1})/N\right) + N, \quad (432)$$

where

$$\mathrm{SS}_{(ml)}(\phi_{1,1}) = H_1^2(1 - \phi_{1,1}^2) + \sum_{t=2}^{N}\left(H_t - \phi_{1,1}H_{t-1}\right)^2.$$

Differentiating Equation (432) with respect to $\phi_{1,1}$ and setting the result equal to 0 yields

$$\frac{\phi_{1,1}\mathrm{SS}_{(ml)}(\phi_{1,1})}{N} - (1 - \phi_{1,1}^2)\left(\sum_{t=2}^{N} H_t H_{t-1} - \phi_{1,1} \sum_{t=2}^{N-1} H_t^2\right) = 0.$$

This is a cubic equation in $\phi_{1,1}$. The estimator $\hat{\phi}_{(ml)}$ is thus equal to the root of this equation that minimizes $l(\phi_{1,1} \mid \mathbf{H})$.

For $p > 2$, we cannot obtain the mle's so easily. We must resort to a nonlinear optimizer in order to numerically determine $\hat{\mathbf{\Phi}}_{(ml)}$. Jones (1980) describes in detail a scheme for computing $\hat{\mathbf{\Phi}}_{(ml)}$ that uses a transformation of variables to facilitate numerical optimization. An interesting feature of his scheme is the use of a Kalman filter to compute the reduced likelihood at each step in the numerical optimization. This approach has two important advantages: first, it can easily deal with time series for which some of the observations are missing; and, second, it can handle the case in which an AR(p) process is observed in the presence of additive noise (this is discussed in more detail in Section 10.14).

Maximum likelihood estimation has not seen much use in parametric spectral analysis for the following two reasons. First, in a pure parametric approach, we need to be able to routinely fit fairly high-order AR models, particularly in the initial stages of a data analysis (p in the range of 10 and higher is common). This is because high-order models are often a necessity in order to adequately capture all of the important features of an sdf. The computational burden of the maximum likelihood method can become unbearable as p increases beyond even 4 or 5. Second, if we regard parametric spectral analysis as merely a way of designing low-order prewhitening filters (the approach we personally favor – see Section 9.10), then, even though the model order might now be small enough to allow use of the maximum likelihood method, other – more easily computed – estimators such as the Burg or forward/backward least squares estimators work perfectly well in practice. Any imperfections in the prewhitening filter can be compensated for in the subsequent nonparametric spectral analysis of the prewhitened series.

Comments and Extensions to Section 9.8

[1]   The large sample distribution of the mle's for the parameters of an AR(p) process has been worked out (see, for example, Newton, 1988, or Brockwell and Davis, 1991). It can be shown that, for large N, the estimator $\hat{\boldsymbol{\Phi}}_{(ml)}$ is approximately distributed as a multivariate Gaussian rv with mean $\boldsymbol{\Phi}_p$ and covariance matrix $\sigma_p^2 \Gamma_p^{-1}/N$, where Γ_p is the covariance matrix defined in Equation (394d); moreover, $\hat{\sigma}_{(ml)}^2$ approximately follows a Gaussian distribution with mean σ_p^2 and variance $2\sigma_p^4/N$ and is approximately independent of $\hat{\boldsymbol{\Phi}}_{(ml)}$. The matrix $\sigma_p^2 \Gamma_p^{-1}$ (sometimes called the *Schur matrix*) can be computed using Equation (412b), with N set to p; however, it can be shown (Pagano, 1973; Godolphin and Unwin, 1983; Newton, 1988) that we also have

$$\sigma_p^2 \Gamma_p^{-1} = A^T A - B^T B = AA^T - BB^T, \tag{433}$$

where A and B are the $p \times p$ lower triangular Toeplitz matrices whose first columns are, respectively, the vectors $[1, -\phi_{1,p}, \ldots, -\phi_{p-1,p}]^T$ and $[\phi_{p,p}, \phi_{p-1,p}, \ldots, \phi_{1,p}]^T$. This formulation for $\sigma_p^2 \Gamma_p^{-1}$ is more convenient to work with at times than that of Equation (412b).

We also note that the Yule–Walker, Burg, forward ls, backward ls and forward/backward ls estimators of the AR(p) parameters *all* share the *same* large sample distribution as the mle's! This result suggests that, given enough data, there is no difference in the performance of all the estimators we have studied; however, small sample studies have repeatedly shown, for example, that the Yule–Walker estimator can be quite poor compared to either the Burg or forward/backward ls estimators and that, for certain narrow-band processes, the Burg estimator is inferior to the forward/backward ls estimator. This discrepancy points out the limitations of large sample results (see Lysne and Tjøstheim, 1987, for further discussion on these points and also on a second-order large sample theory that explains some of the differences between the Yule–Walker and ls estimators).

[2]   The fact that we must resort to a nonlinear optimizer in order to use the maximum likelihood method for AR(p) parameter estimation limits its use in practice. Much effort therefore has gone into finding good approximations to mle's that are easy to compute. These alternative estimators are useful not only by themselves, but also as the initial guesses at the AR parameters that are required by numerical optimization routines.

Three of the estimators that we have previously discussed, namely, the forward ls, backward ls and forward/backward ls estimators, can all be regarded as approximate mle's. To see this in the case of the forward ls estimator, we first simplify the reduced likelihood of Equation (431d) by dropping the $\sum \log(\lambda_j)$ term to obtain

$$l(\boldsymbol{\Phi}_p \mid \mathbf{H}) \approx N \log\left(\mathrm{SS}_{(ml)}(\boldsymbol{\Phi}_p)/N\right) + N.$$

A justification for this simplification is that the deleted term becomes negligible as N gets large. The value of $\boldsymbol{\Phi}_p$ that minimizes the right-hand side above is identical to the value that minimizes the sum of squares $\text{SS}_{(ml)}(\boldsymbol{\Phi}_p)$. If we in turn approximate $\text{SS}_{(ml)}(\boldsymbol{\Phi}_p)$ by dropping the summation of the first p squared prediction errors in Equation (431b) (again negligible as N get large), we obtain an approximate mle given by the value of $\boldsymbol{\Phi}_p$ that minimizes $\sum_{t=p+1}^{N} \epsilon_t^2$. However, this latter sum of squares is identical to the sum of squares $\text{SS}_{(f)}(\cdot)$ of Equation (426), so our approximate estimator turns out to be just the forward ls estimator.

The backward ls estimator is obtained by noting that the log likelihood function in Equation (429) for $\mathbf{H} = [H_1, \ldots, H_N]^T$ is *identical* to the log likelihood function for the 'time reversed' series $\tilde{\mathbf{H}} \equiv [H_N, \ldots, H_1]^T$. This follows from the fact that $\tilde{\mathbf{H}}^T \Gamma_N^{-1} \tilde{\mathbf{H}} = \mathbf{H}^T \Gamma_N^{-1} \mathbf{H}$ (the proof is left as an exercise for the reader). Repeating the argument of the previous paragraph for this 'time reversed' series leads to the backward ls estimator as an approximate mle. Finally, if we rewrite Equation (429) as

$$l(\boldsymbol{\Phi}_p, \sigma_p^2 \mid \mathbf{H}) = \log\left(|\Gamma_N|\right) + \frac{\mathbf{H}^T \Gamma_N^{-1} \mathbf{H} + \tilde{\mathbf{H}}^T \Gamma_N^{-1} \tilde{\mathbf{H}}}{2}, \qquad (434)$$

we can obtain the forward/backward ls estimator as an approximate mle (this is Exercise [9.7]). Due to the symmetric nature of the likelihood function, this particular estimator seems a more natural approximation for Gaussian processes than the other two estimators, both of which are direction dependent. For more details on directionality and reversibility of autoregressive processes, see Weiss (1975) and Lawrance (1991).

9.9 Order Selection for AR(p) Processes

In discussing the various estimators of the parameters for an AR(p) process, we have implicitly assumed that the model order p is known in advance, an assumption that is rarely valid in practice. To make a reasonable choice of p, various order selection criteria have been proposed and studied in the literature. We describe briefly here a few of the most commonly used ones. Two comments are in order, however, before we proceed.

First, any order selection method we use should be appropriate for what we intend to do with the fitted AR model. As we discuss in more detail in the next section, we usually fit AR models in the context of spectral analysis either to directly obtain an estimate of the sdf (a pure parametric approach) or to produce a prewhitening filter (a combined parametric/nonparametric approach). Most of the commonly used order selection criteria are geared toward selecting a low-order AR model that does well for one-step-ahead predictions. These criteria thus seem to be more appropriate for producing a prewhitening filter than for pure parametric spectral estimation. Unfortunately, selection of an appropriate order is vital for pure parametric spectral estimation: if p is too large, the resulting sdf estimate tends to exhibit spurious peaks; if p is too small, structure in the sdf can be smoothed over.

Second, there are some subtle interactions between various order selection criteria and different AR parameter estimators. For example, the FPE criterion we discuss below was originally derived based upon the statistical properties of the forward least squares estimator. Ulrych and Bishop (1975) and Jones (1976) found that, when used in conjunction with Burg's algorithm, this criterion tends to pick out spuriously high-order models. The need to match criteria and estimators properly is often overlooked – much more research is needed in this area.

In the description of the four order selection criteria that follows, we take $\hat{\phi}_{j,k}$ and $\hat{\sigma}_k^2$ to be any of the estimators of $\phi_{j,k}$ and σ_k^2 we have discussed so far in this chapter.

[1] For an AR process of order p, the partial autocorrelation sequence $\{\phi_{k,k}\}$ is nonzero for $k = p$ and zero for $k > p$. In other words, the partial autocorrelation sequence of a pth order AR process has a cutoff after lag p. It is known that, for a Gaussian AR(p) process, the $\hat{\phi}_{k,k}$ terms for $k > p$ are approximately independently distributed with zero mean and a variance of approximately $1/N$ (see Kay and Makhoul, 1983, and references therein). Thus a rough procedure for testing $\phi_{k,k} = 0$ is to examine whether $\hat{\phi}_{k,k}$ lies between $\pm 2/\sqrt{N}$. By plotting $\hat{\phi}_{k,k}$ versus k, we can thus set p to a value beyond which $\phi_{k,k}$ can be regarded as being zero.

[2] For an AR(p) process, we have noted in Equation (409b) that

$$\sigma_{p-1}^2 > \sigma_p^2 = \sigma_{p+1}^2 = \sigma_{p+2}^2 = \cdots,$$

where σ_k^2 for $k > p$ is defined via the augmented Yule–Walker equations of order k. Thus one criterion for selecting p would be to plot $\hat{\sigma}_k^2$ versus k and set p equal to that value of k such that

$$\hat{\sigma}_{k-1}^2 > \hat{\sigma}_k^2 \approx \hat{\sigma}_{k+1}^2 \approx \hat{\sigma}_{k+2}^2 \approx \cdots. \qquad (435)$$

Recall, however, that, for the Yule–Walker and Burg estimation procedures,

$$\hat{\sigma}_k^2 = \hat{\sigma}_{k-1}^2 \left(1 - \hat{\phi}_{k,k}^2\right),$$

showing that, if $\hat{\sigma}_{k-1}^2 > 0$, then $\hat{\sigma}_k^2 < \hat{\sigma}_{k-1}^2$ unless $\hat{\phi}_{k,k}$ happens to be identically zero. The sequence $\{\hat{\sigma}_k^2\}$ is thus nonincreasing, making it problematic at times to determine where the pattern of Equation (435) occurs. The underlying problem is that $\hat{\sigma}_k^2$ tends to underestimate σ_k^2, i.e., to be biased toward zero. In the context of the forward least squares estimator, Akaike (1970) proposed a *final prediction error* (FPE) criterion that attempts to correct for this bias. The FPE for a kth order AR model is defined to be

$$\text{FPE}(k) \equiv \left(\frac{N + k + 1}{N - k - 1} \right) \hat{\sigma}_k^2.$$

This form of the FPE assumes that the process mean is unknown (the usual case in practical applications) and hence that we have recentered our time series by subtracting the sample mean prior to the estimation of the AR parameters (if the process mean is in fact known, the term in parentheses becomes $(N + k)/(N - k)$). Note that, as required, $\text{FPE}(k)$ is an inflated version of $\hat{\sigma}_k^2$. The FPE order selection criterion is to set p equal to the value of k that minimizes $\text{FPE}(k)$.

[3] In the context of the maximum likelihood estimator, Akaike (1974) proposed another order selection criterion, known as *Akaike's information criterion* (AIC). For a kth order AR process with a known process mean, the AIC is defined as

$$\text{AIC}(k) \equiv -2 \log \{\text{maximized likelihood}\} + 2k.$$

However, this criterion, which is based on cross-entropy ideas, is very general and applicable in more than just a time series context. For a Gaussian $\text{AR}(k)$ process $\{H_t\}$, Equations (431d) and (431c) tell us that we can write

$$-2 \log \{\text{maximized likelihood}\} = \sum_{j=0}^{k-1} \log(\hat{\lambda}_j) + N \log(\hat{\sigma}_{(ml)}^2) + N,$$

where $\hat{\sigma}_{(ml)}^2$ is the mle of σ_k^2, while $\hat{\lambda}_j$ depends on the mle's of the $\phi_{j,k}$ terms and not on $\hat{\sigma}_{(ml)}^2$. The first term on the right-hand side becomes negligible for large N, while the third term is just a constant. If we drop these two terms and allow the use of other estimators $\hat{\sigma}_k^2$ of σ_k^2 besides the mle, we obtain the usual form that is quoted in the literature as the AIC for AR processes (see, for example, de Gooijer *et al.*, 1985; Rosenblatt, 1985; or Kay, 1988):

$$\text{AIC}(k) = N \log(\hat{\sigma}_k^2) + 2k$$

(alternatively, we can justify this formulation by using the conditional likelihood of $H_1, \ldots, H_N$, given $H_0 = H_{-1} = \cdots = H_{-k+1} = 0$). The AIC order selection criterion is to set p equal to the value of k that minimizes $\text{AIC}(k)$.

[4] If we are primarily interested in using the fitted AR model to produce a prewhitening filter, the following subjective criterion works
well in practice. First, as described in Chapter 6, we compute a direct spectral estimate $\hat{S}^{(d)}(\cdot)$ using a data taper that provides good
leakage protection. We then fit a sequence of relatively low-order
AR models (starting with, say, $k = 2$ or 4) to our time series, compute the corresponding AR sdf estimates via Equation (392b), and
compare these estimates with $\hat{S}^{(d)}(\cdot)$. We select our model order p
as the smallest value k such that the corresponding AR sdf estimate
generally captures the overall shape of $\hat{S}^{(d)}(\cdot)$. An example of this
procedure is given in the next section.

The order selection criteria we have outlined above are by no means
the only ones that have been proposed or are in extensive use. The
reader is referred to the review article by de Gooijer *et al.* (1985) and
to Choi (1992) for more comprehensive discussions.

How well do these order selection criteria work in practice? Using
simulated data, Landers and Lacoss (1977) found that both the FPE
and AIC – used with Burg's algorithm – selected a model order that
was insufficient to resolve spectral details in an sdf with sharp peaks.
By increasing the model order by a factor 3, they obtained adequate
spectral estimates. Ulrych and Clayton (1976) found that neither the
FPE nor the AIC works well with short time series, prompting Ulrych
and Ooe (1983) to recommend that p be chosen between $N/3$ and $N/2$
for such series. Kay and Marple (1981) report that the results from using
the FPE and AIC criteria have been mixed, particularly with actual
time series rather than simulated AR processes. They conclude that
subjective judgment is still needed to select the AR order for actual time
series. Nonetheless, it is useful to examine the various order selection
criteria to get some idea for our choice of p.

9.10 Prewhitened Spectral Estimators and An Example

In the previous sections we have concentrated on the problems of estimating the parameters of an AR(p) process and selecting the model
order p. In this section we discuss how to use the results of these efforts
to produce an estimator of the sdf for a time series. There are two well-
known ways of doing so. First, we can use a *pure parametric approach*,
in which we simply substitute the estimated AR parameters for the corresponding theoretical parameters in Equation (392b). Second, we can
employ a *prewhitening approach*, in which we use the estimates of the
$\phi_{j,p}$ terms to form a prewhitening filter. We have already introduced
the idea of prewhitening in Section 6.5. Here we give details on how to
implement prewhitening using an estimated AR model.

Suppose we are given a time series that is a realization of a portion
$X_1, \ldots, X_N$ of a stationary process with zero mean and sdf $S_X(\cdot)$. We

carry out the following steps.

[1] We begin by fitting an AR(p) model to the time series using, say, Burg's algorithm to obtain $\bar{\phi}_{1,p}, \ldots, \bar{\phi}_{p,p}$ and $\bar{\sigma}_p^2$ (as noted in the example at the end of this section, the particular estimator that we use can be important; we have found the Burg method to work well in most practical situations). We set the order p to be just large enough to capture the general structure of the sdf $S_X(\cdot)$. In practice, an appropriate p can be determined by comparing plots of AR sdf's of different orders with a direct spectral estimator $\hat{S}_X^{(d)}(\cdot)$ of $S_X(\cdot)$ employing a data taper that offers good protection against broad-band leakage (see item [4] of the previous section). In any case, p should not be large compared to the sample size N.

[2] We then use the fitted AR(p) model as a prewhitening filter. The output from this filter is a time series of length $N - p$ given by

$$\overrightarrow{e_t}(p) \equiv X_t - \bar{\phi}_{1,p}X_{t-1} - \cdots - \bar{\phi}_{p,p}X_{t-p}, \qquad t = p+1,\ldots,N,$$

where, as before, $\overrightarrow{e_t}(p)$ is the forward prediction error observed at time t. If we let $S_e(\cdot)$ denote the sdf for $\{\overrightarrow{e_t}(p)\}$, linear filtering theory tells us that the relationship between $S_e(\cdot)$ and $S_X(\cdot)$ is given by

$$S_e(f) = \left| 1 - \sum_{j=1}^{p} \bar{\phi}_{j,p} e^{-i2\pi fj\,\Delta t} \right|^2 S_X(f).$$

[3] If the prewhitening filter has been chosen correctly, the sdf $S_e(\cdot)$ for the observed forward prediction errors will have a smaller dynamic range than $S_X(\cdot)$. We can thus produce a direct spectral estimate $\hat{S}_e^{(d)}(\cdot)$ for $S_e(\cdot)$ with good bias properties using very little or no tapering. If we let $\{\hat{s}_{k,e}^{(d)}\}$ represent the corresponding acvs estimate, we can then produce an estimate of $S_e(\cdot)$ with decreased variability by applying a lag window $\{w_{k,m}\}$ to obtain

$$\hat{S}_e^{(lw)}(f) \equiv \Delta t \sum_{k=-(N-p-1)}^{N-p-1} w_{k,m}\hat{s}_{k,e}^{(d)} e^{-i2\pi fk\,\Delta t}.$$

[4] Finally we estimate $S_X(\cdot)$ by 'postcoloring' $\hat{S}_e^{(lw)}(\cdot)$ to obtain

$$\hat{S}_X^{(pc)}(f) \equiv \frac{\hat{S}_e^{(lw)}(f)}{\left| 1 - \sum_{j=1}^{p} \bar{\phi}_{j,p} e^{-i2\pi fj\,\Delta t} \right|^2}. \qquad (438)$$

There are two obvious variations on this technique that have seen some use. First, we can make use of the observed backward prediction errors to produce a second estimator similar to $\hat{S}_e^{(lw)}(\cdot)$. These two estimators can be averaged together prior to postcoloring in step [4]. Second, we can obtain the first p values of the acvs corresponding to $\hat{S}_X^{(pc)}(\cdot)$ and use them in the Yule–Walker method to produce a refined prewhitening filter for iterative use in step [2].

Let us note some important points about this combined parametric/nonparametric approach to spectral estimation. First, the chief difference between this approach and a pure parametric approach is that we no longer regard the observed prediction errors $\{\overrightarrow{e_t}(p)\}$ as white noise, but rather we use a nonparametric approach to estimate their sdf. Second, the problem of selecting p is lessened because any imperfections in the prewhitening filter can be compensated for in the nonparametric portion of this combined approach. Third, since all lag window spectral estimators correspond to an acvs estimator that is identically zero after a finite lag q (see Equation (238b)), the numerator of $\hat{S}_X^{(pc)}(\cdot)$ has the form of an MA(q) sdf with $q \leq N - p - 1$. Hence the estimator $\hat{S}_X^{(pc)}(\cdot)$ is the sdf for some ARMA(p, q) model. Our combined approach is thus a way of implicitly fitting an ARMA model to a time series (see the next section). Fourth, the combined approach only makes sense in situations where tapering is normally required; i.e., the sdf $S_X(\cdot)$ has a high dynamic range (see the discussion in Section 6.5). Fifth, even if we use a pure parametric approach, it is useful to carefully examine the observed prediction errors $\{\overrightarrow{e_t}(p)\}$ because this is a valuable way to detect outliers in a time series (Martin and Thomson, 1982).

As an example, let us return to the ocean wave data considered previously in Sections 6.18 and 7.5. We begin by finding a low-order AR model that can serve as a prewhitening filter. The dots in the top plot of Figure 440 show a direct spectral estimate $\hat{S}_X^{(d)}(\cdot)$ for this series using an $NW = 2/\Delta t$ dpss data taper (these are the same as the dots in all three plots of Figure 301 – recall that, based upon a study of Figure 296, we argued that this estimate is leakage-free). After some experimentation, we found that a $p = 5$ order AR model captures the overall spectral structure indicated by $\hat{S}_X^{(d)}(\cdot)$. The sdf corresponding to this fitted model is shown by the thick curves in the two plots of Figure 440. The parameters for this model were estimated using Burg's algorithm. For comparison, we also used the Yule–Walker method. The corresponding sdf is given by the thin curve in the upper plot. Note that, whereas the Burg estimator generally does a good job of tracking $\hat{S}_X^{(d)}(\cdot)$ across all frequencies, the same cannot be said for the Yule–Walker estimator – it considerably overestimates the power in the high frequency region.

Since the FPE order selection criterion picked out the model order $p = 27$ as appropriate for the ocean wave data, we also estimated an

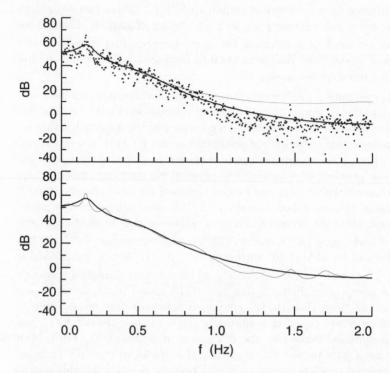

Figure 440. Determination of an AR prewhitening filter and illustration of
the effect of different choices for a parameter estimator and model order. The
dots in the upper plot show the direct spectral estimator of the ocean wave
data that is also shown as dots in the plots of Figure 301. The thick curves
in both plots show the estimated sdf corresponding to an AR(5) model fit to
the data using Burg's algorithm. The thin curve in the upper plot shows the
estimated sdf for the same model, but now with parameters estimated using
the Yule–Walker method. The thin curve in the lower plot shows the estimated
sdf corresponding to an AR(27) model fit using Burg's algorithm.

AR(27) model using Burg's algorithm and plotted the corresponding sdf
estimate as the thin curve in the bottom plot of Figure 440. This spectral
estimate is what we would obtain using a pure parametric approach.
Note that it has much more structure than the $p = 5$ spectral estimate
(the thick curve) and that it agrees quite well in its fine details with
$\hat{S}_X^{(d)}(\cdot)$.

Figure 441(a) shows the forward prediction errors $\{\overrightarrow{e_t}(5)\}$ obtained
when we use the fitted AR(5) model to form a prewhitening filter. Since
there were $N = 1024$ data points in the original ocean wave time series,
there are $N - p = 1019$ points in the forward prediction error series.
Figure 441(b) shows the periodogram for $\{\overrightarrow{e_t}(5)\}$. A comparison of this

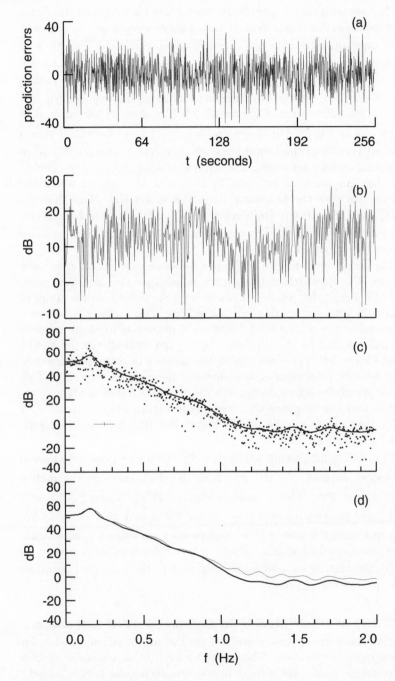

Figure 441. Illustration of prewhitening applied to ocean wave data.

periodogram with direct spectral estimates using a variety of dpss data tapers indicates that the periodogram is leakage-free, so we can take it to be the approximately unbiased direct spectral estimate $\hat{S}_e^{(d)}(\cdot)$ that we are required to obtain in step [3]. We now smooth $\hat{S}_e^{(d)}(\cdot)$ using a Parzen lag window $\{w_{k,m}\}$ with $m = 55$ to match what we did in the middle plot of Figure 301. Postcoloring this lag window estimate yields $\hat{S}_X^{(pc)}(\cdot)$, shown by the thick curve of Figure 441(c). The dots are the same as in the top plot of Figure 440 and show that $\hat{S}_X^{(pc)}(\cdot)$ can be regarded as a smoothed version of the leakage-free direct spectral estimate $\hat{S}_X^{(d)}(\cdot)$ of the original ocean wave series. The amount of smoothing is appropriate, given that our main objective is to determine the rate at which the sdf decreases over the frequency range 0.2 to 1.0 Hz. Note that the AR(27) spectral estimate (thin curve in the bottom plot of Figure 440) is quite comparable to $\hat{S}_X^{(pc)}(\cdot)$, the main difference being that the latter is somewhat smoother.

The crisscross in Figure 441(c) has the same interpretation as those in Figure 301. Its width and height are based upon the statistical properties of a Parzen lag window with $m = 55$ applied to a time series of length 1019 with no tapering (see Table 269). This yields a lag window spectral estimator with $\nu = 68.7$ equivalent degrees of freedom, whereas we found $\nu = 35.2$ for the similarly constructed estimator in the middle plot of Figure 301. The increase in the number of degrees of freedom occurs because prewhitening has obviated the need for tapering. This increase translates into a decrease in the length of a 95% confidence interval so that the height of the crisscross in Figure 441(c) is noticeably smaller than in the middle plot of Figure 301 (the heights are 2.9 dB and 4.1 dB, respectively).

Finally it is interesting to compare $\hat{S}_X^{(pc)}(\cdot)$ with the $m = 55$ Parzen lag window estimate $\hat{S}_X^{(lw)}(\cdot)$ shown as the thick curve in the middle plot of Figure 301. This is done in Figure 441(d), where $\hat{S}_X^{(pc)}(\cdot)$ and $\hat{S}_X^{(lw)}(\cdot)$ are plotted, respectively, as the thick and thin curves. We noted in Chapter 6 that $\hat{S}_X^{(lw)}(\cdot)$ suffers from smoothing window leakage at the high frequencies. There is no evidence of such leakage in $\hat{S}_X^{(pc)}(\cdot)$, illustrating an additional advantage to the combined parametric/nonparametric approach.

9.11 Use of Other Models for Parametric SDF Estimation

So far we have concentrated entirely on the use of AR(p) models for parametric sdf estimation. There are, however, other classes of models that could be used. We discuss briefly one such class here – namely, the class of moving average processes of order q (MA(q)) – because it provides an interesting contrast to AR(p) processes.

A MA(q) process $\{X_t\}$ with zero mean is defined by

$$X_t = \epsilon_t - \theta_{1,q}\epsilon_{t-1} - \theta_{2,q}\epsilon_{t-2} - \cdots - \theta_{q,q}\epsilon_{t-q},$$

where $\theta_{1,q}, \theta_{2,q}, \ldots, \theta_{q,q}$ are q fixed coefficients, and $\{\epsilon_t\}$ is a white noise process with zero mean and variance σ_ϵ^2 (see Equation (43a)). The above process is stationary no matter what values the $\theta_{j,q}$ terms assume. Its sdf is given by

$$S(f) = \sigma_\epsilon^2 \, \Delta t \left| 1 - \sum_{j=1}^{q} \theta_{j,q} e^{-i2\pi f j \, \Delta t} \right|^2, \qquad |f| \le f_{(N)}.$$

There are $q + 1$ parameters – the $\theta_{j,q}$ terms and σ_ϵ^2 – that we need to estimate to produce an MA(q) parametric sdf estimate.

It is interesting to reconsider the six-fold rationale for using AR(p) models discussed in Section 9.2 and to contrast it with the MA(q) case.

[1] As in the case of AR(p) processes, any continuous sdf $S(\cdot)$ can be approximated arbitrarily well by an MA(q) sdf if q is chosen large enough (Anderson, 1971, p. 411). In this sense the class of MA processes is as rich as that of AR processes, but we argue below that in another sense it is *not* as rich.

[2] Algorithms for explicitly fitting MA(q) models to time series tend to be computationally intensive compared with those needed to fit AR(p) models. However, as we see below, if our interest is primarily in spectral estimation so that we do not require explicit estimates of $\theta_{j,q}$, there is an analog to the Yule–Walker approach that is computationally efficient but unfortunately has poor statistical properties.

[3] From the discussion following Equation (424), we know that, for a Gaussian process with known variance s_0 and known cepstral values $c_1, \ldots, c_q$, the maximum entropy spectrum is identical to that of an MA(q) process (if any such process exists satisfying these $q + 1$ constraints).

[4] Fitting an MA(q) model to a time series leads to an IIR filter, which – because of the problem of turn-on transients – is not as easy to work with as a FIR filter for prewhitening a time series of finite length.

[5] Physical arguments sometimes suggest that an MA model is appropriate for a particular time series. For example, we have already noted in Section 2.6 that Spencer-Smith and Todd (1941) argued thus for the thickness of textile slivers as a function of displacement along a sliver.

[6] Pure sinusoidal variations cannot be expressed as an MA(q) model.

Two examples of realizations of MA(1) processes are shown in Figure 44.

As in the case of $\mathrm{AR}(p)$ processes, we need to estimate the $\theta_{j,q}$ terms and the variance σ_ϵ^2 from available data. If we attempt to use the method of matching lagged moments that led to the Yule–Walker equations in the $\mathrm{AR}(p)$ case, we find that

$$s_k = \sigma_\epsilon^2 \sum_{j=0}^{q-k} \theta_{j,q}\theta_{j+k,q}, \qquad k = 0, 1, \ldots, q,$$

where we define $\theta_{0,q} \equiv -1$ (cf. Equation (43b)). These are nonlinear in the MA coefficients, but an additional difficulty is that a valid solution need not exist. As an example, for an MA(1) process, we have

$$s_0 = \sigma_\epsilon^2 \left(1 + \theta_{1,1}^2\right) \text{ and } s_1 = -\sigma_\epsilon^2 \theta_{1,1},$$

which can only be solved for a valid σ_ϵ^2 and $\theta_{1,1}$ if $|s_1/s_0| \leq 1/2$. Thus, whereas there is always an $\mathrm{AR}(p)$ process whose theoretical acvs agrees out to lag p with $\hat{s}_0^{(p)}, \ldots, \hat{s}_p^{(p)}$ computed for any time series, the same is not true of MA processes. In this sense we can claim that the class of AR processes is 'richer' than the class of MA processes.

There is, however, a second way of looking at the Yule–Walker estimation procedure for $\mathrm{AR}(p)$ processes as far as sdf estimation is concerned. If we fit an $\mathrm{AR}(p)$ process to a time series by that procedure, the acvs of the fitted model necessarily agrees *exactly* up to lag p with the estimated acvs. The values of the acvs after lag p are generated recursively in accordance with an $\mathrm{AR}(p)$ model, namely,

$$s_k = \sum_{j=1}^{p} \phi_{j,p} s_{k-j}, \qquad k = p+1, p+2, \ldots$$

(see Equation (394b)). The estimated sdf can thus be regarded as given by

$$\tilde{S}(f) = \Delta t \sum_{k=-\infty}^{\infty} \tilde{s}_k e^{-i2\pi f k \, \Delta t},$$

where

$$\tilde{s}_k = \begin{cases} \hat{s}_k^{(p)}, & |k| \leq p; \\ \sum_{j=1}^{p} \tilde{\phi}_{j,p} \tilde{s}_{|k|-j}, & |k| > p, \end{cases}$$

with $\tilde{\phi}_{j,p}$ defined to be the Yule–Walker estimator of $\phi_{j,p}$. Thus the Yule–Walker sdf estimator can be regarded as accepting $\hat{s}_k^{(p)}$ up to lag p and then estimating the acvs beyond that lag based upon the $\mathrm{AR}(p)$ model.

If we follow this procedure for an MA(q) model, we would set the acvs beyond lag q equal to 0 (see Equation (43b)). Our sdf estimate would thus be just

$$\hat{S}(f) = \Delta t \sum_{k=-q}^{q} \hat{s}_k^{(p)} e^{-i2\pi f k \, \Delta t}.$$

This is identical to the truncated periodogram of Equation (327a). Although this estimator was an early example of a lag window estimator, its smoothing window is such that it is now generally felt to be inferior to other estimators of the lag window class (such as those utilizing the Parzen or Papoulis lag windows).

To conclude, the choice of the class of models to be used in parametric spectral analysis is obviously quite important. The class of AR models has a number of advantages when compared to MA models, not the least of which is ease of computation. A considerable amount of current research is being devoted to using the class of ARMA models for parametric spectral analysis. Unfortunately, algorithms for explicitly fitting ARMA models are much more involved than those for fitting AR models; on the other hand, we have noted that the spectral estimator of Equation (438) is in fact an ARMA sdf, so it is certainly possible to obtain *implicit* ARMA spectral estimates in a computationally efficient manner.

9.12 Confidence Intervals Using AR Spectral Estimators

Until recently, relatively little work has been done on the asymptotic theory of AR spectral estimators and, in particular, on the computation of confidence bounds for autoregressive spectra. As Kaveh and Cooper (1976) point out, since 'the AR spectral estimate is obtained through nonlinear operations on the autocorrelation samples, analytical derivation of its statistical properties is, in general, formidable.'

Early work in this area was done by Kromer (1969), Berk (1974), Baggeroer (1976) and Reid (1979). Recently Newton and Pagano (1984) used both inverse autocovariances and Scheffé projections to find simultaneous confidence bands, while Koslov and Jones (1985) and Burshtein and Weinstein (1987, 1988) developed very similar approaches. Swanepoel and Van Wyk (1986) used the bootstrap technique to obtain confidence bands. We describe here an approach that is most similar to that of Burshtein and Weinstein, although differing quite substantially in mathematical and statistical details. It tries to keep possibly unfamiliar methodology to a minimum.

As we noted in item [1] of the Comments and Extensions to Section 9.8, all of the AR parameter estimators we have discussed in this chapter have the same large sample distribution. To be specific, let

$\hat{\boldsymbol{\Phi}}_p \equiv [\hat{\phi}_{1,p}, \hat{\phi}_{2,p}, \ldots, \hat{\phi}_{p,p}]^T$ and $\hat{\sigma}_p^2$ be any of these estimators of $\boldsymbol{\Phi}_p \equiv [\phi_{1,p}, \phi_{2,p}, \ldots, \phi_{p,p}]^T$ and σ_p^2 based upon a sample $H_1, H_2, \ldots, H_N$ from a Gaussian AR(p) process with zero mean. As $N \to \infty$, we have

$$\sqrt{N}(\hat{\boldsymbol{\Phi}}_p - \boldsymbol{\Phi}_p) \stackrel{\mathrm{d}}{=} \mathcal{N}(\mathbf{0}, \sigma_p^2 \Gamma_p^{-1}) \text{ and } \sqrt{N}(\hat{\sigma}_p^2 - \sigma_p^2) \stackrel{\mathrm{d}}{=} \mathcal{N}(0, 2\sigma_p^4), \quad (446a)$$

where $\mathcal{N}(\mathbf{0}, \sigma_p^2 \Gamma_p^{-1})$ denotes a p-dimensional Gaussian distribution with zero mean vector and covariance matrix $\sigma_p^2 \Gamma_p^{-1}$, and Γ_p is the $p \times p$ covariance matrix of Equation (394d); moreover, $\hat{\sigma}_p^2$ is independent of all the elements of $\hat{\boldsymbol{\Phi}}_p$ (Newton, 1988; Brockwell and Davis, 1991). If we let $\hat{s}_k$ stand for the estimator of s_k derived from $\hat{\boldsymbol{\Phi}}_p$ and $\hat{\sigma}_p^2$, we can obtain a consistent estimator $\hat{\Gamma}_p$ of Γ_p by replacing s_k in Γ_p with $\hat{s}_k$. These results are the key to the methods of calculating asymptotically correct confidence intervals for AR spectra that we now present.

Let us rewrite the expression in Equation (392b) for the sdf of a stationary AR(p) process as

$$S(f) = \frac{\sigma_p^2 \Delta t}{|1 - \mathbf{e}^H(f)\boldsymbol{\Phi}_p|^2}, \qquad |f| \le f_{(N)}, \qquad (446b)$$

where H denotes complex-conjugate (Hermitian) transpose and

$$\mathbf{e}(f) \equiv [e^{i2\pi f\,\Delta t}, e^{i4\pi f\,\Delta t}, \ldots, e^{i2p\pi f\,\Delta t}]^T.$$

From this definition of $\mathbf{e}(f)$, we now define the $p \times 2$ matrix $\mathcal{E}$ via

$$\mathcal{E}^T \equiv [\mathbf{e}_\Re(f) \mid \mathbf{e}_\Im(f)]^T = \begin{bmatrix} \cos(2\pi f\,\Delta t) & \ldots & \cos(2p\pi f\,\Delta t) \\ \sin(2\pi f\,\Delta t) & \ldots & \sin(2p\pi f\,\Delta t) \end{bmatrix},$$

where $\mathbf{e}_\Re(f)$ and $\mathbf{e}_\Im(f)$ are vectors containing the real and imaginary parts of $\mathbf{e}(f)$, respectively. We assume for now that $f \neq 0$ or $\pm f_{(N)}$. It follows from Exercise [2.8] and Equation (446a) that the 2×1 vector

$$\mathbf{B} \equiv \mathcal{E}^T(\hat{\boldsymbol{\Phi}}_p - \boldsymbol{\Phi}_p) \qquad (446c)$$

has, for large N, a variance given by $\mathcal{E}^T(\sigma_p^2/N)\Gamma_p^{-1}\mathcal{E}$ and hence asymptotically

$$\mathbf{B} \stackrel{\mathrm{d}}{=} \mathcal{N}(\mathbf{0}, \sigma_p^2 \mathcal{Q}), \text{ where } \mathcal{Q} \equiv \mathcal{E}^T(\Gamma_p^{-1}/N)\mathcal{E}. \qquad (446d)$$

Under the conditions specified above, we know from the Mann–Wald theorem (Mann and Wald, 1943; see also Bruce and Martin, 1989) that, asymptotically,

$$\frac{\mathbf{B}^T \hat{\mathcal{Q}}^{-1} \mathbf{B}}{\hat{\sigma}_p^2} \stackrel{\mathrm{d}}{=} \chi_2^2, \text{ where } \hat{\mathcal{Q}} \equiv \mathcal{E}^T(\hat{\Gamma}_p^{-1}/N)\mathcal{E}. \qquad (446e)$$

From the definition of $\mathbf{B}$ in (446c), we can write

$$\frac{\mathbf{B}^T \hat{\mathcal{Q}}^{-1} \mathbf{B}}{\hat{\sigma}_p^2} = \frac{(\mathcal{E}^T \hat{\boldsymbol{\Phi}}_p - \mathcal{E}^T \boldsymbol{\Phi}_p)^T \hat{\mathcal{Q}}^{-1}(\mathcal{E}^T \hat{\boldsymbol{\Phi}}_p - \mathcal{E}^T \boldsymbol{\Phi}_p)}{\hat{\sigma}_p^2} \stackrel{\mathrm{d}}{=} \chi_2^2$$

when $f \neq 0$ or $\pm f_{(N)}$. Hence,

$$\mathbf{P}\left[\frac{(\mathcal{E}^T \hat{\boldsymbol{\Phi}}_p - \mathcal{E}^T \boldsymbol{\Phi}_p)^T \hat{\mathcal{Q}}^{-1}(\mathcal{E}^T \hat{\boldsymbol{\Phi}}_p - \mathcal{E}^T \boldsymbol{\Phi}_p)}{\hat{\sigma}_p^2} \leq Q_2(1-\alpha)\right] = 1 - \alpha,$$
(447a)

where $Q_2(\alpha)$ is the $\alpha \times 100\%$ percentage point of the χ_2^2 distribution (see Table 256). Let $\mathcal{A}_0$ represent the event displayed between the brackets above. An equivalent description for $\mathcal{A}_0$ is that the true value of the 2×1 vector $\mathcal{E}^T \boldsymbol{\Phi}_p$ lies inside the ellipsoid defined as the set of vectors, $\mathcal{E}^T \tilde{\boldsymbol{\Phi}}_p$ say, satisfying

$$(\mathcal{E}^T \hat{\boldsymbol{\Phi}}_p - \mathcal{E}^T \tilde{\boldsymbol{\Phi}}_p)^T \mathcal{M}(\mathcal{E}^T \hat{\boldsymbol{\Phi}}_p - \mathcal{E}^T \tilde{\boldsymbol{\Phi}}_p) \leq 1,$$

where $\mathcal{M} \equiv \hat{\mathcal{Q}}^{-1}/(\hat{\sigma}_p^2 Q_2(1-\alpha))$. Scheffé (1959, p. 407) has shown that $\mathcal{E}^T \tilde{\boldsymbol{\Phi}}_p$ is in this ellipsoid if and only if

$$\left|\mathbf{a}^T(\mathcal{E}^T \hat{\boldsymbol{\Phi}}_p - \mathcal{E}^T \boldsymbol{\Phi}_p)\right|^2 \leq \mathbf{a}^T \mathcal{M}^{-1} \mathbf{a}$$
(447b)

for *all* two-dimensional vectors $\mathbf{a}$, giving us another equivalent description for $\mathcal{A}_0$. Hence, specializing to two specific values for $\mathbf{a}$, the occurrence of the event $\mathcal{A}_0$ implies the occurrence of the following two events:

$$\left|[1,0](\mathcal{E}^T \hat{\boldsymbol{\Phi}}_p - \mathcal{E}^T \boldsymbol{\Phi}_p)\right|^2 = \left|\mathbf{e}_{\Re}^T(f)(\hat{\boldsymbol{\Phi}}_p - \boldsymbol{\Phi}_p)\right|^2 \leq [1,0]\mathcal{M}^{-1}[1,0]^T$$

and

$$\left|[0,1](\mathcal{E}^T \hat{\boldsymbol{\Phi}}_p - \mathcal{E}^T \boldsymbol{\Phi}_p)\right|^2 = \left|\mathbf{e}_{\Im}^T(f)(\hat{\boldsymbol{\Phi}}_p - \boldsymbol{\Phi}_p)\right|^2 \leq [0,1]\mathcal{M}^{-1}[0,1]^T.$$

These two events in turn imply the occurrence of the event $\mathcal{A}_1$, defined by

$$\left|\mathbf{e}_{\Re}^T(f)(\hat{\boldsymbol{\Phi}}_p - \boldsymbol{\Phi}_p)\right|^2 + \left|\mathbf{e}_{\Im}^T(f)(\hat{\boldsymbol{\Phi}}_p - \boldsymbol{\Phi}_p)\right|^2$$
$$\leq [1,0]\mathcal{M}^{-1}[1,0]^T + [0,1]\mathcal{M}^{-1}[0,1]^T.$$

Let us rewrite both sides of this inequality. For the left-hand side, we can write

$$\left|\mathbf{e}_{\Re}^T(f)(\hat{\boldsymbol{\Phi}}_p - \boldsymbol{\Phi}_p)\right|^2 + \left|\mathbf{e}_{\Im}^T(f)(\hat{\boldsymbol{\Phi}}_p - \boldsymbol{\Phi}_p)\right|^2$$
$$= \left|\mathbf{e}_{\Re}^T(f)(\hat{\boldsymbol{\Phi}}_p - \boldsymbol{\Phi}_p) - i\mathbf{e}_{\Im}^T(f)(\hat{\boldsymbol{\Phi}}_p - \boldsymbol{\Phi}_p)\right|^2 = \left|\mathbf{e}^H(f)(\hat{\boldsymbol{\Phi}}_p - \boldsymbol{\Phi}_p)\right|^2.$$

For the right-hand side, we let $\mathbf{d}^H \equiv [1, -i]$ and use the fact that the 2×2 matrix $\mathcal{M}$ is symmetric to obtain

$$[1,0]\mathcal{M}^{-1}[1,0]^T + [0,1]\mathcal{M}^{-1}[0,1]^T = \mathbf{d}^H\mathcal{M}^{-1}\mathbf{d}.$$

Hence the event $\mathcal{A}_1$ is equivalent to

$$\left|\mathbf{e}^H(f)(\hat{\boldsymbol{\Phi}}_p - \boldsymbol{\Phi}_p)\right|^2 \leq \mathbf{d}^H\mathcal{M}^{-1}\mathbf{d}, \quad \text{or,} \quad \frac{\left|\mathbf{e}^H(f)(\hat{\boldsymbol{\Phi}}_p - \boldsymbol{\Phi}_p)\right|^2}{\mathbf{d}^H(\hat{\mathcal{Q}}\hat{\sigma}_p^2)\mathbf{d}} \leq Q_2(1-\alpha).$$

Because the occurrence of the event $\mathcal{A}_0$ implies the occurrence of the event $\mathcal{A}_1$, it follows that $\mathcal{A}_0$ is a subset of $\mathcal{A}_1$. We thus must have $\mathbf{P}[\mathcal{A}_0] \leq \mathbf{P}[\mathcal{A}_1]$. Since Equation (447a) states that $\mathbf{P}[\mathcal{A}_0] = 1 - \alpha$, it follows that $\mathbf{P}[\mathcal{A}_1] \geq 1 - \alpha$, i.e., that

$$\mathbf{P}\left[\left|\mathbf{e}^H(f)(\hat{\boldsymbol{\Phi}}_p - \boldsymbol{\Phi}_p)\right|^2 \leq Q_2(1-\alpha)\mathbf{d}^H(\hat{\mathcal{Q}}\hat{\sigma}_p^2)\mathbf{d}\right] \geq 1 - \alpha. \qquad (448a)$$

Let us define

$$G(f) \equiv 1 - \mathbf{e}^H(f)\boldsymbol{\Phi}_p \quad \text{so that} \quad |G(f)|^2 = \frac{\sigma_p^2\,\Delta t}{S(f)},$$

because of Equation (446b). With $\hat{G}(f) \equiv 1 - \mathbf{e}^H(f)\hat{\boldsymbol{\Phi}}_p$, we have

$$\left|\mathbf{e}^H(f)(\hat{\boldsymbol{\Phi}}_p - \boldsymbol{\Phi}_p)\right|^2 = \left|\hat{G}(f) - G(f)\right|^2.$$

Recalling that $\mathbf{d}^H \equiv [1, -i]$ and using the definition for $\hat{\mathcal{Q}}$ in Equation (446e), we obtain

$$\mathbf{d}^H\hat{\mathcal{Q}}\mathbf{d} = \mathbf{e}^H(f)(\hat{\Gamma}_p^{-1}/N)\mathbf{e}(f)$$

since $\mathbf{d}^H\mathcal{E}^T = \mathbf{e}^H(f)$. We can now rewrite Equation (448a) as

$$\mathbf{P}\left[|\hat{G}(f) - G(f)|^2 \leq Q_2(1-\alpha)\hat{\sigma}_p^2\mathbf{e}^H(f)(\hat{\Gamma}_p^{-1}/N)\mathbf{e}(f)\right] \geq 1 - \alpha,$$

from which we conclude that asymptotically the transfer function $G(f)$ of the true AR filter resides within a circle of radius

$$r_2(f) \equiv [Q_2(1-\alpha)\hat{\sigma}_p^2\mathbf{e}^H(f)(\hat{\Gamma}_p^{-1}/N)\mathbf{e}(f)]^{1/2}$$

with probability at least $1 - \alpha$. Equivalently, as can readily be demonstrated geometrically,

$$\mathbf{P}\left[\left\{|\hat{G}(f)| + r_2(f)\right\}^{-2} \leq |G(f)|^{-2} \leq \left\{|\hat{G}(f)| - r_2(f)\right\}^{-2}\right] \geq 1 - \alpha,$$
$$(448b)$$

which gives confidence intervals for $|G(f)|^{-2} = S(f)/\sigma_p^2$, the normalized spectrum. Note that the quantity $\hat{\sigma}_p^2 \hat{\Gamma}_p^{-1}$ appearing in $r_2(f)$ can be readily computed using the sampling version of Equation (433) (other relevant computational details can be found in Burshtein and Weinstein, 1987).

We assumed previously that the frequency of interest $f \neq 0$ or $\pm f_{(N)}$. If in fact f is equal to zero or the Nyquist frequency, then $e_\Im(f) = 0$, and the rank of the quadratic form $\mathbf{B}^T \hat{Q}^{-1} \mathbf{B}/\hat{\sigma}_p^2$ is 1 (rather than 2 as previously), so that now

$$\frac{\mathbf{B}^T \hat{Q}^{-1} \mathbf{B}}{\hat{\sigma}_p^2} \stackrel{\mathrm{d}}{=} \chi_1^2, \qquad f = 0 \text{ or } \pm f_{(N)}.$$

All that is necessary is then to replace $Q_2(1-\alpha)$ by $Q_1(1-\alpha)$, i.e., use the $(1-\alpha) \times 100\%$ percentage point of the χ_1^2 distribution.

Comments and Extensions to Section 9.12

[1]   For a Gaussian stationary process $\{G_t\}$, and a spectrum that is sufficiently smooth, Kromer (1969) showed that, provided $p \to \infty$, $p/N \to \infty$ as $N \to \infty$, then

$$\frac{\hat{S}_G(f) - S_G(f)}{(2p/N)^{1/2} S_G(f)} \stackrel{\mathrm{d}}{=} \begin{cases} \mathcal{N}(0,1), & \text{if } f \neq 0 \text{ or } \pm f_{(N)}; \\ \mathcal{N}(0,2), & \text{if } f = 0 \text{ or } \pm f_{(N)}. \end{cases} \qquad (449)$$

Comparison with the results from the asymptotic distribution of lag window spectral estimators (Section 6.10) reveals that, for $f \neq 0$ or $\pm f_{(N)}$, the value N/p in AR spectral estimation plays the role of ν, the equivalent number of degrees of freedom, in lag window spectral estimation.

A comparison of (448b) with Kromer's result (449) is not particularly easy. The reader is referred to Burshtein and Weinstein (1987, p. 508) for one possible approach, which leads to the conclusion that in both cases the width of the confidence interval at f is proportional to the spectral level at the frequency; however, the constant of proportionality is somewhat different in the two cases.

[2]   *Simultaneous* confidence intervals for $|G(f)|^{-2}$ can be derived by an approach similar to that leading to (448b). Previously, the distributional result (446d) for the 2×1 vector $\mathbf{B}$ led, through the Mann–Wald theorem, to the distributional result (446e). This time, we use (446a) for the $p \times 1$ vector $(\hat{\mathbf{\Phi}}_p - \mathbf{\Phi}_p)$ and the Mann–Wald theorem to obtain the asymptotic result:

$$\frac{N(\hat{\mathbf{\Phi}}_p - \mathbf{\Phi}_p)^T \hat{\Gamma}_p (\hat{\mathbf{\Phi}}_p - \mathbf{\Phi}_p)}{\hat{\sigma}_p^2} \stackrel{\mathrm{d}}{=} \chi_p^2.$$

The equivalent of (447b) is now

$$\left| \mathbf{a}^T (\hat{\mathbf{\Phi}}_p - \mathbf{\Phi}_p) \right|^2 \leq \mathbf{a}^T \mathcal{M}^{-1} \mathbf{a} \tag{450}$$

for all p-dimensional vectors $\mathbf{a}$ with $\mathcal{M} \equiv N\hat{\Gamma}_p/[\hat{\sigma}_p^2 Q_p(1 - \alpha)]$, where $Q_p(\alpha)$ is the $\alpha \times 100\%$ percentage point of the χ_p^2 distribution (see Table 256). We can take $\mathbf{a}_1^T = \mathbf{e}_{\Re}^T(f)$ and $\mathbf{a}_2^T = \mathbf{e}_{\Im}^T(f)$ for all f such that $0 < f < f_{(N)}$ since (450) holds for all p-dimensional vectors $\mathbf{a}$. We thus obtain a simultaneous confidence band for $|G(f)|^{-2}$, with at least a $1 - \alpha$ probability, of the form

$$\mathbf{P} \left[\left\{ |\hat{G}(f)| + r_p(f) \right\}^{-2} \leq |G(f)|^{-2} \leq \left\{ |\hat{G}(f)| - r_p(f) \right\}^{-2} \right] \geq 1 - \alpha,$$

where

$$r_p(f) \equiv [Q_p(1 - \alpha)\hat{\sigma}_p^2 \mathbf{e}^H(f)(\hat{\Gamma}_p^{-1}/N)\mathbf{e}(f)]^{1/2}.$$

9.13 Comments on Complex-Valued Time Series

We have assumed throughout this chapter that our time series of interest is real-valued. If we wish to fit an $\text{AR}(p)$ model to a complex-valued series that is a portion of a realization of a complex-valued stationary process $\{Z_t\}$ with zero mean, we need to make some modifications in the definitions of our various AR parameter estimators (the form of the $\text{AR}(p)$ model itself is the same as indicated by Equation (392a), except that $\phi_{j,p}$ is now in general complex-valued). Most of the changes are due to the following fundamental difference between real-valued and complex-valued processes. For a real-valued stationary process $\{X_t\}$ with acvs $\{s_{k,X}\}$, the time reversed process $\{X_{-t}\}$ also has acvs $\{s_{k,X}\}$, whereas, for a complex-valued process $\{Z_t\}$ with acvs $\{s_{k,Z}\}$, the *complex conjugate* time reversed process $\{Z_{-t}^*\}$ has acvs $\{s_{k,Z}\}$ (the proof is an easy exercise). One immediate implication is that, if the best linear predictor of Z_t, given $Z_{t-1}, \ldots, Z_{t-k}$, is

$$\overrightarrow{Z}_t(k) \equiv \sum_{j=1}^{k} \phi_{j,k} Z_{t-j},$$

then the best linear 'predictor' of Z_t, given $Z_{t+1}, \ldots, Z_{t+k}$, has the form

$$\overleftarrow{Z}_t(k) \equiv \sum_{j=1}^{k} \phi_{j,k}^* Z_{t+j}.$$

Given these facts, a rederivation of the Levinson–Durbin recursions shows that we must make the following minor changes to obtain the

proper Yule–Walker and Burg parameter estimators for complex-valued
time series. For the Yule–Walker estimator, Equations (404b) and (404c)
now become

$$\tilde{\phi}_{j,k} = \tilde{\phi}_{j,k-1} - \tilde{\phi}_{k,k}\tilde{\phi}^*_{k-j,k-1}, \quad 1 \le j \le k-1$$

$$\tilde{\sigma}^2_k = \tilde{\sigma}^2_{k-1}(1 - |\tilde{\phi}_{k,k}|^2)$$

(Equation (404a) remains the same). As before, we initialize the re-
cursions using $\tilde{\phi}_{1,1} = \hat{s}^{(p)}_1 / \hat{s}^{(p)}_0$, but now we set $\tilde{\sigma}^2_1 = \hat{s}^{(p)}_0 (1 - |\tilde{\phi}_{1,1}|^2)$.
In lieu of Equation (416d), the Burg estimator of the kth order partial
autocorrelation coefficient now takes the form

$$\bar{\phi}_{k,k} = \frac{2\sum_{t=k+1}^{N} \overrightarrow{e}_t(k-1) \overleftarrow{e}^*_{t-k}(k-1)}{\sum_{t=k+1}^{N} |\overrightarrow{e}_t(k-1)|^2 + |\overleftarrow{e}_{t-k}(k-1)|^2}.$$

The recursive formula for the forward prediction errors is still given by
Equation (415a), but the formula for the backward prediction errors is
now given by

$$\overleftarrow{e}_{t-k}(k) = \overleftarrow{e}_{t-k}(k-1) - \bar{\phi}^*_{k,k}\overrightarrow{e}_t(k-1), \qquad k+1 \le t \le N,$$

instead of by Equation (415b).

Finally, we note that

[1] the various least squares estimators are now obtained by minimizing
an appropriate sum of squared moduli of prediction errors rather
than sum of squares; and

[2] once we have obtained the AR parameter estimates, the parametric
spectral estimate is obtained in the same manner as in the real-
valued case, namely, by substituting the estimated parameters into
Equation (392b).

For further details on parametric spectral estimation using complex-
valued time series, see Kay (1988).

9.14 Summary of Parametric Spectral Analysis

We assume that $X_1, X_2, \ldots, X_N$ is a sample of length N from a real-
valued stationary process $\{X_t\}$ with zero mean and unknown sdf $S_X(\cdot)$
defined over the interval $[-f_{(N)}, f_{(N)}]$, where $f_{(N)} \equiv 1/(2\,\Delta t)$ is the
Nyquist frequency and Δt is the sampling interval between observations. (If $\{X_t\}$ has an unknown mean, then we need to replace X_t with
$X'_t \equiv X_t - \bar{X}$ in all computational formulae, where $\bar{X} \equiv \sum_{t=1}^{N} X_t/N$ is
the sample mean.) The first step in parametric spectral analysis is to
estimate the $p+1$ parameters – namely, $\phi_{1,p}, \ldots, \phi_{p,p}$ and σ^2_p – of an

AR(p) model (see Equation (392a)) using one of the following estimators (the choice of the order p is discussed in Section 9.9).

[1] *Yule–Walker estimators $\tilde{\phi}_{1,p}, \ldots, \tilde{\phi}_{p,p}$ and $\tilde{\sigma}_p^2$*

We first form the usual biased estimator $\{\hat{s}_k^{(p)}\}$ of the acvs using $X_1, \ldots, X_N$ (see Equation (196b)). With $\tilde{\sigma}_0^2 \equiv \hat{s}_0^{(p)}$, we obtain the Yule–Walker estimators by recursively computing, for $k = 1, \ldots, p$,

$$\tilde{\phi}_{k,k} = \frac{\hat{s}_k^{(p)} - \sum_{j=1}^{k-1} \tilde{\phi}_{j,k-1}\hat{s}_{k-j}^{(p)}}{\tilde{\sigma}_{k-1}^2}; \qquad \text{(see (404a))}$$

$$\tilde{\phi}_{j,k} = \tilde{\phi}_{j,k-1} - \tilde{\phi}_{k,k}\tilde{\phi}_{k-j,k-1}, \quad 1 \le j \le k-1; \quad \text{(see (404b))}$$

$$\tilde{\sigma}_k^2 = \tilde{\sigma}_{k-1}^2(1 - \tilde{\phi}_{k,k}^2) \qquad \text{(see (404c))}$$

(for $k = 1$, we obtain $\tilde{\phi}_{1,1} = \hat{s}_1^{(p)}/\hat{s}_0^{(p)}$ and $\tilde{\sigma}_1^2 = \hat{s}_0^{(p)}(1 - \tilde{\phi}_{1,1}^2)$). An equivalent formulation that makes direct use of $X_1, \ldots, X_N$ rather than $\{\hat{s}_k^{(p)}\}$ is outlined in the discussion surrounding Equation (420). A variation on the Yule–Walker method is to use the acvs estimator corresponding to a direct spectral estimator $\hat{S}^{(d)}(\cdot)$ other than the periodogram (see the discussion concerning Figures 407 and 408). The Yule–Walker parameter estimators are guaranteed to correspond to a stationary AR process.

[2] *Burg estimators $\bar{\phi}_{1,p}, \ldots, \bar{\phi}_{p,p}$ and $\bar{\sigma}_p^2$*

With $\overrightarrow{e}_t(0) \equiv X_t$, $\overleftarrow{e}_t(0) \equiv X_t$ and $\bar{\sigma}_0^2 \equiv \hat{s}_0^{(p)}$, we obtain the Burg estimators by recursively computing, for $k = 1, \ldots, p$,

$$A_k = \sum_{t=k+1}^{N} \overrightarrow{e}_t^{\,2}(k-1) + \overleftarrow{e}_{t-k}^{\,2}(k-1) \qquad \text{(see (416b))}$$

$$B_k = 2\sum_{t=k+1}^{N} \overrightarrow{e}_t(k-1)\overleftarrow{e}_{t-k}(k-1) \qquad \text{(see (416c))}$$

$$\bar{\phi}_{k,k} = B_k/A_k \qquad \text{(see (416d))}$$

$$\overrightarrow{e}_t(k) = \overrightarrow{e}_t(k-1) - \bar{\phi}_{k,k}\overleftarrow{e}_{t-k}(k-1), \quad k+1 \le t \le N; \quad \text{(see (415a))}$$

$$\overleftarrow{e}_{t-k}(k) = \overleftarrow{e}_{t-k}(k-1) - \bar{\phi}_{k,k}\overrightarrow{e}_t(k-1), \quad k+1 \le t \le N; \quad \text{(see (415b))}$$

$$\bar{\phi}_{j,k} = \bar{\phi}_{j,k-1} - \bar{\phi}_{k,k}\bar{\phi}_{k-j,k-1}, \quad 1 \le j \le k-1;$$

$$\bar{\sigma}_k^2 = \bar{\sigma}_{k-1}^2(1 - \bar{\phi}_{k,k}^2). \qquad \text{(see (416f))}$$

The term A_k can be recursively computed using Equation (454). The Burg parameter estimators are guaranteed to correspond to a stationary AR process and have been found to be superior to the Yule–Walker estimators in numerous Monte Carlo experiments.

[3] *Least squares (ls) estimators*

Here we use the resemblance between an AR(p) model and a regression model to formulate ls parameter estimators. As detailed in Section 9.7, there are several possible formulations, leading to the forward ls, backward ls and forward/backward ls estimators. Calculation of these estimators can be done using standard least squares techniques (Golub and Van Loan, 1989; Press *et al.*, 1986) or specialized algorithms that exploit the structure of the AR model (Marple, 1987; Kay, 1988). None of the three ls parameter estimators mentioned above is guaranteed to correspond to a stationary process. In terms of spectrum estimation, Monte Carlo studies show that the forward/backward ls estimator is generally superior to the Yule–Walker estimator and outperforms the Burg estimator in some cases.

[4] *Maximum likelihood estimators (mle's)*

Under the assumption of a Gaussian process, mle's for the AR parameters were derived in Section 9.8. Unfortunately, these estimators require the use of a nonlinear optimization routine and hence are computationally intensive compared to the other estimators mentioned above (particularly for large model orders p). The various ls estimators can be regarded as approximate mle's.

Let $\hat{\phi}_{1,p}, \ldots, \hat{\phi}_{p,p}$ and $\hat{\sigma}_p^2$ represent any of the above estimators. There are two ways we can use these estimators to produce an estimator of the sdf $S_X(\cdot)$.

[1] *Pure parametric approach*

Here we merely substitute the AR parameter estimators for the corresponding theoretical parameters in Equation (392b) to obtain the sdf estimator

$$\hat{S}_X(f) = \frac{\hat{\sigma}_p^2 \, \Delta t}{\left| 1 - \sum_{j=1}^p \hat{\phi}_{j,p} e^{-i2\pi f j \, \Delta t} \right|^2}, \qquad |f| \le f_{(N)}.$$

An approach for obtaining confidence intervals for the normalized sdf based upon this estimator is given in Section 9.12.

[2] *Prewhitening approach*

Here we use the estimated coefficients $\hat{\phi}_{1,p}, \ldots, \hat{\phi}_{p,p}$ to create a prewhitening filter for our time series. This combined parametric/nonparametric approach is discussed in detail in Section 9.10 – its features are generally more appealing than the pure parametric approach.

9.15 Exercises

[9.1] Here we prove a useful corollary to the orthogonality principle. Let $\{X_t\}$ be any stationary process with zero mean and with an acvs $\{s_k\}$ that is positive definite. Define

$$d_t(k) \equiv X_t - \sum_{l=1}^{k} \psi_l X_{t-l},$$

where the ψ_l terms are any set of constants. Suppose that

$$E\{d_t(k)X_{t-j}\} = 0, \qquad 1 \le j \le k;$$

i.e., the $d_t(k)$ terms mimic the property stated in Equation (398c) for $\overrightarrow{\epsilon}_t(k)$. Show that we must have $d_t(k) = \overrightarrow{\epsilon}_t(k)$. Hint: note that the above system of equations is identical to that in Equation (398d).

[9.2] a) Use Equations (402d) and (403) to show that $P_0 = s_0$; i.e., the zeroth-order mean square linear prediction error is just the process variance. Why is this a reasonable interpretation for P_0?

   b) Show how the Levinson–Durbin recursions can be initialized starting with $k = 0$ rather than $k = 1$ as is done in Equation (402d).

   c) Use part a) and Equation (403) to show that

$$P_k = s_0 \prod_{j=1}^{k} (1 - \phi_{j,j}^2).$$

[9.3] Verify Equation (409a), which states that $\phi_{k,k}$ can be interpreted as a correlation coefficient. Hint: express $\overleftarrow{\epsilon}_{t-k}(k-1)$ in terms of $\{X_t\}$; evoke the orthogonality principle (Equation (398c)); and use Equation (402a).

[9.4] Suppose that we are given the following acvs values: $s_0 = 3$, $s_1 = 2$ and $s_2 = 1$. Use the Levinson–Durbin recursions to show that $\phi_{1,1} = 2/3$, $\phi_{1,2} = 4/5$, $\phi_{2,2} = -1/5$, $\sigma_1^2 = 5/3$ and $\sigma_2^2 = 24/15$. Construct L_N^{-1} and D_N, and verify that $L_N^{-1}\Gamma_N L_N^{-T} = D_N$ (see Equation (412b); this example is taken from Therrien, 1983).

[9.5] a) Show that Equation (419a) holds (hint: use Equation (416a)).

   b) Show the related result that A_k of Equation (416b) obeys the following recursive relationship:

$$A_{k+1} = \left(1 - \bar{\phi}_{k,k}^2\right)\left(A_k - \overrightarrow{e}_k^{\,2}(k-1) - \overleftarrow{e}_{N-k+1}^{\,2}(k-1)\right)$$

$$(454)$$

(Andersen, 1978). This recursion is helpful in computing $\bar{\phi}_{k,k}$ in Equation (416d).

[9.6] Any one of the following three sets of $p + 1$ quantities completely characterizes the second-order properties of the AR(p) process of Equation (392a):

   i) $\phi_{1,p}, \phi_{2,p}, \ldots, \phi_{p,p}$ and s_0;

   ii) $\phi_{1,1}, \phi_{2,2}, \ldots, \phi_{p,p}$ and s_0;

   iii) $s_0, s_1, \ldots, s_p$.

Show that this assertion is true by showing how, given any one of these three sets, we can obtain the other two.

[9.7] Show how the forward/backward ls estimator of the AR(p) coefficients can be regarded as an approximation to the maximum likelihood estimator. Hint: starting with Equation (434), produce the analog of Equation (431d), and then use approximations similar to those that yielded the forward ls estimator as an approximation to the maximum likelihood estimator.

[9.8] Suppose that we wish to fit an AR(1) model to a time series that is a realization of a portion $X_1, X_2, \ldots, X_N$ of a stationary process with zero mean.

   a) Show that the Yule–Walker, Burg, forward ls, and backward ls estimators of the coefficient $\phi_{1,1}$ are given by, respectively,

$$\frac{\sum_{t=2}^{N} X_t X_{t-1}}{\sum_{t=1}^{N} X_t^2}; \quad \frac{2\sum_{t=2}^{N} X_t X_{t-1}}{X_1^2 + 2\sum_{t=2}^{N-1} X_t^2 + X_N^2};$$
$$\frac{\sum_{t=2}^{N} X_t X_{t-1}}{\sum_{t=2}^{N} X_t^2}; \quad \text{and} \quad \frac{\sum_{t=1}^{N-1} X_t X_{t+1}}{\sum_{t=1}^{N-1} X_t^2}.$$

   b) Show that the forward/backward ls estimator of $\phi_{1,1}$ is identical to that of the Burg estimator (this relationship does not hold for $p > 1$).

   c) It is instructive to consider the case $N = 2$ for the AR(1) estimators we found above (Jones, 1985, p. 226). Show that, if $X_1 \neq X_2$ the Yule–Walker estimator must be less than $1/2$ in magnitude; that the Burg estimator – and hence the forward/backward ls estimator also – must be less than 1 in magnitude; and that the forward ls estimator and the backward ls estimator can both be arbitrarily large in magnitude. With regard to the stationarity requirement for an AR(1) process, which ones of the five estimators does this example show to be most reasonable?

10

Harmonic Analysis

10.0 Introduction

If a stationary process has a purely continuous spectrum, it is natural to estimate its spectral density function (sdf) since this function is easier to interpret than the integrated spectrum. Estimation of the sdf has occupied our attention in Chapters 6, 7 and 9. However, if we are given a sample of a time series drawn from a process with a purely discrete spectrum (i.e., a 'line' spectrum for which the integrated spectrum is a step function), our estimation problem is quite different: we must estimate the location and magnitude of the jumps in the integrated spectrum. This requires estimation techniques that differ – to some degree at least – from what we have already studied. It is more common, however, to come across processes whose spectra are a mixture of lines and an sdf stemming from a so-called 'background continuum.' In Section 4.4 we distinguished two cases. If the sdf for the continuum is that of white noise, we said that the process has a discrete spectrum – as opposed to a *purely* discrete spectrum, which has only a line component; on the other hand, if the sdf for the continuum differs from that of white noise (sometimes called 'colored' noise), we said that the process has a mixed spectrum (see Figures 142 and 143).

In this chapter we shall use some standard concepts from tidal analysis to motivate and illustrate these models. We shall begin with a discrete parameter harmonic process that has a purely discrete spectrum.

10.1 Harmonic Processes with Purely Discrete Spectra

A real-valued discrete parameter harmonic process with a purely discrete spectrum can be written as

$$X_t = \mu + \sum_{l=1}^{L} D_l \cos\left(2\pi f_l t \, \Delta t + \phi_l\right), \qquad (457a)$$

where μ, D_l and f_l are in general unknown real-valued constants; $L \geq 1$ is an integer; and the ϕ_l terms are independent real-valued random variables (rv's) – representing random phase angles – with a rectangular (uniform) distribution on $[-\pi, \pi]$ (this is Equation (46b) with the insertion of the sampling time Δt so that the frequencies f_l have physically meaningful units of cycles per unit time). Since $\{X_t\}$ is a discrete parameter process, we can consider each f_l to lie in the interval $[-f_{(N)}, f_{(N)}]$ due to the aliasing effect, where $f_{(N)} \equiv 1/(2\,\Delta t)$ is the Nyquist frequency (see Sections 3.8 and 4.5).

This model is in one sense not really a statistical model: for each realization of the process the ϕ_l terms are fixed quantities, so each realization is essentially deterministic and free from what is commonly thought of as random variation. For this reason the process is often dismissed as of no interest; however, its simple properties can prove very useful in practice as we will demonstrate by looking at ocean tide prediction.

The theory of gravitational tidal potential predicts that at a fixed location the hourly height of the ocean tide at the hour with index t can be written as

$$X_t = \mu + \sum_{l=1}^{L'} \alpha_{t,l} H_l \cos\left(2\pi f_l t \, \Delta t + v_{t,l} + u_{t,l} - g_l\right) \text{ with } \Delta t = 1 \text{ hour,}$$
$$(457b)$$

where the sum is over L' primary tidal constituent frequencies, and the harmonic constants H_l and g_l depend only upon the location of interest and must be estimated. (These primary constituent frequencies derive from consideration of a frictionless and inertia-free ocean covering the whole globe; this crude model is inadequate for the real ocean, various hydrographic effects such as the locations of land and rivers making it necessary to introduce the location-dependent amplitudes H_l and phases g_l.) The $\alpha_{t,l}$ and $u_{t,l}$ terms are the so-called nodal parameters (these are essentially corrections for the amplitudes H_l and phases g_l), and $v_{t,l}$ is the phase of the lth constituent of the potential over Greenwich at $t = 0$ (these three quantities can be found from, e.g., Doodson and Warburg, 1941). For locations such as Honolulu, where the tides are virtually 'linear' (Munk and Cartwright, 1966), the above development is sufficient, but at many British ports and elsewhere the distortion

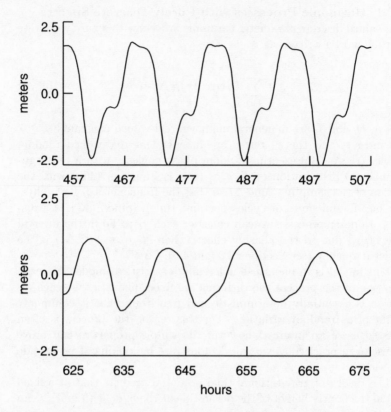

Figure 458. Predicted spring and neap tidal curves at Southampton (January, 1977). Tidal elevations relative to the mean sea level (in meters) are plotted versus an hour count starting from the beginning of 1977. The upper and lower plots show, respectively, spring and neap tides.

of tides by shallow-water effects often cannot be ignored. Nonlinear interactions are incorporated into harmonic tidal prediction by including constituents with frequencies that are sums and differences of the L' primary constituents. The total number of constituents required, say L, can exceed 100 in complicated shallow water areas such as Anchorage, Alaska, or Southampton, England.

In practice the terms of known form, namely the nodal parameters $\alpha_{t,l}$ and $u_{t,l}$ – which are continuously but slowly varying – are updated at fixed periods when computing $\{X_t\}$. Often sufficient accuracy is obtained by regarding them as fixed over a span of a year and updating them annually, although every 60 days is preferred (whichever span is used, the $v_{t,l}$ terms are also updated by 'resetting' the index t to 0 at the start of each span). This implies that in Equation (457b) both the amplitudes and phases of the cosines vary slowly and slightly with time.

This is a good example of what Thomson (1982) called the 'convenient fiction' of pure line components. To illustrate the complicated forms that can occur, predicted heights of the tide at Southampton, England, are given in Figure 458 for two different segments of time in 1977. This uses $L = 102$ frequencies and 60 day updating of $\alpha_{t,l}$, $u_{t,l}$ and $v_{t,l}$ (see Walden, 1982).

With these considerations in mind, we can approximate Equation (457b) by (457a); i.e., we assume the amplitudes $D_l \equiv \alpha_{t,l} H_l$ and phases $\phi_l \equiv v_{t,l} + u_{t,l} - g_l$ are both constant – at least over a one year period – and that the phases are realizations of independent rectangularly distributed rv's. Since Equation (457a) consists in this case of $L = 102$ components, it would seem reasonable to expect X_t to be Gaussian distributed from some version of the central limit theorem. However, this theorem only applies if no single component is dominant. Since for British ports the main semidiurnal constituent (known as M_2) is in fact dominant, Walden and Prescott (1983) suggested instead that the dominance of M_2 should be reflected in the model by writing

$$X_t = \mu + D_1 \cos\left(2\pi f_1 t \,\Delta t + \phi_1\right) + \sum_{l=2}^{L} D_l \cos\left(2\pi f_l t \,\Delta t + \phi_l\right), \quad (459a)$$

where D_1, f_1 and ϕ_1 are the amplitude, frequency and phase of the M_2 constituent. The probability density function of X_t is (Walden and Prescott, 1983)

$$f(x) = \frac{1}{\pi \left(2\pi\psi_0\right)^{1/2}} \int_0^{\pi} \exp\left(-\left[x - D_1 \cos\left(\lambda\right)\right]^2 / 2\psi_0\right) \, d\lambda, \quad (459b)$$

where

$$\psi_0 \equiv \mathrm{var}\left\{\sum_{l=2}^{L} D_l \cos\left(2\pi f_l t \,\Delta t + \phi_l\right)\right\} = \sum_{l=2}^{L} D_l^2 / 2 = \sum_{l=2}^{L} \left(\alpha_{t,l} H_l\right)^2 / 2$$

(see Equation (47b)).

The top plot of Figure 460 compares the theoretical frequency of hourly tidal heights predicted using Equation (459b) with a histogram of X_t values constructed using Equation (457b) for Newlyn, Cornwall, England, in 1979. Both the tails and the modes of the histogram are well fitted by the model density function. By way of contrast, the bottom plot shows the same for Portsmouth, England in 1985. Here the extremes and modes are poorly modeled. The reason for the poorer agreement for Portsmouth is thought to be due to the fact that in practice higher harmonics are correlated with their fundamentals (i.e., the phases are *not* independent). Since the tides at Portsmouth are more affected by harmonics (i.e., their variances contribute more to the total), this effect is more important there than at Newlyn. This amply demonstrates the subtleties that can occur when statistical assumptions are not wholly valid.

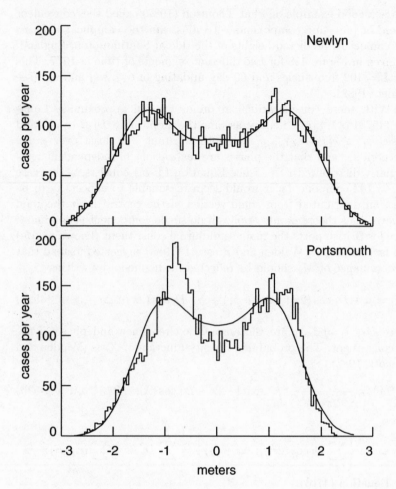

Figure 460. Comparison of model frequency curves and empirical frequency histograms for Newlyn (top plot) and Portsmouth (bottom). The numbers of expected and observed cases occurring in a year are plotted versus the elevations of the tides relative to the mean sea level in meters. In both plots the smooth and jagged curves represent, respectively, the model frequency curve and the empirical frequency histogram.

10.2 Harmonic Processes with Discrete Spectra

A more realistic and useful model than (457a) is given by

$$X_t = \mu + \sum_{l=1}^{L} D_l \cos\left(2\pi f_l t \, \Delta t + \phi_l\right) + \epsilon_t, \tag{460}$$

which differs from (457a) only in that ϵ_t has been added to represent observational error (always present in any sort of physical measurements).

The background continuum $\{\epsilon_t\}$ is assumed to be a real-valued white noise process with zero mean and variance σ_ϵ^2. It is also assumed to be independent of the ϕ_l terms. Equation (460) now describes a true statistical model.

We believe that Equation (460) arises more from analytical convenience than realism – many measuring instruments are band-limited, so that $\{\epsilon_t\}$ is not truly white. We looked at predicted tidal heights in the previous section. Observed tidal heights – i.e., predictions plus noise – certainly have a mixed spectrum with colored rather than white noise (see Section 10.4). However, as most analysis has been done using model (460), we shall also concentrate on it, since – if nothing else – it is an approximation to the mixed process case.

Let us assume initially that $L = 1$ and that the frequency $f \equiv f_1$ is known. The unknown parameters of model (460) are now just $D \equiv D_1$ and σ_ϵ^2. The model can then be rewritten as

$$X_t = \mu + A\cos\left(2\pi ft\,\Delta t\right) + B\sin\left(2\pi ft\,\Delta t\right) + \epsilon_t, \qquad (461a)$$

where $A \equiv D\cos\left(\phi_1\right)$ and $B \equiv -D\sin\left(\phi_1\right)$. Note that, for a given realization, A and B are just constants so that (461a) is a linear regression model. For a time series of length N that is assumed to be a realization of $X_1, X_2, \ldots, X_N$, we can estimate μ, A and B by least squares (Miller, 1973; Bloomfield, 1976). For real sinusoids, this amounts to minimizing

$$\begin{aligned}
SS(\mu, A, B) &\equiv \|\mathbf{X} - H\beta\|^2 \\
&= \sum_{t=1}^{N}\left(X_t - \mu - A\cos\left(2\pi ft\,\Delta t\right) - B\sin\left(2\pi ft\,\Delta t\right)\right)^2,
\end{aligned}$$

$$(461b)$$

where $\mathbf{X}^T \equiv [X_1, X_2, \ldots, X_N]$; $\beta^T \equiv [\mu, A, B]$;

$$H^T \equiv \begin{bmatrix} 1 & 1 & \cdots & 1 \\ \cos\left(2\pi f\,\Delta t\right) & \cos\left(4\pi f\,\Delta t\right) & \cdots & \cos\left(2N\pi f\,\Delta t\right) \\ \sin\left(2\pi f\,\Delta t\right) & \sin\left(4\pi f\,\Delta t\right) & \cdots & \sin\left(2N\pi f\,\Delta t\right) \end{bmatrix};$$

and $\|\mathbf{V}\|$ refers to the Euclidean norm of the vector $\mathbf{V}$. Since

$$\frac{\partial SS}{\partial \beta} = -H^T\left(\mathbf{X} - H\beta\right),$$

we can set the above equal to the zero vector to obtain the following normal equations (here $\hat{\mu}$, $\hat{A}$ and $\hat{B}$ represent the solutions to these equations):

$$\sum_{t=1}^{N} X_t = N\hat{\mu} + \hat{A}\sum_{t=1}^{N}\cos\left(2\pi ft\,\Delta t\right) + \hat{B}\sum_{t=1}^{N}\sin\left(2\pi ft\,\Delta t\right);$$

$$\sum_{t=1}^{N} X_t \cos\left(2\pi f t\,\Delta t\right) = \hat\mu \sum_{t=1}^{N} \cos\left(2\pi f t\,\Delta t\right) + \hat A \sum_{t=1}^{N} \cos^2\left(2\pi f t\,\Delta t\right)$$

$$+ \hat B \sum_{t=1}^{N} \cos\left(2\pi f t\,\Delta t\right)\sin\left(2\pi f t\,\Delta t\right); \qquad (462a)$$

$$\sum_{t=1}^{N} X_t \sin\left(2\pi f t\,\Delta t\right) = \hat\mu \sum_{t=1}^{N} \sin\left(2\pi f t\,\Delta t\right) + \hat B \sum_{t=1}^{N} \sin^2\left(2\pi f t\,\Delta t\right)$$

$$+ \hat A \sum_{t=1}^{N} \cos\left(2\pi f t\,\Delta t\right)\sin\left(2\pi f t\,\Delta t\right). \qquad (462b)$$

We can now use Exercises [1.3c] and [1.4] to reduce these three equations somewhat. For example, Equation (462a) becomes

$$\sum_{t=1}^{N} X_t \cos\left(2\pi f t\,\Delta t\right) = \hat\mu \frac{\cos\left([N+1]\pi f\,\Delta t\right)\sin\left(N\pi f\,\Delta t\right)}{\sin\left(\pi f\,\Delta t\right)}$$

$$+ \hat A \left\{ \frac{N}{2} + \frac{\sin\left(2N\pi f\,\Delta t\right)\cos\left(2[N+1]\pi f\,\Delta t\right)}{2\sin\left(2\pi f\,\Delta t\right)} \right\}$$

$$+ \hat B \left\{ \frac{\sin\left(2N\pi f\,\Delta t\right)\sin\left(2[N+1]\pi f\,\Delta t\right)}{2\sin\left(2\pi f\,\Delta t\right)} \right\},$$

with similar expressions for the other two normal equations.

It can be argued (see Exercise [10.1]) that the three expressions

$$\left|\hat\mu - \bar X\right|, \ \left|\hat A - \frac{2}{N}\sum_{t=1}^{N} X_t \cos\left(2\pi f t\,\Delta t\right)\right|, \ \left|\hat B - \frac{2}{N}\sum_{t=1}^{N} X_t \sin\left(2\pi f t\,\Delta t\right)\right|$$

are all close to zero (at least for large N), where, as usual, $\bar X \equiv \sum_{t=1}^{N} X_t/N$, the sample mean. Thus we have, to a good approximation,

$$\hat\mu \approx \bar X; \ \hat A \approx \frac{2}{N}\sum_{t=1}^{N} X_t \cos\left(2\pi f t\,\Delta t\right); \ \hat B \approx \frac{2}{N}\sum_{t=1}^{N} X_t \sin\left(2\pi f t\,\Delta t\right).$$

$$(462c)$$

If in fact $f = k/(N\,\Delta t)$, where k is an integer such that $1 \leq k < N/2$, it follows from Exercise [1.4] that the approximations in (462c) are in fact equalities. As we have noted in Chapters 3 and 6, this special set of frequencies – along with the frequencies 0 and $f_{(N)}$ if N is even – is known as the *Fourier frequencies* or *standard frequencies*. Since any frequency f in the interval $[-f_{(N)}, f_{(N)}]$ is at most a distance of $1/(N\,\Delta t)$ away from a Fourier frequency not equal to 0 or $f_{(N)}$, it is intuitively reasonable that the approximations in (462c) are good.

For the remainder of this section, we shall take $\mu = 0$ and introduce several frequencies so that our model is

$$X_t = \sum_{l=1}^{L} \left(A_l \cos\left(2\pi f_l t\, \Delta t\right) + B_l \sin\left(2\pi f_l t\, \Delta t\right)\right) + \epsilon_t, \qquad (463a)$$

where we assume for the moment that each f_l is one of the Fourier frequencies not equal to 0 or $f_{(N)}$. The deductions already made are equally valid here, and so

$$\hat{A}_l = \frac{2}{N} \sum_{t=1}^{N} X_t \cos\left(2\pi f_l t\, \Delta t\right) \text{ and } \hat{B}_l = \frac{2}{N} \sum_{t=1}^{N} X_t \sin\left(2\pi f_l t\, \Delta t\right)$$
$$(463b)$$

are the *exact* least squares estimators of A_l and B_l, respectively. Let us consider some of the statistical properties of these estimators *under the assumption that each ϕ_l is a constant* – this implies that A_l and B_l are constants and hence that Equation (463a) is a standard multiple linear regression model. First, note that

$$E\{\hat{A}_l\} = \frac{2}{N} \sum_{t=1}^{N} E\{X_t\} \cos\left(2\pi f_l t\, \Delta t\right)$$

$$= \frac{2}{N} \sum_{t=1}^{N} \sum_{k=1}^{L} \Big(A_k \cos\left(2\pi f_k t\, \Delta t\right) \cos\left(2\pi f_l t\, \Delta t\right)$$

$$+ B_k \sin\left(2\pi f_k t\, \Delta t\right) \cos\left(2\pi f_l t\, \Delta t\right)\Big)$$

since $E\{\epsilon_t\} = 0$. By interchanging the summations and using the orthogonality relationships

$$\sum_{t=1}^{N} \cos\left(2\pi f_k t\, \Delta t\right) \cos\left(2\pi f_l t\, \Delta t\right) = \sum_{t=1}^{N} \sin\left(2\pi f_k t\, \Delta t\right) \cos\left(2\pi f_l t\, \Delta t\right) = 0,$$

$$\sum_{t=1}^{N} \cos^2\left(2\pi f_l t\, \Delta t\right) = N/2,$$

$k \neq l$, we find that

$$E\{\hat{A}_l\} = A_l \text{ and, likewise, } E\{\hat{B}_l\} = B_l.$$

From the definitions of $\hat{A}_l$ and $\hat{B}_l$ and the fact that the X_t terms are uncorrelated, we have

$$\operatorname{var}\{\hat{A}_l\} = \frac{4}{N^2} \sum_{t=1}^{N} \operatorname{var}\{X_t\} \cos^2\left(2\pi f_l t\, \Delta t\right) = \frac{2\sigma_\epsilon^2}{N} = \operatorname{var}\{\hat{B}_l\}$$

when each ϕ_l is fixed. It also follows from the orthogonality relationships that, for $k \neq l$,

$$\text{cov}\{\hat{A}_k, \hat{B}_l\} = \text{cov}\{\hat{A}_l, \hat{B}_l\} = \text{cov}\{\hat{A}_k, \hat{A}_l\} = \text{cov}\{\hat{B}_k, \hat{B}_l\} = 0.$$

We can estimate σ_ϵ^2 by the usual formula used in multiple linear regression, namely,

$$\hat{\sigma}_\epsilon^2 = \frac{1}{N - 2L} \sum_{t=1}^{N} \left[X_t - \sum_{l=1}^{L} \left(\hat{A}_l \cos\left(2\pi f_l t \, \Delta t\right) + \hat{B}_l \sin\left(2\pi f_l t \, \Delta t\right) \right) \right]^2$$

$$= \frac{1}{N - 2L} \left\| \mathbf{X} - H\hat{\beta} \right\|^2, \tag{464}$$

where now

$$H^T \equiv \begin{bmatrix} \cos\left(2\pi f_1 \Delta t\right) & \cos\left(4\pi f_1 \Delta t\right) & \dots & \cos\left(2N\pi f_1 \Delta t\right) \\ \sin\left(2\pi f_1 \Delta t\right) & \sin\left(4\pi f_1 \Delta t\right) & \dots & \sin\left(2N\pi f_1 \Delta t\right) \\ \vdots & \vdots & \ddots & \vdots \\ \cos\left(2\pi f_L \Delta t\right) & \cos\left(4\pi f_L \Delta t\right) & \dots & \cos\left(2N\pi f_L \Delta t\right) \\ \sin\left(2\pi f_L \Delta t\right) & \sin\left(4\pi f_L \Delta t\right) & \dots & \sin\left(2N\pi f_L \Delta t\right) \end{bmatrix}$$

and $\hat{\beta}^T \equiv \left[\hat{A}_1, \hat{B}_1, \dots, \hat{A}_L, \hat{B}_L \right]$. The divisor $N - 2L$ is due to the fact that we have estimated $2L$ parameters from the data.

If the frequencies f_l are not all of the form $k/(N \, \Delta t)$, the estimators $\hat{A}_l$ and $\hat{B}_l$ can be regarded as approximate least squares estimates of A_l and B_l. It can be shown that in this case

$$E\{\hat{A}_l\} = A_l + O\left(\frac{1}{N}\right) \quad \text{and} \quad E\{\hat{B}_l\} = B_l + O\left(\frac{1}{N}\right),$$

which means that, for example, there exists a constant c (independent of N) such that

$$\left| E\{\hat{A}_l\} - A_l \right| < \frac{c}{N}$$

for all N (note, however, that we do not claim that c is a small number!).

To summarize, in order to estimate A_l and B_l, we have assumed that each ϕ_l is fixed for a realization and have derived some statistical properties of the corresponding estimators ($\hat{A}_l$ and $\hat{B}_l$). However, with respect to a spectral representation, we have done two strange things: first, we have let $E\{X_t\}$ vary with time so that $\{X_t\}$ is nonstationary, and, second, we have treated

$$\sum_{l=1}^{L} A_l \cos\left(2\pi f_l t \, \Delta t\right) + B_l \sin\left(2\pi f_l t \, \Delta t\right)$$

as deterministic so that $\{X_t\}$ is an uncorrelated process; i.e., the only stochastic part is $\{\epsilon_t\}$. These apparent contradictions are resolved by reintroducing the variation of the ϕ_l terms when thinking about spectral estimation rather than just amplitude estimation.

Comments and Extensions to Section 10.2

[1]   Here we will look at the complex-valued counterpart of Equation (460), namely,

$$Z_t = \mu + \sum_{l=1}^{L} D'_l e^{i(2\pi f_l t \, \Delta t + \phi_l)} + \epsilon_t, \qquad (465a)$$

where now $\{Z_t\}$ and μ are both complex-valued and $\{\epsilon_t\}$ is a complex-valued white noise process with zero mean and variance σ_ϵ^2 (note, however, that D'_l is still a real-valued amplitude). Although the representation (460) is much more useful for physical applications, the complex-valued form has three advantages:

  a) it leads to more compact mathematical expressions;
  b) it brings out the connection with the periodogram very clearly; and
  c) it is the form most often used in the electrical engineering literature.

It is worth pointing out that, for a complex-valued zero mean random variable such as ϵ_t in (465a), $\text{var}\{\epsilon_t\} = E\{|\epsilon_t|^2\}$. To obtain σ_ϵ^2 it is usually specified that the variance of the real and imaginary parts be $\sigma_\epsilon^2/2$ (this latter quantity is sometimes called the *semivariance* – see, for example, Lang and McClellan, 1980).

The equivalent of model (461a) is now

$$Z_t = \mu + C e^{i2\pi f t \, \Delta t} + \epsilon_t, \qquad (465b)$$

where $C \equiv D' \exp{(i\phi)}$ is the complex-valued amplitude of the complex exponential $\exp{(i2\pi f t \, \Delta t)}$. Again, if we regard ϕ as a constant for a given realization, we can estimate the parameters μ and C by least squares:

$$\text{SS}(\mu, C) = \|\mathbf{Z} - H\beta\|^2 = \sum_{t=1}^{N} \left| Z_t - \mu - C e^{i2\pi f t \, \Delta t} \right|^2,$$

where $\mathbf{Z}^T \equiv [Z_1, Z_2, \ldots, Z_N]$; $\beta^T \equiv [\mu, C]$; and

$$H^T \equiv \begin{bmatrix} 1 & 1 & \cdots & 1 \\ e^{i2\pi f \, \Delta t} & e^{i4\pi f \, \Delta t} & \cdots & e^{i2N\pi f \, \Delta t} \end{bmatrix}.$$

If we set

$$\frac{\partial \text{SS}}{\partial \beta} = -H^T \left(\mathbf{Z}^* - H^* \beta^* \right)$$

to zero and take the complex conjugate of the resulting expressions (the asterisk denotes this operation), we obtain the following normal equations:

$$\sum_{t=1}^{N} Z_t = N\hat{\mu} + \hat{C} \sum_{t=1}^{N} e^{i2\pi ft\,\Delta t},$$

$$\sum_{t=1}^{N} Z_t e^{-i2\pi ft\,\Delta t} = \hat{\mu} \sum_{t=1}^{N} e^{-i2\pi ft\,\Delta t} + N\hat{C}.$$

From Exercise [1.3c], we know that

$$\sum_{t=1}^{N} e^{i2\pi f_k t\,\Delta t} = 0 \text{ for } f_k \equiv \frac{k}{N\,\Delta t}, \quad 1 \le k < N.$$

If we restrict f to this set of frequencies, the following are exact least squares estimators:

$$\hat{\mu} = \frac{1}{N} \sum_{t=1}^{N} Z_t \text{ and } \hat{C} = \frac{1}{N} \sum_{t=1}^{N} Z_t e^{-i2\pi ft\,\Delta t}. \tag{466a}$$

Note that, if $f = 0$, $\hat{C}$ reduces to $\hat{\mu}$; i.e., C refers to the amplitude associated with zero frequency as does μ (this is sometimes called the 'DC' – 'direct current' – component in the electrical engineering literature).

If we take $\mu = 0$ and introduce several frequencies, our complex-valued model now becomes

$$Z_t = \sum_{l=1}^{L} C_l e^{i2\pi f_l t\,\Delta t} + \epsilon_t, \tag{466b}$$

where we assume that each f_l is one of the Fourier frequencies not equal to 0 or $f_{(N)}$. Under the same assumptions and restrictions as for their real sinusoidal counterpart, the following results hold:

$$E\{\hat{C}_l\} = C_l; \quad \text{var}\{\hat{C}_l\} = \frac{\sigma_\epsilon^2}{N}; \quad \text{and} \quad \text{cov}\{\hat{C}_k, \hat{C}_l\} = 0, \quad k \ne l,$$

where $\hat{C}_l$ is the least squares estimate of C_l (see Exercise [10.2]). In order to estimate σ_ϵ^2, we use the complex analog of the usual formula in linear regression (Miller, 1973, Theorem 7.3):

$$\hat{\sigma}_\epsilon^2 = \frac{1}{N-L} \left\| \mathbf{Z} - H\hat{\beta} \right\|^2, \tag{466c}$$

where

$$H^T \equiv \begin{bmatrix} e^{i2\pi f_1\,\Delta t} & e^{i4\pi f_1\,\Delta t} & \cdots & e^{i2N\pi f_1\,\Delta t} \\ \vdots & \vdots & \ddots & \vdots \\ e^{i2\pi f_L\,\Delta t} & e^{i4\pi f_L\,\Delta t} & \cdots & e^{i2N\pi f_L\,\Delta t} \end{bmatrix}$$

and $\hat{\beta}^T \equiv \left[\hat{C}_1, \hat{C}_2, \ldots, \hat{C}_L \right]$. The divisor $N - L$ is due to the fact that we have estimated L complex-valued parameters C_l from our complex-valued data.

10.3 Spectral Representation of Discrete and Mixed Spectra

We know from the Lebesgue decomposition theorem for integrated spectra (Section 4.4) that a combination of spectral lines plus white noise (a 'discrete' spectrum) or a combination of spectral lines plus colored noise (a 'mixed' spectrum) will have an integrated spectrum $S^{(I)}(\cdot)$ that can be written as

$$S^{(I)}(f) = S_1^{(I)}(f) + S_2^{(I)}(f),$$

where $S_1^{(I)}(\cdot)$ is absolutely continuous and $S_2^{(I)}(\cdot)$ is a step function. Consider the real-valued process

$$X_t = \sum_{l=1}^{L} D_l \cos\left(2\pi f_l t\, \Delta t + \phi_l\right) + \eta_t, \qquad (467a)$$

which is the same as model (460) with two substitutions: (1) the process mean μ is set to 0, and (2) the process $\{\eta_t\}$ is not necessarily white noise, but it is independent of each ϕ_l. The spectral representation theorem states that

$$X_t = \int_{-f_{(N)}}^{f_{(N)}} e^{i2\pi f t\, \Delta t}\, dZ(f)$$

$$= \int_{-f_{(N)}}^{f_{(N)}} e^{i2\pi f t\, \Delta t}\, dZ_1(f) + \int_{-f_{(N)}}^{f_{(N)}} e^{i2\pi f t\, \Delta t}\, dZ_2(f),$$

where $\{Z_1(f)\}$ and $\{Z_2(f)\}$ are each orthogonal processes;

$$E\{|dZ_1(f)|^2\} = S_\eta(f)\, df \quad \text{and} \quad E\{|dZ_2(f)|^2\} = S_2^{(I)}(f);$$

$S_\eta(\cdot)$ is the sdf of $\{\eta_t\}$; and $E\{dZ_1^*(f)dZ_2(f')\} = 0$ for all f and f'. Recall from Section 4.1 that by putting

$$C_l \equiv D_l e^{i\phi_l}/2 \quad \text{and} \quad C_{-l} \equiv C_l^*, \qquad l = 1, \ldots, L,$$

we can write

$$X_t = \sum_{l=-L}^{L} C_l e^{i2\pi f_l t\, \Delta t} + \eta_t, \qquad (467b)$$

where $C_0 \equiv 0$; $f_0 \equiv 0$; and $f_{-l} = -f_l$. Furthermore, $E\{X_t\} = 0$ and

$$\text{var}\,\{X_t\} = \sum_{l=-L}^{L} E\{|C_l|^2\} + \text{var}\,\{\eta_t\} = \sum_{l=-L}^{L} D_l^2/4 + \sigma_\eta^2.$$

Corresponding to this result we have

$$E\{|dZ_2(f)|^2\} = \begin{cases} E\{|C_l|^2\} = D_l^2/4, & f = \pm f_l; \\ 0, & f \neq \pm f_l. \end{cases}$$

It follows from the argument leading to Equation (47a) that

$$\text{cov}\,\{X_t, X_{t+\tau}\} = \sum_{l=1}^{L} D_l^2 \cos\left(2\pi f_l \tau\, \Delta t\right)/2 + \text{cov}\,\{\eta_t, \eta_{t+\tau}\}. \qquad (467c)$$

Comments and Extensions to Section 10.3

[1]   For complex exponentials, we have

$$Z_t = \sum_{l=1}^{L} D_l' e^{i(2\pi f_l t \,\Delta t + \phi_l)} + \eta_t$$

(this complex-valued process $\{Z_t\}$ should not be confused with the orthogonal increments process $\{dZ(\cdot)\}$ – the notation is unfortunately similar). This is the same as (465a) except that (1) the process mean $\mu = 0$ and (2) the complex-valued process $\{\eta_t\}$ need not be a white noise process. If we set $C_l \equiv D_l' \exp(i\phi_l)$ so that the above becomes

$$Z_t = \sum_{l=1}^{L} C_l e^{i2\pi f_l t \,\Delta t} + \eta_t, \qquad (468a)$$

we see that (468a) is a one-sided version of (467b); i.e., l is only positive in (468a). Thus $E\{Z_t\} = 0$ and

$$\operatorname{var}\{Z_t\} = \sum_{l=1}^{L} E\{|C_l|^2\} + \operatorname{var}\{\eta_t\} = \sum_{l=1}^{L}(D_l')^2 + \sigma_\eta^2.$$

For this complex-valued model we also have

$$E\{|dZ_2(f)|^2\} = \begin{cases} E\{|C_l|^2\} = (D_l')^2, & f = f_l; \\ 0, & f \neq f_l, \end{cases}$$

and

$$\operatorname{cov}\{Z_t, Z_{t+\tau}\} = \sum_{l=1}^{L}(D_l')^2 e^{i2\pi f_l \tau \,\Delta t} + \operatorname{cov}\{\eta_t, \eta_{t+\tau}\}. \qquad (468b)$$

Proof of this final result is the subject of Exercise [10.3].

10.4   An Example from Tidal Analysis

To illustrate some of the ideas introduced so far in this chapter, we will look at the observed heights of sea levels as recorded by tide gauges. It should be noted that sea level as defined here excludes the effect of individual waves – tidal gauges do *not* measure the instantaneous height of the sea at any particular time but rather the average level about which the waves are oscillating. We can thus write

$$Y_t = X_t + \eta_t,$$

where Y_t is the observed height of the sea level at time t; X_t is the predicted height as given by (457b) – with L' increased to L – under the assumption that $\alpha_l \equiv \alpha_{t,l}$, $v_l \equiv v_{t,l}$ and $u_l \equiv u_{t,l}$ are varying slowly enough that they can be assumed to be constant over the span of data under analysis; and η_t is the unexplained component. Thus (Murray, 1964, 1965)

$$
\begin{aligned}
Y_t &= \mu + \sum_{l=1}^{L} \alpha_l H_l \cos\left(2\pi f_l t\, \Delta t + v_l + u_l - g_l\right) + \eta_t \\[4pt]
&= \mu + \sum_{l=1}^{L} \alpha_l H_l \big\{\cos\left(g_l\right) \cos\left(2\pi f_l t\, \Delta t + v_l + u_l\right) \\
&\qquad\qquad + \sin\left(g_l\right) \sin\left(2\pi f_l t\, \Delta t + v_l + u_l\right)\big\} + \eta_t \\[4pt]
&= \mu + \sum_{l=1}^{L} H_l \cos\left(g_l\right) \left\{\alpha_l \cos\left(2\pi f_l t\, \Delta t + v_l + u_l\right)\right\} \\
&\qquad\qquad + H_l \sin\left(g_l\right) \left\{\alpha_l \sin\left(2\pi f_l t\, \Delta t + v_l + u_l\right)\right\} + \eta_t \\[4pt]
&= \mu + \sum_{l=1}^{L} A_l \left\{\alpha_l \cos\left(2\pi f_l t\, \Delta t + v_l + u_l\right)\right\} \\
&\qquad\qquad + B_l \left\{\alpha_l \sin\left(2\pi f_l t\, \Delta t + v_l + u_l\right)\right\} + \eta_t,
\end{aligned}
$$

where $A_l \equiv H_l \cos\left(g_l\right)$ and $B_l \equiv H_l \sin\left(g_l\right)$. This is a slightly more complicated version of (463a), and the process $\{\eta_t\}$ need not be white noise. The unknowns H_l and g_l can be found from A_l and B_l. The frequencies f_l are not in general related to the Fourier frequencies. The vertical lines in the top plot of Figure 470 indicate the locations of the $L = 102$ frequencies used in the model – the frequency $M_1 = 0.0402557$ cycle/hour and its multiples $M_k = k \cdot M_1$ for $k = 2, 3, 4, 5, 6$ and 8 are indicated by thicker and longer lines than the other 95 frequencies.

It is convenient to write

$$
Y_t = \sum_{l=0}^{2L} \beta_l \theta_{l,t} + \eta_t,
$$

where

$$
\beta_0 \equiv \mu; \quad \beta_{2l-1} \equiv H_l \cos\left(g_l\right); \quad \beta_{2l} \equiv H_l \sin\left(g_l\right);
$$
$$
\theta_{0,t} \equiv 1; \quad \theta_{2l-1,t} = \alpha_l \cos\left(2\pi f_l t\, \Delta t + v_l + u_l\right);
$$
$$
\theta_{2l,t} = \alpha_l \sin\left(2\pi f_l t\, \Delta t + v_l + u_l\right).
$$

Least squares then gives

$$
\sum_t Y_t \theta_{m,t} = \sum_{l=0}^{2L} \beta_l \sum_t \theta_{l,t} \theta_{m,t}
$$

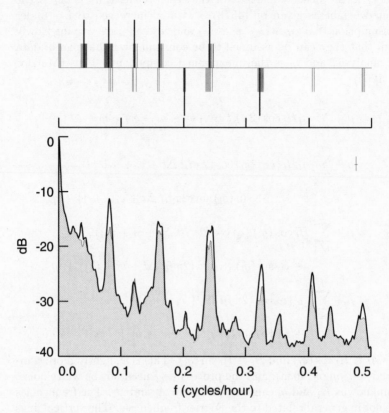

Figure 470. Tidal frequencies (upper plot) and estimated power spectra for Portsmouth residuals for 1971 (lower plot). In the upper plot, each frequency in the $L = 102$ model is indicated by a vertical line originating from the thin horizontal line – there are thus 102 vertical lines in all, but their widths are such that most of them are not distinctly visible. The vertical lines going up from the horizontal line indicate those frequencies that are in both the $L = 102$ and $L = 60$ models; the lines going down indicate frequencies in the $L = 102$ model only. There are also seven lines that are thicker and twice as long as the others – five going up and two going down. These represent, from left to right, the tidal frequencies M_1, M_2, M_3, M_4, M_5, M_6 and M_8. In the lower plot, the thick curve is the spectrum for the 60 constituent model, while the thin curve at the top of the shaded area gives the spectrum for the 102 constituent model. A Parzen lag window with $m = 400$ was used. The horizontal length of the cross in the upper right-hand corner of this plot indicates the smoothing window bandwidth B_W of the lag window spectral estimates, while the vertical length indicates the width of a 95% confidence interval.

for each of the $2L + 1$ parameters. These equations can be written in matrix form as

$$\Theta\beta = \mathbf{c},$$

where

$$\mathbf{c} \equiv \begin{bmatrix} \mu \\ \sum_t Y_t \alpha_1 \cos\left(2\pi f_1 t\, \Delta t + v_l + u_1\right) \\ \vdots \\ \sum_t Y_t \alpha_1 \sin\left(2\pi f_L t\, \Delta t + v_L + u_L\right) \end{bmatrix}$$

and the matrix Θ has $\Theta_{i,j} \equiv \sum_t \theta_{j-1,t}\theta_{i-1,t}$ for its (i,j)th element. With β estimated by $\hat{\beta}$, H_l and g_l are found from

$$\hat{H}_l = \left(\hat{\beta}_{2l}^2 + \hat{\beta}_{2l-1}^2\right)^{1/2} \quad \text{and} \quad \hat{g}_l = \tan^{-1}\left(\hat{\beta}_{2l}/\hat{\beta}_{2l-1}\right), \quad l = 1, \ldots, L.$$

In standard linear regression, the noise process is required to be uncorrelated. In our current application, even before examining estimates of η_t (formed from $Y_t - \hat{Y}_t$, where $\hat{Y}_t$ is the least squares estimate), we would expect the η_t terms to be correlated. For example, seasonal groupings of meteorological effects and, on a shorter scale, surge-generated groupings of residuals will be present. In general, for correlated noise, the maximum likelihood solution to the parameter estimation corresponds to *weighted* rather than the usual unweighted least squares. Remarkably, the unweighted least squares estimates are asymptotically efficient for a polynomial regression model such as is under consideration here (see, e.g., Section 7.7 of Priestley, 1981, and also the Comments and Extensions to this section).

Walden (1982) calculated the harmonic constants H_l and g_l for Portsmouth, England, using three years of carefully edited consecutive hourly height data (1967–9) – over 26 000 points in all. The harmonic constants were estimated using least squares. These estimates were used to 'predict' tidal elevations for each of the years 1962 and 1964–74. These predictions were in turn subtracted from the corresponding observed sea levels to produce residual series of hourly heights. This process was carried out for the so-called orthodox constituents ($L = 60$ harmonic terms comprising the primary frequencies plus some shallow-water frequencies) and for orthodox plus shallow-water terms ($L = 102$ harmonic terms) – see Figure 470. The lower plot of Figure 470 shows the power spectra of the 1971 Portsmouth residuals (note that, although the harmonic constants were computed using three years of data, each residual sequence is one year in length; i.e., $N = 365 \times 24 = 8760$). The thick curve and the thin curve tracing the top of the shaded area show, respectively, spectral estimates based upon the residuals for the $L = 60$ and 102 constituent models. A Parzen lag window was used for both estimates. After some testing, the smoothing parameter m for this window was chosen as 400; from Table 269 the window bandwidth is $1.85/(m\,\Delta t) = 0.0046$ cycle/hour, and the equivalent degrees of freedom ν is $3.71N/m = 81.25$ (note that $C_h = 1$ since the rectangular data taper was used). While the window bandwidth is not sufficient to separate individual constituents, increasing m – i.e., decreasing the window bandwidth – led to no visual improvement in the form of the spectral estimate. With $L = 102$, there is a diminution – as compared to the $L = 60$ analysis – of the contribution in the frequency bands close to integer multiples of 0.0805 cycle/hour. Since this frequency is equivalent to 1 cycle/12.42 hours (a semi-diurnal term), multiples of 2, 3 and so forth correspond to quarter-diurnal, sixth-diurnal and so forth terms. These are approximately the positions of the majority of the additional frequencies in the $L = 102$ model – see Figure 470. Note that, even for the $L = 102$ model, there is still considerable low-frequency power in the residual process so that it cannot be reasonably approximated as white noise.

How can we explain the residual power at low frequencies and in the tidal bands (i.e., the frequency ranges encompassing semi-, quarter-, eighth-diurnal, etc., tides)? The prediction model (457b) does not include nongravitational effects. For example, pressure fluctuations due to weather systems – which can induce a sea level response – have a continuous noise spectrum. Storm surges will show primarily in the band 0.03 to 0.08 cycle/hour. Land and sea breezes will cause discernible sea level changes and have associated spectral *lines* at 1 and 2 cycles/day as well as a continuous noise component (Munk and Cartwright, 1966). The overall spectrum is of course a mixed spectrum. At higher frequencies, residual power can arise through the omission of shallow-water constituents or through interaction between the tide and local mean sea level. Aliasing could also be involved. Suppose there is nonnegligible power in the *continuous time* residual series at harmonics of the main semidiurnal M_2 constituent at frequencies corresponding to M_{14} and M_{16}, i.e., approximately 0.5636 and 0.6441 cycle/hour. Sampling every hour gives a Nyquist frequency of $f_{(N)} = 1/2$ cycle/hour. Since power from frequency $2f_{(N)} - f$ would be aliased with f under this sampling scheme, we note that

0.5636 cycle/hour aliases to 0.4364 cycle/hour, while

0.6441 cycle/hour aliases to 0.3559 cycle/hour.

Thus, the residual power at M_{14} and M_{16} in the continuous series would appear at 0.436 and 0.356 cycle/hour in the sample spectrum. For Portsmouth there is a peak close to both of these frequencies (for other data – such as Southampton for 1955 – there is a noticeable peak about the first frequency only).

Comments and Extensions to Section 10.4

[1] If the sdf of the background continuum is colored, there is no efficiency to be gained *asymptotically* by including information on the covariance structure of the continuum. Hannan (1973), in examining a single sinusoid with frequency f_1, explained that the regression component is by nature a 'signal' sent at a single frequency so that only the sdf of the background continuum at that frequency matters. Ultimately, as the sample size increases, we can gain nothing from knowledge of the covariance structure of the background continuum. He also pointed out that, for finite sample sizes and a continuum with a very irregular sdf near f_1, the influence of the covariance structure of the continuum might be appreciable even though its influence vanishes as the sample size increases.

10.5 A Special Case of Unknown Frequencies

Up to this point we have assumed that the frequencies f_l in Equation (460) are known in advance (as in the example of Section 10.4). If they are not known, then the models are no longer multiple linear regressions. The usual procedure is to somehow estimate the f_l terms and then to use these estimates in finding the other unknowns.

The classic procedure for estimating f_l – which dates from the 1890s – is based upon examining the periodogram, which we have already met in Chapter 6. The basic idea behind periodogram analysis in the current context is as follows. From Equation (460), note that D_l^2 represents the squared amplitude of the sinusoid with frequency f_l. We also have $D_l^2 = A_l^2 + B_l^2$ from reexpressing Equation (460) – with μ assumed to be 0 – as Equation (463a). If $f_l \neq 0$ or $f_{(N)}$, we can estimate D_l^2 by using the estimators for A_l and B_l in Equation (463a) to obtain

$$\hat{D}_l^2 = \left(\frac{2}{N} \sum_{t=1}^{N} X_t \cos\left(2\pi f_l t\, \Delta t\right) \right)^2 + \left(\frac{2}{N} \sum_{t=1}^{N} X_t \sin\left(2\pi f_l t\, \Delta t\right) \right)^2$$

$$= \frac{4}{N^2} \left| \sum_{t=1}^{N} X_t e^{-i2\pi f_l t\, \Delta t} \right|^2. \tag{473}$$

It follows from Section 10.3 that the estimated contribution to the power spectrum at $f = \pm f_l$ is $\hat{D}_l^2 / 4$. The periodogram at f_l is

$$\hat{S}^{(p)}(f_l) = \frac{\Delta t}{N} \left| \sum_{t=1}^{N} X_t e^{-i2\pi f_l t\, \Delta t} \right|^2$$

(see Equation (196c)), so the scheme is to plot $\hat{S}^{(p)}(f)$ versus f and look for 'large' values of $\hat{S}^{(p)}(f)$. Note that, for any f not equal to an f_l in

Equation (460), we can imagine adding an $(L + 1)$st term to the model with $f_{L+1} = f$ and $D_{L+1} = 0$. The estimator $\hat{D}^2_{L+1}$ should thus be an estimator of zero and consequently should be small.

Now let us assume that each f_l in Equation (460) can be written in the form $k_l/(N \Delta t)$, where k_l is an integer (unique for each different f_l) such that $0 \leq k_l \leq N/2$; i.e., each f_l is one of the Fourier frequencies $k/(N \Delta t)$, $k = 0, 1, \ldots, \lfloor N/2 \rfloor$ – as usual, $\lfloor N/2 \rfloor$ refers to the largest integer less than or equal to $N/2$. If $f_l \neq 0$ or $f_{(N)}$, we know from Section 10.2 that

$$
\begin{aligned}
E\{\hat{S}^{(p)}(f_l)\} &= \frac{N \Delta t}{4} E\{D^2_l\} = \frac{N \Delta t}{4} \left(E\{\hat{A}^2_l\} + E\{\hat{B}^2_l\} \right) \\
&= \frac{N \Delta t}{4} \left(\text{var}\{\hat{A}_l\} + \left[E\{\hat{A}_l\} \right]^2 + \text{var}\{\hat{B}_l\} + \left[E\{\hat{B}_l\} \right]^2 \right) \\
&= \frac{N \Delta t}{4} \left(\frac{2\sigma^2_\epsilon}{N} + A^2_l + \frac{2\sigma^2_\epsilon}{N} + B^2_l \right) \\
&= \left(N \frac{D^2_l}{4} + \sigma^2_\epsilon \right) \Delta t. \tag{474}
\end{aligned}
$$

The first term in the last line above is $N \Delta t$ times the contribution to the power spectrum from the line at $f = \pm f_l$, while the second is the value of the white noise contribution to the spectrum. If $f_l \neq 0$ or $f_{(N)}$ is one of the Fourier frequencies that is *not* in Equation (460), then

$$
E\{\hat{S}^{(p)}(f_l)\} = \sigma^2_\epsilon \Delta t
$$

by the artifact of adding a fictitious term to the model with zero amplitude.

Thus, in the case where each f_l is of a special nature (i.e., a Fourier frequency), plotting $\hat{S}^{(p)}(f_l)$ versus f_l on the grid of standard frequencies and searching this plot for large values is a reasonable way of finding which frequencies f_l belong in the model. The sampling properties for $\hat{S}^{(p)}(f_l)$ were given in Section 6.6.

Comments and Extensions to Section 10.5

[1]   For the complex exponential model of (465a), the equivalent of Equation (474) is

$$
E\{\hat{S}^{(p)}(f_l)\} = \left[N(D'_l)^2 + \sigma^2_\epsilon \right] \Delta t.
$$

10.6 General Case of Unknown Frequencies

For the general case where f_l is not necessarily one of the Fourier frequencies (and $\{X_t\}$ in (460) is not necessarily a Gaussian process), we have the following result, valid for all $|f| \le f_{(N)}$:

$$E\{\hat{S}^{(p)}(f)\} = \sigma_\epsilon^2 \, \Delta t + \sum_{l=1}^{L} \frac{D_l^2}{4} \left[\mathcal{F}(f + f_l) + \mathcal{F}(f - f_l) \right], \qquad (475)$$

where $\mathcal{F}(\cdot)$ is Fejér's kernel – see Equation (198b) for its definition and also Figure 200. The proof of (475) is Exercise [10.4], which can be most easily seen by expressing the increments of $S_2^{(I)}(\cdot)$ in Section 10.3 in terms of the Dirac delta function (cf. Equation (144)):

$$dS_2^{(I)}(f) = D_l^2 \left[\delta(f - f_l) + \delta(f + f_l) \right]/4.$$

What does Equation (475) tell us about the usefulness of the periodogram for identifying f_l in Equation (460)? Consider the case where there is only one frequency, f_1 (i.e., $L = 1$ in Equation (460)). Equation (475) reduces to a constant term, $\sigma_\epsilon^2 \, \Delta t$, plus the sum of $D_1^2/4$ times two Fejér kernels centered at frequencies $\pm f_1$. If $f_1 = k_1/(N \, \Delta t)$ for some integer k_1 and if we evaluate $E\{\hat{S}^{(p)}(f)\}$ at the Fourier frequencies, Equation (475) is in agreement with our earlier analysis: a plot of $E\{\hat{S}^{(p)}(f)\}$ versus f has a single large value at f_1 against a 'background' of level $\sigma_\epsilon^2 \, \Delta t$. If, however, f_1 is not of the form $k_1/(N \, \Delta t)$, the plot is more complicated since we are now sampling Fejér's kernel at points other than the 'null points.' For example, if f_1 falls exactly halfway between two of the Fourier frequencies and if we only plot $E\{\hat{S}^{(p)}(\cdot)\}$ at the Fourier frequencies, we would find the two largest values at $f_1 \pm 1/(2N \, \Delta t)$; moreover, these values would have approximately the same amplitude due to the symmetry of the Fejér kernel (the qualifier 'approximately' is needed due to the contribution from the second Fejér kernel centered at $-f_1$). It can be shown that the size of these two values would be about 40% of the value of $E\{\hat{S}^{(p)}(f_1)\}$. Thus, we need to plot $\hat{S}^{(p)}(\cdot)$ at more than just the $\lfloor N/2 \rfloor + 1$ Fourier frequencies. In practice a grid twice as fine as that of the Fourier frequencies suffices – at least initially – to aid us in searching for periodic components. An additional rationale for this advice is that $\hat{S}^{(p)}(\cdot)$ is a trigonometric polynomial of degree $N - 1$ and hence is uniquely determined by its values at N points. (There is, however, a potential penalty for sampling twice as finely as the Fourier frequencies – the loss of independence between the periodogram ordinates. This presents no problem when we are merely examining plots of the periodogram ordinates to assess appropriate terms for Equation (460), but it is a problem in constructing valid statistical tests based upon periodogram ordinates.)

Let us consider the relationship between the periodogram and least squares estimation for the single frequency model

$$X_t = A\cos\left(2\pi f_1 t\,\Delta t\right) + B\sin\left(2\pi f_1 t\,\Delta t\right) + \epsilon_t.$$

This is the model described by (461a) with $\mu = 0$ now assumed to be known (and f relabeled as f_1). The exact least squares estimators of A and B, say $\hat{A}$ and $\hat{B}$, could be explicitly found from the normal equations

$$\sum_{t=1}^{N} X_t \cos\left(2\pi f_1 t\,\Delta t\right) = \hat{A}\sum_{t=1}^{N}\cos^2\left(2\pi f_1 t\,\Delta t\right)$$

$$+ \hat{B}\sum_{t=1}^{N}\cos\left(2\pi f_1 t\,\Delta t\right)\sin\left(2\pi f_1 t\,\Delta t\right)$$

and

$$\sum_{t=1}^{N} X_t \sin\left(2\pi f_1 t\,\Delta t\right) = \hat{A}\sum_{t=1}^{N}\cos\left(2\pi f_1 t\,\Delta t\right)\sin\left(2\pi f_1 t\,\Delta t\right)$$

$$+ \hat{B}\sum_{t=1}^{N}\sin^2\left(2\pi f_1 t\,\Delta t\right)$$

(cf. Equations (462a) and (462b)). The explicit expressions for $\hat{A}$ and $\hat{B}$ are somewhat unwieldy but can be simplified considerably by approximating f_1 on the right-hand side of the normal equations by its nearest Fourier frequency, say, $f_{k_1} = k_1/(N\,\Delta t)$. If we assume that f_{k_1} is not 0 or $f_{(N)}$, we have

$$\sum_{t=1}^{N}\cos^2\left(2\pi f_{k_1} t\,\Delta t\right) = \sum_{t=1}^{N}\sin^2\left(2\pi f_{k_1} t\,\Delta t\right) = \frac{N}{2}$$

and

$$\sum_{t=1}^{N}\cos\left(2\pi f_{k_1} t\,\Delta t\right)\sin\left(2\pi f_{k_1} t\,\Delta t\right) = 0$$

(see Exercise [1.4]). With this approximation the normal equations become

$$\sum_{t=1}^{N} X_t \cos\left(2\pi f_1 t\,\Delta t\right) \approx \hat{A}\frac{N}{2} \quad\text{and}\quad \sum_{t=1}^{N} X_t \sin\left(2\pi f_1 t\,\Delta t\right) \approx \hat{B}\frac{N}{2}.$$

Thus we can regard

$$\tilde{A} \equiv \frac{2}{N}\sum_{t=1}^{N} X_t \cos\left(2\pi f_1 t\,\Delta t\right) \quad\text{and}\quad \tilde{B} \equiv \frac{2}{N}\sum_{t=1}^{N} X_t \sin\left(2\pi f_1 t\,\Delta t\right)$$

as approximate least squares estimators of A and B. The residual sum of squares resulting from these approximate least squares estimators is

$$\text{SS} \equiv \sum_{t=1}^{N} \left[X_t - \left(\tilde{A} \cos\left(2\pi f_1 t\, \Delta t\right) + \tilde{B} \sin\left(2\pi f_1 t\, \Delta t\right) \right) \right]^2.$$

If we substitute for $\tilde{A}$ and $\tilde{B}$ and simplify somewhat, we obtain

$$\text{SS} = \sum_{t=1}^{N} X_t^2 - \frac{2}{\Delta t} \hat{S}^{(p)}(f_1). \tag{477a}$$

We assumed that f_1 was known. In general it will not be, and in this case we can see from (477a) that, if we choose f_1 to *maximize* the periodogram, then this will correspond to a *minimum* of the residual sum of squares for the approximate least squares estimators. As N gets large, any f_1 (not equal to 0 or $f_{(N)}$) will become closer and closer to a Fourier frequency, and so any results based upon maximizing the periodogram will hold for minimizing the residual sum of squares for the exact least squares estimators.

In the case where

[1] the error terms $\{\epsilon_t\}$ in the single frequency model are an independent and identically distributed sequence of random variables with zero mean and finite variance and

[2] $\hat{f}_1$ represents the frequency of the maximum value of the periodogram,

Whittle (1952) and Walker (1971) showed that

$$E\{\hat{f}_1\} = f_1 + O\left(N^{-1}\right) \quad \text{and} \quad \text{var}\{\hat{f}_1\} \approx \frac{3}{N^3 R(\pi\, \Delta t)^2}, \tag{477b}$$

where $R \equiv \left(A^2 + B^2\right)/(2\sigma_\epsilon^2)$ is the signal-to-noise ratio; i.e., since

$$\text{var}\{X_t\} = \frac{A^2 + B^2}{2} + \sigma_\epsilon^2$$

for a *randomly* phased sinusoid, R is just the ratio of the two terms on the right-hand side, the first due to the sinusoid (the 'signal'), and the second to the white noise process (the 'noise'). For *nonrandomly* phased sinusoids, since $A^2 + B^2 = D^2$ is the squared amplitude of the sinusoidal component in $\{X_t\}$, we can intuitively interpret R as a comparison of the variation in the sinusoid to the expected variation in $\{\epsilon_t\}$.

The $O\left(N^{-3}\right)$ behavior of $\text{var}\{\hat{f}_1\}$ is somewhat surprising (a rate of decrease of $O\left(N^{-1}\right)$ is more common, as in Equation (188) for example), but Bloomfield (1976, p. 29) gives the following argument to lend credence to it. Suppose R is reasonably large so that, by inspecting part of a realization of this process, say, $x_1, x_2, \ldots, x_N$, we can see that there are, say, between M and $M + 1$ complete cycles of the sinusoid. Therefore we know that

$$\frac{M}{N\,\Delta t} \leq f_1 \leq \frac{(M+1)}{N\,\Delta t}. \tag{478}$$

It follows that $|\tilde{f}_1 - f_1| \leq 1/(N\,\Delta t)$ for any estimator $\tilde{f}_1$ satisfying the above. This implies that

$$\text{var}\,\{\tilde{f}_1\} \leq \frac{1}{(N\,\Delta t)^2} = O\left(N^{-2}\right).$$

If we can estimate f_1 to within order $O\left(N^{-2}\right)$ with only the constraint that $\tilde{f}_1$ satisfies (478), it is plausible that the fancier periodogram estimator is $O\left(N^{-3}\right)$. More recently Rice and Rosenblatt (1988) have shown that the product of the amplitude of the sinusoid and the sample size must be quite large for the asymptotic theory to be meaningful.

If there is more than just a single frequency in Equation (460) (i.e., $L > 1$), the form of $E\{\hat{S}^{(p)}(\cdot)\}$ in Equation (475) can be quite complicated. Since it is now the superposition of $2L$ Fejér kernels centered at the points $\{\pm f_l\}$ together with the constant term $\sigma_\epsilon^2\,\Delta t$, there can be 'interference' between the superimposed kernels. For example, if two of the f_l terms are close together and N is small enough, a plot of $E\{\hat{S}^{(p)}(\cdot)\}$ might only indicate a single broad peak. Likewise, a frequency with a small D_l^2 might be effectively masked by a 'sidelobe' from Fejér's kernel due to a nearby frequency with a large amplitude (see Section 10.8 and Figure 487).

In practical applications, we only observe $\hat{S}^{(p)}(\cdot)$, which can differ substantially from $E\{\hat{S}^{(p)}(\cdot)\}$ due to the presence of the white noise component. Our problem of identifying frequency components is much more difficult due to the distortions that arise from sampling variations. One effect of the white noise component is to introduce 'spurious' peaks into the periodogram. Recall that, in the null case when $L = 0$ and the white noise process is Gaussian, the $\hat{S}^{(p)}(f)$ terms at the Fourier frequencies are independent χ^2 rv's with one or two degrees of freedom (see Section 6.6). Because they are independent rv's, in portions of the spectrum where the white noise component dominates, we can expect to see a rather choppy behavior with many 'peaks' and 'valleys.' Furthermore, the χ_1^2 and χ_2^2 distributions have heavy tails; in a random sample from such a distribution we can expect one or more observations that appear unusually large. These large observations could be mistaken for peaks due to a harmonic component. This fact points out the need for statistical tests to help us decide if a peak in the periodogram can be considered statistically significant (see Sections 10.9 and 10.11).

Comments and Extensions to Section 10.6

[1]   Hannan (1973) looked at the case where the additive noise in Equation (460) is a stationary process and examined also the special cases where $f_1 = 0$ or $f_{(N)}$. He found that for $0 < f_1 < f_{(N)}$

$$\lim_{N \to \infty} N \left(\hat{f}_1 - f_1 \right) = 0 \text{ almost surely,}$$

while for $f_1 = 0$ or $f_{(N)}$ a stronger result holds: there is an integer N_0 – determined by the realization – such that $\hat{f}_1 = f_1$ for all $N \geq N_0$ with $\mathbf{P}\left[N_0 < \infty \right] = 1$.

[2]   Suppose that $f'_k = k/(N' \Delta t)$ maximizes the periodogram $\hat{S}^{(p)}(\cdot)$ over a grid of equally spaced frequencies defined by $f'_j = j/(N' \Delta t)$ for $1 \leq j < N'/2$ (the choice $N' = N$ yields the grid of Fourier frequencies, but $N' > N$ is also of interest if we pad $N' - N$ zeros to our time series for use with an FFT algorithm – see [1] of the Comments and Extensions to Section 3.10). Suppose, however, that we are interested in obtaining the frequency maximizing $\hat{S}^{(p)}(\cdot)$ over *all* frequencies. From Equation (196c), we can write

$$\hat{S}^{(p)}(f) = \Delta t \sum_{\tau=-(N-1)}^{N-1} \hat{s}_\tau^{(p)} e^{-i2\pi f \tau \Delta t}$$

$$= \Delta t \left(\hat{s}_0^{(p)} + 2 \sum_{\tau=1}^{N-1} \hat{s}_\tau^{(p)} \cos \left(2\pi f \tau \Delta t \right) \right),$$

where $\{ \hat{s}_\tau^{(p)} \}$ is the biased estimator of the acvs. If we let $\omega \equiv 2\pi f \Delta t$ and

$$g(\omega) \equiv \sum_{\tau=1}^{N-1} \hat{s}_\tau^{(p)} \cos \left(\omega \tau \right),$$

then a peak in $g(\cdot)$ at ω corresponds to a peak in $\hat{S}^{(p)}(\cdot)$ at $f = \omega/(2\pi \Delta t)$. The first and second derivatives of $g(\cdot)$ are

$$g'(\omega) = - \sum_{\tau=1}^{N-1} \tau \hat{s}_\tau^{(p)} \sin \left(\omega \tau \right) \text{ and } g''(\omega) = - \sum_{\tau=1}^{N-1} \tau^2 \hat{s}_\tau^{(p)} \cos \left(\omega \tau \right).$$

$$(479)$$

Note that a peak in $g(\cdot)$ corresponds to a root in $g'(\cdot)$. If $\omega^{(0)} \equiv 2\pi f'_k \Delta t$ is not too far from the true location of the peak in $g(\cdot)$, we can apply the Newton–Raphson method to recursively compute

$$\omega^{(j)} = \omega^{(j-1)} - \frac{g'(\omega^{(j-1)})}{g''(\omega^{(j-1)})} \text{ for } j = 1, 2, \ldots, J,$$

stopping when $|\omega^{(J)} - \omega^{(J-1)}|$ is smaller than some specified tolerance. This yields a value of $\omega^{(J)}/(2\pi\,\Delta t)$ for the location of the peak in $\hat{S}^{(p)}(\cdot)$ (Newton and Pagano, 1983; Newton, 1988, Section 3.9.2).

In practice this scheme can fail if f_k' is not sufficiently close to the true peak location. Section 9.4 of Press *et al.* (1986) describes a useful method for finding the root of $g'(\cdot)$, call it $\omega^{(r)}$, using a Newton–Raphson scheme in combination with bisection. This method can be applied here if we can find two values of ω – call them $\omega^{(l)}$ and $\omega^{(u)}$ – that bracket the root; i.e., we have $\omega^{(l)} < \omega^{(r)} < \omega^{(u)}$ with $g'(\omega^{(l)}) > 0$ and $g'(\omega^{(u)}) < 0$. If N' is sufficiently large (usually $N' = 2N$ does the trick), these bracketing values can be taken to be $2\pi f_{k-1}'\,\Delta t$ and $2\pi f_{k+1}'\,\Delta t$. (There can also be problems if the equation for $g''(\omega)$ in (479) cannot be accurately computed, in which case *Brent's method* (Press *et al.*, 1986, Section 9.3) can be used to find the bracketed root of $g'(\cdot)$.)

10.7 An Artificial Example from Kay and Marple

Let us illustrate some of the points in the previous sections by looking at an artificial example due to Kay and Marple (1981, p. 1386). Figure 481 shows three different sinusoids and their weighted summation

$$X_t = 0.9\cos\left(2\pi t/7.5\right) + 0.9\cos\left(2\pi t/5.3 + \pi/2\right) + \cos\left(2\pi t/3\right) \quad (480)$$

at indices $t = 1, 2, \ldots, 16$ (the 16 asterisks). This time series can be regarded as a realization of a noise-free model (457a) with $\mu = 0$, $L = 3$ and $\Delta t = 1$. The frequencies of the three sinusoids are $1/7.5 = 0.133$ (thin solid curve), $1/5.3 = 0.189$ (dashed) and $1/3$ (thick solid). In our sample of length 16 of $\{X_t\}$ we have about 2+ cycles of the first sinusoid, 3 cycles of the second and 5+ cycles of the third.

In Figure 482 we study the effect of grid size on our ability to detect the presence of harmonic components when examining the periodogram $\hat{S}^{(p)}(\cdot)$ for $X_1, \ldots, X_{16}$. The asterisks in the top plot show the value of $10\log_{10}\hat{S}^{(p)}(\cdot)$ at the nine Fourier frequencies $f_k = k/N = k/16$ for $k = 0, 1, \ldots, 8$. The three thin vertical lines indicate the locations of the frequencies of the three sinusoids that form $\{X_t\}$. There are two broad peaks visible, the one with the lower frequency being larger in magnitude. Note that this plot fails to resolve the two lowest frequencies and that the apparent location of the highest frequency is shifted to the left. From this plot we might conclude that there are only two sinusoidal components in $\{X_t\}$.

Figure 482(b) shows $10\log_{10}\hat{S}^{(p)}(\cdot)$ at the frequencies $k/32$ for $k = 0, 1, \ldots, 16$. There are still two major broad peaks visible (with maximum heights at points $k = 6$ and $k = 11$, respectively) and some minor bumps that are an order of magnitude smaller. Note now that

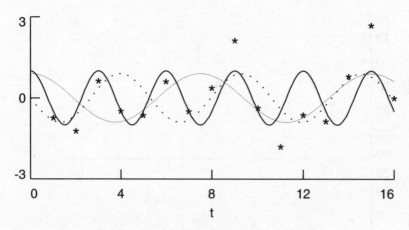

Figure 481. Sinusoids with periods 7.5 (thin solid curve), 5.3 (dashed) and 3 (thick solid) along with their summation at $t = 1, \ldots, 16$ (asterisks).

the peak with the higher frequency is slightly larger than the one with the lower frequency. Moreover, there is a hint of some structure in the first broad bump (there is a slight dip between points $k = 4$ and 6). By going to a finer grid than that provided by the Fourier frequencies, we are getting a better idea of the sinusoidal components in $\{X_t\}$.

In Figure 482(c) we have evaluated the periodogram on a still finer grid of 513 frequencies. The difference between adjacent frequencies is $1/1024 \approx 0.001$. This plot clearly shows two major broad peaks, the first of which (roughly between $f = 0.1$ and 0.2) has a small dip in the middle. The smaller peaks are evidently sidelobes due to the superposition of Fejér kernels. The appearance of nonzero terms in the periodogram at frequencies other than those of sinusoids in the harmonic process is due to leakage – see Sections 3.7 and 6.3 and the next section. Leakage can affect our ability to use the periodogram to determine what frequencies make up $\{X_t\}$. For example, the three highest peaks in the periodogram in Figure 482(c) occur at frequencies 0.117, 0.181 and 0.338 as compared to the true frequencies of 0.133, 0.189 and 0.333. These discrepancies are here entirely due to distortions caused by the six superimposed Fejér kernels (see Equation (475)) since there is *no* observational noise present. Note, however, that the discrepancies are not large, particularly when we recall we are only dealing with 16 points from a function composed of three sinusoids.

We next study the effect of adding white noise to $X_1, \ldots, X_{16}$ to produce a realization of Equation (460). As in the previous section, we

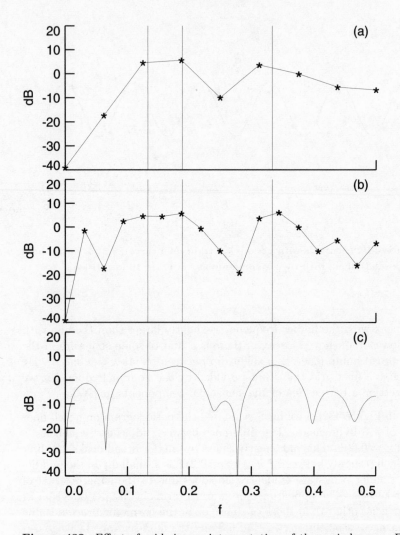

Figure 482. Effect of grid size on interpretation of the periodogram. For a time series of length 16, the plots show the periodogram (on a decibel scale) versus frequency at the nine Fourier frequencies (asterisks in top plot, connected by lines), on a grid of frequencies twice as fine as the Fourier frequencies (17 asterisks in middle plot), and on a finely spaced grid of 513 frequencies (bottom).

can use the decomposition

$$\operatorname{var}\{X_t\} = \sum_{l=1}^{L} D_l^2/2 + \sigma_\epsilon^2 \ \text{ to define } \ R \equiv \frac{\sum_{l=1}^{L} D_l^2}{2\sigma_\epsilon^2},$$

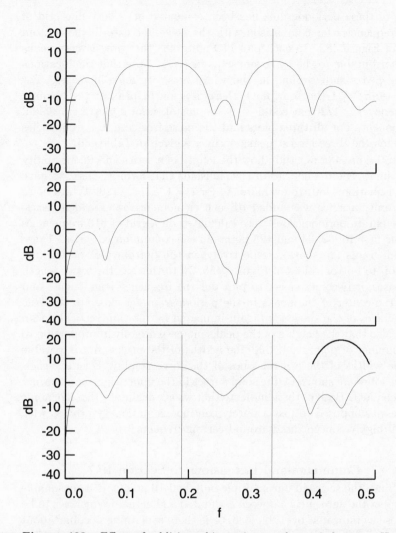

Figure 483. Effect of additive white noise on the periodogram. Here we add white noise to $X_1, \ldots, X_{16}$ of Equation (480) and compute the resulting periodogram $\hat{S}^{(p)}(\cdot)$ over a finely spaced grid of 513 frequencies. The variances of the additive white noise are 0.01, 0.1 and 1, respectively, in the top to bottom plots. The periodogram for the noise-free case is shown in the bottom plot of Figure 482.

where we consider R as a signal-to-noise ratio (see Equations (467c) and (477b)). The plots in Figure 483 show the effect on the periodogram of adding white noise with variances of 0.01, 0.1 and 1.0 (top to bottom plots, respectively), yielding signal-to-noise ratios of 131, 13.1 and 1.31.

Each of these periodograms has been evaluated on a very fine grid of 513 frequencies for comparison with the noise-free case in the bottom plot of Figure 482. In each plot of Figure 483 the three largest peaks are attributable to the three sinusoids in $\{X_t\}$. Note that the locations of the peaks shift around as the white noise variance increases. For the case $\sigma_\epsilon^2 = 1$ (bottom plot), there is a fourth peak at the Nyquist frequency $f = 1/2$ that could easily be mistaken for a fourth sinusoidal component. For all three plots and the noise-free case, the frequencies at which the three largest peaks occur are given in Table 486.

In Figure 485 we study how the length of a series affects our ability to estimate the frequencies of the sinusoids that form it. Here we have used Equation (480) to evaluate X_t for $t = 1, 2, \ldots, 128$. White noise with unit variance was added to each element of the extended series, and separate periodograms were calculated on a grid of 513 frequencies for the first 16, 32, 64 and 128 points in the contaminated series. These periodograms are shown, respectively, in the bottom plot of 483 and the top to bottom plots of Figures 485. Note that, as the series length increases, it becomes easier to pick out the frequency components and that the center of the peaks in the periodogram get closer to the true frequencies of the sinusoids (again indicated by the thin vertical lines). Note also that the heights of the peaks increase directly in proportion to the number of points and that the widths of the peaks are comparable to the widths of the central lobes of the corresponding Fejér's kernels (these lobes are shown by the solid curves in the upper right-hand corner of each plot). Both of these indicate that we are dealing with a harmonic process as opposed to just a purely continuous stationary process with an sdf highly concentrated around certain frequencies.

Comments and Extensions to Section 10.7

[1]   Since the series in our example only had 16 points, is it reasonable to expect the agreement between estimated and actual frequencies to be as close as indicated by Table 486, or is there something peculiar about our example? Let us see what approximation (477b) gives us for the variance of the estimated frequency for the sinusoid with the highest frequency in our four time series of length 16. Strictly speaking, this approximation is based on the assumption that there is only a single frequency in our model instead of three; however, this approximation can be taken as valid if the frequencies are spaced 'sufficiently' far apart (Priestley, 1981, p. 410). If we thus ignore the other two frequencies, the signal-to-noise ratio for the $f_3 = 1/3$ component alone is $R = 1/(2\sigma_\epsilon^2)$ since $D_3^2 = 1$. This implies that

$$\text{var}\,\{\hat{f}_3\} \approx \frac{6\sigma_\epsilon^2}{N^3\pi^2} = 0.00015 \times \sigma_\epsilon^2 \text{ for } N = 16.$$

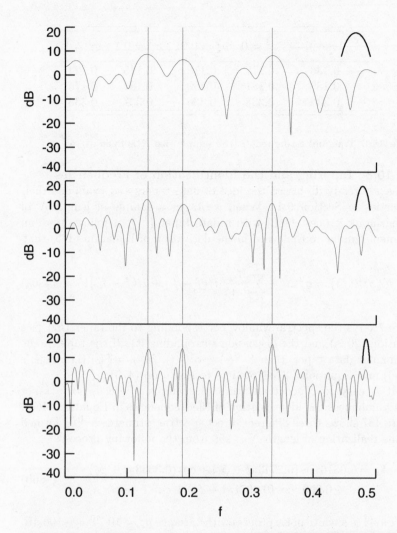

Figure 485. Effect of sample size on the periodogram. Here we add white noise with unit variance to $X_1, \ldots, X_{128}$ of Equation (480) and compute the resulting periodogram $\hat{S}^{(p)}(\cdot)$ using just the first 32 and 64 values of the series (top and middle plots) and finally all 128 values (bottom plot). The periodogram for just the first 16 values is shown in the bottom plot of Figure 483.

For $\sigma_\epsilon^2 = 0.01$, 0.1 and 1, the standard deviations of $\hat{f}_3$ are 0.001, 0.004 and 0.012. From the last row of Table 486, we see that $\left| \hat{f}_3 - f_3 \right|$ is 0.003, 0.004 and 0.007 for these three cases. Thus the observed deviations are of the order of magnitude expected by statistical theory.

l	True f_l	$\sigma_\epsilon^2 = 0$	$\sigma_\epsilon^2 = 0.01$	$\sigma_\epsilon^2 = 0.1$	$\sigma_\epsilon^2 = 1$
1	0.133	0.117	0.118	0.115	0.124
2	0.189	0.181	0.180	0.191	0.176
3	0.333	0.338	0.336	0.337	0.330

Table 486. True and estimated f_l (the sample size N is fixed at 16).

10.8 Tapering and the Identification of Frequencies

We have already discussed the idea of data tapers and examined their properties in Section 6.4. When we taper a sample of length N of the harmonic process (460) with a data taper $\{h_t\}$, it follows from an argument similar to that used in the derivation of Equation (475) that

$$E\{\hat{S}^{(d)}(f)\} = \sigma_\epsilon^2 \, \Delta t + \sum_{l=1}^{L} \frac{D_l^2}{4} \left[\mathcal{H}(f + f_l) + \mathcal{H}(f - f_l) \right], \qquad (486a)$$

where $\mathcal{H}(\cdot)$ is the spectral window corresponding to the taper $\{h_t\}$ (see Equation (207a) and the discussion surrounding it). If the taper is the rectangular data taper, then $\hat{S}^{(d)}(\cdot) = \hat{S}^{(p)}(\cdot)$, $\mathcal{H}(\cdot) = \mathcal{F}(\cdot)$ (i.e., Fejér's kernel) and Equation (486a) reduces to Equation (475).

It is easy to concoct an artificial example where use of a taper greatly enhances the identification of the frequencies in Equation (460). Figure 487 shows three different direct spectral estimates $\hat{S}^{(d)}(\cdot)$ formed for one realization of length $N = 256$ from the following process:

$$X_t = 0.0316 \cos\left(0.2943\pi t + \phi_1\right) + \cos\left(0.3333\pi t + \phi_2\right)$$
$$+ 0.0001 \cos\left(0.3971\pi t + \phi_3\right) + \epsilon_t, \qquad (486b)$$

where $\{\epsilon_t\}$ is a white noise process with variance $\sigma_\epsilon^2 = 10^{-10} = -100$ dB. For the three plots in the figure, the corresponding data tapers are, from top to bottom, a rectangular data taper (cf. left-hand plots of Figure 209 for $N = 64$), a dpss data taper with $NW = 2$ (right-hand plots of Figure 211 for $N = 64$) and a dpss data taper with $NW = 4$ (left-hand plots of Figure 212 for $N = 64$). The thick curve in the upper right-hand corner of each plot shows the shape of the central lobe of the associated spectral window $\mathcal{H}(\cdot)$. The thin vertical lines in each plot indicate the locations of the three frequencies in Equation (486b). For the rectangular taper (top plot), the sidelobes of Fejér's kernel that is associated with the middle term in (486b) completely mask the other two sinusoids with smaller amplitudes. Thus the periodogram reveals only the presence of a single harmonic component. Use of a dpss taper

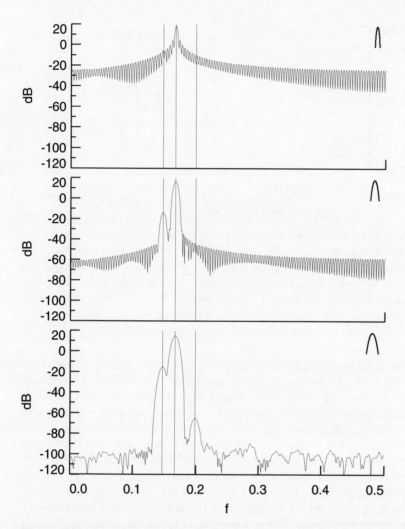

Figure 487. Effect of tapering on detection of harmonic components. From top to bottom, the direct spectral estimates used, respectively, a rectangular taper, a dpss taper with $NW = 2$ and a dpss taper with $NW = 4$. Each spectral estimate is plotted over a finely spaced grid of 513 frequencies.

with $NW = 2$ (middle plot) yields a spectral window with sidelobes reduced enough to reveal the presence of a second sinusoid, but the third sinusoid is still hidden (this dpss data taper is roughly comparable to the popular Hanning data taper). Finally, all three sinusoidal components are revealed when a dpss taper with $NW = 4$ is used (bottom plot).

Note, however, that, in comparison to the $NW = 2$ dpss data taper,

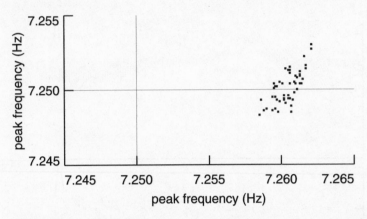

Figure 488. Scatter plot of locations of peak frequencies computed from a direct spectral estimator with $NW = 2/\Delta t$ (vertical axis) versus locations computed from the periodogram (horizontal axis). Each of the 50 points in the plot indicates the locations computed for a single realization of the model of Equation (489), which has a single sinusoidal term at 7.25 Hz and a high signal-to-noise ratio of 37 dB. There is substantial bias in the periodogram-based estimator of peak location. This bias is alleviated by tapering.

the broader central lobe of the $NW = 4$ dpss taper somewhat degrades the resolution of the two dominant sinusoidal components. An example could be constructed in which the $NW = 4$ dpss taper would fail to resolve two closely spaced frequencies, while a taper with a narrower central lobe (say, a dpss taper with $NW = 1$ or 2 or a 100% cosine taper) would succeed in doing so. The selection of a proper taper is clearly dependent upon the true model – a fact that accounts for the large number of different data tapers that have been suggested as appropriate for different problems (Harris, 1978). Rife and Vincent (1970) present an interesting study on the estimation of the magnitude and frequencies of two 'tones' when using different families of tapers with differing central lobe and sidelobe properties.

Tapering is sometimes useful even when there is but a single sinusoidal term in a harmonic process. In this case, Equation (475) tells us that

$$E\{\hat{S}^{(p)}(f)\} = \sigma_\epsilon^2 \, \Delta t + \frac{D_1^2}{4} \left[\mathcal{F}(f + f_1) + \mathcal{F}(f - f_1) \right].$$

For f close to f_1, the second of the two Fejér kernels above is generally the dominant term in the expectation, but the presence of the first kernel can distort $E\{\hat{S}^{(p)}(\cdot)\}$ such that its peak value is shifted away from f_1. This distortion can be lessened by tapering. As a simple example, we generated 50 realizations, each of length $N = 101$ from the process described by

$$\cos\left(2\pi f_1 t\,\Delta t + \pi/4\right) + \epsilon_t, \quad t = 0, \ldots, 100, \tag{489}$$

where $\{\epsilon_t\}$ is a zero mean white noise process with variance 0.0001, $f_1 = 7.25$ Hz and $\Delta t = 0.01$ seconds (yielding a Nyquist frequency of $f_{(N)} = 50$ Hz). This process has a very high signal-to-noise ratio of 37 dB. For each realization, we computed both the periodogram $\hat{S}^{(p)}(\cdot)$ and a direct spectral estimator $\hat{S}^{(d)}(\cdot)$ using a dpss data taper with $NW = 2/\Delta t$. We then found the locations of the frequencies at which $\hat{S}^{(p)}(\cdot)$ and $\hat{S}^{(d)}(\cdot)$ attained their peak values (see item [2] of the Comments and Extensions to Section 10.6). A scatter plot of the locations of the peak frequencies for $\hat{S}^{(d)}(\cdot)$ versus those for $\hat{S}^{(p)}(\cdot)$ is shown in Figure 488 for the 50 realizations. The thin horizontal and vertical lines mark the location of the true frequency $f_1 = 7.25$ Hz of the sinusoidal component. We see that, whereas the peak frequencies derived from $\hat{S}^{(d)}(\cdot)$ are clustered nicely about f_1, those from $\hat{S}^{(p)}(\cdot)$ are biased high in frequency. The mean square error in estimating f_1 via the periodogram is dominated by this bias and is considerably greater than that associated with $\hat{S}^{(d)}(\cdot)$. (We will return to this example later when we discuss the parametric approach to harmonic analysis – see the discussion concerning Figure 519 in Section 10.14.)

10.9 Tests for Periodicity – White Noise Case

It should be clear from the previous sections that we often need some statistical tests to help us determine when a spectral peak is significantly larger than is likely to arise if there were no genuine periodic components. We shall consider two basic cases, where the decision is

[1] between white noise and spectral lines plus white noise (i.e., white noise versus a discrete spectrum) and

[2] between colored noise and spectral lines plus colored noise (i.e., colored noise versus a mixed spectrum).

Typical tests for the first case make use of the fact that, under the null hypothesis that $\{X_t\}$ is (Gaussian) white noise, the periodogram ordinates calculated at the Fourier frequencies (excluding the zero and Nyquist frequencies) are distributed as independent and identically distributed χ_2^2 rv's times a multiplicative constant (see Section 6.6). The tests then look for periodogram ordinates of a size incompatible with the null hypothesis. Implicitly it is assumed that any spectral lines occur at Fourier frequencies, a rather strong assumption we reconsider at the end of this section.

Tests for case [2] – the mixed spectra case – also mostly use periodogram ordinates, although a technique based upon the autocovariance function can be constructed (Priestley, 1981, Sections 8.3–4). For this case we wish to draw attention in the next section to an appealing technique given in Thomson (1982).

For the remainder of this section we look at case [1] by using Equation (460) as our model for the observed time series, where the process $\{\epsilon_t\}$ is assumed to be Gaussian white noise. We also assume – purely for mathematical convenience – that the length of the realization N is odd; i.e., we can write $N = 2m + 1$ for some positive integer m (note that $\lfloor N/2 \rfloor = m$). The null hypothesis is

$$D_1 = \cdots = D_L = 0.$$

We consider the periodogram terms $\hat{S}^{(p)}(f_k)$ with $f_k \equiv k/(N \Delta t)$ for $k = 1, \ldots, m$. Under the null hypothesis,

$$\frac{2\hat{S}^{(p)}(f_k)}{\sigma_\epsilon^2 \Delta t} \stackrel{\mathrm{d}}{=} \chi_2^2$$

(see Section 6.6), and thus it has a probability density function given by

$$f(u) = e^{-u/2}/2, \qquad 0 \le u < \infty$$

(see Figure 228). For any $u_0 \ge 0$,

$$\mathbf{P}\left[\frac{2\hat{S}^{(p)}(f_k)}{\sigma_\epsilon^2 \Delta t} \le u_0\right] = \mathbf{P}\left[\chi_2^2 \le u_0\right] = \int_0^{u_0} f(u)\, du = 1 - e^{-u_0/2}.$$

Let

$$\gamma = \max_{1 \le k \le m} \frac{2\hat{S}^{(p)}(f_k)}{\sigma_\epsilon^2 \Delta t}.$$

Under the null hypothesis, γ is the maximum of the m independent and identically distributed χ_2^2 rv's; thus, for any u_0,

$$\mathbf{P}\left[\gamma > u_0\right] = 1 - \mathbf{P}\left[\frac{2\hat{S}^{(p)}(f_k)}{\sigma_\epsilon^2 \Delta t} \le u_0 \text{ for } 1 \le k \le m\right]$$

$$= 1 - \left(1 - e^{-u_0/2}\right)^m.$$

This equation gives us the distribution of γ under the null hypothesis. Under the alternative hypothesis that $D_l > 0$ for one or more terms in Equation (460), γ will be large, so we use a one-sided test with a critical region of the form $\{\gamma : \gamma > u_0\}$, where u_0 is chosen so that $\mathbf{P}[\gamma > u_0] = \alpha$, the chosen level of the test.

The chief disadvantage of this test (due to Schuster, 1898) is that we must know σ_ϵ^2 in advance to be able to calculate γ. Since σ_ϵ^2 usually can only be estimated, Fisher (1929) derived an exact test for the null hypothesis based upon the statistic

$$g \equiv \frac{\max_{1 \leq k \leq m} \hat{S}^{(p)}(f_k)}{\sum_{j=1}^{m} \hat{S}^{(p)}(f_j)}. \qquad (491a)$$

From Exercise [10.6] we can write the above denominator as

$$\sum_{j=1}^{m} \hat{S}^{(p)}(f_j) = \frac{\Delta t}{2} \sum_{t=1}^{N} \left(X_t - \bar{X} \right)^2.$$

Note that, except for a scaling factor, the right-hand side is an estimator of σ_ϵ^2 under the null hypothesis. We can regard g as the maximum of the sum of squares due to a single frequency f_k over the total sum of squares. Note also that σ_ϵ^2 acts as a proportionality constant in the distribution of both $\max \hat{S}^{(p)}(f_k)$ and $\sum \hat{S}^{(p)}(f_j)$, so that σ_ϵ^2 'ratios out' of Fisher's g statistic. Fisher showed that the exact distribution of g under the null hypothesis is given by

$$\mathbf{P}\left[g > g_0\right] = \sum_{j=1}^{M} (-1)^{j-1} \binom{m}{j} (1 - jg_0)^{m-1}, \qquad (491b)$$

where M is the largest integer satisfying both $M < 1/g_0$ and $M \leq m$. If, as a simplification, we use just the $j = 1$ term in this summation, we can determine the value g_0 for which $\mathbf{P}\left[g > g_0\right] = 0.05$ to within an error of 0.1%; moreover, this error is in the direction of decreasing the probability for the case. The error in this approximation is about the same order for common values for the size of the test α. Nowroozi (1967) gives tables of critical values, say g_F, for Fisher's test for $\alpha = 0.01, 0.02, 0.05$ and 0.1 using just the first term of the summation, i.e.,

$$g_F \approx 1 - (\alpha/m)^{1/(m-1)}, \qquad (491c)$$

an approximation that improves with increasing m (see also Shimshoni, 1971).

It is noted in Anderson (1971) that Fisher's test is the uniformly most powerful symmetric invariant decision procedure against *simple* periodicities, i.e., where the alternative hypothesis is that there exists a periodicity at only one Fourier frequency. But how well does it perform when there is a *compound* periodicity, i.e., spectral lines at several frequencies? Siegel (1980) studied this problem. He suggested a test statistic based on all large values of the rescaled periodogram

$$\tilde{S}^{(p)}(f_k) \equiv \frac{\hat{S}^{(p)}(f_k)}{\sum_{j=1}^{m} \hat{S}^{(p)}(f_j)} \qquad (491d)$$

instead of only their maximum. For a value $0 < g_0 \leq g_F$ (where, as before, g_F is the critical value for Fisher's test) and for each $\tilde{S}^{(p)}(f_k)$ that exceeds g_0, Siegel sums the excess of $\tilde{S}^{(p)}(f_k)$ above g_0. This forms the statistic

$$\sum_{k=1}^{m} \left(\tilde{S}^{(p)}(f_k) - g_0 \right)_+ ,$$

where $(a)_+ = \max(a, 0)$ is the positive-part function. The value of this statistic will depend on the choice of g_0, and this can be related to g_F by writing $g_0 = \lambda g_F$ with $0 < \lambda \leq 1$ to give

$$T_\lambda \equiv \sum_{k=1}^{m} \left(\tilde{S}^{(p)}(f_k) - \lambda g_F \right)_+ .$$

Note that, when $\lambda = 1$, the event $g > g_F$ is identical to the event $T_1 > 0$; i.e., we have Fisher's test. Suppose that we choose g_F to correspond to the $100(1 - \alpha)\%$ significance level of Fisher's test so that

$$\mathbf{P}\left[T_1 > 0\right] = \mathbf{P}\left[\sum_{k=1}^{m} \left(\tilde{S}^{(p)}(f_k) - g_F \right)_+ > 0\right] = \alpha.$$

Stevens (1939) showed that

$$\mathbf{P}\left[T_1 > 0\right] = \sum_{j=1}^{M} (-1)^{j-1} \binom{m}{j} (1 - jg_F)^{m-1},$$

exactly as is required by Equation (491b).

We need to compute $\mathbf{P}\left[T_\lambda > t\right]$ for general λ in the range 0 to 1 so that we can find the critical points for significance tests. Siegel (1979) gives the formula

$$\mathbf{P}\left[T_\lambda > t\right] =$$
$$\sum_{j=1}^{m} \sum_{k=0}^{j-1} (-1)^{j+k+1} \binom{m}{j} \binom{j-1}{k} \binom{m-1}{k} t^k (1 - j\lambda g_F - t)_+^{m-k-1}.$$

(492)

Siegel (1980) gives critical values t_λ for T_λ for significance levels 0.01 and 0.05 with m ranging from 5 to 50 (corresponding to values of N from 11 to 101) and with $\lambda = 0.4$, 0.6 and 0.8. For a value of $\lambda = 0.6$, Siegel found that his test statistic $T_{0.6}$ proved only slightly less powerful than Fisher's g (equivalently, T_1) against an alternative of simple periodicity, but that it outperformed Fisher's test against an alternative of *compound periodicity* (Siegel looked at periodic activity at both two and three frequencies).

	$\alpha = 0.05$			$\alpha = 0.01$		
m	Exact	$c\chi_o^2(\beta)$	$0.6g_F$	Exact	$c\chi_o^2(\beta)$	$0.6g_F$
10	0.181	0.178	0.267	0.214	0.217	0.322
20	0.116	0.114	0.162	0.134	0.134	0.198
30	0.0880	0.0872	0.119	0.0993	0.0998	0.145
40	0.0721	0.0715	0.0944	0.0799	0.0802	0.115
50	0.0616	0.0612	0.0788	0.0673	0.0676	0.0957
100	0.0373	0.0372	0.0443	0.0389	0.0391	0.0533
500	0.0115	0.0115	0.0110	0.0106	0.0107	0.0129
1000	0.00695	0.00695	0.00592	0.00607	0.00608	0.00687
2000	0.00421	0.00421	0.00317	0.00348	0.00348	0.00365

Table 493. Critical values $t_{0.6}$ for $T_{0.6}$ with sample sizes $N = 2m + 1$.

Table 493 gives exact critical values for $T_{0.6}$ for $\alpha = 0.05$ and 0.01 and sizes of m up to 2000 (thus extending Siegel's table). For computing (492) for such large m, we rewrite it as

$$\mathbf{P}\left[T_\lambda > t\right] = \sum_{j=1}^{m}\sum_{k=0}^{j-1}(-1)^{j+k+1}e^{Q_{j,k}}, \qquad (493a)$$

where

$$Q_{j,k} \equiv \log\binom{m}{j} + \log\binom{j-1}{k} + \log\binom{m-1}{k}$$
$$+ k\log(t) + (m-k-1)\log(1 - j\lambda g_F - t)_+$$

(Walden, 1992). If we recall that

$$\log\binom{a}{b} = \log(a!) - \log((a-b)!) - \log(b!)$$
$$= \log(\Gamma(a+1)) - \log(\Gamma(a-b+1)) - \log(\Gamma(b+1)),$$

we can use standard approximations to the log of a gamma function with a large argument (see, for example, Press *et al.*, 1986, p. 157). The argument of the exponent is kept under control if double precision arithmetic is used. Computation of these critical values for m in the range 20 to 2000 allows the construction of very accurate interpolation formulae for critical values for other values of m. For $\alpha = 0.05$ the interpolation formula is

$$t_{0.6} = 1.033m^{-0.72356}, \qquad (493b)$$

and for $\alpha = 0.01$ it is

$$t_{0.6} = 1.4987m^{-0.79695}. \qquad (493c)$$

An alternative way of finding the critical values for m large was given by Siegel (1979). He notes that

$$T_\lambda \stackrel{\mathrm{d}}{=} c\chi_0^2(\beta) \tag{494}$$

asymptotically, where $\chi_0^2(\beta)$ is what Siegel calls the noncentral chi-square distribution with zero degrees of freedom and noncentrality parameter β. The first and second moments of the distribution in (494) are

$$E\{T_\lambda\} = (1 - \lambda g_F)^m$$

and

$$E\{T_\lambda^2\} = \left[2(1 - \lambda g_F)^{m+1} + (m-1)(1 - 2\lambda g_F)_+^{m+1}\right]/(m+1),$$

and the parameters c and β in (494) are related by

$$c = \frac{\mathrm{var}\,\{T_\lambda\}}{4E\{T_\lambda\}} = \frac{E\{T_\lambda^2\} - (E\{T_\lambda\})^2}{4E\{T_\lambda\}} \quad \text{and} \quad \beta = E\{T_\lambda\}/c.$$

Hence it is merely necessary to find the critical values of $c\chi_0^2(\beta)$ with c and β given by the above. This can be done using the algorithm of Farebrother (1987). Critical values found in this way are given in Table 493 alongside of those found using Equation (493a).

The tests we have looked at above assume that any spectral lines occur at Fourier frequencies. Fisher's and Siegel's tests look particularly useful in theory for the case of large samples where the Fourier frequencies are closely spaced and thus are likely to be very close to any true lines (for large enough N, it will become obvious from the periodogram that there are significant peaks in the spectrum so that in practice the tests lose some of their value). What happens, however, if a spectral line occurs midway between two Fourier frequencies? Equation (475) can be employed to show that the expected value of the periodogram at such an intermediate frequency can be substantially less than that obtained at a Fourier frequency (cf. Equation (474); see also Whittle, 1952). Even though there must be some degradation in performance, Priestley (1981, p. 410) argues that Fisher's test will still be reasonably powerful, provided the signal-to-noise ratio $D_l^2/(2\sigma_\epsilon^2)$ is large (he also gives a summary of some other tests for periodicity, none of which appear particularly satisfactory).

Finally, we note that, although we have assumed an odd sample size N for mathematical convenience, Fisher's (1939) method for accommodating even values of N can be applied here (see also the appendix of Shimshoni, 1971). An example of analyzing an even length series is given next.

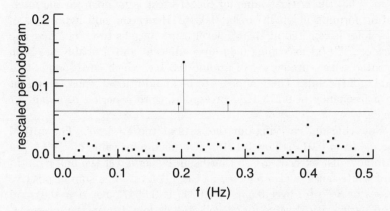

Figure 495. Application of Fisher's and Siegel's tests to ocean noise data.

10.10 Periodicity Tests with Ocean Noise Data

In Section 1.2 we noted that the sample acs of the ocean noise data in the bottom plot of Figure 7 shows a tendency to oscillate with a period of 5 seconds. What do the tests for simple or compound periodicity in white noise tell us about this series? First, it is convenient to use the fast Fourier transform to calculate $\hat{S}^{(p)}(f_k)$ for the series of length 128. We thus make the modification that now $m = (N - 2)/2$, giving $m = 63$ in this case. For an even length series the Nyquist frequency is also a Fourier frequency; however the m values of $\hat{S}^{(p)}(f_k)$ that we use exclude both the zero and Nyquist frequencies. It is readily shown (see Exercise [10.7]) that the term $\sum_{j=1}^{m} \hat{S}^{(p)}(f_j)$ in the expression for the rescaled periodogram $\tilde{S}^{(p)}(\cdot)$ of Equation (491d) is now given by

$$\sum_{j=1}^{m} \hat{S}^{(p)}(f_j) = \frac{\Delta t}{2} \left(\sum_{t=1}^{N} (X_t - \bar{X})^2 - \frac{1}{N} \left[\sum_{t=1}^{N} X_t (-1)^t \right]^2 \right).$$

Of course, $\sum_{j=1}^{m} \hat{S}^{(p)}(f_j)$ can just be calculated directly in the frequency domain, so that it is not strictly necessary to know this time domain equivalent.

Each $\tilde{S}^{(p)}(f_k)$ is plotted as a small solid square in Figure 495. The critical value g_F for Fisher's test ($\alpha = 0.05$) found using Equation (491c) is marked as the upper horizontal line (its value is 0.1087); the lower horizontal line marks $0.6g_F$ used in Siegel's test. The maximum value of $\tilde{S}^{(p)}(f_k)$ is 0.1336 and substantially exceeds g_F, so that the null hypothesis of white noise is rejected at the 5% level using Fisher's test. Siegel's test statistic $T_{0.6}$ is formed from the sum of the positive excesses of the $\tilde{S}^{(p)}(f_k)$ terms over $0.6g_F = 0.0653$. These excesses are shown by the three thin vertical lines in Figure 495; their sum is 0.0907. For $\alpha = 0.05$

and $m = 63$, the critical value for Siegel's test is found from the interpolation formula in (493b) to be 0.0515. Hence the null hypothesis of white noise is rejected at the 5% level using Siegel's test. The first two values of $\tilde{S}^{(p)}(f_k)$ exceeding $0.6g_F$ are adjacent and probably are both associated with a frequency of around 0.2 Hz, which corresponds to a period of 5 seconds. There appears to be an additional sinusoid present with a frequency of 0.273 Hz, corresponding to a period of about 3.7 seconds.

Now consider carrying out the tests at the 1% level. The critical value g_F ($\alpha = 0.01$) is now 0.1316. The maximum value of $\tilde{S}^{(p)}(f_k)$ of 0.1336 now just exceeds g_F, so that using Fisher's test the null hypothesis of white noise is also rejected at the 1% level. The sum of positive excesses of $\tilde{S}^{(p)}(f_k)$ over $0.6g_F = 0.0790$ is 0.0547. For $\alpha = 0.01$ and $m = 63$, the critical value for Siegel's test is found from the interpolation formula in (493c) to be 0.0552. Hence the null hypothesis of white noise is not rejected (but only just!) at the 1% level using Siegel's test.

The practical conclusion from these tests is that there is almost certainly one, and probably two sinusoids, of periods about 5 and 3.7 seconds, present in the ocean noise data, and a physical explanation should be sought to explain their presence.

10.11 Tests for Periodicity – Colored Noise Case

We now turn our attention to case [2] described at the beginning of Section 10.9 – tests for spectral lines in *colored* noise. The multitaper-based method given in Thomson (1982) is worthy of study. For convenience we assume the following version of Equation (467a) involving just a single frequency f_1:

$$X_t = D_1 \cos\left(2\pi f_1 t\,\Delta t + \phi_1\right) + \eta_t, \qquad (496a)$$

where $\{\eta_t\}$ (the background continuum) is a zero mean stationary process with sdf $S_\eta(\cdot)$ that is not necessarily constant; i.e., $\{\eta_t\}$ can be colored noise. We also assume in this section that $\{\eta_t\}$ has a Gaussian distribution.

In the usual formulation, ϕ_1 in Equation (496a) would be considered as an rv, but in Thomson's approach it (along with D_1) is regarded as an unknown constant. This assumption turns $\{X_t\}$ into a process with a time varying mean value:

$$E\{X_t\} = D_1 \cos\left(2\pi f_1 t\,\Delta t + \phi_1\right) = C_1 e^{i2\pi f_1 t\,\Delta t} + C_1^* e^{-i2\pi f_1 t\,\Delta t},$$

where $C_1 \equiv D_1 e^{i\phi_1}/2$. Note that we can write

$$X_t = E\{X_t\} + \eta_t. \qquad (496b)$$

As in Chapter 7, let us consider

$$J_k(f) \equiv (\Delta t)^{1/2} \sum_{t=1}^{N} h_{t,k} X_t e^{-i2\pi f t\,\Delta t},$$

where $\{h_{t,k}\}$ is a kth order dpss data taper with parameter NW and associated spectral window

$$\mathcal{H}_k(f) \equiv \frac{1}{\Delta t} |H_k(f)|^2, \quad \text{where } H_k(f) \equiv \Delta t \sum_{t=1}^{N} h_{t,k} e^{-i2\pi ft\,\Delta t}.$$

We recall that $|J_k(f)|^2 = \hat{S}_k^{(mt)}(f)$, the kth eigenspectrum in the multi-taper scheme. The expected value of the complex-valued quantity $J_k(f)$ is

$$E\{J_k(f)\} = (\Delta t)^{1/2} \sum_{t=1}^{N} h_{t,k} \left(C_1 e^{i2\pi f_1 t\,\Delta t} + C_1^* e^{-i2\pi f_1 t\,\Delta t} \right) e^{-i2\pi ft\,\Delta t}$$

$$= \frac{1}{(\Delta t)^{1/2}} \left[C_1 H_k(f - f_1) + C_1^* H_k(f + f_1) \right].$$

The frequency domain analog of Equation (496b) is thus

$$J_k(f) = E\{J_k(f)\} + (\Delta t)^{1/2} \sum_{t=1}^{N} h_{t,k} \eta_t e^{-i2\pi ft\,\Delta t}.$$

Note that the squared modulus of the term with η_t is a direct spectral estimator of $S_\eta(f)$ using the data taper $\{h_{t,k}\}$. At $f = f_1$, the above becomes

$$J_k(f_1) = E\{J_k(f_1)\} + (\Delta t)^{1/2} \sum_{t=1}^{N} h_{t,k} \eta_t e^{-i2\pi f_1 t\,\Delta t},$$

where

$$E\{J_k(f_1)\} = \frac{1}{(\Delta t)^{1/2}} \left[C_1 H_k(0) + C_1^* H_k(2f_1) \right]. \qquad (497a)$$

Now, since the squared modulus of $H_k(\cdot)$ is highly concentrated in the interval $[-W, W]$, we have, for $2f_1 > W$,

$$E\{J_k(f_1)\} \approx \frac{1}{(\Delta t)^{1/2}} C_1 H_k(0) = (\Delta t)^{1/2} C_1 \sum_{t=1}^{N} h_{t,k}. \qquad (497b)$$

Because a kth order dpss data taper is skew-symmetric about its mid-point for odd k (see the left-hand plots of Figures 336 and 338), the above summation is in fact zero for odd k, but it is real-valued and positive for even k.

We now assume that $2f_1 > W$ and take the approximation in Equation (497b) to be an equality in what follows. We can write the following model for $J_k(f_1)$:

$$J_k(f_1) = C_1 \frac{H_k(0)}{(\Delta t)^{1/2}} + \tilde{\epsilon}_k, \quad k = 0, \ldots, K-1, \qquad (498)$$

where

$$\tilde{\epsilon}_k \equiv (\Delta t)^{1/2} \sum_{t=1}^{N} h_{t,k} \eta_t e^{-i2\pi f_1 t \Delta t},$$

and K is the number of orthogonal data tapers used in the multitaper scheme (recall that usually $K < 2NW$). Equation (498) thus defines a complex-valued regression model: $J_k(f_1)$ is the observed dependent variable; C_1 is an unknown parameter; $H_k(0)/(\Delta t)^{1/2}$ is the kth independent variable; and $\tilde{\epsilon}_k$ is the error term. In order to use the results of the complex-valued regression theory we appeal to below, the $\tilde{\epsilon}_k$ terms must be uncorrelated complex-valued Gaussian rv's with zero means and a common variance $\sigma_{\tilde{\epsilon}}^2$; moreover, the real and imaginary components of $\tilde{\epsilon}_k$ must be uncorrelated, and each must have the same semivariance, namely, $\sigma_{\tilde{\epsilon}}^2/2$ (Section 2.5 has the definition of a complex-valued Gaussian rv). Because, by assumption, $\{\eta_t\}$ is a zero mean real-valued Gaussian stationary process, it is clear that $\tilde{\epsilon}_k$ must follow a complex Gaussian distribution with zero mean. Since $\tilde{\epsilon}_k$ is just the kth eigencoefficient in a multitaper estimator of $S_\eta(f_1)$, we can appeal to the arguments of Chapter 7 to claim that the $\tilde{\epsilon}_k$ terms are approximately pairwise uncorrelated if the sdf $S_\eta(\cdot)$ is slowly varying in the interval $[f_1 - W, f_1 + W]$ – this condition is an *assumption* that we now make. Moreover, we also have

$$\sigma_{\tilde{\epsilon}}^2 = \text{var}\{\tilde{\epsilon}_k\} = E\{|\tilde{\epsilon}_k|^2\} = S_\eta(f_1),$$

so each $\tilde{\epsilon}_k$ has the same variance. Finally, the additional assumption that the real and imaginary components of $\tilde{\epsilon}_k$ are uncorrelated and have equal variance is reasonable as long as f_1 is not too close to either zero or the Nyquist frequency. With these stipulations, $\tilde{\epsilon}_k$ matches the requirements of complex-valued regression theory.

We can now obtain an estimator for C_1 by employing a version of least squares valid for complex-valued quantities. This amounts to estimating C_1 by finding the value $\hat{C}_1$ that minimizes

$$\text{SS}(\hat{C}_1) \equiv \sum_{k=0}^{K-1} \left| J_k(f_1) - \hat{C}_1 \frac{H_k(0)}{(\Delta t)^{1/2}} \right|^2.$$

It follows from Miller (1973, Equation (3.2)) that

$$\hat{C}_1 = (\Delta t)^{1/2} \frac{\sum_{k=0}^{K-1} J_k(f_1) H_k(0)}{\sum_{k=0}^{K-1} H_k^2(0)} = (\Delta t)^{1/2} \frac{\sum_{k=0,2,\dots}^{K-1} J_k(f_1) H_k(0)}{\sum_{k=0,2,\dots}^{K-1} H_k^2(0)}$$

(499a)

(we can eliminate $k = 1, 3, \dots$ from the summations because $H_k(0) = 0$ for odd k). Moreover, under the conditions on $\tilde{\epsilon}_k$ stated in the previous paragraph, we can state the following results, all based on Theorem 8.1 of Miller (1973). First, $\hat{C}_1$ is a complex Gaussian rv with mean C_1 and variance $\sigma_{\tilde{\epsilon}}^2 \Delta t / \sum_{k=0}^{K-1} H_k^2(0)$. Second, an estimator of $\sigma_{\tilde{\epsilon}}^2$ is given by

$$\hat{\sigma}_{\tilde{\epsilon}}^2 \equiv \frac{1}{K} \sum_{k=0}^{K-1} \left| J_k(f_1) - \hat{J}_k(f_1) \right|^2, \qquad (499b)$$

where $\hat{J}_k(f_1)$ is the fitted value for $J_k(f_1)$ given by

$$\hat{J}_k(f_1) \equiv \hat{C}_1 \frac{H_k(0)}{(\Delta t)^{1/2}}.$$

Third, the rv $2K\hat{\sigma}_{\tilde{\epsilon}}^2/\sigma_{\tilde{\epsilon}}^2$ follows a chi-square distribution with $2K - 2$ degrees of freedom. Finally, $\hat{C}_1$ and $2K\hat{\sigma}_{\tilde{\epsilon}}^2/\sigma_{\tilde{\epsilon}}^2$ are independent of each other.

Thomson's test for a periodicity in colored noise is based upon the usual F-test for the significance of a regression parameter. Under the null hypothesis that $C_1 = 0$, the rv $\hat{C}_1$ has a complex Gaussian distribution with zero mean and variance $\sigma_{\tilde{\epsilon}}^2 \Delta t / \sum_{k=0}^{K-1} H_k^2(0)$; moreover, because $J_k(f_1) = \tilde{\epsilon}_k$ under the null hypothesis, it follows from our assumptions about $\tilde{\epsilon}_k$ that the real and imaginary components of $\hat{C}_1$ are uncorrelated and have the same variance. Since $|\hat{C}_1|$ is thus the sum of squares of two uncorrelated real-valued Gaussian rv's with zero means and equal variances, we have

$$\frac{2 \left| \hat{C}_1 \right|^2 \sum_{k=0}^{K-1} H_k^2(0)}{\sigma_{\tilde{\epsilon}}^2 \Delta t} \stackrel{\mathrm{d}}{=} \chi_2^2.$$

Because $\hat{C}_1$ and the χ_{2K-2}^2 rv $2K\hat{\sigma}_{\tilde{\epsilon}}^2/\sigma_{\tilde{\epsilon}}^2$ are independent, it follows that the above χ_2^2 rv is independent of $2K\hat{\sigma}_{\tilde{\epsilon}}^2/\sigma_{\tilde{\epsilon}}^2$ also. The ratio of independent χ^2 rv's with 2 and $2K - 2$ degrees of freedom – each divided by their respective degrees of freedom – is F-distributed with 2 and $2K - 2$ degrees of freedom. Hence, we obtain (after some reduction)

$$\frac{(K - 1) \left| \hat{C}_1 \right|^2 \sum_{k=0}^{K-1} H_k^2(0)}{\Delta t \sum_{k=0}^{K-1} \left| J_k(f_1) - \hat{J}_k(f_1) \right|^2} \stackrel{\mathrm{d}}{=} F_{2,2K-2}. \qquad (499c)$$

If in fact $C_1 \neq 0$, the above statistic should give a value exceeding a high percentage point of the $F_{2,2K-2}$ distribution (the 95% point, say). This test statistic was given by Thomson (1982, Equation (13.10)); see also Park *et al.* (1987a) for an application of this approach to decaying sinusoids.

If the F-test at $f = f_1$ is significant (i.e., the null hypothesis that $C_1 = 0$ is rejected), Thomson (1982) suggests reshaping the spectrum around f_1 to give a better estimate of $S_\eta(\cdot)$, the sdf of the background continuum $\{\eta_t\}$. Since $\sigma_\epsilon^2 = S_\eta(f_1)$, Equation (499b) can be rewritten as

$$\hat{S}_\eta(f_1) = \frac{1}{K} \sum_{k=0}^{K-1} \left| J_k(f_1) - \hat{J}_k(f_1) \right|^2 = \frac{1}{K} \sum_{k=0}^{K-1} \left| J_k(f_1) - \hat{C}_1 \frac{H_k(0)}{(\Delta t)^{1/2}} \right|^2.$$

For f in the neighborhood of f_1 (i.e., $f_1 - W \leq f \leq f_1 + W$), the above generalizes to

$$\hat{S}_\eta(f) = \frac{1}{K} \sum_{k=0}^{K-1} \left| J_k(f) - \hat{C}_1 \frac{H_k(f - f_1)}{(\Delta t)^{1/2}} \right|^2. \tag{500}$$

Some comments should be made on the practical application of Thomson's method. Since the eigencoefficients $J_k(\cdot)$ are typically computed using an FFT, it has been implicitly assumed that f_1 coincides with a standard frequency after padding with zeros to obtain a fine grid. Taking f_1 to be the standard frequency at which $\hat{C}_1$ is a maximum defines the frequency at which to carry out the F-test. (Note that the regression model in Equation (498) is defined pointwise at each frequency, so that there is no assumption of uncorrelated eigencoefficients across frequencies.) If another line occurs at a frequency f_2 outside $[f_1 - W, f_1 + W]$ – i.e., outside the design bandwidth – then a large value of $\hat{C}_2$ should result, and another F-test can be carried out at f_2. (The assumption of a slowly varying spectrum around f_1 must also now hold around f_2.) This approach extends to more lines satisfying the same requirements.

Comments and Extensions to Section 10.11

[1]   Let us consider the special case $f_1 = 0$ so that Equation (496a) becomes $X_t = D_1 + \eta_t$ (we set $\phi_1 = 0$ since the 'phase' of a constant term is undefined). The parameter D_1 is now the expected value of the stationary process $\{X_t\}$; i.e., $E\{X_t\} = D_1$. With $C_1 \equiv D_1/2$, Equation (497a) becomes

$$E\{J_k(0)\} = \frac{1}{(\Delta t)^{1/2}} 2C_1 H_k(0) = \frac{1}{(\Delta t)^{1/2}} D_1 H_k(0),$$

and the least squares estimator of C_1 (Equation (499a)) becomes

$$\hat{C}_1 = \frac{(\Delta t)^{1/2}}{2} \frac{\sum_{k=0}^{K-1} J_k(0) H_k(0)}{\sum_{k=0}^{K-1} H_k^2(0)}, \quad \text{so} \quad \hat{D}_1 = (\Delta t)^{1/2} \frac{\sum_{k=0}^{K-1} J_k(0) H_k(0)}{\sum_{k=0}^{K-1} H_k^2(0)}.$$

If $K = 1$, the above reduces to

$$\hat{D}_1 = (\Delta t)^{1/2} \frac{J_0(0) H_0(0)}{H_0^2(0)} = \frac{\sum_{t=1}^{N} h_{t,0} X_t}{\sum_{t=1}^{N} h_{t,0}}.$$

This is the alternative estimator to the sample mean given previously in Equation (217a). The reshaped sdf estimator that is appropriate for this case is

$$\hat{S}_\eta(f) = \left| J_0(f) - \hat{D}_1 \frac{H_0(0)}{(\Delta t)^{1/2}} \right|^2 = \Delta t \left| \sum_{t=1}^{N} h_{t,0} \left(X_t - \hat{D}_1 \right) e^{-i 2\pi f t \Delta t} \right|^2,$$

in agreement with the 'mean corrected' direct spectral estimator of Equation (217b) with $\hat{D}_1$ substituted for $\tilde{\mu}$.

[2] The upper $(1 - \alpha) \times 100\%$ percentage point of the $F_{2,\nu}$ distribution can be surprisingly easily computed using the formula

$$\frac{\nu(1 - \alpha^{2/\nu})}{2\alpha^{2/\nu}}.$$

10.12 Completing a Harmonic Analysis

In Sections 10.5 and 10.6 we outlined ways of searching for spectral lines immersed in white noise by looking for peaks in the periodogram. Tests for periodicity – both simple and compound – against an alternative hypothesis of white noise were given in Section 10.9; additionally Thomson's multitaper-based test for detecting periodicity in a white or colored background continuum was discussed in the previous section. Suppose that, based on one of these tests, we have concluded that there are one or more line components in a time series. To complete our analysis, we must estimate the amplitudes of the line components and the sdf of the background continuum. If we have evidence that the background continuum is white noise, we need only estimate the variance σ_ϵ^2 of the white noise since its sdf is just $\sigma_\epsilon^2 \Delta t$; for the case of a colored continuum, we must estimate the sdf after somehow removing the line components from the time series. Let us first consider the white noise case in detail.

Let us assume the model given by Equation (463a):

$$X_t = \sum_{l=1}^{L} \left(A_l \cos \left(2\pi f_l t \, \Delta t \right) + B_l \sin \left(2\pi f_l t \, \Delta t \right) \right) + \epsilon_t.$$

Suppose that we have estimated the line frequencies f_l by $\hat{f}_l$. We can then form the *approximate conditional* least squares estimates for A_l and B_l from Equation (463b):

$$\hat{A}_l = \frac{2}{N} \sum_{t=1}^{N} X_t \cos\left(2\pi \hat{f}_l t \, \Delta t\right) \quad \text{and} \quad \hat{B}_l = \frac{2}{N} \sum_{t=1}^{N} X_t \sin\left(2\pi \hat{f}_l t \, \Delta t\right).$$

Here 'conditional' refers to the use of the estimator $\hat{f}_l$ rather than the actual f_l, and 'approximate' refers to the fact that, for non-Fourier frequencies, these estimators will not be exact least squares estimators (see Section 10.2). The size of the jumps at $\pm \hat{f}_l$ in the integrated spectrum for $\{X_t\}$ would be estimated by

$$\frac{\hat{A}_l^2 + \hat{B}_l^2}{4} = \frac{\hat{S}^{(p)}(\hat{f}_l)}{N \, \Delta t}$$

(see the discussion surrounding Equation (473)). The white noise variance σ_ϵ^2 would be estimated from

$$\hat{\sigma}_\epsilon^2 \equiv \frac{\text{SS}(\hat{A}_l, \hat{B}_l, \hat{f}_l)}{N - 2L},$$

where

$$\text{SS}(\hat{A}_l, \hat{B}_l, \hat{f}_l) \equiv \sum_{t=1}^{N} \left(X_t - \sum_{l=1}^{L} \left[\hat{A}_l \cos\left(2\pi \hat{f}_l t \, \Delta t\right) + \hat{B}_l \sin\left(2\pi \hat{f}_l t \, \Delta t\right) \right] \right)^2.$$

The sdf $S_\epsilon(\cdot)$ of the white noise would be estimated by $\hat{S}_\epsilon(f) = \hat{\sigma}_\epsilon^2 \, \Delta t$ for $|f| \leq f_{(N)}$.

A second approach would be to find *exact conditional* least squares estimates for A_l and B_l by determining those values of A_l and B_l that actually minimize $\text{SS}(A_l, B_l, \hat{f}_l)$ as a function of A_l and B_l. These can be obtained by regressing X_t on $\cos\left(2\pi \hat{f}_l t \, \Delta t\right)$ and $\sin\left(2\pi \hat{f}_l t \, \Delta t\right)$. If we let $\tilde{A}_l$ and $\tilde{B}_l$ denote these minimizing values, the obvious estimate for σ_ϵ^2 is

$$\tilde{\sigma}_\epsilon^2 \equiv \frac{\text{SS}(\tilde{A}_l, \tilde{B}_l, \hat{f}_l)}{N - 2L}.$$

A third approach would be to find *exact unconditional* least squares estimates for A_l, B_l and f_l by determining those values of A_l, B_l and f_l that minimize $\text{SS}(A_l, B_l, f_l)$ as a function of A_l, B_l and f_l. If we let $\breve{A}_l$, $\breve{B}_l$ and $\breve{f}_l$ denote the minimizing values, we would now estimate σ_ϵ^2 by

$$\breve{\sigma}_\epsilon^2 \equiv \frac{\text{SS}(\breve{A}_l, \breve{B}_l, \breve{f}_l)}{N - 2L}.$$

In practice, exact unconditional least squares estimates must be found by means of a nonlinear optimization routine, for which initial values of the parameters can be obtained from a periodogram analysis. Bloomfield (1976) has an extensive discussion of these estimates (along with some useful FORTRAN subroutines). For most time series encountered in nature, approximate conditional least squares estimates are adequate (Bloomfield gives a pathological artificial example where exact unconditional least squares estimates do remarkably better than the other two estimation procedures). Rife and Boorstyn (1976) consider a similar simplification of an exact maximum likelihood approach.

Once we have obtained estimates of A_l, B_l and f_l by any of these three approaches, we can examine the adequacy of our fitted model by looking at the *residual process* $\{R_t\}$. For example, if we use approximate conditional least squares estimates, this process is defined by

$$R_t = X_t - \sum_{l=1}^{L} \left[\hat{A}_l \cos\left(2\pi \hat{f}_l t\, \Delta t\right) + \hat{B}_l \sin\left(2\pi \hat{f}_l t\, \Delta t\right) \right].$$

If our fitted model is adequate, the R_t series should look like a realization of a white noise process because they can be regarded as proxies for the unknown ϵ_t series. In particular, if we look at the periodogram or other spectral estimates of the R_t series, we can see – or formally test – whether the residual process can be regarded as white noise. While this procedure works well in many cases, spectral estimates of $\{R_t\}$ tend to indicate a deficiency of power (i.e., deep troughs) around the line frequencies.

Now we consider the second possibility, namely, a colored continuum. Our model is now

$$X_t = \sum_{l=1}^{L} \left(A_l \cos\left(2\pi f_l t\, \Delta t\right) + B_l \sin\left(2\pi f_l t\, \Delta t\right)\right) + \eta_t,$$

where $\{\eta_t\}$ is a stationary process with zero mean and nonwhite sdf $S_\eta(\cdot)$. We assume again that we have estimated the frequencies f_l by $\hat{f}_l$. One approach is to use, say, approximate conditional least squares – as in the case of a white continuum – to estimate the amplitudes of the line components and then to form the residual process $\{R_t\}$. If $\hat{A}_l$, $\hat{B}_l$ and $\hat{f}_l$ are good estimates, then $\{R_t\}$ should be a good approximation to $\{\eta_t\}$. We can thus estimate $S_\eta(\cdot)$ by computing an sdf estimate using these residuals, but again we note that such estimates tend to have a deficiency of power around the line frequencies.

A second approach is to reformulate the model in terms of complex exponentials, i.e., to write

$$X_t = \sum_{l=1}^{L} \left(C_l e^{i2\pi f_l t\, \Delta t} + C_l^* e^{-i2\pi f_l t\, \Delta t}\right) + \eta_t, \tag{503}$$

and then to estimate C_l via the multitaper approach of the previous section. If the $\hat{f}_l$ terms are sufficiently separated and not too close to zero, as discussed, then the estimator of C_l is given by

$$\hat{C}_l = (\Delta t)^{1/2} \frac{\sum_{k=0,2,\ldots}^{K-1} J_k(\hat{f}_l) H_k(0)}{\sum_{k=0,2,\ldots}^{K-1} H_k^2(0)}$$

(cf. Equation (499a)), where $J_k(\hat{f}_l)$ and $H_k(0)$ are as defined previously. The size of the jumps in the integrated spectrum at $\pm \hat{f}_l$ would be estimated by $|\hat{C}_l|^2$. The sdf $S_\eta(\cdot)$ of the background continuum would be estimated by reshaping – in the neighborhoods of each $\hat{f}_l$ – the multitaper sdf estimate based upon $\{X_t\}$ (see Equation (500)). This estimate of $S_\eta(\cdot)$ tends to be much better behaved around the line frequencies than estimates based upon the residual process.

Finally we note that the multitaper approach can also be used with a white noise continuum. Here we need only estimate the variance σ_ϵ^2 of the white noise, and the appropriate estimator is now

$$\hat{\sigma}_\epsilon^2 = \frac{1}{N - 2L} \sum_{t=1}^{N} \left(X_t - \sum_{l=1}^{L} \left[\hat{C}_l e^{i2\pi \hat{f}_l t \Delta t} + \hat{C}_l^* e^{-i2\pi \hat{f}_l t \Delta t} \right] \right)^2.$$

This expression can be reduced somewhat by noting that

$$\hat{C}_l e^{i2\pi \hat{f}_l t \Delta t} + \hat{C}_l^* e^{-i2\pi \hat{f}_l t \Delta t} = 2\Re(\hat{C}_l) \cos\left(2\pi \hat{f}_l t \Delta t\right)$$
$$- 2\Im(\hat{C}_l) \sin\left(2\pi \hat{f}_l t \Delta t\right),$$

where $\Re(z)$ and $\Im(z)$ refer, respectively, to the real and imaginary components of the complex number z. (Rife and Boorstyn, 1974, 1976, present an interesting discussion about the estimation of complex exponentials in white noise.)

10.13 Harmonic Analysis of River Flow Data

As a rather simple example of a harmonic analysis, let us consider a time series related to the flow of the Willamette River at Salem, Oregon (the first 128 values of this series were used as an example in Chapter 1). Figures 505, 508, 510 and 512 summarize our analysis, the details of which we now discuss (the number preceding each paragraph indicates the plot under discussion; for example, 505(c) refers to the third plot in Figure 505).

505(a):   We begin by plotting the $N = 395$ data points in the time series. Each point represents the log of the average daily water flow over a one month period from October, 1950, to August, 1983. The

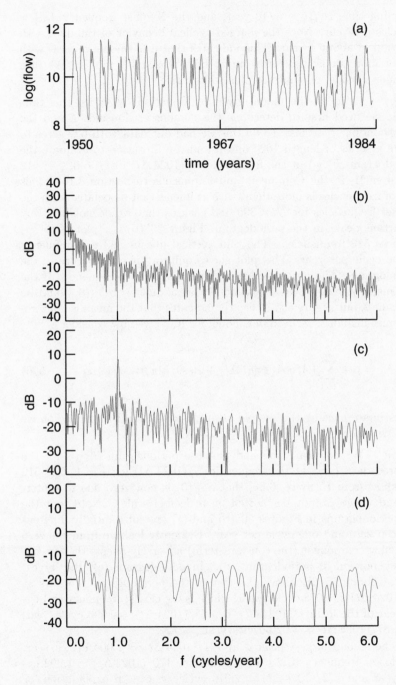

Figure 505. Analysis of Willamette River data, part 1.

sampling time is $\Delta t = 1/12$ year, and the Nyquist frequency $f_{(N)}$ is 6 cycles per year. Note the marked cyclical behavior of the data with a period of about 1 year. This indicates that a harmonic process with one or more sinusoidal terms might be a reasonable model. We thus consider a model of sinusoids plus white noise.

505(b): To fit the model given by Equation (463a) to this time series, we need first to determine the number of sinusoids L and the corresponding f_l terms. To do this we pad our data with 629 zeros to create a series of length 1024 and use an FFT routine to evaluate the periodogram $\hat{S}^{(p)}(\cdot)$ at the frequencies $j/(1024\,\Delta t)$ for $j = 0, \ldots, 512$ (see item [1] in the Comments and Extensions to Section 3.10). This grid of frequencies is more than twice as fine as that associated with the Fourier frequencies for $N = 395$ and ensures that we do not miss any important peaks in the periodogram. Figure 505(b) is a plot of $\hat{S}^{(p)}(\cdot)$ at these 513 frequencies. The thin vertical line indicates a frequency of one cycle per year. This plot shows quite a bit of structure in the periodogram. The largest value of $\hat{S}^{(p)}(\cdot)$ is at zero frequency. If the true mean were zero (as Equation (463a) assumes), $\hat{S}^{(p)}(0)$ would be very small, but Figure 505(a) clearly indicates that the mean is nonzero. We must include a term in our model for it, so we now consider

$$X_t = \mu + \sum_{l=1}^{L} \left(A_l \cos\left(2\pi f_l t\,\Delta t\right) + B_l \sin\left(2\pi f_l t\,\Delta t\right)\right) + \epsilon_t. \qquad (506)$$

We estimate μ using the sample mean $\bar{X} = 9.83$, and from now on we work with the centered data $X_t' \equiv X_t - \bar{X}$.

505(c): We now calculate and plot the periodogram $\hat{S}^{(p)}(\cdot)$ for the centered data $\{X_t'\}$ at the frequencies $j/(1024\,\Delta t)$ for $j = 1, \ldots, 512$ (we know from Exercise [6.5b] that $\hat{S}^{(p)}(0)$ is now zero due to centering and hence cannot be plotted on a decibel scale). Note that the two periodograms in Figures 505(b) and (c) are substantially different between zero and one cycle per year. Evidently leakage from the zero frequency component (the constant term) adversely affects the low frequency portion of periodogram (b). In periodogram (c), the largest peak – about 15 dB above the rest of the periodogram – occurs at $86/(1024\,\Delta t) = 1.008$ cycles/year, which is the closest frequency to 1 cycle/year of the form $j/(1024\,\Delta t)$; i.e., $85/(1024\,\Delta t) = 0.996$ cycle/year, while $87/(1024\,\Delta t) = 1.020$ cycles/year.

The second largest peak is at $171/(1024\,\Delta t) = 2.004$ cycles/year, the closest frequency to 2 cycles/year since $170/(1024\,\Delta t) = 1.992$ cycles/year and $172/(1024\,\Delta t) = 2.016$ cycles/year. An explanation for the presence of this peak is that the river flow data have an annual variation that is periodic but not purely sinusoidal. If we describe this variation by a real-valued periodic function $g(\cdot)$ with a period of 1 year, the results of Section 3.1 say that we can write

$$g(t) = \frac{a_0}{2} + \sum_{l=1}^{\infty} a_l \cos{(2\pi l t)} + b_l \cos{(2\pi l t)}$$

(cf. Equation (57) with $T = 1$). The frequency that is associated with the lth cosine and sine in the summation is just $f_l' \equiv l$ cycles/year. The lowest such frequency, namely $f_1' = 1$ cycle/year, is called the *fundamental frequency* for $g(\cdot)$ and is equal to the reciprocal of its period. All other frequencies in the above representation for $g(\cdot)$ are multiples of the fundamental frequency; i.e., $f_l' = l f_1' = l$ cycles/year. The frequency f_{l+1}' is called the lth harmonic of the fundamental frequency f_1' so that, for example, $f_2' = 2$ cycles/year is the first harmonic. If we now sample $g(\cdot)$ at points $\Delta t = 1/12$ year apart, we obtain a periodic sequence given by

$$g_t \equiv g(t\,\Delta t) = \frac{a_0}{2} + \sum_{l=1}^{\infty} a_l \cos{(2\pi l t\,\Delta t)} + b_l \cos{(2\pi l t\,\Delta t)}. \qquad (507\mathrm{a})$$

Because this choice of Δt produces an aliasing effect, the above can be rewritten as

$$g_t = \mu' + \sum_{l=1}^{6} \left(A_l' \cos{(2\pi l t\,\Delta t)} + B_l' \sin{(2\pi l t\,\Delta t)} \right), \qquad (507\mathrm{b})$$

where μ', A_l' and B_l' are related to a_l and b_l (see Exercise [10.8]). Note that the right-hand side of the above equation resembles the nonrandom portion of (506) if we let $L = 6$ and equate f_l with $l = f_l'$. Given the nature of this time series, it is thus not unreasonable to find a peak in periodogram (c) at 2 cycles/year, the first harmonic for a fundamental frequency of 1 cycle/year. Recall that we also encountered harmonics in the example from tidal analysis in Section 10.4.

505(d):   Besides the peaks corresponding to annual and semiannual periods, there are numerous other lesser peaks in the periodogram that might or might not be due to other sinusoidal components. To see if we can identify components – such as the second or higher harmonics of 1 cycle/year – that might be hidden due to leakage from the dominant peaks, Figure 505(d) shows a direct spectral estimate $\hat{S}^{(d)}(\cdot)$ based upon $\{X_t'\}$ and a dpss data taper with parameter $NW = 4/\Delta t$. While there are again lots of lesser peaks, none of them seems to be particularly prominent. The overall agreement between periodogram (c) and $\hat{S}^{(d)}(\cdot)$ is good, so evidently leakage is not a problem here.

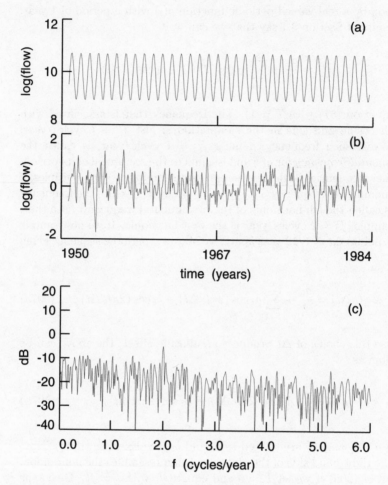

Figure 508. Analysis of Willamette River data, part 2.

508(a): Let us assume for the moment that there is only one frequency
of importance in the time series so that the following simple harmonic
model is appropriate:

$$X_t' = A_1 \cos\left(2\pi f_1 t\,\Delta t\right) + B_1 \sin\left(2\pi f_1 t\,\Delta t\right) + \epsilon_t. \tag{508}$$

Although, based on physical arguments, it would be reasonable to consider that f_1 is known to be 1 cycle/year, let us assume – for illustrative
purposes – that it is unknown and use the periodogram to estimate it.
If we search $\hat{S}^{(p)}(\cdot)$ in Figure 505(c) for its peak value, we find that it
occurs at $\hat{f}_1 = 1.0032$ cycles/year. (In practice, an easy way to do this
is to just append more zeros to our data, use an FFT and interpolate.)

Given this periodogram estimate of f_1, we can now estimate the remaining parameters in the model by approximate conditional least squares (acls) to get

$$\hat{A}_1 = -0.686, \quad \hat{B}_1 = 0.578 \text{ and } \hat{\sigma}_\epsilon^2 = 0.2224$$

(for comparison, the sample variance of this time series is 0.6238). Our estimate of the size of the jump in the integrated spectrum at ± 1.0032 cycles/year is thus $(\hat{A}_1^2 + \hat{B}_1^2)/4 = 0.2012$. Figure 508(a) is a plot of the fitted model for $\{X_t\}$, i.e.,

$$\bar{X} + \hat{A}_1 \cos\left(2\pi \hat{f}_1 t \, \Delta t\right) + \hat{B}_1 \sin\left(2\pi \hat{f}_1 t \, \Delta t\right).$$

508(b): Here we plot the associated residuals from the model, namely,

$$R_t^{(1)} \equiv X_t - \bar{X} - \hat{A}_1 \cos\left(2\pi \hat{f}_1 t \, \Delta t\right) - \hat{B}_1 \sin\left(2\pi \hat{f}_1 t \, \Delta t\right).$$

If the simple one term model were adequate, $\{R_t^{(1)}\}$ would be close to a white noise process.

508(c): To examine whether $\{R_t^{(1)}\}$ can be considered to be white noise, we plot the periodogram of these residuals. Note that the only substantial difference between this periodogram and that of $\{X_t'\}$ in Figure 505(c) is near $f = 1$ cycle/year, as is reasonable to expect. The peak at $f = 2$ cycles/year is the most prominent one in the periodogram of the residuals. Although this peak does occur at a frequency that, given the nature of this series, is reasonable, it is only about 5 dB above the rest of the periodogram. Since we are in a situation where Fisher's g statistic (Equation (491a)) can help decide whether this peak is significant, we compute it for these residuals and obtain $g = 0.0875$. Under the null hypothesis of white noise, the upper 1% critical level for the test statistic is approximately 0.049 (from the approximation in Equation (491c), or Table 1(b) of Shimshoni, 1971, with $m = (N-1)/2 = 197$). Since g exceeds this value, we have evidence that the peak is not just due to random variation. (There is, however, reason to use this statistic here with caution: the periodogram for $\{R_t^{(1)}\}$ decreases about 10 dB as f increases from 0 to 6 cycles/year, indicating that the presumed null hypothesis of white noise is questionable – see Figure 510(d).)

510(a): Let us now assume that there are two frequencies of importance in the time series so that our model becomes

$$X_t = \mu + \sum_{l=1}^{2} A_l \cos\left(2\pi f_l t \, \Delta t\right) + B_l \sin\left(2\pi f_l t \, \Delta t\right) + \epsilon_t,$$

where we use the same estimates of μ, A_1, B_1 and f_1 as before. Again, rather than assuming that f_2 is 2 cycles/year, we estimate it from the

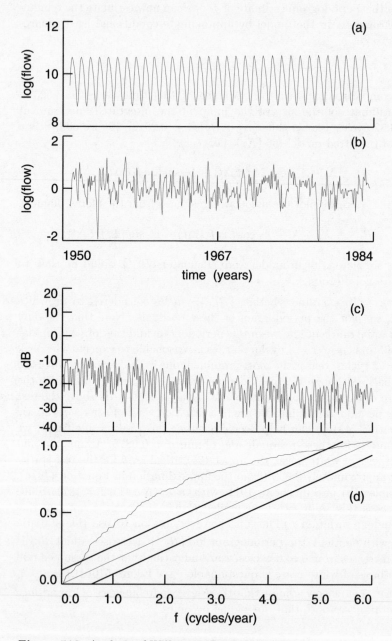

Figure 510. Analysis of Willamette River data, part 3.

periodogram for $\{X_t'\}$ and obtain $\hat{f}_2 = 2.0034$. The acls estimates of A_2 and B_2 are

$$\hat{A}_2 = -0.150, \quad \hat{B}_2 = 0.128, \quad \text{and now} \quad \hat{\sigma}_\epsilon^2 = 0.2036,$$

a reduction in residual variance of about 10% over the single frequency model. Our estimate of the size of the jump in the integrated spectrum at ± 2.0034 cycles/year is $(\hat{A}_2^2 + \hat{B}_2^2)/4 = 0.0097$. Figure 510(a) is a plot of the fitted two frequency model for $\{X_t\}$. There is little visual difference between this plot and Figure 508(a) for the one frequency model.

510(b):   This plot shows the residuals for the two frequency model, namely,

$$R_t^{(2)} \equiv X_t - \bar{X} - \sum_{l=1}^{2} \left(\hat{A}_l \cos\left(2\pi \hat{f}_l t \, \Delta t\right) + \hat{B}_l \sin\left(2\pi \hat{f}_l t \, \Delta t\right) \right).$$

Again, if the two frequency model were adequate, $\{R_t^{(2)}\}$ would be close to a white noise process. (We note in passing that there are two substantive downshoots in the residuals. It would be interesting to find out if these correspond to known periods of drought.)

510(c):   Here we plot the periodogram of $\{R_t^{(2)}\}$ to examine whether it can be considered to be white noise. As is to be expected, the only substantial difference between this periodogram and that for $\{R_t^{(1)}\}$ in Figure 508(c) is near $f = 2$ cycles/year, where there is now a downshoot instead of a peak.

510(d):   We now test whether the residuals $\{R_t^{(2)}\}$ can be considered to be drawn from a white noise process by applying the cumulative periodogram test for white noise (see item [4] of the Comments and Extensions to Section 6.6). The jagged curve in the plot is the normalized cumulative periodogram $\{\mathcal{P}_k\}$ for these residuals (see Equation (233)). The thin line emanating from the origin of the plot indicates the theoretical normalized integrated spectrum for a white noise process. The two thick lines are $L_u(\cdot)$ and $L_l(\cdot)$ of Equation (234) for $\alpha = 0.05$. Since at some frequencies the normalized cumulative periodogram falls outside the region between these two lines, we can reject the null hypothesis of white noise at the 5% level of significance.

512(a):   Let us now consider a model of the form

$$X_t = \mu + \sum_{l=1}^{L} \left(C_l e^{i2\pi f_l t \, \Delta t} + C_l^* e^{-i2\pi f_l t \, \Delta t} \right) + \eta_t, \tag{511}$$

where $\{\eta_t\}$ is a zero mean stationary process with nonwhite sdf $S_\eta(\cdot)$. (cf. Equation (503)). Although we already have a good idea of the

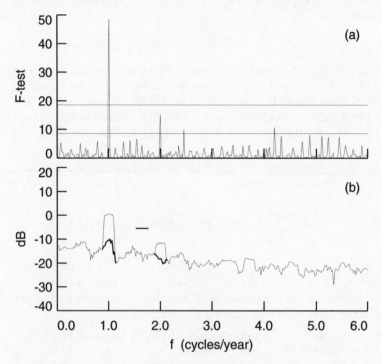

Figure 512. Analysis of Willamette River data, part 4.

appropriate frequencies f_l to use in this model, let us illustrate the use of the F-test for periodicity discussed in Section 10.11. With $NW = 4/\Delta t$ and $K = 5$ dpss data tapers, we compute the F-test of Equation (499c) at frequencies $j/(1024\,\Delta t)$ with $j = 1, \ldots, 512$ and plot it in 512(a). At a given fixed frequency, the F-test is F-distributed with 2 and $2K - 2 = 8$ degrees of freedom under the null hypothesis of no spectral line. We reject the null hypothesis for large values of the F-test. The heights of the two thin horizontal lines on the plot indicate the levels of the upper 99% and 99.9% percentage points of the $F_{2,8}$ distribution (8.6 and 18.6, respectively). The two largest excursions in the F-test occur at the same frequencies as the two largest values of the periodogram in Figure 505(c), namely, 1.008 and 2.004 cycles/year. The largest excursion (at 1.008 cycles/year) exceeds the 99.9% point and is clearly significant; the F-test at 2.004 cycles/year exceeds the 99% point (but not the 99.9%). There are also two other frequencies at which the F-test exceeds the 99% point, namely, 2.461 and 4.195 cycles/year. The spectral estimates in Figures 505, 508 and 510 do not clearly indicate a line component at either of these frequencies, so how can we explain this result? A quote from Thomson (1990a) is pertinent here:

It is important to remember that in typical time-series problems hundreds or thousands of uncorrelated estimates are being dealt with; consequently one will encounter numerous instances of the F-test giving what would normally be considered highly significant test values that, in actuality, will only be sampling fluctuations. A good rule-of-thumb is not to get excited by significance levels less than $1 - 1/N$.

Here $1 - 1/N = 1 - 1/395 = 0.9975$, which translates into a critical value of 13.9. Since only the excursions at 1.008 and 2.004 cycles/year exceed 13.9, the rule-of-thumb says that all the other peaks in the F-test are due to sampling fluctuations.

Based upon the results of the F-test, we now proceed under the assumption that there are line components at the frequencies $f_1 = 1.008$ and $f_2 = 2.004$ cycles/year so that $L = 2$ in Equation (511). The associated complex-valued amplitudes for these frequencies, namely, C_1 and C_2, can be estimated using Equation (499a) to obtain

$$\hat{C}_1 = -0.2912 - i0.3122 \qquad |\hat{C}_1|^2 = 0.1823$$
$$\text{yielding}$$
$$\hat{C}_2 = 0.0232 - i0.0984 \qquad |\hat{C}_2|^2 = 0.0102.$$

The values 0.1823 and 0.0102 are the sizes of the jumps in the integrated spectrum at $\pm f_1 = \pm 1.008$ and $\pm f_2 = \pm 2.004$ cycles/year, respectively. These are in reasonable agreement with the acls estimates previously determined (0.2012 and 0.0097).

512(b):   Here we plot – using a thin curve – the multitaper sdf estimate for $\{X_t'\}$ using the same set of tapers as in the F-test; i.e., $K = 5$, $NW = 4/\Delta t$ and

$$\hat{S}^{(mt)}(f) = \frac{1}{K} \sum_{k=0}^{K-1} \hat{S}_k^{(mt)}(f) = \frac{\Delta t}{K} \sum_{k=0}^{K-1} \left| \sum_{t=1}^{N} h_{t,k} X_t' e^{-i2\pi ft\Delta t} \right|^2,$$

where $\{h_{t,k}\}$ is the kth order dpss data taper. Since $W = 4/(N\,\Delta t) = 4 \times 12/395 = 0.12$, the width of the central lobe of the effective spectral window in the multitaper scheme is approximately $2W = 0.24$, which is indicated by the length of the thick horizontal line on the plot. Note that the peaks at 1 and 2 cycles/year are spread out by about this amount.

To obtain an estimate of $S_\eta(\cdot)$, we reshape $\hat{S}^{(mt)}(\cdot)$ using Equation (500) at all frequencies f such that $|f - f_1| \leq W$ or $|f - f_2| \leq W$. Because f_1 corresponds to $85/(1024\,\Delta t)$ and f_2 corresponds to $170/(1024\,\Delta t)$, this amounts to reshaping $\hat{S}^{(mt)}(\cdot)$ at the frequencies $j/(1024\,\Delta t)$ for $j = 75, \ldots, 95$ and $j = 160, \ldots, 180$. The two short thick curves indicate the reshaped portions of $\hat{S}^{(mt)}(\cdot)$. Our estimate of $S_\eta(\cdot)$ is thus given by these curves in addition to the unreshaped portions of $\hat{S}^{(mt)}(\cdot)$.

To summarize, we started by defining a grid of frequencies more than twice as fine as that associated with the standard Fourier frequencies. Removal of the mean reduced leakage at low frequencies, while periodograms identified the fundamental frequency and its first harmonic. Amplitudes of the sinusoids were estimated using approximate conditional least squares. However, using the cumulative periodogram test, the background continuum was declared nonwhite. Hence the analysis was restarted with the F-test for periodicity developed in Section 10.11. The fundamental frequency and its first harmonic were identified as significant, and an estimate of the background continuum spectrum was found by reshaping the multitaper spectrum estimate to take into account the two sinusoidal terms (Figure 512(b)).

10.14 A Parametric Approach to Harmonic Analysis

Autoregressive (AR) spectral estimation, as discussed in Chapter 9, is thought of by statisticians as a method for estimating sdf's, i.e., a spectrum with only a purely continuous component and no line components. However, this technique has been used on many occasions in applied work to attempt to estimate the frequencies of line components, perhaps immersed in a background continuum (Ulrych, 1972b; Chen and Stegen, 1974; Satorius and Zeidler, 1978). We investigate the AR approach to spectral line estimation in this section.

A *deterministic* real sinusoid of the form $x_t = D \cos\left(2\pi f t \, \Delta t + \phi\right)$ gives rise to the second-order difference equation

$$x_t = 2\cos\left(2\pi f \, \Delta t\right) x_{t-1} - x_{t-2} \tag{514a}$$

with the initial conditions $x_0 = D\cos\left(\phi\right)$ and $x_1 = D\cos\left(2\pi f \, \Delta t + \phi\right)$. Note that the fixed amplitude and phase are present only in the initial conditions. To derive this difference equation, we write

$$x_{t+2} = \varphi_{1,2} x_{t+1} + \varphi_{2,2} x_t, \text{ given } x_t = D\cos\left(2\pi f t \, \Delta t + \phi\right). \tag{514b}$$

From the trigonometric relationship $\cos\left(A + B\right) = \cos\left(A\right)\cos\left(B\right) - \sin\left(A\right)\sin\left(B\right)$, it follows that

$$\begin{aligned}
x_{t+2} &= D\cos\left(2\pi f[t+2]\,\Delta t + \phi\right) = D\cos\left([2\pi f t \,\Delta t + \phi] + 4\pi f \,\Delta t\right) \\
&= D\cos\left(2\pi f t \,\Delta t + \phi\right)\cos\left(4\pi f \,\Delta t\right) \\
&\quad - D\sin\left(2\pi f t \,\Delta t + \phi\right)\sin\left(4\pi f \,\Delta t\right),
\end{aligned}$$

while

$$\begin{aligned}
x_{t+1} &= D\cos\left(2\pi f t \,\Delta t + \phi\right)\cos\left(2\pi f \,\Delta t\right) \\
&\quad - D\sin\left(2\pi f t \,\Delta t + \phi\right)\sin\left(2\pi f \,\Delta t\right)
\end{aligned}$$

and

$$x_t = D \cos \left(2\pi f t \, \Delta t + \phi \right).$$

If we use these last three equations to substitute for the three terms in Equation (514b), and if we then equate terms in $D \cos \left(2\pi f t \, \Delta t + \phi \right)$ and $D \sin \left(2\pi f t \, \Delta t + \phi \right)$, we find

$$\begin{bmatrix} \cos \left(2\pi f \, \Delta t \right) & 1 \\ \sin \left(2\pi f \, \Delta t \right) & 0 \end{bmatrix} \begin{bmatrix} \varphi_{1,2} \\ \varphi_{2,2} \end{bmatrix} = \begin{bmatrix} \cos \left(4\pi f \, \Delta t \right) \\ \sin \left(4\pi f \, \Delta t \right) \end{bmatrix},$$

which has the solution $[\varphi_{1,2}, \varphi_{2,2}]^T = [2 \cos \left(2\pi f \, \Delta t \right), -1]^T$ as required.

A *randomly phased* sinusoid of the form $X_t = D \cos \left(2\pi f t \, \Delta t + \phi \right)$, where now ϕ is an rv uniformly distributed between $-\pi$ and π, also satisfies

$$X_t = 2 \cos \left(2\pi f \, \Delta t \right) X_{t-1} - X_{t-2}, \qquad (515a)$$

where the initial conditions $X_0 = D \cos \left(\phi \right)$ and $X_1 = D \cos \left(2\pi f \, \Delta t + \phi \right)$ involve the rv ϕ and the constant D. We call (515a) a *pseudo*-AR(2) process. This terminology is meant to remind us that (515a) lacks one aspect of the usual AR(2) process – the innovations term ϵ_t. If we write (515a) as

$$X_t = \varphi_{1,2} X_{t-1} + \varphi_{2,2} X_{t-2},$$

it is easy to verify that the roots of the polynomial equation

$$1 - \varphi_{1,2} z^{-1} - \varphi_{2,2} z^{-2} = 0$$

are $z = \exp \left(\pm i 2\pi f \, \Delta t \right)$, both of which are on the unit circle. We know that a randomly phased sinusoid has

$$\operatorname{cov} \left\{ X_t, X_{t+\tau} \right\} = D^2 \cos \left(2\pi f \tau \, \Delta t \right) / 2$$

(cf. Equation (467c) with no noise). The same result can be derived from (515a) (see Exercise [10.9]).

The representation of a single sinusoid by a second-order difference equation extends to the representation of the summation of p sinusoids using the pseudo-AR($2p$) equation

$$X_t = \sum_{k=1}^{2p} \varphi_{k,2p} X_{t-k}. \qquad (515b)$$

The roots $\{z_j\}$ of the polynomial equation corresponding to (515b), namely,

$$1 - \varphi_{1,2p} z^{-1} - \varphi_{2,2p} z^{-2} - \cdots - \varphi_{2p,2p} z^{-2p} = 0, \qquad (515c)$$

are all on the unit circle in conjugate pairs so that $z_j = \exp \left(\pm i 2\pi f_j \, \Delta t \right)$, $j = 1, \ldots, p$, where each f_j is one frequency of the p sinusoids. As an example, let us consider Equation (515b) for the case of two sinusoids with $\Delta t = 1$. By the same approach used for the single sinusoid case, we obtain the matrix equation

$$
\begin{bmatrix}
\cos(6\pi f_1) & \cos(4\pi f_1) & \cos(2\pi f_1) & 1 \\
\sin(6\pi f_1) & \sin(4\pi f_1) & \sin(2\pi f_1) & 0 \\
\cos(6\pi f_2) & \cos(4\pi f_2) & \cos(2\pi f_2) & 1 \\
\sin(6\pi f_2) & \sin(4\pi f_2) & \sin(2\pi f_2) & 0
\end{bmatrix}
\begin{bmatrix}
\varphi_{1,4} \\
\varphi_{2,4} \\
\varphi_{3,4} \\
\varphi_{4,4}
\end{bmatrix}
=
\begin{bmatrix}
\cos(8\pi f_1) \\
\sin(8\pi f_1) \\
\cos(8\pi f_2) \\
\sin(8\pi f_2)
\end{bmatrix},
$$

from which $\varphi_{1,4}, \ldots, \varphi_{4,4}$ can be obtained.

So far we have only considered perfectly observed sinusoids. We now consider randomly phased sinusoids plus white noise. For p sinusoids observed with zero mean white noise $\{\alpha_t\}$, the observed process is the sum of a pseudo-AR($2p$) process and white noise:

$$
\widetilde{X}_t \equiv X_t + \alpha_t = \sum_{k=1}^{2p} \varphi_{k,2p} X_{t-k} + \alpha_t,
\tag{516a}
$$

where $E\{X_t \alpha_{t+\tau}\} = 0$ for all integers τ; i.e., the noise $\{\alpha_t\}$ is assumed to be uncorrelated with the 'signal' $\{X_t\}$. If we substitute $X_{t-k} = \widetilde{X}_{t-k} - \alpha_{t-k}$ into the right-hand side of the above, we get

$$
\widetilde{X}_t - \sum_{k=1}^{2p} \varphi_{k,2p} \widetilde{X}_{t-k} = \alpha_t - \sum_{k=1}^{2p} \varphi_{k,2p} \alpha_{t-k},
\tag{516b}
$$

which is an ARMA($2p, 2p$) process whose AR and MA coefficients are identical. Such a result was first derived by Ulrych and Clayton (1976), but without the stipulation that sinusoids were randomly phased and hence stochastic. Kay and Marple (1981, p. 1403) report a method (essentially due to Pisarenko, 1973) for estimating the parameters $\varphi_{k,2p}$ in Equation (516a).

It is important to clearly distinguish between a pseudo-AR process (i.e., one without an innovations term) plus white noise and a standard AR process plus white noise. We have just seen how the former – for the case of p randomly phased sinusoids plus white noise – yields an ARMA($2p,2p$) process with identical AR and MA coefficients. In the latter case, however, an ARMA process with an equal number of AR and MA coefficients is again obtained but now with *distinct* coefficients (Walker, 1960; Tong, 1975; Friedlander, 1982). To see this, let $\{Y_t\}$ be a standard AR(p) process with innovations variance σ_p^2 as in Chapter 9, and let $\{\xi_t\}$ be a white noise process with variance σ_ξ^2 that is uncorrelated with $\{Y_t\}$. Then the process defined by $\widetilde{Y}_t \equiv Y_t + \xi_t$ has an sdf given by

$$S(f) = \frac{\sigma_p^2 \, \Delta t}{\left|1 - \sum_{k=1}^{p} \phi_{k,p} e^{-i2\pi f k \, \Delta t}\right|^2} + \sigma_\xi^2 \, \Delta t$$

$$= \frac{\sigma_p^2 \, \Delta t + \sigma_\xi^2 \, \Delta t \left|1 - \sum_{k=1}^{p} \phi_{k,p} e^{-i2\pi f k \, \Delta t}\right|^2}{\left|1 - \sum_{k=1}^{p} \phi_{k,p} e^{-i2\pi f k \, \Delta t}\right|^2}.$$

If we let

$$\sigma_p^2 + \sigma_\xi^2 \left|1 - \sum_{k=1}^{p} \phi_{k,p} e^{-i2\pi f k \, \Delta t}\right|^2 = \sigma_\zeta^2 \left|1 - \sum_{k=1}^{p} \theta_{k,p} e^{-i2\pi f k \, \Delta t}\right|^2$$

by appropriately defining σ_ζ^2 and $\theta_{k,p}$, we obtain

$$S(f) = \sigma_\zeta^2 \, \Delta t \frac{\left|1 - \sum_{k=1}^{p} \theta_{k,p} e^{-i2\pi f k \, \Delta t}\right|^2}{\left|1 - \sum_{k=1}^{p} \phi_{k,p} e^{-i2\pi f k \, \Delta t}\right|^2},$$

which is the sdf for an ARMA(p,p) process with AR parameters $\phi_{k,p}$, MA parameters $\theta_{k,p}$ and innovations variance σ_ζ^2. The $\theta_{k,p}$ terms are no longer identical to the $\phi_{k,p}$ terms, but note that the $p+1$ values $\{\theta_{k,p}\}$ and σ_ζ^2 are determined by the $p+2$ values $\{\phi_{k,p}\}$, σ_p^2 and σ_ξ^2. (Proof of the above statements is the subject of Exercise [10.10].)

Since a standard AR(p) process plus white noise is an ARMA(p,p) process, the AR parameters can be estimated from the ARMA(p,p) nature of the observed process. This is done by Tong (1975) and Friedlander (1982). These methods, however, do not make use of the fact that there are only $p+2$ free parameters, but rather they treat the model as if there were $2p+2$ free parameters ($\phi_{1,p}, \ldots, \phi_{p,p}$; $\theta_{1,p}, \ldots, \theta_{p,p}$; σ_p^2 and σ_ξ^2); i.e., the MA parameters are not considered related to the other $p+2$ parameters. Jones (1980) and Tugnait (1986) present algorithms for solving the AR parameters that involve only the $p+2$ free parameters.

To illustrate very simply the AR approach to sinusoid estimation, we consider an example due to Makhoul (1981b) of a deterministic sinusoid with a frequency of $f = 1/6$ Hz and a sampling interval of $\Delta t = 1$ second. With these values substituted into Equation (514a), the appropriate difference equation becomes

$$x_t = 2\cos(\pi/3)x_{t-1} - x_{t-2} = x_{t-1} - x_{t-2}. \tag{517}$$

For convenience let us assume that the amplitude of the sinusoid is $D = 1/\cos(\pi/6)$ and its phase is $\phi = -\pi/2$. The initial conditions are thus

$$x_0 = D\cos(\phi) = D\cos(-\pi/2) = 0$$

and

$$x_1 = D\cos{(2\pi f + \phi)} = D\cos{(-\pi/6)} = 1,$$

and subsequent values are

$$x_2 = x_1 - x_0 = 1, \qquad x_3 = x_2 - x_1 = 0,$$
$$x_4 = x_3 - x_2 = -1, \qquad x_5 = x_4 - x_3 = -1,$$

etc.

Suppose that we have only the samples $x_1, \ldots, x_4$ and that we seek to estimate the parameters $\phi_{1,2}$, $\phi_{2,2}$ and σ_2^2 of a *standard* AR(2) model. If we do so by using the forward/backward least squares estimator, we must find the values of $\phi_{1,2}$ and $\phi_{2,2}$ such that the forward/backward sum of squares

$$\sum_{t=3}^{4} \left[\left(x_t - \phi_{1,2}x_{t-1} - \phi_{2,2}x_{t-2} \right)^2 + \left(x_{t-2} - \phi_{1,2}x_{t-1} - \phi_{2,2}x_t \right)^2 \right] \quad (518)$$

is minimized (cf. Equation (427)). If we call the solution points $\hat{\phi}_{1,2}$ and $\hat{\phi}_{2,2}$, we find that the minimum sum of squares is *zero* when $\hat{\phi}_{1,2} = 1$ and $\hat{\phi}_{2,2} = -1$; i.e., the fitted model is *exactly* the true model (517). Note that the innovations term ϵ_t can be dropped from the fitted model – it has a mean of zero by assumption, and our estimate of the innovations variance would be zero, implying that $\epsilon_t = 0$ for all t.

If now we consider Burg's algorithm instead (Section 9.5), we must first fit an AR(1) model and estimate $\phi_{1,1}$ by minimizing the forward and backward sum of squares

$$\sum_{t=2}^{4} \left[\left(x_t - \phi_{1,1}x_{t-1} \right)^2 + \left(x_{t-1} - \phi_{1,1}x_t \right)^2 \right]$$

(see Equation (416e)). It follows from Exercise [9.8a] that the estimate is

$$\bar{\phi}_{1,1} = \frac{2\sum_{t=2}^{4} x_t x_{t-1}}{x_1^2 + 2\sum_{t=2}^{3} x_t^2 + x_4^2} = 1/2,$$

after substitution of the sample values. We now use the Levinson–Durbin relationship

$$\phi_{1,2} = \phi_{1,1} - \phi_{2,2}\phi_{1,1} = \phi_{1,1} \left(1 - \phi_{2,2} \right).$$

If we now replace $\phi_{1,2}$ in Equation (518) with the right-hand side of the above (with $\bar{\phi}_{1,1}$ replacing $\phi_{1,1}$), the next step in Burg's algorithm is the minimization of

$$\sum_{t=3}^{4} \Bigg(\left[x_t - \left(1 - \phi_{2,2} \right) \bar{\phi}_{1,1}x_{t-1} - \phi_{2,2}x_{t-2} \right]^2$$
$$+ \left[x_{t-2} - \left(1 - \phi_{2,2} \right) \bar{\phi}_{1,1}x_{t-1} - \phi_{2,2}x_t \right]^2 \Bigg)$$

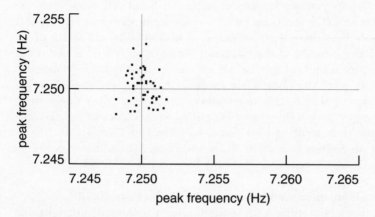

Figure 519. Scatter plot of locations of peak frequencies computed from a direct spectral estimator $\hat{S}^{(d)}(\cdot)$ with $NW = 2/\Delta t$ (vertical axis) versus locations determined from an AR(24) sdf estimator with parameters estimated using the forward/backward least squares method (horizontal axis). Each point in the scatter plot corresponds to location estimators for one of 50 different realizations from the model of Equation (489). The true location of the peak frequency is 7.25 Hz (indicated by the vertical and horizontal lines). (This figure should be compared with Figure 488.)

with respect to $\phi_{2,2}$. With the data substituted we obtain $\bar{\phi}_{2,2} = -1$, and hence $\bar{\phi}_{1,2} = (1 - \bar{\phi}_{2,2})\bar{\phi}_{1,1} = 1$. These are the same estimates as obtained previously. (Makhoul, 1981b, appears to have miscalculated $\bar{\phi}_{1,1}$ and hence derived poor estimates of $\phi_{1,2}$ and $\phi_{2,2}$ from Burg's algorithm.)

As another example of the parametric approach to harmonic analysis, let us reconsider the stochastic process of Equation (489), which we discussed previously in the context of tapering (see Figure 488). This process consists of a single sinusoid plus a small amount of white noise – it has such a high signal-to-noise ratio (37 dB) that plots of individual realizations look indistinguishable from a plot of a sinusoid with a frequency of $f_1 = 7.25$ Hz. We used the forward/backward least squares estimator to fit AR(24) models to the same 50 realizations from this process that were used in Figure 488. For each realization, we determined the frequency of the peak value in the sdf corresponding to the fitted AR model (details on how to find the location of this frequency are given in item [1] of the Comments and Extensions to the next section). In Figure 488, we showed a scatter plot of peak locations as determined from a direct spectral estimator $\hat{S}^{(d)}(\cdot)$ using an $NW = 2/\Delta t$ data taper (vertical axis) and the periodogram $\hat{S}^{(p)}(\cdot)$ (horizontal axis). A similar scatter plot is shown in Figure 519, but now we display the locations of the peak frequencies computed using the fitted AR models on the horizontal axis (the locations for $\hat{S}^{(d)}(\cdot)$ are again shown on the vertical axis). Note that there is no indication of bias in either estimator of f_1, but that the variance of the estimator based upon $\hat{S}^{(d)}(\cdot)$ is about 1.9 times greater than that for the AR estimator. This example illustrates the usefulness of the parametric approach to harmonic analysis. (We note, however, that a data taper other than a dpss taper might work better here because a dpss taper is optimal with respect to leakage control rather than locating peaks (see, for example, Tseng *et al.*, 1981). Also, as we discuss in Section 10.16, the Burg AR estimator performs very poorly on this example – see the bottom row of Table 527.)

Comments and Extensions to Section 10.14

[1]   For readers familiar with the notion of a backward shift operator and its use in expressing stationary nondeterministic ARMA processes (see, for example, Box and Jenkins, 1976), we point out the fact – unfortunately often neglected in the literature – that, since the roots of (515c) are on the unit circle, the operator common to the AR and MA components of Equation (516b) does not have an inverse and hence cannot be canceled out.

10.15   Parametric Approach with River Flow Data

As an application of the parametric approach to harmonic analysis, we reconsider the Willamette River data discussed in Section 10.13 (see the top plot of Figure 505). We fit AR models of orders $p = 1$ to 150 using Burg's algorithm. Figure 521 shows a plot of Akaike's FPE order selection criterion versus these values of p – see Section 9.9 for details. The minimum value over this range of p occurs at $p = 38$, but there is a local minimum at $p = 27$ quite close to the one at 38 (the two thin vertical lines in the plot indicate the locations $p = 27$ and $p = 38$). This suggests that an AR process of order 38 or 27 might be appropriate, but we also consider the case $p = 150$ for comparison. The top three plots in Figure 522 show the resulting spectral estimates. In each of these sdf's, the two largest peaks occur near 1 and 2 cycles/year, in agreement with periodogram-based analysis of Section 10.13. Table 523 lists the observed peak frequencies $\hat{f}_j$, peak values $\hat{S}(\hat{f}_j)$ and peak widths for these sdf's and – for comparison – the periodogram in Figure 505(c). The peak widths at the observed peak frequency are measured by $f' + f''$, where $f' > 0$ and $f'' > 0$ are the smallest values such that

$$10 \log_{10}\left(\frac{\hat{S}(\hat{f}_j)}{\hat{S}(\hat{f}_j + f')} \right) = 10 \log_{10}\left(\frac{\hat{S}(\hat{f}_j)}{\hat{S}(\hat{f}_j - f'')} \right) = 3 \text{ dB}.$$

This table and the plots show a number of interesting qualitative differences in the sdf estimates:

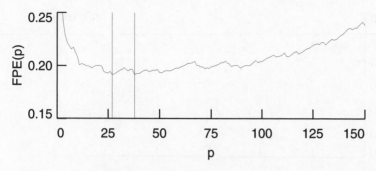

Figure 521. Akaike's FPE order selection criterion for Willamette River data (parameter estimation via Burg's algorithm).

[1] the peak near 2 cycles/year in the AR(27) sdf is displaced in frequency somewhat low compared to the other two AR sdf's, while the peak near 1 cycle/year has about the same location for all three values of p (and is in reasonable agreement with the location given by the periodogram);

[2] the widths of both peaks decrease as p increases;

[3] the widths of the peaks near 1 cycle/year in the AR sdf's are 2 to 6 times smaller than the width of the corresponding peak in the periodogram, while at 2 cycles/year the width is smaller only for order 150;

[4] the widths of the peaks at 1 and 2 cycles/year for a given order p differ by at least a factor of 2, whereas the corresponding periodogram widths are remarkably similar (0.0270 and 0.0269);

[5] the heights of the two peaks in the AR(27) sdf are about an order of magnitude larger than the corresponding peaks in the AR(150) sdf; and

[6] the AR(27) and AR(38) sdf's are considerably smoother looking than the AR(150) sdf (or the periodogram in Figure 505(c)).

Item [1] illustrates the danger of too low an AR order (a loss of accuracy of peak location), while item [6], of too high an AR order (introduction of spurious peaks). Items [2] and [4] indicate an inherent problem with assessing peak widths in AR sdf estimates: whereas the peak widths in the periodogram can be attributed simply to the width of the central lobe of the associated spectral window, there is evidently no corresponding simple relationship for the AR estimates. If we believe that the true frequencies for this time series are at 1 and 2 cycles/year, items [1] and [3] indicate that the narrowness of the AR peaks does not translate into increased accuracy of the peak location. Indeed, as is pointed out by Burg (1975, p. 64) the narrowness of AR peaks is a cause for care in computing peak locations (see the Comments and Extensions below).

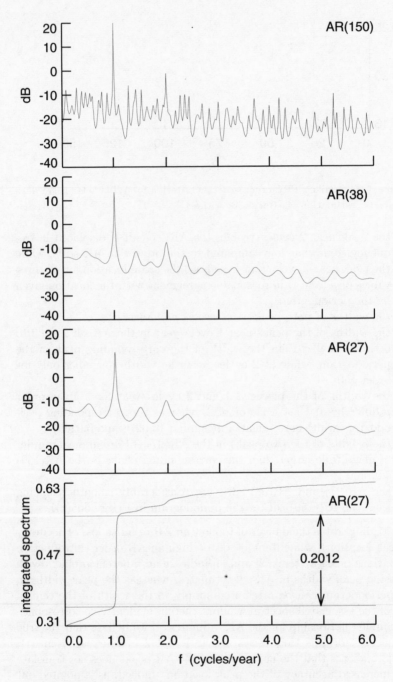

Figure 522. Analysis of Willamette River data, part 5.

p	$\hat{f}_1$	3 dB width	$\hat{S}(\hat{f}_1)$	$\hat{f}_2$	3 dB width	$\hat{S}(\hat{f}_2)$
150	1.0055	0.0045	21.7 dB	2.0055	0.0091	−1.0 dB
38	1.0053	0.0061	13.5 dB	1.9970	0.0622	−7.7 dB
27	1.0042	0.0127	10.4 dB	1.9757	0.1225	−9.8 dB
$\hat{S}^{(p)}(\cdot)$	1.0032	0.0270	8.2 dB	2.0034	0.0269	−5.0 dB

Table 523. Characteristics of autoregressive sdf's for Willamette River data (cf. top three plots of Figure 522). The top three rows of numbers refer to the two dominant peaks in the AR(p) sdf's shown in the top three plots of Figure 522 (the AR parameters were estimated using Burg's algorithm). The first column is the AR order p; the second is the location of the peak near 1 cycle/year; the third is the width of the peak 3 dB down from the peak value; and the fourth is the peak value itself. Columns 5, 6 and 7 show similar values for the peak near 2 cycles/year. The final row gives the corresponding values for the periodogram $\hat{S}^{(p)}(\cdot)$ in Figure 505(c).

Item [5] raises the question of how we can relate the height of a peak in an AR sdf at frequency f_l to the amplitude D_l of a sinusoidal component at that frequency in the assumed model

$$X_t = \sum_{l=1}^{L} D_l \cos\left(2\pi f_l \,\Delta t + \phi_l\right) + \epsilon_t.$$

In the case of the periodogram, we have the approximate relationship

$$E\{\hat{S}^{(p)}(f_l)\} \approx \sigma_\epsilon^2 \,\Delta t + N \,\Delta t D_l^2/4$$

(since $\mathcal{F}(0) = N\,\Delta t$, this follows from Equation (475) under the assumption that f_l is well separated from the other frequencies in the model and is not too close to zero or the Nyquist frequency; see also Figure 485). A similar result for AR sdf's estimated via Burg's algorithm has yet to be established. Recently, Pignari and Canavero (1991) concluded, via both theoretical calculations and Monte Carlo simulations, that '... the power estimation of sinusoidal signals by means of the Burg method is scarcely reliable, even for very high signal-to-noise ratios for which noise corruption of data is unimportant' (in our notation, the power in a sinusoid is proportional to D_l^2). In an earlier work, Lacoss (1971) found that, if we know the acvs *perfectly* to lag p for a single sinusoid plus white noise, the height of the peak at the sinusoidal frequency in a pth order Yule–Walker sdf is proportional to $D_l^4 p^2$, while the 3 dB peak width is inversely proportional to $D_l^2 p^2$. This implies that the product of the peak height and the peak width is proportional to D_l^2, but the relevance of these results to AR sdf estimation using Burg's algorithm is unclear.

Burg (1985) advocated plotting the integrated spectrum $S^{(I)}(\cdot)$ as a way of determining the power in sinusoidal components. Since the integrated spectrum is related to the sdf $S(\cdot)$ by

$$S^{(I)}(f) = \int_{-f_{(N)}}^{f} S(u) \, du,$$

the power in the frequency interval $[f', f'']$ is given by $S^{(I)}(f'') - S^{(I)}(f')$. As an example, the bottom plot in Figure 522 shows the integrated spectrum – obtained via numerical integration – corresponding to the AR(27) sdf estimate for the Willamette River data. The vertical distance between the two thin horizontal lines in this plot is 0.2012, which is the periodogram-based estimate of the size of the jump in the integrated spectrum at 1 cycle/year due to the line component with this frequency. Note that the increase in the AR(27) integrated spectrum around 1 cycle/year is roughly consistent with this value, but that precise determination of the power would be tricky pending future research on exactly how to pick the interval $[f', f'']$ to bracket a peak. (Burg, 1975, p. 64, notes that numerical integration of estimated AR sdf's is a useful way of checking that peak values have been accurately determined.)

Comments and Extensions to Section 10.15

[1]　The locations of the peaks in an AR sdf can be accurately determined in a manner similar to that used to locate the peak frequencies in the periodogram (see item [2] in the Comments and Extensions to Section 10.6). Recall that the sdf for a stationary AR(p) process has the form

$$S(f) = \frac{\sigma_p^2 \, \Delta t}{\left| 1 - \sum_{j=1}^{p} \phi_{j,p} e^{-i 2\pi f j \, \Delta t} \right|^2}, \qquad |f| \le f_{(N)}$$

(this is Equation (392b)). With $\phi_{0,p} \equiv -1$, the denominator of the above can be rewritten as

$$D(f) \equiv \left| \sum_{j=1}^{p+1} \phi_{j-1,p} e^{-i 2\pi f j \, \Delta t} \right|^2 .$$

To within a scale factor, we can regard $D(\cdot)$ as the periodogram for the 'time series' $X_j \equiv \phi_{j-1,p}$ with $j = 1, \ldots, p+1$. By a derivation analogous to that of Equation (196c), we have

$$D(f) = (p+1) \sum_{\tau=-p}^{p} \hat{s}_{\tau,\phi}^{(p)} e^{-i2\pi f \tau \, \Delta t},$$

where

$$\hat{s}_{\tau,\phi}^{(p)} \equiv \frac{1}{p+1} \sum_{j=1}^{p+1-|\tau|} \phi_{j-1,p} \phi_{j-1+|\tau|,p}.$$

Since a maximum (peak) in $S(\cdot)$ corresponds to a minimum (valley) in $D(\cdot)$, we can use the Newton–Raphson method described following Equation (479) – with $\hat{s}_{\tau,\phi}^{(p)}$ substituted for $\hat{s}^{(p)}$ – to determine the location of a valley. This technique requires a good initial estimate of the location. This estimate can usually be obtained by evaluating $D(\cdot)$ on a fine enough grid of frequencies via an FFT of $\phi_{0,p}, \ldots, \phi_{p,p}$ padded with lots of zeros. (As we noted previously in the discussion following Equation (479), the Newton–Raphson method can fail if the initial estimate is poor. We indicated there that, if the location can be bracketed, a bisection technique (combined with Newton–Raphson) or Brent's method can be used in these cases. Since we are now searching for a minimum rather than a maximum, the previously stated bracketing conditions must be modified to be $g'(\omega^{(l)}) < 0$ and $g'(\omega^{(u)}) > 0$.)

10.16 Problems with the Parametric Approach

For a process consisting of a sinusoid in additive noise, two basic problems have been identified with the AR spectral estimates. Chen and Stegen (1974) found the location of the peak in the estimate to depend on the phase of the sinusoid, while Fougere *et al.* (1976) found that the estimate can sometimes contain two adjacent peaks in situations where only a single peak should appear – this is known in the literature as *spectral line splitting* or *spontaneous line splitting*. Causes and cures for these problems have been the subject of many papers – see Kay and Marple (1981, p. 1396) for a nice summary. We give here a perspective on the current status of the literature.

Toman (1965) looked at the effect on the periodogram of short truncation lengths T of the *continuous* parameter deterministic sinusoid $X(t) = D \cos(2\pi f_0 t - \pi/2) = D \sin(2\pi f_0 t)$, where $f_0 = 1$ Hz so that the corresponding period is 1 second. The data segment is taken to be from $t = 0$ to $t = T$, and the periodogram is here defined as

$$\frac{1}{T} \left| \int_0^T X(t) e^{-i2\pi f t} \, dt \right|^2 .$$

Note that, because the period is unity, the ratio of the length of the data segment to the period is just T. Toman demonstrated that, for $0 \leq T \leq 0.58$, the location of the peak value of the periodogram is at $f = 0$; as T increases from 0.58 to 0.716, the location of the peak value of the periodogram increases from 0 to 1 Hz (the latter being the true sinusoidal frequency f_0); for T greater than 0.716, the peak frequency oscillates, giving 1 Hz at the maximum of each oscillation and converging to 1 Hz as $T \to \infty$. For the special case $T = 1$ corresponding to one complete cycle of the sinusoid, the peak occurred at 0.84 Hz.

Toman's experimental results were clearly and concisely explained by Jackson (1967). He noted that, with $X(t) = D \cos{(2\pi f_0 t + \phi)}$,

$$\int_0^T X(t) e^{-i2\pi f t}\, dt = \frac{D}{2} e^{i\phi} e^{-i\pi (f - f_0)T} \frac{\sin{(\pi (f - f_0)T)}}{\pi (f - f_0)}$$
$$+ \frac{D}{2} e^{-i\phi} e^{-i\pi (f + f_0)T} \frac{\sin{(\pi (f + f_0)T)}}{\pi (f + f_0)}.$$

For the special case considered by Toman, namely, $\phi = -\pi/2$, $f_0 = 1$ and $T = 1$, the above reduces to

$$i\frac{D}{2} e^{-i\pi f} \left(\operatorname{sinc}{(f - 1)} - \operatorname{sinc}{(f + 1)}\right),$$

where, as usual, $\operatorname{sinc}{(t)} \equiv \sin{(\pi t)}/(\pi t)$ is the sinc function. The periodogram is thus proportional to $|\operatorname{sinc}{(f - 1)} - \operatorname{sinc}{(f + 1)}|^2$. Jackson noted that the interference of the two sinc functions produces extrema at $\pm 0.84 f$, as observed by Toman. For a single cycle of a cosine, i.e., $\phi = 0$, the maxima occur at $\pm 1.12 f$. Jackson similarly explained Toman's other observations.

Ulrych (1972a) – see also Ulrych (1972b) – thought that such spectral shifts could be prevented by the use of Burg's algorithm. In his example – see Table 527 – neither shifting nor splitting was present even with additive white noise. Chen and Stegen (1974) carried out a more detailed investigation of Burg's algorithm using as input 1 Hz sinusoids sampled at 20 samples per second with white noise superimposed. Some of their results are summarized in Table 527. It turned out that the particular parameter combinations used by Ulrych (1972a) happen to coincide with points in the oscillating location of the spectral peak that agree with the true value of 1 Hz; in general, however, this is not the case, so that Burg's method is indeed susceptible to these phenomena.

Fougere *et al.* (1976) noted that spectral line splitting is most likely to occur when

[1] the signal-to-noise ratio is high;

Frequency (Hz)	Phase	$f_{(N)}$	N	p	Peak shift	Line splitting
Ulrych, 1972a:						
1	$-\pi/2$	10	21	11	no	no
1	0	10	21	11	no	no
1	$-\pi/2$	10	12	11	no	no
Chen and Stegen, 1974:						
1	$-\pi/2$	10	24	2,8	no	no
1	$-\pi/2$	10	24	20	yes	yes
1	$-\pi/2$	10	10–60	8	oscillates	—
1	varied	10	15	8	oscillates	—
1	varied	10	12–65	8	oscillates	—
Fougere *et al.*, 1976:						
1	$-\pi/2\,(\pi/9)\,\pi/2$	10	21	19	no	no
5	$-\pi/2\,(\pi/90)\,\pi/2$	10	6	5	yes	yes
1.25 (2) 49.25	$-\pi/4$	50	101	24	yes (all)	yes (all)

Table 527. Summary of simulation results for a single sinusoid with additive white noise (estimation by Burg's algorithm). The first column gives the frequency of the sinusoid; the second column is phase (with respect to a cosine term); $f_{(N)}$ is the Nyquist frequency; N is the sample size; and p is the order of the autoregressive model. The notation $x\,(a)\,y$ means from x to y in steps of a.

[2] the initial phase of sinusoidal components is some odd multiple of 45°; and

[3] the time duration of the data sequence is such that the sinusoidal components have an odd number of quarter cycles.

Other papers of interest are Fougere (1977, 1985). A summary of some of Fougere's results is given in Table 527. In the 1 Hz example, an AR(19) model with 21 data points produced no anomalous results for a range of input phases; on the other hand, an AR(24) model with 101 data points produced line splitting for a range of frequencies for the input sinusoids. Fougere (1977) concluded that '... splitting does not occur as a result of using an overly large number of filter weights [AR coefficients], as is frequently claimed in the literature.' Chen and Stegen (1974) had reached the *opposite* conclusion based on the second of their examples given in Table 527.

Chen and Stegen (1974) considered that the frequency shifts they observed were due to essentially the same mechanism as observed by

Jackson (1967). However, Fougere (1977) notes that 'Jackson's worst cases were sine waves an even number of cycles long with either 90° or 0° initial phase.' For these two cases Burg's spectra are neither split nor shifted but are extremely accurate.

Fougere (1977) put the problem with line splitting in Burg's algorithm down to the substitution of estimated $\phi_{k,k}$ from lower order fitted AR models into the higher order fitted models (as we have seen, the forward/backward sums of squares from the final model order are not necessarily minimized using Burg's algorithm due to the substitution from lower order fits). Fougere suggested setting $\phi_{k,k} = U \sin(\Phi_k)$ for $k = 1, \ldots, p$, with U slightly less than 1, and solving for the Φ_k terms *simultaneously* using a nonlinear optimization scheme. Once Φ_k has been estimated by $\hat{\Phi}_k$, the corresponding estimates for $\phi_{k,k}$, namely, $\hat{\phi}_{k,k} = U \sin(\hat{\Phi}_k)$, must be less than unity as required for stability. Fougere demonstrated that, at least in the single sinusoid examples examined, line splitting could be eliminated by using this procedure. (Recently, Bell and Percival, 1991, obtained similar results with a 'two step' Burg algorithm, which estimates the reflection coefficients in pairs and hence can be regarded as a compromise between Burg's algorithm and Fougere's method.)

Kay and Marple (1979) diagnosed both a different cause and cure from the suggestions of Fougere. They considered that

> ... spectral line splitting is a result of estimation errors and is not inherent in the autoregressive approach. In particular, the interaction between positive and negative sinusoidal frequency components [Jackson's mechanism] in the Burg reflection coefficient and Yule–Walker autocorrelation estimates and the use of the biased estimator in the Yule–Walker approach are responsible for spectral line splitting.

Their cure (for one sinusoid) was the use of *complex-valued data* and the *unbiased* acvs estimator in the Yule–Walker case. To follow their argument, consider a noise-free randomly phased sinusoid with unit amplitude, $X_t = \cos(2\pi f t \, \Delta t + \phi)$, and let us take our N data points to be $X_1, \ldots, X_N$. Then the biased estimator of the acvs is

$$\hat{s}_\tau^{(p)} \equiv \frac{1}{N} \sum_{t=1}^{N-\tau} X_t X_{t+\tau} \text{ for } \tau \geq 0,$$

since we have $E\{X_t\} = 0$ under the random phase assumption. Hence, for $\tau \geq 0$,

$$\hat{s}_\tau^{(p)} = \frac{1}{N} \sum_{t=1}^{N-\tau} \cos(2\pi f t \, \Delta t + \phi) \cos([2\pi f t \, \Delta t + \phi] + 2\pi f \tau \, \Delta t)$$

$$= \left(\frac{N-\tau}{2N}\right) \cos\left(2\pi f\tau\,\Delta t\right)$$

$$+ \frac{\sin\left(2\pi[N-\tau]f\,\Delta t\right)}{2N\sin\left(2\pi f\,\Delta t\right)} \cos\left(2\phi + 2\pi[N+1]f\,\Delta t\right) \quad (529)$$

(the proof of this result is Exercise [10.11]). This can be written in terms of the true acvs $s_\tau = \cos\left(2\pi f\tau\,\Delta t\right)/2$ as

$$\hat{s}_\tau^{(p)} = \left(\frac{N-\tau}{N}\right) s_\tau + \frac{\sin\left(2\pi[N-\tau]f\,\Delta t\right)}{2N\sin\left(2\pi f\,\Delta t\right)} \cos\left(2\phi + 2\pi[N+1]f\,\Delta t\right);$$

i.e., $\hat{s}_\tau^{(p)}$ is composed of one term that is a biased estimate of s_τ and a second term that is phase dependent. Both these terms would contribute to an inaccurate spectral estimate using the Yule–Walker equations based on $\{\hat{s}_\tau^{(p)}\}$.

Suppose now that N is large enough that the phase dependent term is close to zero. Then

$$\hat{s}_\tau^{(p)} \approx \left(\frac{N-\tau}{N}\right) s_\tau = \left(1 - \frac{\tau}{N}\right) s_\tau.$$

Kay and Marple (1979) argue that the estimator $\hat{s}_\tau^{(p)}$ '... corresponds more nearly to the acvs of two sinusoids which beat together to approximate a linear tapering.' Since an AR(p) spectral estimate fitted by the Yule–Walker method has an acvs that is *identical* with the sample acvs up to lag p, the spectral estimate should have two – rather than only one – spectral peaks. Although this problem might be alleviated by using the unbiased acvs estimator for large N, splitting can still occur when the phase dependent term is nonnegligible. To see this, note that the unbiased estimator can be written as

$$\left(\frac{N}{N-\tau}\right) \hat{s}_\tau^{(p)} = s_\tau + \frac{\sin\left(2\pi[N-\tau]f\,\Delta t\right)}{2(N-\tau)\sin\left(2\pi f\,\Delta t\right)} \cos\left(2\phi + 2\pi[N+1]f\,\Delta t\right).$$

In the case where $\phi = -\pi/2$, $N = 9$ and $f = 1/(4\,\Delta t) = f_{(N)}/2$, it is easy to show that

$$\left(\frac{N}{N-\tau}\right) \hat{s}_\tau^{(p)} = \left(1 - \frac{1}{N-\tau}\right) s_\tau,$$

so that the true acvs is again tapered (although not linearly). In fact, when $p = 8$, line splitting occurs (Kay and Marple, 1979).

Note that the model used by Kay and Marple is a noise-free sinusoid. This is reasonable for investigating line splitting, since the phenomenon has been seen to occur at *high* signal-to-noise ratios.

In Burg's algorithm the numerator of the estimator for $\phi_{k,k}$ is a sum of lagged products of forward and backward prediction errors, and thus it is plausible that it has the same problems as the acvs estimator. Indeed, $\bar{\phi}_{1,1}$ has $2N$ times the Yule–Walker lag one biased acvs estimator as its numerator. It would thus be expected that the information conveyed by $\bar{\phi}_{1,1}$ to a higher order fit would be faulty.

Kay and Marple (1979) point out that the phase-dependent term of Equation (529) can be regarded as the interaction between complex exponentials since this term can be written as

$$\frac{1}{4N} \sum_{t=1}^{N-\tau} \left(e^{i(4\pi ft\,\Delta t + 2\pi f\tau\,\Delta t + 2\phi)} + e^{-i(4\pi ft\,\Delta t + 2\pi f\tau\,\Delta t + 2\phi)} \right).$$

This fact led them to consider the *'analytic' series* associated with $\{X_t\}$. By definition this series is $X_t + i\mathcal{HT}\{X_t\}$, where $\mathcal{HT}$ is the *Hilbert transform* of $\{X_t\}$ (Papoulis, 1991, Section 10–3). For our purposes, we can regard the Hilbert transform of a series as a $-\pi/2$ phase-shifted version of the series – thus, if $X_t = \cos(2\pi ft\,\Delta t + \phi)$, then $\mathcal{HT}\{X_t\} = \sin(2\pi ft\,\Delta t + \phi)$. In our example, the analytic series is thus

$$\cos(2\pi ft\,\Delta t + \phi) + i\sin(2\pi ft\,\Delta t + \phi) = e^{i(2\pi ft\,\Delta t + \phi)} \equiv Z_t,$$

say. For $\tau \geq 0$ the unbiased acvs estimator for these complex-valued data is

$$\frac{N}{N-\tau}\hat{s}_{\tau,Z} = \frac{1}{N-\tau} \sum_{t=1}^{N-\tau} Z_t^* Z_{t+\tau}$$

$$= \frac{1}{N-\tau} \sum_{t=1}^{N-\tau} e^{-i(2\pi ft\,\Delta t + \phi)} e^{i(2\pi[t+\tau]f\,\Delta t + \phi)}$$

$$= \frac{1}{N-\tau} \sum_{t=1}^{N-\tau} e^{i2\pi f\tau\,\Delta t} = e^{i2\pi f\tau\,\Delta t}, \tag{530}$$

which is exactly the acvs of a complex exponential with unit amplitude (cf. Equation (468b)). Since this estimate equals the exact or theoretical acvs, the fact that the AR(p) process fitted by the Yule–Walker method has a theoretical acvs that is identical with the estimated acvs up to lag p will prevent the possibility of line splitting. Thus, for the Yule–Walker method, Kay and Marple (1979) recommend using complex-valued data (i.e., the analytic series) and the unbiased acvs estimator.

Let us now turn to the case of Burg's algorithm. For the analytic series $\{Z_t\}$, the fitting of the initial AR(1) model by this method demands the minimization of

$$\sum_{t=2}^{N} |Z_t - \phi_{1,1}Z_{t-1}|^2 + |Z_{t-1} - \phi_{1,1}^* Z_t|^2,$$

or, identically,

$$\sum_{t=2}^{N} |Z_t - \phi_{1,1} Z_{t-1}|^2 + |Z_{t-1}^* - \phi_{1,1} Z_t^*|^2$$

(see Section 9.13). The Burg estimate of $\phi_{1,1}$ is

$$\bar{\phi}_{1,1} = \frac{2\sum_{t=1}^{N-1} Z_t^* Z_{t+1}}{|Z_1|^2 + 2\sum_{t=2}^{N-1}|Z_t|^2 + |Z_{N-1}|^2} = \frac{2(N-1)e^{i2\pi f \, \Delta t}}{2(N-1)} = e^{i2\pi f \, \Delta t},$$

where we have used the result in Equation (530) and the fact that $|Z_t|^2 = 1$ for all t. It can be shown that the estimates of the higher order reflection coefficients, $\bar{\phi}_{2,2}$, $\bar{\phi}_{3,3}$, etc., are all zero (Kay and Marple, 1979). These results imply that the acvs estimator corresponding to Burg's algorithm is thus $\tilde{s}_{\tau,Z} = \exp{(i2\pi f\tau \, \Delta t)}$, again identical to the theoretical acvs. Since the implied and theoretical acvs's are identical for all τ, line splitting in Burg's algorithm also appears to be cured by the use of a complex exponential, i.e., by making the series 'analytic.'

Kay and Marple (1981) point out that, for *multiple* sinusoids, the performance of Fougere's algorithm is undocumented and that the use of complex-valued data in conjunction with Burg's algorithm can still result in line splitting. However, the unconstrained forward/backward least squares minimization algorithm given by Marple (1980) and discussed in Section 9.7 has shown no evidence of line splitting.

10.17 Singular Value Decomposition Approach

We stated in Equation (515b) that the summation of p *real-valued* sinusoids can be represented by a real-valued pseudo-AR($2p$) equation in which the parameters $\varphi_{k,2p}$ are the coefficients of the polynomial equation (515c) with roots occurring in pairs of the form $\exp{(\pm i2\pi f_j \, \Delta t)}$ for $j = 1, \ldots, p$. As might be anticipated, for the summation of p *complex-valued* sinusoids, i.e., complex exponentials,

$$\tilde{Z}_t \equiv \sum_{l=1}^{p} D_l' e^{i(2\pi f_l t \, \Delta t + \phi_l)},$$

there is a corresponding representation in terms of a complex-valued pseudo-AR(p) equation where the parameters $\varphi_{k,p}$ are the coefficients of the polynomial equation

$$1 - \sum_{k=1}^{p} \varphi_{k,p} z^{-k} = 0 \quad \text{or, equivalently,} \quad z^p - \sum_{k=1}^{p} \varphi_{k,p} z^{p-k} = 0$$

with roots of the form $z_j \equiv \exp{(i2\pi f_j\,\Delta t)}$ for $j = 1, \ldots, p$. To see that we do indeed have

$$\tilde{Z}_t = \sum_{k=1}^{p} \varphi_{k,p}\tilde{Z}_{t-k},$$

we first write

$$\tilde{Z}_t = \sum_{l=1}^{p} C_l e^{i2\pi f_l t\,\Delta t} = \sum_{l=1}^{p} C_l z_l^{t}, \text{ where } C_l \equiv D_l' e^{i\phi_l}.$$

The relationship between the roots $\{z_j\}$ and coefficients $\{\varphi_{k,p}\}$ is given by

$$\prod_{j=1}^{p} (z - z_j) = z^p - \sum_{k=1}^{p} \varphi_{k,p} z^{p-k}. \tag{532}$$

Now $\tilde{Z}_{t-k} = \sum_{l=1}^{p} C_l z_l^{t-k}$, and hence

$$\sum_{k=1}^{p} \varphi_{k,p}\tilde{Z}_{t-k} = \sum_{k=1}^{p} \varphi_{k,p}\left(\sum_{l=1}^{p} C_l z_l^{t-k}\right) = \sum_{l=1}^{p} C_l z_l^{t-p} \sum_{k=1}^{p} \varphi_{k,p} z_l^{p-k}.$$

However, because z_l is a root, we have

$$\prod_{j=1}^{p} (z_l - z_j) = z^p - \sum_{k=1}^{p} \varphi_{k,p} z_l^{p-k} = 0, \text{ i.e., } \sum_{k=1}^{p} \varphi_{k,p} z_l^{p-k} = z^p,$$

so we can now write

$$\sum_{k=1}^{p} \varphi_{k,p}\tilde{Z}_{t-k} = \sum_{l=1}^{p} C_l z_l^{t-p} z^p = \sum_{l=1}^{p} C_l z_l^{t} = \tilde{Z}_t$$

as required.

We now make use of this pseudo-AR(p) representation for the summation of p complex exponentials to design an estimation scheme for the frequencies f_l for a model with additive complex-valued white noise $\{\epsilon_t\}$ with zero mean; i.e.,

$$Z_t \equiv \tilde{Z}_t + \epsilon_t = \sum_{l=1}^{p} C_l e^{i2\pi f_l t\,\Delta t} + \epsilon_t = \sum_{k=1}^{p} \varphi_{k,p}\tilde{Z}_{t-k} + \epsilon_t$$

(cf. Equation (465a) with $\mu = 0$). To determine the frequencies f_l, Tufts and Kumaresan (1982) consider the following model involving a predictive filter of order p' (with $p' > p$ and p assumed unknown) in both the forward and backward directions:

$$Z_t = \sum_{k=1}^{p'} \varphi_{k,p'} Z_{t-k} \text{ and } Z_t^{*} = \sum_{k=1}^{p'} \varphi_{k,p'} Z_{t+k}^{*},$$

which, given data $Z_1, \ldots, Z_N$, yields the equations

$$
\begin{aligned}
Z_{p'+1} &= \varphi_{1,p'} Z_{p'} &&+ \varphi_{2,p'} Z_{p'-1} &&+ \cdots + \varphi_{p',p'} Z_1 \\
&\;\vdots &&\;\vdots &&\;\;\vdots &&\;\ddots &&\;\vdots \\
Z_N &= \varphi_{1,p'} Z_{N-1} &&+ \varphi_{2,p'} Z_{N-2} &&+ \cdots + \varphi_{p',p'} Z_{N-p'} \\
Z_1^* &= \varphi_{1,p'} Z_2^* &&+ \varphi_{2,p'} Z_3^* &&+ \cdots + \varphi_{p',p'} Z_{p'+1}^* \\
&\;\vdots &&\;\vdots &&\;\;\vdots &&\;\ddots &&\;\vdots \\
Z_{N-p'}^* &= \varphi_{1,p'} Z_{N-p'+1}^* &&+ \varphi_{2,p'} Z_{N-p'+2}^* &&+ \cdots + \varphi_{p',p'} Z_N^*
\end{aligned}
$$

For the noiseless case when $Z_t = \tilde{Z}_t$, these equations would be exactly true if $p' = p$. For the moment we assume the noiseless case, but take $p' > p$. We can write the above equations compactly as

$$A\varphi = \mathbf{Z}, \tag{533a}$$

where $\varphi \equiv [\varphi_{1,p'}, \ldots, \varphi_{p',p'}]^T$,

$$
A \equiv \begin{bmatrix}
Z_{p'} & Z_{p'-1} & \cdots & Z_1 \\
\vdots & & \ddots & \vdots \\
Z_{N-1} & Z_{N-2} & \cdots & Z_{N-p'} \\
Z_2^* & Z_3^* & \cdots & Z_{p'+1}^* \\
\vdots & \vdots & \ddots & \vdots \\
Z_{N-p'+1}^* & Z_{N-p'+2}^* & \cdots & Z_N^*
\end{bmatrix}
\quad \text{and} \quad
\mathbf{Z} \equiv \begin{bmatrix}
Z_{p'+1} \\
\vdots \\
Z_N \\
Z_1^* \\
\vdots \\
Z_{N-p'}^*
\end{bmatrix}
$$

(note that A is a $2(N - p') \times p'$ matrix).

As is true for any matrix, we can express A in terms of its singular value decomposition; see, for example, Golub and Kahan (1965) and Businger and Golub (1969). (For real-valued matrices see Jackson, 1972, or Golub and Van Loan, 1989, p. 71.) Two sets of orthogonal vectors $\{\mathbf{u}_j\}$ and $\{\mathbf{v}_j\}$ can be found such that

$$A^H \mathbf{u}_j = \gamma_j \mathbf{v}_j \text{ and } A\mathbf{v}_j = \lambda_j \mathbf{u}_j,$$

from which it follows that

$$
\begin{aligned}
AA^H \mathbf{u}_j &= \gamma_j \lambda_j \mathbf{u}_j, & j &= 1, \ldots, 2(N - p'), \\
A^H A \mathbf{v}_j &= \gamma_j \lambda_j \mathbf{v}_j, & j &= 1, \ldots, p',
\end{aligned} \tag{533b}
$$

(see Exercise [10.12]). Here the superscript 'H' denotes the operation of complex-conjugate transposition. Each $\mathbf{u}_j$ is an eigenvector of AA^H, while $\mathbf{v}_j$ is an eigenvector of $A^H A$. If we rank the γ_j and λ_j terms each in decreasing order of magnitude, it can be shown that, for some integer $r \leq \min\left(2(N-p'), p'\right)$,

$$\gamma_j = \begin{cases} \lambda_j > 0, & j \leq r; \\ 0, & \text{otherwise}; \end{cases} \quad \text{and } \lambda_j = 0 \text{ for } j > r;$$

i.e., there are r eigenvalues $\gamma_j \lambda_j = \lambda_j^2 > 0$ common to the two systems of equations in (533b), and all the others are zero. The $\mathbf{u}_j$ and $\mathbf{v}_j$ for $j = 1,$ $\ldots, r$ are called the *left and right eigenvectors* or *singular vectors* of the matrix A; the λ_j terms are called the *eigenvalues* or *singular values* of A. The integer r is the rank of A (its importance becomes clearer when we look at the two cases of zero and nonzero additive noise). The matrix A can be written as

$$A = U\Lambda V^H, \tag{534a}$$

where U is a $2(N-p') \times r$ matrix whose columns are the eigenvectors $\mathbf{u}_j$; V is a $p' \times r$ matrix whose columns are the eigenvectors $\mathbf{v}_j$; and Λ is the diagonal matrix whose diagonal elements are the eigenvalues λ_j. The right-hand side of the above is called the *singular value decomposition* or SVD of A. By the orthonormality of the eigenvectors,

$$U^H U = V^H V = I_r; \ UU^H = I_r \text{ if } r = 2(N-p'); \ VV^H = I_r \text{ if } r = p', \tag{534b}$$

where I_r is the $r \times r$ identity matrix (Jackson, 1972).

Let us now return to the particular matrix equation of interest, namely, Equation (533a). If we let $R \equiv A^H A$ and $\mathbf{b} \equiv A^H \mathbf{Z}$, then we can premultiply both sides of (533a) by A^H to write

$$A^H A\varphi = A^H \mathbf{Z} \quad \text{as} \quad R\varphi = \mathbf{b},$$

where the (j, k)th element of R is given by

$$\sum_{t=p'+1}^{N} Z_{t-j}^* Z_{t-k} + \sum_{t=1}^{N-p'} Z_{t+j} Z_{t+k}^*$$

and the jth element of $\mathbf{b}$ is given by

$$\sum_{t=p'+1}^{N} Z_{t-j}^* Z_t + \sum_{t=1}^{N-p'} Z_{t+j} Z_t^*.$$

The system $R\varphi = \mathbf{b}$ is the least squares system of equations corresponding to minimizing the forward/backward sum of squares

$$\sum_{t=p'+1}^{N} \left| Z_t - \sum_{k=1}^{p'} \varphi_{k,p'} Z_{t-k} \right|^2 + \sum_{t=1}^{N-p'} \left| Z_t^* - \sum_{k=1}^{p'} \varphi_{k,p'} Z_{t+k}^* \right|^2.$$

If we use the singular value decomposition given in Equation (534a) and the result $U^H U = I_r$ from Equation (534b), we can write

$$R = A^H A = \left(V \Lambda U^H \right) \left(U \Lambda V^H \right) = V \Lambda I_r \Lambda V^H = V \Lambda^2 V^H,$$

where Λ^n refers to the diagonal matrix with diagonal elements λ_j^n. Also, **b** can be written

$$\mathbf{b} = A^H \mathbf{Z} = V \Lambda U^H \mathbf{Z},$$

so that $R\varphi = \mathbf{b}$ becomes

$$V \Lambda^2 V^H \varphi = V \Lambda U^H \mathbf{Z}. \tag{535a}$$

The *generalized inverse* of $R = V \Lambda^2 V^H$ is given by

$$R^{\#} \equiv V \Lambda^{-2} V^H.$$

This generalized inverse always exists and gives the solution (using $V^H V = I_r$ from Equation (534b))

$$\tilde{\varphi} = \left(V \Lambda^{-2} V^H \right) \left(V \Lambda U^H \right) \mathbf{Z} = V \Lambda^{-1} U^H \mathbf{Z}. \tag{535b}$$

That this is indeed a solution to Equation (535a) can be checked by noting that

$$V \Lambda^2 V^H \tilde{\varphi} = V \Lambda^2 V^H V \Lambda^{-1} U^H \mathbf{Z} = V \Lambda^2 I_r \Lambda^{-1} U^H \mathbf{Z} = V \Lambda U^H \mathbf{Z}$$

(again using $V^H V = I_r$). This solution minimizes $\tilde{\varphi}^H \tilde{\varphi} = |\tilde{\varphi}|^2$ (see, for example, Jackson, 1972). It is interesting to note that, if $r = p'$, then $RR^{\#} = R^{\#}R = VV^H = I_{p'}$ from Equation (534b), and in this case (535b) is the usual least squares solution for φ.

When the additive noise $\{\epsilon_t\}$ is zero for all t, the process $\{Z_t\}$ consists of just the summation of p complex exponentials, and $r = p < p'$. From Equation (535b), we have

$$\tilde{\varphi} = [\mathbf{v}_1, \ldots, \mathbf{v}_p] \begin{bmatrix} \lambda_1^{-1} & & & \\ & \lambda_2^{-1} & & \\ & & \ddots & \\ & & & \lambda_p^{-1} \end{bmatrix} \begin{bmatrix} \mathbf{u}_1^H \\ \vdots \\ \vdots \\ \mathbf{u}_p^H \end{bmatrix} \mathbf{Z}$$

$$= \left(\sum_{k=1}^{p} \frac{1}{\lambda_k} \mathbf{v}_k \mathbf{u}_k^H \right) \mathbf{Z} \tag{535c}$$

(Tufts and Kumaresan, 1982). To ensure that the rank of A is $r = p$, Hua and Sarkar (1988) point out that we should choose p' such that

$p \leq p' \leq N - p$. With $\tilde{\varphi}$ determined from Equation (535c), we form the corresponding polynomial equation, namely,

$$z^{p'} - \sum_{k=1}^{p'} \tilde{\varphi}_{k,p'} z^{p'-k} = 0.$$

Since $r = p$ this expression has p roots z_j on the unit circle at the locations of the sinusoidal frequencies, and the other $p'-p$ roots lie inside the unit circle (Tufts and Kumaresan, 1982). The frequency f_j is thus found by inverting $z_j = \exp(i2\pi f_j \Delta t)$; i.e., $f_j = \Im(\log(z_j))/(2\pi \Delta t)$, where $\Im(z)$ is the imaginary part of the complex number z, and we take the solution such that $|f_j| \leq f_{(N)}$.

Of course, a much more realistic situation is when the additive noise $\{\epsilon_t\}$ is nonzero. In this case A has full rank, i.e., $r = \min(2(N - p'), p')$, and, by a derivation identical to that used for Equation (535c), the solution using the generalized inverse is

$$\hat{\varphi} = \left(\sum_{k=1}^{p'} \frac{1}{\hat{\lambda}_k} \hat{\mathbf{v}}_k \hat{\mathbf{u}}_k^H \right) \mathbf{Z}. \tag{536}$$

Since the matrix A and $\mathbf{Z}$ of (533a) here include additive noise, the left and right eigenvectors and eigenvalues are in general different from those in Equation (535c), so we have added a 'hat' to these quantities to emphasize this. The eigenvalues $\hat{\lambda}_{p+1}, \ldots, \hat{\lambda}_{p'}$ are due to the noise and will usually be small. The solution $\hat{\varphi}$ thus becomes dominated by the noise components unless the sum in (536) is truncated at p, so we define

$$\hat{\varphi}' \equiv \left(\sum_{k=1}^{p} \frac{1}{\hat{\lambda}_k} \hat{\mathbf{v}}_k \hat{\mathbf{u}}_k^H \right) \mathbf{Z},$$

for which again the sinusoidal terms are the dominant influence. If prior knowledge of the exact number of sinusoids p is lacking, it might be possible to guess the number by looking for the point k in the eigenvalue sequence

$$\hat{\lambda}_1 \geq \hat{\lambda}_2 \geq \cdots \hat{\lambda}_k \geq \hat{\lambda}_{k+1} \geq \cdots \geq \hat{\lambda}_{p'}$$

at which the value is large before rapidly decreasing at $k+1$. Sometimes such a cutoff is clear, but other times it is not. The polynomial equation

$$z^{p'} - \sum_{k=1}^{p'} \tilde{\varphi}'_{k,p'} z^{p'-k} = 0$$

has in general p roots z'_k close to the unit circle, and these can be used to find the frequencies $f'_k = \Im(\log(z'_k))/(2\pi \Delta t)$.

Comments and Extensions to Section 10.17

[1] Tufts and Kumaresan (1982) consider that the singular value decomposition approach is likely to be most useful when only a short time series is available and the sinusoids are more closely spaced than the reciprocal of the observation time.

[2] If minimization of the forward and backward sum of squares alone is used for the estimation of φ, large values of p' can result in spurious spectral peaks. By using the singular value decomposition, the length of p' can be made considerably greater because of the effective gain in signal-to-noise ratio, thus increasing the resolution of the sinusoids. Tufts and Kumaresan found a good empirical choice to be $p' = 3N/4$. Since this implies that p' complex roots must be determined, it is clear that the method is only practicable for short series, as indeed was intended.

[3] For computational purposes it may be more convenient to reexpress Equation (536) as

$$\hat{\varphi} = \left(\sum_{k=1}^{p'} \frac{1}{\hat{\lambda}_k^2} \hat{\mathbf{v}}_k \hat{\mathbf{v}}_k^H A^H \right) \mathbf{Z},$$

where we have used

$$A\hat{\mathbf{v}}_k = \hat{\lambda}_k \hat{\mathbf{u}}_k \text{ to write } \hat{\mathbf{u}}_k^H = \hat{\mathbf{v}}_k^H A^H / \hat{\lambda}_k.$$

Note from Equation (533b) that $\hat{\mathbf{v}}_k$ is the estimated (right) eigenvector, and $\hat{\lambda}_k^2$ the estimated eigenvalue, of $R = A^H A$.

10.18 SVD Approach with River Flow Data

As an example of the singular value decomposition (SVD) approach to harmonic analysis, we here apply the methods of Section 10.17 to the Willamette River data discussed in Sections 10.13 and 10.15. The description of the SVD method given above is in terms of complex exponentials, whereas in Section 10.15 the analysis was in terms of real-valued sinusoids. We can render the real-valued Willamette River series into a suitable form for the SVD approach by creating the 'analytic' series corresponding to the data $\{X_t\}$. The analytic series is

$$X_t + i\mathcal{HT}\{X_t\},$$

where $\mathcal{HT}$ is the Hilbert transform of $\{X_t\}$ (see Section 10.16 above and Section 10–3 of Papoulis, 1991). By definition, the transfer function (frequency response function) of the Hilbert transform, $G(\cdot)$ say, must have unit gain, a phase angle of $-\pi/2$ for f between 0 and $1/2$ and a phase angle of $\pi/2$ for f between 0 and $-1/2$. Thus $G(\cdot)$ takes the form

$$G(f) \equiv \begin{cases} -i, & 0 \leq f < 1/2; \\ i, & -1/2 \leq f < 0 \end{cases}$$

(Oppenheim and Schafer, 1989). The corresponding impulse response sequence $\{g_t\}$ for the Hilbert transform is

$$g_t = \int_{-1/2}^{1/2} G(f)e^{i2\pi ft}\,df = \begin{cases} 2/(\pi t), & t \text{ odd}; \\ 0, & t \text{ even} \end{cases}$$

A realizable approximation to this ideal (infinitely long) impulse response can be obtained by multiplying the g_t sequence by a set of convergence factors (see Sections 3.7 and 5.8); for convenience, we use a Hanning lag window (see Exercise [6.20]). The convergence factors are used to reduce the impulse response sequence smoothly to a total length of 51 nonzero coefficients (i.e., 25 each side of $t = 0$). This creates no phase error, but slight gain errors do occur, mainly near zero and Nyquist frequencies. The resulting Hilbert transform filter is then renormalized to preserve the variance of the series to which it is to be applied. The sample mean is subtracted from the real-valued Willamette River series (the Hilbert transform filter has zero gain at zero frequency), and the series is convolved with the filter. Due to transient effects of the filtering at the ends of the data, the series is reduced in length; for computational convenience a length of 200 values unaffected by transients was retained. These form the imaginary part of the analytic series and are plotted with the corresponding real part (from the original series) in the top two plots of Figure 539.

In applying the methods of Section 10.17 to the complex (analytic) series a value of p' of 25 is suitable. It would be desirable to make this value larger for improved resolution of the sinusoids, but we must be able to find p' complex roots, and 25 is already a large number to find. The results of the investigations in Section 10.13 were two sinusoids with estimated frequencies of 1.008 and 2.004 cycles/year. Hence, this suggests truncating at $p = 2$ in the SVD method. The resulting estimated roots are plotted in polar form in the bottom plot of Figure 539 and were found using the subroutine ZROOTS given in Press *et al.* (1986, p. 265). The two complex roots z'_k closest to the unit circle are at (0.8611, 0.5039) and (0.4839, 0.8442), having magnitudes 0.9977 and 0.9730, respectively, and phase angles quite close to 30° and 60°. Recalling that $\Delta t = 1/12$ year, the corresponding frequencies f'_k are 1.011 and 2.006 cycles/year, very similar to those obtained in Section 10.13.

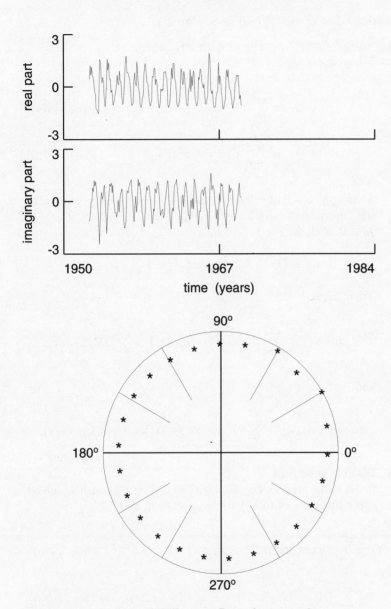

Figure 539. Analysis of Willamette River data, part 6.

10.19 Summary of Harmonic Analysis

In this section we collect together the most useful results of this chapter. It is assumed throughout that $\mu = 0$; D_l, D'_l and f_l are real-valued constants; C_l is a complex-valued constant; and the ϕ_l terms are independent real-valued rv's uniformly distributed on $[-\pi, \pi]$. The spectral classifications used are as defined in Section 4.4.

[1] *Real-valued discrete parameter harmonic processes*
   a) Mixed spectrum:

$$X_t = \sum_{l=1}^{L} D_l \cos\left(2\pi f_l t\, \Delta t + \phi_l\right) + \eta_t \quad \text{(see (467a))}$$

$$= \sum_{l=-L}^{L} C_l e^{i2\pi f_l t\, \Delta t} + \eta_t, \quad \text{(see (467b))}$$

where $\{\eta_t\}$ is a real-valued (possibly colored) noise process with zero mean and variance σ_η^2, independent of each ϕ_l, $C_0 \equiv 0$, $f_0 \equiv 0$, and, for $l = 1, \ldots, L$,

$$C_l \equiv D_l e^{i\phi_l}/2, \quad C_{-l} \equiv C_l^* \quad \text{and} \quad f_{-l} = -f_l.$$

Then, $E\{X_t\} = 0$,

$$\operatorname{var}\{X_t\} = \sum_{l=-L}^{L} E\{|C_l|^2\} + \operatorname{var}\{\eta_t\} = \sum_{l=1}^{L} D_l^2/2 + \sigma_\eta^2$$

and

$$\operatorname{cov}\{X_t, X_{t+\tau}\} = \sum_{l=1}^{L} D_l^2 \cos\left(2\pi f_l \tau\, \Delta t\right)/2 + \operatorname{cov}\{\eta_t, \eta_{t+\tau}\}.$$

$$\text{(see (467c))}$$

   b) Discrete spectrum:
   If the noise process $\{\eta_t\}$ is white, i.e., $\eta_t \equiv \epsilon_t$, then the mixed spectrum reduces to a discrete spectrum, and

$$X_t = \sum_{l=1}^{L} D_l \cos\left(2\pi f_l t\, \Delta t + \phi_l\right) + \epsilon_t \quad \text{(see (460))}$$

$$= \sum_{l=1}^{L} \left(A_l \cos\left(2\pi f_l t\, \Delta t\right) + B_l \sin\left(2\pi f_l t\, \Delta t\right)\right) + \epsilon_t, \text{(see (463a))}$$

where $\{\epsilon_t\}$ is a real-valued white noise process with zero mean and variance σ_ϵ^2, independent of each ϕ_l. We have $E\{X_t\} = 0$,

$$\operatorname{var}\{X_t\} = \sum_{l=1}^{L} D_l^2/2 + \sigma_\epsilon^2$$

and

$$\operatorname{cov}\{X_t, X_{t+\tau}\} = \sum_{l=1}^{L} D_l^2 \cos\left(2\pi f_l \tau \,\Delta t\right)/2, \quad |\tau| > 0.$$

c) Known frequencies, discrete spectrum:
   If each f_l is any of the Fourier frequencies not equal to 0 or $f_{(N)}$,

$$\hat{A}_l = \frac{2}{N} \sum_{t=1}^{N} X_t \cos\left(2\pi f_l t \,\Delta t\right); \quad \hat{B}_l = \frac{2}{N} \sum_{t=1}^{N} X_t \sin\left(2\pi f_l t \,\Delta t\right)$$

(see (463b))

give *exact* least squares estimates of A_l and B_l. If now each ϕ_l is treated as a constant so that A_l and B_l are constants and Equation (463a) is a multiple linear regression model, then

$$E\{\hat{A}_l\} = A_l, \ \ E\{\hat{B}_l\} = B_l \ \text{ and } \ \operatorname{var}\{\hat{A}_l\} = \operatorname{var}\{\hat{B}_l\} = \frac{2\sigma_\epsilon^2}{N}.$$

Further, if $k \neq l$, then $\operatorname{cov}\{\hat{A}_k, \hat{B}_l\}$, $\operatorname{cov}\{\hat{A}_l, \hat{B}_l\}$, $\operatorname{cov}\{\hat{A}_k, \hat{A}_l\}$ and $\operatorname{cov}\{\hat{B}_k, \hat{B}_l\}$ are all zero, and σ_ϵ^2 is estimated by Equation (464). If the frequencies f_l are not all Fourier frequencies, then $\hat{A}_l$ and $\hat{B}_l$ give *approximate* least squares estimates, and

$$E\{\hat{A}_l\} = A_l + O\left(\frac{1}{N}\right) \ \text{ and } \ E\{\hat{B}_l\} = B_l + O\left(\frac{1}{N}\right).$$

d) Unknown frequencies, discrete spectrum:
   When the frequencies f_l are unknown, the standard approach is to look for peaks in the periodogram. If f_l is a Fourier frequency (not equal to 0 or $f_{(N)}$) present in $\{X_t\}$, then

$$E\{\hat{S}^{(p)}(f_l)\} = \left(N\frac{D_l^2}{4} + \sigma_\epsilon^2\right)\Delta t, \qquad \text{(see (474))}$$

while, if f_l is a Fourier frequency not in $\{X_t\}$, then

$$E\{\hat{S}^{(p)}(f_l)\} = \sigma_\epsilon^2 \,\Delta t.$$

When f_l is not necessarily one of the Fourier frequencies, it follows that, for all $|f| \leq f_{(N)}$,

$$E\{\hat{S}^{(p)}(f)\} = \sigma_\epsilon^2 \,\Delta t + \sum_{l=1}^{L} \frac{D_l^2}{4}\left[\mathcal{F}(f + f_l) + \mathcal{F}(f - f_l)\right],$$

(see (475))

where $\mathcal{F}(\cdot)$ is Fejér's kernel. If a data taper $\{h_t\}$ is applied to the process, then the expectation of the direct spectral estimator is given by

$$E\{\hat{S}^{(d)}(f)\} = \sigma_\epsilon^2 \, \Delta t + \sum_{l=1}^{L} \frac{D_l^2}{4} \left[\mathcal{H}(f + f_l) + \mathcal{H}(f - f_l)\right],$$

$$\text{(see (486a))}$$

where $\mathcal{H}(\cdot)$ is the spectral window corresponding to $\{h_t\}$. If the taper is the default rectangular taper, then $\mathcal{H}(\cdot) \equiv \mathcal{F}(\cdot)$.

[2] *Complex-valued discrete parameter harmonic processes*

a) Mixed spectrum:

$$Z_t = \sum_{l=1}^{L} D_l' e^{i(2\pi f_l t \, \Delta t + \phi_l)} + \eta_t = \sum_{l=1}^{L} C_l e^{i2\pi f_l t \, \Delta t} + \eta_t,$$

$$\text{(see (468a))}$$

where $\{\eta_t\}$ is a complex-valued (possibly colored) noise process with zero mean and variance σ_η^2, independent of each ϕ_l, and $C_l \equiv D_l' \exp(i\phi_l)$. Then, $E\{Z_t\} = 0$,

$$\text{var}\,\{Z_t\} = \sum_{l=1}^{L} E\{|C_l|^2\} + \text{var}\,\{\eta_t\} = \sum_{l=1}^{L} (D_l')^2 + \sigma_\eta^2$$

and

$$\text{cov}\,\{Z_t, Z_{t+\tau}\} = \sum_{l=1}^{L} (D_l')^2 e^{i2\pi f_l \tau \, \Delta t} + \text{cov}\,\{\eta_t, \eta_{t+\tau}\}.$$

b) Discrete spectrum:

If the noise process is white, i.e., $\eta_t \equiv \epsilon_t$ then the mixed spectrum reduces to a discrete spectrum, and

$$Z_t = \sum_{l=1}^{L} D_l' e^{i(2\pi f_l t \, \Delta t + \phi_l)} + \epsilon_t, \qquad \text{(see (465a))}$$

where $\{\epsilon_t\}$ is a real-valued white noise process with zero mean and variance σ_ϵ^2, independent of each ϕ_l. We have $E\{Z_t\} = 0$,

$$\text{var}\,\{Z_t\} = \sum_{l=1}^{L} (D_l')^2 + \sigma_\epsilon^2$$

and

$$\text{cov}\,\{Z_t, Z_{t+\tau}\} = \sum_{l=1}^{L} (D_l')^2 e^{i2\pi f_l \tau \, \Delta t}, \qquad |\tau| > 0.$$

c) Known frequencies, discrete spectrum:

Let us assume a process with a discrete spectrum. If each f_l is any of the Fourier frequencies not equal to 0 or $f_{(N)}$,

$$\hat{C}_l = \frac{1}{N} \sum_{t=1}^{N} Z_t e^{-i2\pi f_l t \, \Delta t} \qquad \text{(see Exercise [10.2])}$$

is the exact least squares estimate of C_l. With each ϕ_l treated as a constant,

$$E\{\hat{C}_l\} = C_l, \ \ \text{var}\,\{\hat{C}_l\} = \frac{\sigma_\epsilon^2}{N} \ \ \text{and} \ \ \text{cov}\,\{\hat{C}_k, \hat{C}_l\} = 0, \quad k \neq l.$$

d) Unknown frequencies, discrete spectrum:

If f_l is a Fourier frequency (not equal to 0 or $f_{(N)}$) present in $\{Z_t\}$, then

$$E\{\hat{S}^{(p)}(f_l)\} = \left[N(D_l')^2 + \sigma_\epsilon^2\right] \Delta t,$$

while, if f_l is a Fourier frequency not in $\{Z_t\}$,

$$E\{\hat{S}^{(p)}(f_l)\} = \sigma_\epsilon^2 \, \Delta t.$$

[3] *Tests for periodicity for real-valued process*

a) White noise:

This case is discussed in detail in Section 10.9. The sample size N is usually taken to be odd so that $N = 2m + 1$ for some integer m (Section 10.10 shows how to accommodate an even sample size). The null hypothesis is $D_1 = \cdots = D_L = 0$. Fisher's exact test for *simple* periodicity uses the statistic

$$g \equiv \frac{\max_{1 \leq k \leq m} \hat{S}^{(p)}(f_k)}{\sum_{j=1}^{m} \hat{S}^{(p)}(f_j)}, \qquad \text{(see (491a))}$$

where $\hat{S}^{(p)}(f_k)$ is a periodogram term. The exact distribution of g under the null hypothesis is given by Equation (491b). Critical values g_F for Fisher's test for $\alpha = 0.01, 0.02, 0.05$ and 0.1 can be adequately approximated using Equation (491c). To test for the presence of *compound* periodicities, Siegel derived the statistic based on excesses over a threshold,

$$T_\lambda \equiv \sum_{k=1}^{m} \left(\tilde{S}^{(p)}(f_k) - \lambda g_F\right)_+,$$

where $\tilde{S}^{(p)}(f_k) \equiv \hat{S}^{(p)}(f_k) / \sum_{j=1}^{m} \hat{S}^{(p)}(f_j)$, $0 < \lambda \leq 1$, and g_F is the critical value for Fisher's test. The exact distribution of T_λ under the null hypothesis is given by Equation (492). When $\lambda = 0.6$, critical values for m in the range 20 to 2000 may be approximated using Equation (493b) for $\alpha = 0.05$, and Equation (493c) for $\alpha = 0.01$.

b) Colored noise:

This case is discussed in detail in Section 10.11. To test for periodicity at a particular frequency f_1 for a real-valued process incorporating colored *Gaussian* noise, the recommended statistic is

$$\frac{(K-1)\left|\hat{C}_1\right|^2 \sum_{k=0}^{K-1} H_k^2(0)}{\Delta t \sum_{k=0}^{K-1} \left|J_k(f_1) - \hat{J}_k(f_1)\right|^2}, \qquad \text{(see (499c))}$$

where $H_k(0) \equiv \Delta t \sum_{t=1}^{N} h_{t,k}$,

$$J_k(f_1) \equiv (\Delta t)^{1/2} \sum_{t=1}^{N} h_{t,k} X_t e^{-i2\pi f_1 t \,\Delta t}, \quad \hat{J}_k(f_1) \equiv \hat{C}_1 \frac{H_k(0)}{(\Delta t)^{1/2}}$$

and $\hat{C}_1$ is given by Equation (499a). Here $\{h_{t,k}\}$ is a kth order dpss data taper. This statistic has a simple $F_{2,2K-2}$ distribution under the null hypothesis of no periodicity at the frequency f_1.

10.20 Exercises

[10.1] Verify that the approximations in Equation (462c) are valid.

[10.2] For the complex-valued model given by Equation (466b), show that $\operatorname{cov}\{\hat{C}_k, \hat{C}_l\} = 0$ for $k \neq l$, where $\hat{C}_k$ and $\hat{C}_l$ are the least squares estimators of, respectively, C_k and C_l; i.e.,

$$\hat{C}_k = \frac{1}{N} \sum_{t=1}^{N} Z_t e^{-i2\pi f_k t \,\Delta t} \text{ and } \hat{C}_l = \frac{1}{N} \sum_{t=1}^{N} Z_t e^{-i2\pi f_l t \,\Delta t}.$$

[10.3] Prove that

$$\operatorname{cov}\{Z_t, Z_{t+\tau}\} = \sum_{l=1}^{L} (D_l')^2 e^{i2\pi f_l \tau \,\Delta t} + \operatorname{cov}\{\eta_t, \eta_{t+\tau}\}$$

for $\{Z_t\}$ given by Equation (468a).

[10.4] Verify Equation (475).

[10.5] Substitute appropriate estimates for the unknown quantities in the expression for the variance of a periodogram estimate of frequency in Equation (477b) to determine roughly whether the estimated frequencies $\hat{f}_1 = 1.0032$ cycles/year and and $\hat{f}_2 = 2.0034$ cycles/year for the Willamette River data include – within their two standard deviation limits – the values 1 cycle/year and 2 cycles/year (see Section 10.13). Are the corresponding AR(p) estimates in Table 523 also within these two standard deviation limits?

[10.6] Prove that, for N odd,

$$\sum_{j=1}^{(N-1)/2} \hat{S}^{(p)}(f_j) = \frac{\Delta t}{2} \sum_{t=1}^{N} (X_t - \bar{X})^2.$$

[10.7] Prove that, for N even,

$$\sum_{j=1}^{(N-2)/2} \hat{S}^{(p)}(f_j) = \frac{\Delta t}{2} \left(\sum_{t=1}^{N}(X_t - \bar{X})^2 - \frac{1}{N} \left[\sum_{t=1}^{N} X_t(-1)^t \right]^2 \right).$$

[10.8] Show that Equation (507a) can be rewritten as Equation (507b). What happens if $\Delta t = 1/5$ instead of $1/12$?

[10.9] Using Equation (515a) and the initial conditions $X_0 = D \cos(\phi)$ and $X_1 = D \cos(2\pi f \, \Delta t + \phi)$, show that

$$\text{cov}\,\{X_t, X_{t+\tau}\} = D^2 \cos(2\pi f \tau \, \Delta t)/2.$$

[10.10] Show that an AR(p) sdf plus a white noise sdf yields an ARMA(p, p) sdf. As a concrete example, find the coefficients for the ARMA(2,2) process obtained by adding white noise with unit variance to the AR(2) process defined by Equation (45). Plot the sdf's for the AR(2) and ARMA(2,2) processes for comparison.

[10.11] Verify the result stated in Equation (529) by using the trigonometric identity

$$\sum_{k=0}^{M-1} \cos(x + ky) = \cos\left(x + \frac{M-1}{2}y\right) \frac{\sin(My/2)}{\sin(y/2)}.$$

[10.12] With $m \equiv 2(N - p')$ and $n = p'$, derive Equation (534a) by considering the eigenvectors and eigenvalues of the $(m + n) \times (m + n)$ matrix

$$B \equiv \begin{bmatrix} 0 & A \\ A^H & 0 \end{bmatrix}.$$

References

Abramowitz, M. and Stegun, I. A. (1964) *Handbook of Mathematical Functions.* Washington, D. C.: U. S. Government Printing Office (reprinted in 1968 by Dover Publications, New York).

Akaike, H. (1970) Statistical Predictor Identification. *Annals of the Institute of Statistical Mathematics*, 22, 203–17.

—— (1974) A New Look at the Statistical Model Identification. *IEEE Transactions on Automatic Control*, 19, 716–22.

Andersen, N. O. (1978) Comments on the Performance of Maximum Entropy Algorithms. *Proceedings of the IEEE*, 66, 1581–2.

Anderson, T. W. (1971) *The Statistical Analysis of Time Series.* New York: John Wiley & Sons.

Baggeroer, A. B. (1976) Confidence Intervals for Regression (MEM) Spectral Estimates. *IEEE Transactions on Information Theory*, 22, 534–45 (reprinted in Childers, 1978).

Barnes, J. A., Chi, A. R., Cutler, L. S., Healey, D. J., Leeson, D. B., McGunigal, T. E., Mullen, J. A., Jr., Smith, W. L., Sydnor, R. L., Vessot, R. F. C. and Winkler, G. M. R. (1971) Characterization of Frequency Stability. *IEEE Transactions on Instrumentation and Measurement*, 20, 105–20.

Bartlett, M. S. (1950) Periodogram Analysis and Continuous Spectra. *Biometrika*, 37, 1–16.

—— (1955) *An Introduction to Stochastic Processes.* Cambridge, England: Cambridge University Press.

Bartlett, M. S. and Kendall, D. G. (1946) The Statistical Analysis of Variance-Heterogeneity and the Logarithmic Transformation. *Supplement to the Journal of the Royal Statistical Society*, 8, 128–38.

546

Beauchamp, K. G. (1984) *Applications of Walsh and Related Functions.* London: Academic Press.

Bell, B. M. and Percival, D. B. (1991) A Two Step Burg Algorithm. *IEEE Transactions on Signal Processing*, 39, 185–9.

Bell, B. M., Percival, D. B. and Walden, A. T. (1993) Calculating Thomson's Spectral Multitapers by Inverse Iteration. *Journal of Computational and Graphical Statistics* (to appear).

Berk, K. N. (1974) Consistent Autoregressive Spectral Estimates. *Annals of Statistics*, 2, 489–502.

Best, D. J. and Roberts, D. E. (1975) The Percentage Points of the χ^2 Distribution: Algorithm AS 91. *Applied Statistics*, 24, 385–8.

Blackman, R. B. and Tukey, J. W. (1958) *The Measurement of Power Spectra.* New York: Dover Publications.

Bloomfield, P. (1976) *Fourier Analysis of Time Series: An Introduction.* New York: John Wiley & Sons.

Bogert, B. P., Healy, M. J. and Tukey, J. W. (1963) The Quefrency Alanysis of Time Series for Echoes: Cepstrum, Pseudo-Autocovariance, Cross-Cepstrum and Saphe Cracking. In *Proceedings of the Symposium on Time Series Analysis*, edited by M. Rosenblatt, New York: John Wiley & Sons, 209–43 (reprinted in Brillinger, 1984a).

Bohman, H. (1961) Approximate Fourier Analysis of Distribution Functions. *Arkiv för Matematik*, 4, 99–157.

Box, G. E. P. and Jenkins, G. M. (1976) *Time Series Analysis: Forecasting and Control* (Revised Edition). San Francisco: Holden-Day.

Bracewell, R. N. (1978) *The Fourier Transform and Its Applications* (Second Edition). New York: McGraw-Hill.

Brillinger, D. R. (1981a) *Time Series: Data Analysis and Theory* (Expanded Edition). San Francisco: Holden-Day.

——— (1981b) The Key Role of Tapering in Spectrum Estimation. *IEEE Transactions on Acoustics, Speech, and Signal Processing*, 29, 1075–6.

Brillinger, D. R., editor (1984a) *The Collected Works of John W. Tukey, Volume I, Time Series: 1949–1964.* Belmont, California: Wadsworth.

——— (1984b) *The Collected Works of John W. Tukey, Volume II, Time Series: 1965–1984.* Belmont, California: Wadsworth.

Brockwell, P. J. and Davis, R. A. (1991) *Time Series: Theory and Methods* (Second Edition). New York: Springer-Verlag.

Bronez, T. P. (1985) Nonparametric Spectral Estimation of Irregularly-Sampled Multidimensional Random Processes. Ph.D. dissertation, Department of Electrical Engineering, Arizona State University.

——— (1986) Nonparametric Spectral Estimation with Irregularly Sampled Data. In *Proceedings of the Third IEEE ASSP Workshop on Spectrum Estimation and Modeling*, Boston, 133–6.

——— (1988) Spectral Estimation of Irregularly Sampled Multidimensional Processes by Generalized Prolate Spheroidal Sequences. *IEEE Transactions on Acoustics, Speech, and Signal Processing*, 36, 1862–73.

——— (1992) On the Performance Advantage of Multi-taper Spectral Analysis. *IEEE Transactions on Signal Processing*, 40, to appear.

Bruce, A. G. and Martin, R. D. (1989) Leave-k-out Diagnostics for Time Series. *Journal of the Royal Statistical Society, Series B*, 51, 363–424.

Bunch, J. R. (1985) Stability of Methods for Solving Toeplitz Systems of Equations. *SIAM Journal on Scientific and Statistical Computing*, 6, 349–64.

Burg, J. P. (1967) Maximum Entropy Spectral Analysis. In *Proceedings of the 37th Meeting of the Society of Exploration Geophysicists* (reprinted in Childers, 1978).

——— (1968) A New Analysis Technique for Time Series Data. In *NATO Advanced Study Institute on Signal Processing with Emphasis on Underwater Acoustics* (reprinted in Childers, 1978).

——— (1975) Maximum Entropy Spectral Analysis. Ph.D. dissertation, Department of Geophysics, Stanford University.

——— (1985) Absolute Power Density Spectra. In *Maximum-Entropy and Bayesian Methods in Inverse Problems*, edited by C. R. Smith and W. T. Grandy, Jr., Dordrecht: D. Reidel Publishing Company, 273–86.

Burshtein, D. and Weinstein, E. (1987) Confidence Intervals for the Maximum Entropy Spectrum. *IEEE Transactions on Acoustics, Speech, and Signal Processing*, 35, 504–10.

——— (1988) Corrections to 'Confidence Intervals for Maximum Entropy Spectrum'. *IEEE Transactions on Acoustics, Speech, and Signal Processing*, 36, 826.

Businger, P. A. and Golub, G. H. (1969) Algorithm 358: Singular Value Decomposition of a Complex Matrix. *Communications of the ACM*, 12, 564–5.

Carter, G. C. (1987) Coherence and Time Delay Estimation. *Proceedings of the IEEE*, 75, 236–55.

Chambers, J. M., Cleveland, W. S., Kleiner, B. and Tukey, P. A. (1983) *Graphical Methods for Data Analysis*. Boston: Duxbury Press.

Champeney, D. C. (1987) *A Handbook of Fourier Theorems*. Cambridge, England: Cambridge University Press.

Chatfield, C. (1984) *The Analysis of Time Series: An Introduction* (Third Edition). London: Chapman and Hall.

Chave, A. D., Thomson, D. J. and Ander, M. E. (1987) On the Robust Estimation of Power Spectra, Coherences, and Transfer Functions. *Journal of Geophysical Research*, 92, 633–48.

Chen, W. Y. and Stegen, G. R. (1974) Experiments with Maximum Entropy Power Spectra of Sinusoids. *Journal of Geophysical Research*, 79, 3019–22.

Childers, D. G., editor (1978) *Modern Spectrum Analysis*. New York: IEEE Press.

Choi, B. S. (1992) *ARMA Model Identification*. New York: Springer-Verlag.

Chung, K. L. (1974) *A Course in Probability Theory* (Second Edition). New York: Academic Press.

Cleveland, W. S. and Parzen, E. (1975) The Estimation of Coherence, Frequency Response, and Envelope Delay. *Technometrics*, 17, 167–72.

Coates, D. S. and Diggle, P. J. (1986) Tests for Comparing Two Estimated Spectral Densities. *Journal of Time Series Analysis*, 7, 7–20.

Conover, W. J. (1980) *Practical Nonparametric Statistics* (Second Edition). New York: John Wiley & Sons.

Conradsen, K. and Spliid, H. (1981) A Seasonal Adjustment Filter for Use in Box-Jenkins Analysis of Seasonal Time Series. *Applied Statistics*, 30, 172–7.

Courant, R. and Hilbert, D. (1953) *Methods of Mathematical Physics, Volume I*. New York: Interscience Publishers.

Cramér, H. (1942) On Harmonic Analysis in Certain Functional Spaces. *Arkiv för Matematik, Astronomi och Fysik*, 28B, 1–7.

Daniell, P. J. (1946) Discussion on the Papers by Bartlett, Foster, Cunningham and Hynd. *Supplement to the Journal of the Royal Statistical Society*, 8, 88–90.

de Gooijer, J. G., Abraham, B., Gould, A. and Robinson, L. (1985) Methods for Determining the Order of an Autoregressive-Moving Average Process: A Survey. *International Statistical Review*, 53, 301–29.

Diggle, P. J. (1990) *Time Series: A Biostatistical Introduction*. Oxford: Oxford University Press.

Doodson, A. T. and Warburg, H. D. (1941) *Admiralty Manual of Tides*. London: H. M. Stationery Office.

Farebrother, R. W. (1987) The Distribution of a Noncentral χ^2 Variable with Nonnegative Degrees of Freedom. *Applied Statistics*, 36, 402–5.

Fisher, R. A. (1929) Tests of Significance in Harmonic Analysis. *Proceedings of the Royal Society of London, Series A*, 125, 54–9.

——— (1939) The Sampling Distribution of Some Statistics Obtained from Non-Linear Equations. *Annals of Eugenics*, 9, 238–49.

Fougere, P. F. (1977) A Solution to the Problem of Spontaneous Line Splitting in Maximum Entropy Power Spectrum Analysis. *Journal of Geophysical Research*, 82, 1051–4.

——— (1985) A Review of the Problem of Spontaneous Line Splitting in

Maximum Entropy Power Spectral Analysis. In *Maximum-Entropy and Bayesian Methods in Inverse Problems*, edited by C. R. Smith and W. T. Grandy, Jr., Dordrecht: D. Reidel Publishing Company, 303–15.

Fougere, P. F., Zawalick, E. J. and Radoski, H. R. (1976) Spontaneous Line Splitting in Maximum Entropy Power Spectrum Analysis. *Physics of the Earth and Planetary Interiors*, 12, 201–7.

Franke, J. (1985) ARMA Processes Have Maximal Entropy Among Time Series with Prescribed Autocovariances and Impulse Responses. *Advances in Applied Probability*, 17, 810–40.

Friedlander, B. (1982) System Identification Techniques for Adaptive Signal Processing. *IEEE Transactions on Acoustics, Speech, and Signal Processing*, 30, 240–6.

Fuller, W. A. (1976) *Introduction to Statistical Time Series*. New York: John Wiley & Sons.

Geçkinli, N. C. and Yavuz, D. (1978) Some Novel Windows and a Concise Tutorial Comparison of Window Families. *IEEE Transactions on Acoustics, Speech, and Signal Processing*, 26, 501–7.

Godolphin, E. J. and Unwin, J. M. (1983) Evaluation of the Covariance Matrix for the Maximum Likelihood Estimator of a Gaussian Autoregressive-Moving Average Process. *Biometrika*, 70, 279–84.

Golub, G. H. and Kahan, W. (1965) Calculating the Singular Values and Pseudo-Inverse of a Matrix. *SIAM Journal on Numerical Analysis*, 2, 205–24.

Golub, G. H. and Van Loan, C. F. (1989) *Matrix Computations* (Second Edition). Baltimore: Johns Hopkins University Press.

Gourlay, A.R. and Watson, G.A. (1973) *Computational Methods for Matrix Eigenproblems*. London: John Wiley & Sons.

Granger, C. W. J. (1966) The Typical Spectral Shape of an Economic Variable. *Econometrica*, 34, 150–61.

Graybill, F. A. (1983) *Matrices with Applications in Statistics* (Second Edition). Belmont, California: Wadsworth.

Greene, D. H. and Knuth, D. E. (1990) *Mathematics for the Analysis of Algorithms* (Third Edition). Boston: Birkhäuser.

Greenhall, C. A. (1990) Orthogonal Sets of Data Windows Constructed from Trigonometric Polynomials. *IEEE Transactions on Acoustics, Speech, and Signal Processing*, 38, 870–2.

Grenander, U. (1951) On Empirical Spectral Analysis of Stochastic Processes. *Arkiv för Matematik*, 1, 503–31.

Grenander, U. and Rosenblatt, M. (1984) *Statistical Analysis of Stationary Time Series* (Second Edition). New York: Chelsea Publishing Company.

Grenander, U. and Szegő, G. (1984) *Toeplitz Forms and Their Applications* (Second Edition). New York: Chelsea Publishing Company.

Griffiths, P. and Hill, I. D., editors (1985) *Applied Statistics Algorithms.* Chichester, England: Ellis Horwood Ltd.

Grünbaum, F. A. (1981) Eigenvectors of a Toeplitz Matrix: Discrete Version of the Prolate Spheroidal Wave Functions. *SIAM Journal on Algebraic and Discrete Methods*, 2, 136–41.

Hamming, R. W. (1983) *Digital Filters* (Second Edition). Englewood Cliffs, New Jersey: Prentice-Hall.

Hannan, E. J. (1973) The Estimation of Frequency. *Journal of Applied Probability*, 10, 510–9.

Hardin, J. C. (1986) An Additional Source of Uncertainty and Bias in Digital Spectral Estimates Near the Nyquist Frequency. *Journal of Sound and Vibration*, 110, 533–7.

Harris, F. J. (1978) On the Use of Windows for Harmonic Analysis with the Discrete Fourier Transform. *Proceedings of the IEEE*, 66, 51–83 (reprinted in Kesler, 1986).

Hasan, T. (1983) Complex Demodulation: Some Theory and Applications. In *Handbook of Statistics 3: Time Series in the Frequency Domain*, edited by D. R. Brillinger and P. R. Krishnaiah, Amsterdam: North-Holland, 125–56.

Hua, Y. and Sarkar, T. K. (1988) Perturbation Analysis of TK Method for Harmonic Retrieval Problems. *IEEE Transactions on Acoustics, Speech, and Signal Processing*, 36, 228–40.

Ihara, S. (1984) Maximum Entropy Spectral Analysis and ARMA Processes. *IEEE Transactions on Information Theory*, 30, 377–80.

Isserlis, L. (1918) On a Formula for the Product-Moment Coefficient of Any Order of a Normal Frequency Distribution in Any Number of Variables. *Biometrika*, 12, 134–9.

Jackson, D. D. (1972) Interpretation of Inaccurate, Insufficient and Inconsistent Data. *Geophysical Journal of the Royal Astronomical Society*, 28, 97–109.

Jackson, P. L. (1967) Truncations and Phase Relationships of Sinusoids. *Journal of Geophysical Research*, 72, 1400–3.

Jacovitti, G. and Scarano, G. (1987) On a Property of the PARCOR Coefficients of Stationary Processes Having Gaussian-Shaped ACF. *Proceedings of the IEEE*, 75, 960–1.

Jaynes, E. T. (1982) On the Rationale of Maximum-Entropy Methods. *Proceedings of the IEEE*, 70, 939–52.

Jenkins, G. M. (1961) General Considerations in the Analysis of Spectra. *Technometrics*, 3, 133–66.

Jenkins, G. M. and Watts, D. G. (1968) *Spectral Analysis and Its Applications.* San Francisco: Holden-Day.

Jones, N. B., Lago, P. J. and Parekh, A. (1987) Principal Component Analysis of the Spectra of Point Processes – An Application in Electromyography. In *Mathematics in Signal Processing*, edited

by T. S. Durrani, J. B. Abbiss, J. E. Hudson, R. N. Madan, J. G. Mc-
Whirter and T. A. Moore, Oxford: Clarendon Press, 147–64.

Jones, R. H. (1971) Spectrum Estimation with Missing Observations.
Annals of the Institute of Statistical Mathematics, 23, 387–98.

—— (1976) Autoregression Order Selection. *Geophysics*, 41, 771–3
(reprinted in Childers, 1978).

—— (1980) Maximum Likelihood Fitting of ARMA Models to Time
Series with Missing Observations. *Technometrics*, 22, 389–95.

—— (1985) Time Series Analysis – Time Domain. In *Probability,
Statistics, and Decision Making in the Atmospheric Sciences*, edited
by A. H. Murphy and R. W. Katz, Boulder, Colorado: Westview
Press, 223–59.

Jones, R. H., Crowell, D. H., Nakagawa, J. K. and Kapuniai, L. (1972)
Statistical Comparisons of EEG Spectra Before and During Stimu-
lation in Human Neonates. In *Computers in Biomedicine, a Supplement to the Proceedings of the Fifth Hawaii International Conference on System Sciences*, North Hollywood, California: Western
Periodicals Company, 4 pages.

Kaiser, J. F. (1966) Digital Filters. In *System Analysis by Digital Computer*, edited by F. F. Kuo and J. F. Kaiser, New York: John Wiley
& Sons, 218–85.

—— (1974) Nonrecursive Digital Filter Design Using the $I_0 - SINH$
Window Function. In *Proceedings – 1974 IEEE International Symposium on Circuits and Systems*, 20–3.

Kaveh, M. and Cooper, G. R. (1976) An Empirical Investigation of the
Properties of the Autoregressive Spectral Estimator. *IEEE Transactions on Information Theory*, 22, 313–23 (reprinted in Childers,
1978).

Kay, S. M. (1981a) The Effect of Sampling Rate on Autocorrelation
Estimation. *IEEE Transactions on Acoustics, Speech, and Signal
Processing*, 29, 859–67.

—— (1981b) Efficient Generation of Colored Noise. *Proceedings of
the IEEE*, 69, 480–1.

—— (1988) *Modern Spectral Estimation: Theory and Application*.
Englewood Cliffs, New Jersey: Prentice-Hall.

Kay, S. M. and Demeure, C. (1984) The High-Resolution Spectrum
Estimator – A Subjective Entity. *Proceedings of the IEEE*, 72, 1815–
6 (reprinted in Kesler, 1986).

Kay, S. M. and Makhoul, J. (1983) On the Statistics of the Estimated
Reflection Coefficients of an Autoregressive Process. *IEEE Transactions on Acoustics, Speech, and Signal Processing*, 31, 1447–55
(reprinted in Kesler, 1986).

Kay, S. M. and Marple, S. L., Jr. (1979) Sources of and Remedies
for Spectral Line Splitting in Autoregressive Spectrum Analysis. In

Proceedings of the 1979 IEEE International Conference on Acoustics, Speech, and Signal Processing, 151–4.

———— (1981) Spectrum Analysis – A Modern Perspective. *Proceedings of the IEEE*, 69, 1380–419 (reprinted in Kesler, 1986).

Kesler, S. B., editor (1986) *Modern Spectrum Analysis, II*. New York: IEEE Press.

Kikkawa, S. and Ishida, M. (1988) Number of Degrees of Freedom, Correlation Times, and Equivalent Bandwidths of a Random Process. *IEEE Transactions on Information Theory*, 34, 151–5.

King, M. E. (1990) Multiple Taper Spectral Analysis of Earth Rotation Data. Ph.D. dissertation, Scripps Institute of Oceanography, University of California, San Diego.

Koopmans, L. H. (1974) *The Spectral Analysis of Time Series*. New York: Academic Press.

Koslov, J. W. and Jones, R. H. (1985) A Unified Approach to Confidence Bounds for the Autoregressive Spectral Estimator. *Journal of Time Series Analysis*, 6, 141–51.

Kromer, R. E. (1969) Asymptotic Properties of the Autoregressive Spectral Estimator. Ph.D. dissertation, Department of Statistics, Stanford University.

Kung, S. Y. and Arun, K. S. (1987) Singular-Value-Decomposition Algorithms for Linear System Approximation and Spectrum Estimation. In *Advances in Statistical Signal Processing* (Volume 1), edited by H. V. Poor, Greenwich, Connecticut: JAI Press, 203–50.

Kuo, C., Lindberg, C. and Thomson, D. J. (1990) Coherence Established Between Atmospheric Carbon Dioxide and Global Temperature. *Nature*, 343, 709–14.

Lacoss, R. T. (1971) Data Adaptive Spectral Analysis Methods. *Geophysics*, 36, 661–75 (reprinted in Childers, 1978).

Lagunas-Hernández, M. A., Santamaría-Perez, M. E. and Figueiras-Vidal, A. R. (1984) ARMA Model Maximum Entropy Power Spectral Estimation. *IEEE Transactions on Acoustics, Speech, and Signal Processing*, 32, 984–90.

Landers, T. E. and Lacoss, R. T. (1977) Some Geophysical Applications of Autoregressive Spectral Estimates. *IEEE Transactions on Geoscience Electronics*, 15, 26–32 (reprinted in Childers, 1978).

Lang, S. W. and McClellan, J. H. (1980) Frequency Estimation with Maximum Entropy Spectral Estimators. *IEEE Transactions on Acoustics, Speech, and Signal Processing*, 28, 716–24 (reprinted in Kesler, 1986).

Lanning, E. N. and Johnson, D. M. (1983) Automated Identification of Rock Boundaries: An Application of the Walsh Transform to Geophysical Well-Log Analysis. *Geophysics*, 48, 197–205.

Lawrance, A. J. (1991) Directionality and Reversibility in Time Series.

International Statistical Review, 59, 67–79.

Lindberg, C. R. and Park, J. (1987) Multiple-Taper Spectral Analysis of Terrestrial Free Oscillations: Part II. *Geophysical Journal of the Royal Astronomical Society*, 91, 795–836.

Liu, T.-C. and Van Veen, B. (1992) Multiple Window Based Minimum Variance Spectrum Estimation for Multidimensional Random Fields. *IEEE Transactions on Signal Processing*, 40, 578–89.

Loupas, T. and McDicken, W. N. (1990) Low-Order Complex AR Models for Mean and Maximum Frequency Estimation in the Context of Doppler Color Flow Mapping. *IEEE Transactions on Ultrasonics, Ferroelectrics, and Frequency Control*, 37, 590–601.

Lysne, D. and Tjøstheim, D. (1987) Loss of Spectral Peaks in Autoregressive Spectral Estimation. *Biometrika*, 74, 200–6.

Makhoul, J. (1981a) On the Eigenvectors of Symmetric Toeplitz Matrices. *IEEE Transactions on Acoustics, Speech, and Signal Processing*, 29, 868–72.

——— (1981b) Lattice Methods in Spectral Estimation. In *Applied Time Series Analysis II*, edited by D. F. Findley, New York: Academic Press, 301–25.

——— (1986) Maximum Confusion Spectral Analysis. In *Proceedings of the Third IEEE ASSP Workshop on Spectrum Estimation and Modeling*, Boston, 6–9.

——— (1990) Volume of the Space of Positive Definite Sequences. *IEEE Transactions on Acoustics, Speech, and Signal Processing*, 38, 506–11.

Mann, H. B. and Wald, A. (1943) On the Statistical Treatment of Linear Stochastic Difference Equations. *Econometrica*, 11, 173–220.

Marple, S. L., Jr. (1980) A New Autoregressive Spectrum Analysis Algorithm. *IEEE Transactions on Acoustics, Speech, and Signal Processing*, 28, 441–54 (reprinted in Kesler, 1986).

——— (1987) *Digital Spectral Analysis with Applications*. Englewood Cliffs, New Jersey: Prentice-Hall.

Martin, R. D. and Thomson, D. J. (1982) Robust-Resistant Spectrum Estimation. *Proceedings of the IEEE*, 70, 1097–115.

McLeod, A. I. and Jiménez, C. (1984) Nonnegative Definiteness of the Sample Autocovariance Function. *The American Statistician*, 38, 297–8.

——— (1985) Reply to Discussion by Arcese and Newton. *The American Statistician*, 39, 237–8.

Miller, K. S. (1973) Complex Linear Least Squares. *SIAM Review*, 15, 706–26.

——— (1974) *Complex Stochastic Processes: An Introduction to Theory and Application*. Reading, Massachusetts: Addison-Wesley.

Monro, D. M. and Branch, J. L. (1977) The Chirp Discrete Transform of General Length: Algorithm AS 117. *Applied Statistics*, 26, 351–61.

Morettin, P. A. (1981) Walsh Spectral Analysis. *SIAM Review*, 23, 279–91.

——— (1984) The Levinson Algorithm and its Applications in Time Series Analysis. *International Statistical Review*, 52, 83–92.

Mullis, C. T. and Scharf, L. L. (1991) Quadratic Estimators of the Power Spectrum. In *Advances in Spectrum Analysis and Array Processing,* Volume I, edited by S. Haykin, Englewood Cliffs, New Jersey: Prentice-Hall, 1–57.

Munk, W. and Cartwright, D. (1966) Tidal Spectroscopy and Prediction. *Philosophical Transactions of the Royal Society of London, Series A*, 259, 533–81.

Murray, M. T. (1964) A General Method for the Analysis of Hourly Heights of Tide. *International Hydrographic Review*, 41, 91–102.

——— (1965) Optimization Processes in Tidal Analysis. *International Hydrographic Review*, 42, 73–82.

Newton, H. J. (1988) *TIMESLAB: A Time Series Analysis Laboratory.* Pacific Grove, California: Wadsworth & Brooks/Cole.

Newton, H. J. and Pagano, M. (1983) A Method for Determining Periods in Time Series. *Journal of the American Statistical Association*, 78, 152–7.

——— (1984) Simultaneous Confidence Bands for Autoregressive Spectra. *Biometrika*, 71, 197–202.

Nitzberg, R. (1979) Spectral Estimation: An Impossibility? *Proceedings of the IEEE*, 67, 437–8.

Nowroozi, A. A. (1967) Table for Fisher's Test of Significance in Harmonic Analysis. *Geophysical Journal of the Royal Astronomical Society*, 12, 517–20.

Nuttall, A. H. and Carter, G. C. (1982) Spectral Estimation Using Combined Time and Lag Weighting. *Proceedings of the IEEE*, 70, 1115–25.

Oppenheim, A. V. and Schafer, R. W. (1989) *Discrete-Time Signal Processing.* Englewood Cliffs, New Jersey: Prentice-Hall.

Pagano, M. (1973) When is an Autoregressive Scheme Stationary? *Communications in Statistics*, 1, 533–44.

Papoulis, A. (1973) Minimum-Bias Windows for High-Resolution Spectral Estimates. *IEEE Transactions on Information Theory*, 19, 9–12.

——— (1985) Levinson's Algorithm, Wold's Decomposition, and Spectral Estimation. *SIAM Review*, 27, 405–41.

——— (1991) *Probability, Random Variables, and Stochastic Processes* (Third Edition). New York: McGraw-Hill.

Park, J., Lindberg, C. R. and Thomson, D. J. (1987a) Multiple-Taper Spectral Analysis of Terrestrial Free Oscillations: Part I. *Geophysical*

Journal of the Royal Astronomical Society, 91, 755–94.

Park, J., Lindberg, C. R. and Vernon, F. L., III. (1987b) Multitaper Spectral Analysis of High-Frequency Seismograms. *Journal of Geophysical Research*, 92, 12 675–84.

Parlett, B. N. (1980) *The Symmetric Eigenvalue Problem*. Englewood Cliffs, New Jersey: Prentice-Hall.

Parzen, E. (1957) On Choosing an Estimate of the Spectral Density Function of a Stationary Time Series. *Annals of Mathematical Statistics*, 28, 921–32.

—— (1961) Mathematical Considerations in the Estimation of Spectra. *Technometrics*, 3, 167–90.

—— (1983) Autoregressive Spectral Estimation. In *Handbook of Statistics 3: Time Series in the Frequency Domain*, edited by D. R. Brillinger and P. R. Krishnaiah, Amsterdam: North-Holland, 221–47.

Pignari, S. and Canavero, F. G. (1991) Amplitude Errors in the Burg Spectrum Estimation of Sinusoidal Signals. *Signal Processing*, 22, 107–12.

Pisarenko, V. F. (1973) The Retrieval of Harmonics from a Covariance Function. *Geophysical Journal of the Royal Astronomical Society*, 33, 347–66 (reprinted in Kesler, 1986).

Press, H. and Tukey, J. W. (1956) Power Spectral Methods of Analysis and Application in Airplane Dynamics. In *AGARD Flight Test Manual, Vol. IV, Instrumentation*, edited by E. J. Durbin, Paris: North Atlantic Treaty Organization, Advisory Group for Aeronautical Research and Development, C:1–C:41 (reprinted in Brillinger, 1984a).

Press, W. H., Flannery, B. P., Teukolsky, S. A. and Vetterling, W. T. (1986) *Numerical Recipes: The Art of Scientific Computing*. Cambridge, England: Cambridge University Press.

Priestley, M. B. (1981) *Spectral Analysis and Time Series*. London: Academic Press.

Rabiner, L. R. and Gold, B. (1975) *Theory and Application of Digital Signal Processing*. Englewood Cliffs, New Jersey: Prentice-Hall.

Rabiner, L. R. and Schafer, R. W. (1978) *Digital Processing of Speech Signals*. Englewood Cliffs, New Jersey: Prentice-Hall.

Ramsey, F. L. (1974) Characterization of the Partial Autocorrelation Function. *Annals of Statistics*, 2, 1296–301.

Reid, J. S. (1979) Confidence Limits and Maximum Entropy Spectra. *Journal of Geophysical Research*, 84, 5289–301.

Rice, J. A. and Rosenblatt, M. (1988) On Frequency Estimation. *Biometrika*, 75, 477–84.

Rice, S. O. (1945) Mathematical Analysis of Random Noise, Part III: Statistical Properties of Random Noise Currents. *Bell System Technical Journal*, 24, 46–156.

Rife, D. C. and Boorstyn, R. R. (1974) Single-Tone Parameter Estimation from Discrete-Time Observations. *IEEE Transactions on Information Theory*, 20, 591–8.

—— (1976) Multiple Tone Parameter Estimation from Discrete-Time Observations. *Bell System Technical Journal*, 55, 1389–410.

Rife, D. C. and Vincent, G. A. (1970) Use of the Discrete Fourier Transform in the Measurement of Frequencies and Levels of Tones. *Bell System Technical Journal*, 49, 197–228.

Rosenblatt, M. (1985) *Stationary Sequences and Random Fields*. Boston: Birkhäuser.

Samarov, A. and Taqqu, M. S. (1988) On the Efficiency of the Sample Mean in Long-Memory Noise. *Journal of Time Series Analysis*, 9, 191–200.

Satorius, E. H. and Zeidler, J. R. (1978) Maximum Entropy Spectral Analysis of Multiple Sinusoids in Noise. *Geophysics*, 43, 1111–8 (reprinted in Kesler, 1986).

Scheffé, H. (1959) *The Analysis of Variance*. New York: John Wiley & Sons.

Schuster, A. (1898) On the Investigation of Hidden Periodicities with Application to a Supposed 26 Day Period of Meteorological Phenomena. *Terrestrial Magnetism*, 3, 13–41.

Shimshoni, M. (1971) On Fisher's Test of Significance in Harmonic Analysis. *Geophysical Journal of the Royal Astronomical Society*, 23, 373–7.

Siegel, A. F. (1979) The Noncentral Chi-Squared Distribution with Zero Degrees of Freedom and Testing for Uniformity. *Biometrika*, 66, 381–6.

—— (1980) Testing for Periodicity in a Time Series. *Journal of the American Statistical Association*, 75, 345–8.

Singleton, R. C. (1969) An Algorithm for Computing the Mixed Radix Fast Fourier Transform. *IEEE Transactions on Audio and Electroacoustics*, 17, 93–103.

Sjoholm, P. F. (1989) Statistical Optimization of the Log Spectral Density Estimate. In *Twenty-third Asilomar Conference on Signals, Systems & Computers, Volume 1*, San Jose, California: Maple Press, 355–9.

Slepian, D. (1976) On Bandwidth. *Proceedings of the IEEE*, 64, 292–300.

—— (1978) Prolate Spheroidal Wave Functions, Fourier Analysis, and Uncertainty – V: The Discrete Case. *Bell System Technical Journal*, 57, 1371–430.

—— (1983) Some Comments on Fourier Analysis, Uncertainty and Modeling. *SIAM Review*, 25, 379–93.

Slepian, D. and Pollak, H. O. (1961) Prolate Spheroidal Wave Functions,

Fourier Analysis and Uncertainty – I. *Bell System Technical Journal*, 40, 43–63.

Sloane, E. A. (1969) Comparison of Linearly and Quadratically Modified Spectral Estimates of Gaussian Signals. *IEEE Transactions on Audio and Electroacoustics*, 17, 133–7.

Smith, B. T., Boyle, J. M., Dongarra, J. J., Garbow, B. S., Ikebe, Y., Klema, V. C. and Moler, C. B. (1976) *Matrix Eigensystem Routines – EISPACK Guide* (Second Edition). Berlin: Springer-Verlag.

Spencer-Smith, J. L. and Todd, H. A. C. (1941) A Time Series Met With in Textile Research. *Supplement to the Journal of the Royal Statistical Society*, 7, 131–45.

Stephens, M. A. (1974) EDF Statistics for Goodness of Fit and Some Comparisons. *Journal of the American Statistical Association*, 69, 730–7.

Stevens, W. L. (1939) Solution to a Geometrical Problem in Probability. *Annals of Eugenics*, 9, 315–20.

Stoffer, D. S. (1991) Walsh–Fourier Analysis and Its Statistical Applications. *Journal of the American Statistical Association*, 86, 461–79.

Stoffer, D. S., Scher, M. S., Richardson, G. A., Day, N. L. and Coble, P. A. (1988) A Walsh–Fourier Analysis of the Effects of Moderate Maternal Alcohol Consumption on Neonatal Sleep-State Cycling. *Journal of the American Statistical Association*, 83, 954–63.

Swanepoel, J. W. H. and Van Wyk, J. W. J. (1986) The Bootstrap Applied to Power Spectral Density Function Estimation. *Biometrika*, 73, 135–41.

Taylor, A. E. and Mann, W. R. (1972) *Advanced Calculus* (Second Edition). Lexington, Massachusetts: Xerox College Publishing.

Therrien, C. W. (1983) On the Relation Between Triangular Matrix Decomposition and Linear Prediction. *Proceedings of the IEEE*, 71, 1459–60.

Thomson, D. J. (1977) Spectrum Estimation Techniques for Characterization and Development of WT4 Waveguide – I. *Bell System Technical Journal*, 56, 1769–815.

—— (1982) Spectrum Estimation and Harmonic Analysis. *Proceedings of the IEEE*, 70, 1055–96.

—— (1990a) Time Series Analysis of Holocene Climate Data. *Philosophical Transactions of the Royal Society of London, Series A*, 330, 601–16.

—— (1990b) Quadratic-Inverse Spectrum Estimates: Applications to Palaeoclimatology. *Philosophical Transactions of the Royal Society of London, Series A*, 332, 539–97.

Thomson, D. J. and Chave, A. D. (1991) Jackknifed Error Estimates for Spectra, Coherences, and Transfer Functions. In *Advances in Spectrum Analysis and Array Processing*, Volume I, edited by S. Haykin,

Englewood Cliffs, New Jersey: Prentice-Hall, 58–113.

Titchmarsh, E. C. (1939) *The Theory of Functions* (Second Edition). Oxford: Oxford University Press.

Toman, K. (1965) The Spectral Shifts of Truncated Sinusoids. *Journal of Geophysical Research*, 70, 1749–50.

Tong, H. (1975) Autoregressive Model Fitting with Noisy Data by Akaike's Information Criterion. *IEEE Transactions on Information Theory*, 21, 476–80 (reprinted in Childers, 1978).

Tseng, F. I., Sarkar, T. K. and Weiner, D. D. (1981) A Novel Window for Harmonic Analysis. *IEEE Transactions on Acoustics, Speech, and Signal Processing*, 29, 177–88.

Tufts, D. W. and Kumaresan, R. (1982) Estimation of Frequencies of Multiple Sinusoids: Making Linear Prediction Perform Like Maximum Likelihood. *Proceedings of the IEEE*, 70, 975–89.

Tugnait, J. K. (1986) Recursive Parameter Estimation for Noisy Autoregressive Signals. *IEEE Transactions on Information Theory*, 32, 426–30.

Tukey, J. W. (1980) Can We Predict Where 'Time Series' Should Go Next? In *Directions in Time Series Analysis*, edited by D. B. Brillinger and G. C. Tiao, Hayward, California: Institute of Mathematical Statistics, 1–31 (reprinted in Brillinger, 1984b).

Ulrych, T. J. (1972a) Maximum Entropy Power Spectrum of Truncated Sinusoids. *Journal of Geophysical Research*, 77, 1396–400.

—— (1972b) Maximum Entropy Power Spectrum of Long Period Geomagnetic Reversals. *Nature*, 235, 218–9.

Ulrych, T. J. and Bishop, T. N. (1975) Maximum Entropy Spectral Analysis and Autoregressive Decomposition. *Reviews of Geophysics and Space Physics*, 13, 183–200 (reprinted in Childers, 1978).

Ulrych, T. J. and Clayton, R. W. (1976) Time Series Modelling and Maximum Entropy. *Physics of the Earth and Planetary Interiors*, 12, 188–200.

Ulrych, T. J. and Ooe, M. (1983) Autoregressive and Mixed Autoregressive-Moving Average Models and Spectra. In *Nonlinear Methods of Spectral Analysis* (Second Edition), edited by S. Haykin, Berlin: Springer-Verlag, 73–125.

Van Schooneveld, C. and Frijling, D. J. (1981) Spectral Analysis: On the Usefulness of Linear Tapering for Leakage Suppression. *IEEE Transactions on Acoustics, Speech, and Signal Processing*, 29, 323–9.

Van Veen, B. D. and Scharf, L. L. (1990) Estimation of Structured Covariance Matrices and Multiple Window Spectrum Analysis. *IEEE Transactions on Acoustics, Speech, and Signal Processing*, 38, 1467–72.

Wahba, G. (1980) Automatic Smoothing of the Log Periodogram. *Journal of American Statistical Association*, 75, 122–32.

Walden, A. T. (1982) The Statistical Analysis of Extreme High Sea Levels Utilizing Data from the Solent Area. Ph.D. dissertation, University of Southampton.

—— (1989) Accurate Approximation of a 0th Order Discrete Prolate Spheroidal Sequence for Filtering and Data Tapering. *Signal Processing*, 18, 341–8.

—— (1990a) Variance and Degrees of Freedom of a Spectral Estimator Following Data Tapering and Spectral Smoothing. *Signal Processing*, 20, 67–79.

—— (1990b) Improved Low-Frequency Decay Estimation Using the Multitaper Spectral Analysis Method. *Geophysical Prospecting*, 38, 61–86.

—— (1992) Asymptotic Percentage Points for Siegel's Test Statistic for Compound Periodicities. *Biometrika*, 79, 438–40.

Walden, A. T. and Prescott, P. (1983) Statistical Distributions for Tidal Elevations. *Geophysical Journal of the Royal Astronomical Society*, 72, 223–36.

Walden, A. T. and White, R. E. (1984) On Errors of Fit and Accuracy in Matching Synthetic Seismograms and Seismic Traces. *Geophysical Prospecting*, 32, 871–91.

—— (1990) Estimating the Statistical Bandwidth of a Time Series. *Biometrika*, 77, 699–707.

Walker, A. M. (1960) Some Consequences of Superimposed Error in Time Series Analysis. *Biometrika*, 47, 33–43.

—— (1971) On the Estimation of a Harmonic Component in a Time Series with Stationary Independent Residuals. *Biometrika*, 58, 21–36.

Walker, J. (1985) Searching for Patterns of Rainfall in a Storm. *Scientific American*, 252 (1), 112–9.

Weiss, G. (1975) Time-Reversibility of Linear Stochastic Processes. *Journal of Applied Probability*, 12, 831–6.

Welch, P. D. (1967) The Use of Fast Fourier Transform for the Estimation of Power Spectra: A Method Based on Time Averaging Over Short, Modified Periodograms. *IEEE Transactions on Audio and Electroacoustics*, 15, 70–3 (reprinted in Childers, 1978).

White, R. E. (1980) Partial Coherence Matching of Synthetic Seismograms with Seismic Traces. *Geophysical Prospecting*, 28, 333–58.

Whittle, P. (1952) The Simultaneous Estimation of a Time Series' Harmonic Components and Covariance Structure. *Trabajos de Estadistica y de Investigacion Operativa*, 3, 43–57.

Wiener, N. (1949) *Extrapolation, Interpolation, and Smoothing of Stationary Time Series*. Cambridge, Massachusetts: MIT Press.

Wilson, R. (1987) Finite Prolate Spheroidal Sequences and Their Applications I: Generation and Properties. *IEEE Transactions on Pattern Analysis and Machine Intelligence*, 9, 787–95.

Wilson, R. and Spann, M. (1988) Finite Prolate Spheroidal Sequences and Their Applications II: Image Feature Description and Segmentation. *IEEE Transactions on Pattern Analysis and Machine Intelligence*, 10, 193–203.

Yaglom, A. M. (1987) *Correlation Theory of Stationary and Related Random Functions, Volume I: Basic Results.* New York: Springer-Verlag.

Yuen, C. K. (1979) Comments on Modern Methods for Spectrum Estimation. *IEEE Transactions on Acoustics, Speech, and Signal Processing*, 27, 298–9.

Zhang, H.-C. (1992) Reduction of the Asymptotic Bias of Autoregressive and Spectral Estimators by Tapering. *Journal of Time Series Analysis*, 13, 451–69.

Author Index

Subject Index